Lecture Notes in Economics and Mathematical Systems

539

Founding Editors:

M. Beckmann
H. P. Künzi

Managing Editors:

Prof. Dr. G. Fandel
Fachbereich Wirtschaftswissenschaften
Fernuniversität Hagen
Feithstr. 140/AVZ II, 58084 Hagen, Germany

Prof. Dr. W. Trockel
Institut für Mathematische Wirtschaftsforschung (IMW)
Universität Bielefeld
Universitätsstr. 25, 33615 Bielefeld, Germany

Editorial Board:

A. Basile, A. Drexl, W. Güth, K. Inderfurth, W. Kürsten, U. Schittko

Springer
Berlin
Heidelberg
New York
Hong Kong
London
Milan
Paris
Tokyo

Nikolaus Hautsch

Modelling
Irregularly Spaced
Financial Data

Theory and Practice
of Dynamic Duration Models

 Springer

Author

Nikolaus Hautsch
Institute of Economics
University of Copenhagen
Studiestraede 6
1455 Copenhagen K
Denmark

Library of Congress Control Number: 2004103616

ISSN 0075-8442
ISBN 3-540-21134-9 Springer-Verlag Berlin Heidelberg New York

Springer-Verlag is a part of Springer Science+Business Media

springeronline.com

Typesetting: Camera ready by author
Cover design: *Erich Kirchner*, Heidelberg

Printed on acid-free paper   55/3142/du   5 4 3 2 1 0

To Christiane

Preface

This book has been written as a doctoral dissertation at the Department of Economics at the University of Konstanz. I am indebted to my supervisor Winfried Pohlmeier for providing a stimulating and pleasant research environment and his continuous support during my doctoral studies. I strongly benefitted from inspiring discussions with him, his valuable advices and helpful comments regarding the contents and the exposition of this book.

I am grateful to Luc Bauwens for refereeing my work as a second supervisor. Moreover, I wish to thank him for offering me the possibility of a research visit at the Center of Operations Research and Econometrics (CORE) at the Université Catholique de Louvain. Important parts of this book have been conceived during this period.

Similarly, I am grateful to Tony Hall who invited me for a research visit at the University of Technology, Sydney, and provided me access to an excellent database from the Australian Stock Exchange. I would like to thank him for his valuable support and the permission to use this data for empirical studies in this book.

I wish to thank my colleagues at the University of Konstanz Frank Gerhard, Dieter Hess, Joachim Inkmann, Markus Jochmann, Stefan Klotz, Sandra Lechner and Ingmar Nolte who offered me advice, inspiration, friendship and successful co-operations. Moreover, I am grateful to the student research assistants at the Chair of Econometrics at the University of Konstanz, particularly Magdalena Ramada Sarasola, Danielle Tucker and Nadine Warmuth who did a lot of editing work.

Last but not least I would like to thank my family for supporting me in any period of life. I am exceptionally indebted to my wonderful wife. Without her love, her never-ending mental support and her invaluable encouragement this book could not have been written.

Konstanz,   *Nikolaus Hautsch*
December 2003

Contents

1

Introduction

Recent theoretical and empirical research in econometrics and statistics witnessed a growing interest in the modelling of high-frequency data. The ultimate limiting case is reached when *all* single events are recorded. Engle (2000) calls this limiting frequency "*ultra-high frequency*". Especially in financial economics and econometrics, the examination of ultra-high-frequency data is an extremely active field of research. This is a clear consequence of the availability of intraday databases consisting of detailed information on the complete trading process involving in the limit case all single transactions and orders on a financial market.

A key property of transaction data is the irregular spacing in time. The question of how this salient feature should be treated in an econometric model is one of the most disputed issues in high-frequency data econometrics. Clearly, researchers can circumvent this problem by aggregating the data up to fixed (discrete) intervals. However, such a procedure naturally comes along with a loss of information and raises the question of an optimal aggregation level. Moreover, it is quite unclear what kind of bias is induced by choosing a discrete sampling scheme. In this context, two major aspects are particularly important. Firstly, the time interval between subsequent events has itself informational content and is a valuable economic variable that serves as a measure of trading activity and might affect price and volume behavior. By aggregating the data up to equi-distant intervals, this information is discarded. The second aspect is more technical, however not less crucial. Ignoring the irregular spacing of data in an econometric model can cause substantial misspecifications. As illustrated by Aït-Sahalia and Mykland (2003), two effects have to be taken into account in this context: the effects of sampling discreteness and sampling randomness. Effects of sampling discreteness are associated with the implications when irregularly spaced data is sampled at discrete, equi-distant time intervals. The latter is related to the additional effect that the randomness of the sampling intervals has when a continuous-time model is estimated. By investigating these effects, Aït-Sahaila and Mykland illustrate that in many situations the effect of the sampling randomness is greater than the effect of the sampling discreteness. They point out that the consideration of the sampling randomness is essential when using irregularly spaced data.

The explicit consideration of the irregularity of time intervals necessitates to consider the data statistically as *point processes*. Point processes characterize the random occurrence of single events along the time axis in dependence of observable characteristics (marks) and of the process history. The importance of point process models in financial econometrics has been discussed for the first time by the 2003 Nobel laureate Robert F. Engle on the 51th European Meeting of the Econometric Society in Istanbul, 1996. His paper[1] can be regarded as the starting point for a fast growing body of research in high-frequency financial econometrics literature.

From an econometric point of view, the major objective of this monograph is to provide a methodological framework to model *dynamic point processes*, i.e. point processes that are subjected to serial dependencies. This book provides a thorough review of recent developments in the econometric and statistical literature, puts forward existing approaches and opens up new directions. We present alternative ways to model dynamic point processes, illustrate their theoretical properties and discuss their ability to account for specific features of point process data, like the occurrence of time-varying covariates, censoring mechanisms and multivariate structures.

In this book, we explicitly focus on dynamic *duration* and *intensity* models[2] that allow us to describe point processes in a discrete-time and continuous-time framework. Using a duration framework and modelling the distribution of waiting times between subsequent events is the most common way to characterize point processes in *discrete time*. We illustrate that in such a framework, the inclusion of dynamic structures is straightforward since traditional time series models, like autoregressive moving average (ARMA) models (see Box and Jenkins, 1976) or generalized autoregressive conditional heteroscedasticity (GARCH) type specifications (see Engle, 1982, Bollerslev, 1986) can be adopted. However, the main drawback of a discrete-time approach is that it is extremely difficult to account for events arriving during a duration spell. This becomes particularly important in the context of multivariate data or time-varying covariates.

The latter issue is addressed by modelling point processes in a *continuous-time* framework. A convenient and powerful way to describe a point process in continuous time is to specify the (stochastic) intensity function. The intensity function is a central concept in the theory of point processes and is defined as the instantaneous rate of occurrence given the process history and observable

---

[1] It has been published later under Engle (2000).

[2] A third possibility to model point processes is to model the number of points in equi-distant time intervals. Since the focus on this book explicitly lies on the analysis of *irregularly* spaced data, the discussion of (dynamic) count data models is beyond its scope. Interested readers are referred to the textbooks by Winkelmann (1997), Cameron and Trivedi (1998) or the paper by Davis, Rydberg, Shephard, and Streett (2001), among others.

factors. The major advantage of the modelling of the intensity function is that it allows us to account for events that occur in *any* point in time like in the case of time-varying covariates or at the event arrival of other processes. As illustrated in this book, dynamic intensity processes can be specified either as so-called *self-exciting* intensity processes where the intensity is driven by functions of the backward recurrence time to all previous points (see e.g. Hawkes, 1971) or as time series specifications (see e.g. Russell, 1999) where the intensity function follows a dynamic structure that is updated at each occurrence of a new point.

From an economic point of view, the focus of this book lies on the analysis of so-called *financial point processes*. Financial point processes are defined as stochastic processes that characterize the arrival time of certain events in the trading process on a financial market. A salient feature of financial event data is the occurrence of strong clustering structures, i.e., events on a financial markets typically arrive in clusters leading to strong serial dependencies in the waiting time between consecutive points. Hence, the modelling of such data requires the aforementioned dynamic point process approaches. By selecting the points of a process in a certain way, specific types of financial point processes are generated. These have different economic implications and allow us to analyze market activity from various viewpoints. We illustrate that the analysis of trade durations, defined as the time between consecutive trades, allows us to gain deeper insights into market dynamics on a trade-to-trade level and into market microstructure relationships. Alternatively, by defining financial events in terms of the price and/or the volume process, we show how to construct valuable risk and liquidity measures. In this context, we illustrate several applications of the econometric framework on the basis of different databases.

The book has the following structure: The first part of the book is devoted to the provision of the methodological and economic background (Chapters 2 and 3), as well as a detailed description of the databases (Chapter 4). In the second part (Chapters 5 through 7) we present the different econometric models and illustrate their application to various financial data.

Chapter 2 introduces in the theory of point processes and provides the methodological background. This chapter serves as a starting point for the following chapters of the book. The main focus lies on the intensity function and the integrated intensity function as the key concepts to describe point processes in a continuous-time framework. We briefly review their statistical properties and illustrate how to perform intensity-based inference. By restricting our attention to a non-dynamic framework, we illustrate the basic statistical concepts and discuss different ways to model point processes. In this context, we provide a thorough review of the different possibilities to classify point process models, namely the distinction between *proportional intensity (PI) models* and *accelerated failure time (AFT) models*, as well as the classification in intensity models, duration models and count data models. The relationships and differences between the individual types of point process models are crucial for an understanding of the particular dynamic extensions considered in the remainder of the book.

In *Chapter 3*, we discuss alternative financial duration concepts and focus on their economic implications. Here, we consider the (multivariate) point process describing the complete trading process of a particular financial asset over a given time span. By selecting systematically individual points of this process, different types of financial duration processes are generated. The most common type of financial duration is the *trade duration* that is defined as the time between subsequent transactions. It measures the speed of the market, and thus is strongly related to the intensity for liquidity demand. Furthermore, it is a major source of intraday volatility and plays an important role in market microstructure theory. We particularly discuss the relationship between trade durations and other market microstructure variables, like the bid-ask spread, the transaction price, as well as the transaction volume and derive several testable economic hypotheses. A further focus of Chapter 3 lies on the consideration of aggregated trade durations, like *price durations* and *volume durations*. A price duration is defined as the waiting time until a certain cumulative absolute price change is observed, thus it corresponds to a first passage time associated with the price process. As discussed in more detail in this chapter, price durations play an important role in constructing alternative volatility and risk measures. Volume durations are associated with the time until a given cumulative volume is traded on the market. Since they capture the time and volume component of the trading process, they measure important determinants of liquidity. A refinement of this concept are the so-called *excess volume durations* that measure the time until a predetermined buy or sell excess volume is traded. As argued in Hautsch (2003), they serve as natural indicators for the existence of information on the market and allow to assess liquidity suppliers' risk with respect to asymmetric information and inventory problems. The use of excess volume durations and price durations in constructing measures for particular dimensions of liquidity is illustrated in more detail.

After discussing the economic implications of alternative financial duration concepts, *Chapter 4* provides deeper insights into the statistical properties of financial durations. Here, we present the transaction databases that are used for several empirical applications in this book consisting of data from the New York Stock Exchange (NYSE), the German Stock Exchange, the London International Financial Futures Exchange (LIFFE), the EUREX trading platform in Frankfurt, as well as from the Australian Stock Exchange (ASX). One particular section is devoted to issues of data preparation and specific pitfalls that have to be taken into account when transaction data sets are constructed. Furthermore, we illustrate the main statistical properties of different types of financial durations based on various assets and exchanges.

In *Chapter 5*, we present the most common type of autoregressive duration model, namely, the *autoregressive conditional duration (ACD) model* proposed by Engle and Russell (1998). The main idea behind the ACD model is a dynamic parameterization of the conditional duration mean. It allows for a modelling of autocorrelated durations in a discrete-time framework and combines elements of a GARCH specification with features of a duration model. It is the most common type of autoregressive point process model since it is easy to estimate and can be used as a building block for models of the price process (see Engle, 2000). However, several studies illustrate that the basic form of the ACD model is too restrictive to correctly describe financial duration processes. This has given birth to a new field of research and a plethora of papers that focussed on extensions of the ACD model in several directions. In this context, we provide a review of several extended ACD models and discuss new suggestions and directions.

A main objective of this chapter is devoted to specification tests for ACD models. While several studies focus on tests against violations of distributional assumptions, there is still some need to test for misspecifications of the functional form. This issue is related to the question whether the conditional mean restriction, which is the essential assumption behind the ACD model, actually holds. We address this issue by applying the framework of (integrated) conditional moment tests to test against violations of the conditional mean restriction. The evaluation of different ACD specifications based on trade and price durations shows that in fact more simple models imply violations of the conditional mean restriction, and thus nonlinear ACD specifications are required. Applications of the ACD model are related to empirical tests of market microstructure hypotheses, as well as the quantification of illiquidity risks on the basis of excess volume durations. The latter study allows us to gain deeper insights into the dynamics and determinants of the time-varying excess demand intensity and its relationship to liquidity and market depth.

In *Chapter 6*, we present dynamic semiparametric proportional intensity models as an alternative and direct counterpart to the class of autoregressive AFT models considered in Chapter 5. In the semiparametric PI model, it is assumed that the intensity function is specified based on a multiplicative relationship between some non-specified baseline intensity that characterizes the underlying distribution of the durations and a function of covariates. The important advantage of this model is that no explicit distributional assumptions are needed since the baseline intensity is estimated semiparametrically. The so-called *autoregressive conditional proportional intensity (ACPI) model* proposed by Gerhard and Hautsch (2002a) exploits the close relationship between ordered response models and models for categorized durations. The main idea behind the model is to categorize the durations and formulate the semiparametric PI model in terms of a specific type of ordered response model which is augmented by an observation driven dynamic. In this sense, the ACPI model implies a dynamic parameterization of the integrated intensity function. This framework leads to a model that allows for a consistent estimation of the dynamic parameters without requiring explicit distributional assumptions for the baseline intensity. Moreover, discrete points of the baseline survivor function can be estimated simultaneously with the dynamic parameters. The ACPI model is specified in discrete time, however embodies characteristics of pure duration models, as well as of intensity approaches. A particular strength of the model is that it is a valuable approach to account for censoring structures. Censoring occurs if there exist intervals in which the point process cannot be observed directly. This might be, for example, due to non-trading periods, like nights, weekends or holidays. In this case, the exact timing of particular points of the process cannot be observed and can only be isolated by a corresponding interval. Such effects cause problems in an autoregressive framework when the information about the exact timing of a point is needed for the sequel of the time series. Because of the relationship between the integrated intensity and the conditional survivor function, an autoregressive model for the integrated intensity is a natural way to account for censoring mechanisms in a dynamic framework. We illustrate the application of this framework for estimating price change volatilities based on censored price durations. In this context, we re-investigate the study of Gerhard and Hautsch (2002b) by modelling price durations based on the Bund future trading at LIFFE. By focussing explicitly on highly aggregated price durations, the need to account for censoring mechanisms becomes apparent.

Chapter 7 addresses univariate and multivariate autoregressive intensity models. This chapter deals with dynamic parameterizations of the intensity function itself and allows for a modelling of point processes in a continuous-time framework. We focus on two general types of models: *autoregressive conditional intensity models* and *self-exciting intensity models*. The first class of models is proposed by Russell (1999) and is based on an autoregressive structure for the intensity function that is updated at each occurrence of a new point. In the latter class of models, the intensity is driven by a function of the backward recurrence time to all previous points. Hawkes (1971) introduces a particular type of linear self-exciting process based on an exponentially decaying backward recurrence function. For both types of models we illustrate the extension to the multivariate case and discuss their theoretical properties, as well as estimation issues.

As a generalized framework for the modelling of point processes we consider *dynamic latent factor (LFI) models* for intensity processes as proposed by Bauwens and Hautsch (2003). The main idea is to assume that the conditional intensity given the (observable) history of the process is not deterministic but stochastic and follows itself a dynamic process. Hence, the LFI model embodies characteristics of a doubly stochastic Poisson process (see, for example, Grandell, 1976, or Cox and Isham, 1980). We illustrate the implementation of a latent dynamic factor in both types of the aforementioned intensity models, leading to a latent factor autoregressive conditional intensity (LF-ACI) model and a latent factor Hawkes (LF-Hawkes) model and discuss estimation issues, as well as theoretical properties.

Several applications of dynamic intensity models to financial duration data show the potential of this framework. Here, we illustrate the use of multivariate Hawkes models to estimate multivariate instantaneous price change volatilities. Moreover, a central objective addresses the analysis of the limit order book of the ASX. By applying bivariate ACI models, we model the relationship between the simultaneous buy/sell arrival intensity and the current state of the market characterized by market depth, tightness and the cumulated volume in the particular queues of the order book. Furthermore, we illustrate applications of univariate and multivariate LFI models for the modelling of XETRA trade durations. Finally, in *Chapter 8* we conclude and summarize the main findings of this book.

Note that it is not necessary to read all chapters in a strict order. For readers who are mainly interested in methodological issues, it is recommended to read Chapter 2 and then to focus on Chapters 5, 6 and 7 which can be read separately. Readers who are interested in economic and statistical implications of financial durations, as well as issues of data preparation, are referred to the Chapters 3 and 4. Notice that for a better understanding of the notation, the Appendix contains an alphabetical list of all variables used in this book.

Point Processes

This chapter provides the methodological background for the specification and estimation of point processes. In this sense, it is the starting point for the following chapters of the book. We give a brief introduction to the fundamental statistical concepts and the basic ways to model point processes. For ease of introduction we restrict our attention to non-dynamic point processes. In Section 2.1, we illustrate the most important theoretical concepts. Here, the focus lies on the idea of the intensity function as a major concept in the theory of point processes. In Section 2.2, different types of point processes are discussed while Section 2.3 gives an overview of alternative ways to model and to estimate such processes. Section 2.4 is concerned with the treatment of censoring mechanisms and time-varying covariates that might also occur in financial point processes. Section 2.5 provides an overview of different possibilities for dynamic extensions of basic point process models. This section serves as the starting point for the various autoregressive point process models considered in the sequel of the book.

2.1 Basic Concepts of Point Processes

In this section, we discuss the central concepts of point processes. Textbook treatments of point processes are given by Cox and Isham (1980), Kalbfleisch and Prentice (1980), Brémaud (1981), Karr (1991), Snyder and Miller (1991), or Lancaster (1997) among others.

2.1.1 Fundamental Definitions

Point processes

Let t denote the physical (calendar) time. Furthermore, let $\{t_i\}_{i \in \{1,2,\ldots\}}$ be a sequence of nonnegative random variables on some probability space $(\Omega, \mathfrak{F}, \mathcal{P})$ associated with random arrival times $0 \leq t_i \leq t_{i+1}$. Then, the sequence $\{t_i\}$

is called a *point process* on $[0, \infty)$. If $t_i < t_{i+1} \; \forall \; i$, the process is called a *simple point process*. Thus, the definition of a simple point process excludes the possibility of the occurrence of events simultaneously. During the remainder of the book, we restrict our consideration to simple point processes. Furthermore, denote n as the sample range and t_n as the last observed point of the process. Then, $\{t_i\}_{i=1}^n$ is the complete observable process.

Marks, (Time-Varying) Covariates

Let $\{z_i\}_{i \in \{1,2,\dots\}}$ be a sequence of random vectors corresponding to the characteristics associated with the arrival times $\{t_i\}_{i \in \{1,2,\dots\}}$. Then, $\{z_i\}_{i \in \{1,2,\dots\}}$ is called the sequence of *marks* while the double sequence $\{t_i, z_i\}_{i \in \{1,2,\dots\}}$ is called a (simple) *marked point process*. Marks can be interpreted as *time-invariant* covariates (i.e. covariates that do not change between two subsequent points) or as the realizations of a *time-varying* covariate process at the corresponding points t_i. In order to consider such processes, $z(t)$ is defined as a vector of time-varying covariates corresponding to t. Then, $z(t_i) := z_i$.

Counting Processes

The process $N(t)$ with $N(t) := \sum_{i \geq 1} \mathbb{1}_{\{t_i \leq t\}}$ is called a *right-continuous counting process* associated with $\{t_i\}$. $N(t)$ is a right-continuous step function with upward jumps (at each t_i) of magnitude 1. Furthermore, the process $\check{N}(t)$ with $\check{N}(t) := \sum_{i \geq 1} \mathbb{1}_{\{t_i < t\}}$ is called a *left-continuous counting process* associated with $\{t_i\}$. $\check{N}(t)$ is a left-continuous step function and counts the number of events that occur *before* t.

Durations and Backward Recurrence Times

Define x_i as the waiting time between two successive points, defined as

$$
x_i := \begin{cases} t_i - t_{i-1}, & i = 2, 3, \dots \\ t_i, & i = 1 \quad (\text{with } t_0 := 0). \end{cases}
$$

Then, $\{x_i\}_{i \in \{1,2,\dots\}}$ is called the *duration process* associated with $\{t_i\}$. Moreover, the process $x(t)$ with $x(t) := t - t_{\check{N}(t)}$ is called the *backward recurrence time* at t. The backward recurrence time is the time elapsed since the previous point and is a left-continuous function that grows linearly through time with discrete jumps back to zero *after* each arrival time t_i. Note that $x(t_i) = t_i - t_{i-1} = x_i$.

2.1.2 The Homogeneous Poisson Process

The simplest type of point process is given by the homogeneous Poisson process. The Poisson process of rate λ is defined by the requirements that for all t, as the time interval $\Delta \downarrow 0$,

$$\text{Prob}\left[(N(t + \Delta) - N(t)) = 1 \,|\, \mathfrak{F}_t\right] = \lambda\Delta + o(\Delta), \qquad (2.1)$$
$$\text{Prob}\left[(N(t + \Delta) - N(t)) > 1 \,|\, \mathfrak{F}_t\right] = o(\Delta), \qquad (2.2)$$

leading to

$$\text{Prob}\left[(N(t + \Delta) - N(t)) = 0 \,|\, \mathfrak{F}_t\right] = 1 - \lambda\Delta + o(\Delta), \qquad (2.3)$$

where $o(\Delta)$ denotes a remainder term with the property $o(\Delta)/\Delta \to 0$ as $\Delta \to 0$ and $\mathfrak{F}_t$ denotes the history of the process up to (and inclusive) t. One major property of the Poisson process is that the probability for the occurrence of an event in $(t, t + \Delta]$ is independent from $\mathfrak{F}_t$, i.e., it does not depend on the number of points observed before (and also exactly at) t. Equations (2.1) and (2.2) are associated with the *intensity representation* of a Poisson process.

Further key properties of the Poisson process are related to the distribution of events in a fixed time interval. Following Lancaster (1997), it is easy to see that the probability for the occurrence of j events *before* $t + \Delta$ is obtained by

$$\text{Prob}\left[\check{N}(t + \Delta) = j\right] = \text{Prob}\left[\check{N}(t) = j\right] \cdot \text{Prob}\left[(N(t + \Delta) - N(t)) = 0\right]$$

$$+ \text{Prob}\left[\check{N}(t) = j - 1\right] \cdot \text{Prob}\left[(N(t + \Delta) - N(t)) = 1\right]$$

$$+ \text{Prob}\left[\check{N}(t) = j - 2\right] \cdot \text{Prob}\left[(N(t + \Delta) - N(t)) = 2\right]$$

$$\vdots$$

$$= \text{Prob}\left[\check{N}(t) = j\right] \cdot (1 - \lambda\Delta)$$

$$+ \text{Prob}\left[\check{N}(t) = j - 1\right] \cdot \lambda\Delta + o(\Delta). \qquad (2.4)$$

By rearranging the terms and dividing by Δ, one obtains

$$\Delta^{-1}\left(\text{Prob}\left[\check{N}(t + \Delta) = j\right] - \text{Prob}\left[\check{N}(t) = j\right]\right)$$

$$= -\lambda\text{Prob}\left[\check{N}(t) = j\right] + \lambda\text{Prob}\left[\check{N}(t) = j - 1\right] + o(\Delta)\Delta^{-1} \qquad (2.5)$$

and thus, for $\Delta \downarrow 0$

$$\frac{d}{dt}\text{Prob}\left[\check{N}(t) = j\right] = -\lambda\text{Prob}\left[\check{N}(t) = j\right] + \lambda\text{Prob}\left[\check{N}(t) = j - 1\right]. \qquad (2.6)$$

For $j = 0$, (2.6) becomes

$$\frac{d}{dt}\text{Prob}\left[\check{N}(t)=0\right]=-\lambda\text{Prob}\left[\check{N}(t)=0\right], \qquad (2.7)$$

because $\text{Prob}\left[\check{N}(t)=j-1\right]=0$ for $j=0$. By accounting for the initial condition $\text{Prob}\left[\check{N}(0)=0\right]=1$, we obtain

$$\text{Prob}\left[\check{N}(t)=0\right]=\exp(-\lambda t), \qquad (2.8)$$

which corresponds to the survivor function of an exponential distribution. Thus, the waiting time until the first event is exponentially distributed. Analogously it is shown that (2.8) is also the probability that no events occur in an interval of length t starting at any arbitrary point on the time axis, like, at the occurrence of the previous event. Therefore, the duration between subsequent points is independent of the length of all other spells. Hence, it can be concluded that the durations are independently exponentially distributed with probability density function (p.d.f.) and cumulative distribution function (c.d.f.) given by

$$f(x)=\lambda\exp\left(-\lambda x\right) \qquad (2.9)$$
$$F(x)=1-\exp\left(-\lambda x\right). \qquad (2.10)$$

This property is associated with the *duration representation* of a Poisson process.

The third key property of the Poisson process is obtained by solving the differential equation (2.6) successively for $j=1,2,\ldots$ given the solution of (2.6) for $j=0$. It is easily shown that this differential equation has the solution

$$\text{Prob}\left[\check{N}(t)=j\right]=\frac{\exp(-\lambda t)(\lambda t)^j}{j!}, \qquad j=1,2,\ldots. \qquad (2.11)$$

Therefore, the number of events during an interval of length t is Poisson distributed with parameter λt. Note that (2.11) does not depend on the starting point on the time axis, thus, the property holds for *any* time interval. Hence, the number of events in the interval $(t',t]$, $N(t)-N(t')$, is Poisson distributed with parameter $\lambda(t-t')$. This relationship is called the *counting representation* of a Poisson process.

2.1.3 The Intensity Function and its Properties

A central concept in the theory of point processes is the intensity function[1] that is defined as following:

---

[1] Note that the notation is not consistent in all papers or textbooks. For example, Brémaud (1981) and Karr (1991) denote it as "stochastic intensity function". Here, we follow the notation of, for example, Aalen (1978) and Snyder and Miller (1991).

Definition 2.1. *Let $N(t)$ be a simple point process on $[0, \infty)$ that is adapted to some history $\mathfrak{F}_t$ and assume that $\lambda(t; \mathfrak{F}_t)$ is a positive process with sample paths that are left-continuous and have right-hand limits. Then, the process*

$$\lambda(t; \mathfrak{F}_t) := \lim_{\Delta \downarrow 0} \frac{1}{\Delta} \mathrm{E}\left[N(t + \Delta) - N(t)| \mathfrak{F}_t\right], \quad \lambda(t; \mathfrak{F}_t) > 0, \ \forall \, t, \qquad (2.12)$$

is called the $\mathfrak{F}_t$-intensity process of the counting process $N(t)$.

Hence, the $\mathfrak{F}_t$-intensity process characterizes the evolution of the point process $N(t)$ conditional on some history $\mathfrak{F}_t$. Typically one considers the case $\mathfrak{F}_t = \mathfrak{F}_t^N$, where $\mathfrak{F}_t^N$ denotes the *internal* filtration

$$\mathfrak{F}_t^N = \sigma(t_{N(t)}, z_{N(t)}, t_{N(t)-1}, z_{N(t)-1}, \ldots, t_1, z_1)$$

consisting of the complete (observable) history of the point process up to t. However, the intensity concept allows also to condition on a wider filtration $\mathfrak{F}_t \supset \mathfrak{F}_t^N$ including, for example, unobservable factors or the path of time-varying covariates between two subsequent points.

Since the assumption of a simple point process implies (2.2), the intensity function can be alternatively written as

$$\lambda(t; \mathfrak{F}_t) = \lim_{\Delta \downarrow 0} \frac{1}{\Delta} \mathrm{Prob}\left[(N(t + \Delta) - N(t)) > 0 \,|\mathfrak{F}_t\right], \qquad (2.13)$$

which can be associated, roughly speaking, with the conditional probability per unit time to observe an event in the next instant, given the conditioning information.

Typically, the $\mathfrak{F}_t$-intensity is also expressed in terms of the conditional expectation of $N(t) - N(t')$ given $\mathfrak{F}_{t'}$ for all t, t' with $t > t'$:

$$\mathrm{E}[N(t) - N(t')|\mathfrak{F}_{t'}] = \mathrm{E}\left[\left.\int_{t'}^{t} \lambda(s)ds \right| \mathfrak{F}_{t'}\right]. \qquad (2.14)$$

Thus, the expected number of events in the interval $(t', t]$, given $\mathfrak{F}_{t'}$, is computed as the conditional expectation of the integrated intensity function. Note that (2.14) implies a martingale representation. In particular, under fairly weak regularity conditions (see, for example, Chapter 2 of Karr, 1991), the expression

$$N(t) - \int_0^t \lambda(s)ds$$

is a martingale. This relationship can be exploited to derive diagnostic tests for intensity models.

The intensity function is the counterpart to the *hazard function*, which is a central concept in traditional duration or survival analysis[2]. However, in

---

[2] See, for example, Kalbfleisch and Prentice (1980), Kiefer (1988) or Lancaster (1997).

classical duration literature, one typically analyzes cross-sectional duration data. Hence, in such a framework, there typically exists no history of the process before the beginning of a spell, and thus, one typically considers the hazard function, defined as

$$\tilde{\lambda}(s) := \lim_{\Delta \downarrow 0} \frac{1}{\Delta} \text{Prob} \left[s \leq x_i < s + \Delta \,|\, x_i \geq s \right]. \qquad (2.15)$$

Clearly, the hazard and intensity function are alternative formulations of the same concept. However, the only distinction is that the hazard function is typically defined in terms of the durations x_i and is used for cross-sectional data. In contrast, the intensity function is defined in continuous time and is used in the framework of point processes. Clearly, both approaches coincide when the hazard concept is extended to allow not only to condition on the time elapsed since the beginning of the spell, but on the *complete* information set at each point t. Alternatively, when there exists no history of the process before the beginning of the spell, then, $\lambda(t; x(t)) = \tilde{\lambda}(x(t))$.

The convenient and well known properties of a homogeneous Poisson process (see Section 2.1.2) are often exploited to construct diagnostics for point processes. In this context, the *integrated intensity function*, defined as

$$\Lambda(t_{i-1}, t_i) := \int_{t_{i-1}}^{t_i} \lambda(s; \mathfrak{F}_s) ds \qquad (2.16)$$

plays a key role. The following theorem establishes distributional properties of the integrated intensity function that are crucial for the derivation of diagnostic tests.

Theorem 2.2 (Theorem T16 of Brémaud, 1981). *Let $N(t)$ be a simple point process on $[0, \infty)$. Suppose that $N(t)$ has the intensity function $\lambda(t; \mathfrak{F}_t)$ that satisfies*

$$\int_0^\infty \lambda(t; \mathfrak{F}_t) dt = \infty. \qquad (2.17)$$

Define for all t, the stopping-time τ_t as the solution to

$$\int_0^{\tau_t} \lambda(s; \mathfrak{F}_s) ds = t. \qquad (2.18)$$

Then, the point process $\tilde{N}(t) = N(\tau_t)$ is a homogenous Poisson process with intensity $\lambda = 1$.

Proof. See the proof of Theorem T16 of Brémaud (1981). □

Note that the only condition needed is given by (2.17) and is easily satisfied in the context of transaction data since it is assumed zero probability for the occurrence of no more events after some points in time. Based on this theorem, it can be shown that $\tilde{t}_i - \tilde{t}_{i-1} = \int_{t_{i-1}}^{t_i} \lambda(s; \mathfrak{F}_s)ds = \Lambda(t_{i-1}, t_i)$, where $\{\tilde{t}_i\}_{i \in \{1,2,...\}}$ denotes the sequence of points associated with $\tilde{N}(t)$ (for a small formal proof, see Bowsher, 2002). Then it follows that

$$\Lambda(t_{i-1}, t_i) \sim \text{ i.i.d. } Exp(1). \tag{2.19}$$

Note that (2.18) is associated with a transformation of the time scale transforming any non-Poisson process into a homogenous Poisson process. Moreover, $\Lambda(t_{i-1}, t_i)$ establishes the link between the intensity function and the duration until the occurrence of the next point. For example, in a simple case where the intensity function is constant during two points, the duration x_i is given by $x_i = \Lambda(t_{i-1}, t_i)/\lambda(t_i; \mathfrak{F}_{t_i})$. Furthermore, $\Lambda(t_{i-1}, t_i)$ can be interpreted as a generalized error (for example, in the spirit of Cox and Snell, 1968) and indicates whether the path of the conditional intensity function under-predicts ($\Lambda(t_{i-1}, t_i) > 1$) or over-predicts ($\Lambda(t_{i-1}, t_i) < 1$) the number of events between t_i and t_{i-1}.

The relationship between the intensity function $\lambda(t; \mathfrak{F}_t)$ and the conditional survivor function is obtained by the so called "exponential formula" (see Yashin and Arjas, 1988)

$$S(x_i; \mathfrak{F}_{t_i}) = \exp\left(-\Lambda(t_{i-1}, t_i)\right). \tag{2.20}$$

When there is no conditioning, this relationship is simply derived by solving the differential equation

$$\frac{d}{dx_i} S(x_i) = -\lambda(t_{i-1} + x_i)S(x_i) \tag{2.21}$$

subject to the initial condition $S(0) = 1$. As illustrated by Yashin and Arjas (1988), this relationship also holds if the conditioning is based on a fixed σ-algebra. However, it does not necessarily hold when the conditioning is "dynamically", i.e., depending on time-dependent random factors (e.g. time-varying covariates). More concretely, Yashin and Arjas (1988) prove that the "exponential formula", (2.20), is only valid if $S(x_i; \mathfrak{F}_{t_{i-1}+x_i})$ is absolutely continuous in x_i. Note that his assumption excludes jumps of the conditional survivor function induced by changes of the information set during a spell. This can be an issue in the framework of time-varying covariates and multivariate processes.

2.1.4 Intensity-Based Inference

Karr (1991) (see Karr's Theorem 5.2) proves the existence of a unique probability measure such that $N(t)$ has the $\mathfrak{F}_t$-intensity $\lambda(t; \mathfrak{F}_t)$. This theorem is fundamental for intensity-based statistical inference since it implies that a statistical model can be completely specified in terms of the $\mathfrak{F}_t$-intensity. A

valuable conclusion from Karr's Theorem 5.2 is the establishment of a likelihood function in terms of the intensity. Assume that the intensity process
satisfies the condition

$$\int_0^{t_n} \lambda(s; \mathfrak{F}_s)ds < \infty$$

and denote W as the data matrix, consisting of the observed points $\{t_i\}_{i=1}^n$
and possible covariates. Then, the log-likelihood function associated with the
parameter vector θ is given by

$$\ln \mathcal{L}\,(W; \theta) = \int_0^{t_n} (1 - \lambda(s; \mathfrak{F}_s))ds + \int_{(0,t_n]} \ln \lambda(s; \mathfrak{F}_s)dN(s)$$

$$= \int_0^{t_n} (1 - \lambda(s; \mathfrak{F}_s))ds + \sum_{i \geq 1} \mathbb{1}_{\{t_i \leq t_n\}} \ln \lambda(t_i; \mathfrak{F}_{t_i}), \qquad (2.22)$$

where

$$\int_{(0,t]} U(s)dN(s) = \sum_{i \geq 1} \mathbb{1}_{\{t_i \leq t\}} U(t_i) \qquad (2.23)$$

defines the stochastic Stieltjes integral[3] of some measurable point process $U(t)$
with respect to $N(s)$.

2.2 Types of Point Processes

2.2.1 Poisson Processes

The most simple point process is the *homogeneous Poisson process*, (2.1)-(2.2).
As presented in Section 2.1.2, a key property of the homogeneous Poisson
process is that the intensity is constant. Thus, the data generating process
(DGP) is fully described by the intensity $\lambda(t) = \lambda$.

By allowing the intensity rate λ to depend on some observable factors, we
obtain the general class of *non-homogenous Poisson processes*. For example, if
the Poisson process depends on marks which are observable at the beginning
of the spell, then the DGP is fully described by the $z_{\check{N}(t)}$-intensity $\lambda(t; z_{\check{N}(t)})$.[4]
One special case is the so-called *non-stationary Poisson process*, where λ is
a function of time. It is described by (2.1)-(2.2) but with λ replaced by $\lambda(t)$.

---

[3] See, for example, Karr (1991), p. 57.

[4] Care is needed with respect to the specification of the intensity function in dependence of observable factors. Note that the intensity function is left-continuous
with right-hand limits. This property implies that $\lambda(t; \mathfrak{F}_t)$ cannot depend on factors which are not observable at least instantaneously *before* t. See also Section
2.3.1.

Clearly, a non-stationary Poisson process typically does not imply the independence of consecutive durations.

A further type of non-homogenous Poisson process is obtained by the class of *doubly stochastic Poisson processes*. In this framework, it is assumed that the intensity is driven by some unobserved stochastic process $\lambda^*(t)$. This leads to a DGP that is characterized by the intensity $\lambda(t; \mathfrak{F}_t^*)$, where $\mathfrak{F}_t^*$ denotes the history of the unobserved process up to t. Such processes will be discussed in more detail in Chapter 7. For more details concerning specific types of non-homogeneous Poisson processes, see Cox and Isham (1980).

2.2.2 Renewal Processes

Another way to generalize Poisson processes is to relax the assumption that the durations between subsequent points are exponentially distributed. Such processes are called *renewal processes*. In contrast to Poisson processes, here the intensity depends on the backward recurrence time $x(t)$ leading to non-constant intensity shapes during a spell. Hence, the DGP of a renewal process is described by the intensity or, respectively, hazard function $\lambda(t; x(t)) = \tilde{\lambda}(x(t))$. Therefore, the durations x_i between successive events are no longer exponentially distributed, but follow more flexible distributions with density $f(\cdot)$. For renewal processes, in contrast to Poisson processes, the time since the previous point has an influence on the further development of the process. Because of this property, one has to take care of the initial conditions. In general, the *forward recurrence time* x_1 from $t = 0$ to the first subsequent point does not have the density $f(\cdot)$ except for the case when there is a point in $t = 0$. In the latter case, the process is called an *ordinary* renewal process, otherwise it is a *modified* renewal process. More details are given in Cox and Isham (1980). The shape of the intensity function determines the so-called *duration dependence* which is associated with the sign of $d\tilde{\lambda}(s)/ds$.[5] If $d\tilde{\lambda}(s)/ds > (<) 0$, the process is said to exhibit positive (negative) duration dependence, i.e., the hazard rate increases (decreases) over the length of the spell. Note that in a renewal process, the durations between subsequent points are assumed to be independent. Thus, the process history before t_{i-1} has no impact on the current intensity. Therefore, a renewal process is typically used to model cross-sectional duration data.

2.2.3 Dynamic Point Processes

A generalized renewal process is obtained by allowing for some Markovian structure in the sequence of duration distributions. A so called *Semi-Markov process* is obtained by specifying a stochastic sequence of distribution functions for the duration sequence. By supposing R distribution functions

---

[5] This expression has to be used carefully since one might confuse it with some dynamic dependence in a time series context.

$F^{(1)}(\cdot), \ldots, F^{(R)}(\cdot)$ associated with R particular states, the process is said to be in state j in t if the current distribution function of $x_{\tilde{N}(t)+1}$ is $F^{(j)}(\cdot)$.[6] The transitions are determined by a transition matrix P^* where the ijth element is the transition probability from state i to state j. For more details, see Cox and Isham (1980).

By relaxing the Markovian structure of the Semi-Markov process, more general dynamic point processes are obtained. One particular type of dynamic point process is the class of *self-exciting processes*. Hawkes (1971) introduces a special class of linear self-exciting processes where the intensity is a weighted function of the backward recurrence time to all previous points. The advantage of this approach is to estimate the nature of the dependence of the intensity without imposing a strong a priori time series structure. For more details and applications of Hawkes processes to financial data, see Chapter 7.

An alternative to self-exciting processes is the class of *autoregressive point processes*. In this context, renewal processes are augmented by time series structures that may be specified in terms of the intensity function, the integrated intensity function, the duration between subsequent points, or in terms of the counting function. Detailed illustrations of such types of point processes will be given in the sequel of this book.

2.3 Non-Dynamic Point Process Models

In this section, we will discuss briefly three basic ways to model point processes:

- models for intensity processes
- models for duration processes
- models for counting processes

In this section, we restrict our considerations to the non-dynamic case. In the following chapters of the book, duration and intensity concepts will be reconsidered in more detail and will be extended to a dynamic framework.

2.3.1 Intensity-Based Models

The so-called proportional hazard model (PH) and accelerated failure time model (AFT) have their origins in cross-sectional survival analysis in bio-

---

[6] Cox and Isham (1980) refer the term "*Semi*-Markov" to the property that the individual state-dependent distribution functions $F^{(j)}(\cdot)$ are associated with renewal processes. Hence, the Semi-Markov process is a generalization of a Markov process in which the time spent in a particular state is no longer exponentially distributed.

statistics[7] and labor economics[8]. Thus, they are typically formulated in terms of the hazard function $\tilde{\lambda}(x(t))$. However, for reasons of consistency, we accomplish the following representations in terms of the intensity formulation. Furthermore, we restrict our attention to the case of time-invariant covariates (marks) $z_{\check{N}(t)}$ without any censoring structures. The phenomena of censoring and time-varying covariates will be considered in Sections 2.4.1 and 2.4.2.

Proportional Intensity (PI) Models

The PI model is specified as the product of a baseline intensity function $\lambda_0(t) > 0$ and a strictly positive function of a vector of covariates $z_{\check{N}(t)}$ (observable at the beginning of the spell) with coefficients γ. It is given by

$$\lambda(t; z_{\check{N}(t)}) = \lambda_0(t) \exp(-z'_{\check{N}(t)}\gamma). \qquad (2.24)$$

The baseline intensity $\lambda_0(t)$ corresponds to $\lambda(t; z_{\check{N}(t)} = 0)$. I.e., if the regressors are centered, $\lambda_0(t)$ has an interpretation as the intensity function for the mean values of $z_{\check{N}(t)}$. The key property of the PI model is that $\partial \ln \lambda(t; z_{\check{N}(t)})/\partial z_{\check{N}(t)} = \gamma$, thus, γ is interpreted as the constant proportional effect of $z_{\check{N}(t)}$ on the conditional rate of occurrence.

The most common examples of a (parametric) PI model are the exponential and Weibull regression model. In these models, the baseline intensity $\lambda_0(t)$ is specified according to the exponential or Weibull hazard function. While the Weibull model allows only for monotonic increasing or decreasing hazard shapes, more flexible parameterizations are obtained, for instance, by the gamma, generalized gamma, Burr or generalized F distribution (see Appendix A). However, an alternative is to specify the baseline intensity as a linear spline function. Assume that the observed durations are partitioned into K categories, where $\bar{x}_k$, $k = 1, \ldots, K - 1$ denote the chosen category bounds. Then, a semiparametric baseline intensity is obtained by

$$\lambda_0(t) = \exp\left[\nu_0 + \sum_{k=1}^{K-1} \nu_{0,k} \mathbb{1}_{\{x_{N(t)} \leq \bar{x}_k\}}(\bar{x}_k - x_{N(t)})\right], \qquad (2.25)$$

where ν_0 is a constant and $\nu_{0,k}$ are coefficients associated with the nodes of the spline function. Parametric PI models are easily estimated by maximum likelihood (ML) adapting the log likelihood function (2.22). Note that in a

---

[7] See, for example, Kalbfleisch and Prentice (1980), Cox and Oakes (1984) or the recent survey by Oakes (2001).

[8] A well known example is the analysis of the length of unemployment spells which is studied by a wide range of theoretical and empirical papers, see e.g. Lancaster (1979), Nickell (1979), Heckmann and Singer (1984), Moffitt (1985), Honoré (1990), Meyer (1990), Han and Hausman (1990), Gritz (1993), McCall (1996) or van den Berg and van der Klaauw (2001) among many others.

full information ML approach, a consistent estimation of γ requires the correct specification of $\lambda_0(t)$. However, often parametric distributions are not sufficient to model the hazard shape. In such a case, a full information ML estimation of the PI model leads to inconsistent estimates of γ.

A more flexible PI model is proposed by Cox (1972). In this framework, the baseline intensity $\lambda_0(t)$ remains completely unspecified and can be estimated non-parametrically. Cox (1975) illustrates that the estimation of γ does not require a specification of the baseline intensity, and thus, the estimation of γ and $\lambda_0(t)$ can be separated. In order to estimate γ and $\lambda_0(t)$, he suggests a two-step approach where γ is consistently estimated using a *partial likelihood* approach while the estimation of the baseline intensity follows from a modification of the estimator by Kaplan and Meier (1958) as proposed by Breslow (1972 and 1974).[9]

Since the partial likelihood approach proposed by Cox (1975) is based on the order statistics of the durations, this method is of limited use in the case of ties, i.e., when several observations have the same outcome. Then, only approximative procedures can be used (see, for example, Breslow, 1974). In this case, a valuable alternative way to estimate the PI model, is to apply a categorization approach as proposed by Han and Hausman (1990) and Meyer (1990). The main idea behind this procedure is to exploit the relationship between ordered response specifications and models for grouped durations (see Sueyoshi, 1995). Since a categorization approach will play an important role in the context of (censored) autoregressive proportional intensity models (see Chapter 6), it will be discussed here in more detail.

The main idea is to write the PI model in terms of the (log) integrated baseline intensity. According to the implications of Theorem 2.2, the integrated intensity function $\Lambda(t_{i-1}, t_i)$ is i.i.d. standard exponentially distributed. Thus, the PI model is rewritten as

$$\Lambda_0(t_{i-1}, t_i) = \exp(z'_{i-1}\gamma)\Lambda(t_{i-1}, t_i), \qquad (2.26)$$

where $\Lambda_0(t_{i-1}, t_i) = \int_{t_{i-1}}^{t_i} \lambda_0(s)ds$ denotes the *integrated baseline intensity*. By assuming the validity of the "exponential formula", (2.20), the integrated baseline intensity is written as $\Lambda_0(t_{i-1}, t_i) = -\ln S_0(x_i)$, where $S_0(\cdot)$ denotes the baseline survivor function associated with the baseline intensity $\lambda_0(\cdot)$. Hence, (2.26) can be formulated in terms of the baseline survivor function as

$$\ln S_0(x_i) = -\exp(z'_{i-1}\gamma)\Lambda(t_{i-1}, t_i). \qquad (2.27)$$

Since $\Lambda(t_{i-1}, t_i)$ is standard exponentially distributed, it is straightforward to verify that

$$E[\Lambda_0(t_{i-1}, t_i)] = \exp(z'_{i-1}\gamma) \qquad (2.28)$$

$$\mathrm{Var}[\Lambda_0(t_{i-1}, t_i)] = \exp(2z'_{i-1}\gamma). \qquad (2.29)$$

---

[9] For more details, see Kalbfleisch and Prentice (1980) or in the survey of Kiefer (1988).

Equivalently, the PI model can be expressed in terms of the log integrated intensity leading to

$$\ln \Lambda_0(t_{i-1}, t_i) = \ln(-\ln S_0(x_i)) = z'_{i-1}\gamma + \epsilon_i^*, \quad i = 1, \ldots, n, \qquad (2.30)$$

where $\epsilon_i^* := \ln \Lambda(t_{i-1}, t_i)$ follows an i.i.d. standard extreme value type I distribution (standard Gumbel (minimum) distribution)[10] with mean $E[\epsilon_i^*] = -0.5772$, variance $\text{Var}[\epsilon_i^*] = \pi^2/6$ and density function

$$f_{\epsilon^*}(s) = \exp(s - \exp(s)). \qquad (2.31)$$

Note that this approach does not require to specify the baseline intensity λ_0, however it requires a complete parameterization of the *log integrated* baseline intensity. Thus, the PI model boils down to a regression model for $\ln \Lambda_0(t_{i-1}, t_i)$. The function $\Lambda_0(t_{i-1}, t_i)$ can be interpreted as a transformation of the underlying duration $x_i := t_i - t_{i-1}$, where the transformation is known when $\lambda_0(t)$ is completely specified, otherwise it is unknown. For example, by assuming a standard Weibull specification, i.e. $\lambda_0(t) = ax(t)^{a-1}$, the PI model yields

$$\ln x_i = z'_{i-1}\frac{\gamma}{a} + \frac{1}{a}\epsilon_i^*, \quad i = 1, \ldots, n. \qquad (2.32)$$

In this case, the PI model is a log linear model with standard extreme value distributed errors. The coefficient vector γ/a can be consistently (however, not efficiently) estimated by OLS[11]. Alternatively, the (fully parametric) PI model is consistently and efficiently estimated by ML.

When $\lambda_0(t)$ is non-specified, $\Lambda_0(t_{i-1}, t_i)$ is unknown and can be interpreted as a latent variable. Han and Hausman (1990) propose to treat this model as a special type of ordered response model leading to a semiparametric estimation of $\lambda_0(t)$. The central idea is to introduce a categorization of the duration x_i and to consider (2.30) as a latent process which is observable only at discrete points associated with the chosen category bounds. Then, by adapting the categorization introduced above, we define

$$\mu_k^* := \ln \Lambda_0(t_{i-1}, t_{i-1} + \bar{x}_k), \qquad k = 1, \ldots, K - 1, \qquad (2.33)$$

as the value of the latent variable $\ln \Lambda_0(\cdot)$ at the (observable) category bound $\bar{x}_k$. In this formulation, the PI model is interpreted as an ordered response model based on a standard extreme value distribution. Hence, the thresholds of the latent model correspond to the log integrated baseline intensity calculated at the points of the corresponding categorized durations. The direct relationship between the latent thresholds and the log integrated baseline intensity is one of the main advantages of this approach. Thus, the unknown

---

[10] See Appendix A.

[11] In this case, an intercept term has to be included in (2.32) since ϵ_i^* has a nonzero mean.

baseline survivor function $S_0(\cdot)$ can be estimated at the $K-1$ discrete points by a nonlinear function of the estimated thresholds μ_k^*, thus

$$S_0(\bar{x}_k) = \exp(-\exp(\mu_k^*)), \quad k = 1, \ldots, K-1. \tag{2.34}$$

Based on the discrete points of the baseline survivor function, we can estimate a *discrete* baseline intensity $\bar{\lambda}_0(t)$, corresponding to the conditional failure probability given the elapsed time since the last event,

$$\bar{\lambda}_0(t_{i-1} + \bar{x}_k) := \mathrm{Prob}\left[\bar{x}_k \le x_i < \bar{x}_{k+1} \,|\, x_i \ge \bar{x}_k\right]$$
$$= \frac{S_0(\bar{x}_k) - S_0(\bar{x}_{k+1})}{S_0(\bar{x}_k)}, \quad k = 0, \ldots, K-2, \tag{2.35}$$

where $\bar{x}_0 := 0$. This formulation serves as an approximation of the baseline intensity if divided by the length of the discretization interval. If the length of the intervals goes to zero, the approximation converges to the original definition, (2.12), thus

$$\lambda_0(t_{i-1} + \bar{x}_k) \approx \frac{\bar{\lambda}_0(t_{i-1} + \bar{x}_k)}{\bar{x}_{k+1} - \bar{x}_k}. \tag{2.36}$$

Note that the important result of Cox (1975) also holds in this discretization framework. Hence, γ can be estimated consistently without an explicit specification of $\lambda_0(t)$, which means that the consistency of $\hat{\gamma}$ does not depend on the chosen categorization and the number of the categories.[12] This result will be exploited in a dynamic framework in Chapter 6.

The log likelihood function associated with the PI model represented by (2.30), (2.31) and (2.33) has the well known form of an ordered response specification based on the standard extreme value distribution

$$\ln \mathcal{L}(W; \theta) = \sum_{i=1}^{n} \sum_{k=1}^{K-1} \mathbb{1}_{\{x_i \in (\bar{x}_{k-1}, \bar{x}_k]\}} \ln \int_{\mu_{k-1}^* - z_{i-1}'\gamma}^{\mu_k^* - z_{i-1}'\gamma} f_{\epsilon^*}(s)ds. \tag{2.37}$$

Accelerated Failure Time Models

As illustrated above, in the PI model, explanatory variables act multiplicatively with the baseline intensity. In contrast, in the AFT model, it is assumed that $z_{\check{N}(t)}$ accelerates (or decelerates) the time to failure. Thus, the covariates alter the rate at which one proceeds along the time axis, leading to an intensity function that is given by

$$\lambda(t; z_{\check{N}(t)}) = \lambda_0\left(t \exp(-z_{\check{N}(t)}'\gamma)\right) \exp(-z_{\check{N}(t)}'\gamma), \tag{2.38}$$

---

[12] Nevertheless, the efficiency of the estimator is affected by the chosen categorization.

which can be alternatively written as[13]

$$\ln x_i = z'_{i-1}\gamma + \varepsilon_i, \quad i = 1, \dots, n, \tag{2.39}$$

where ε_i is an error term that follows some continuous distribution. Hence, while the PI model implies a linear model with an unknown left-hand variable and a standard extreme value distributed error term, the AFT model leads to a log-linear representation with x_i as the left-hand variable and an unknown distribution of the error term. Thus, while in the PI model, the covariates act multiplicatively with the intensity function, in the AFT model, the covariates act multiplicatively with the log duration. For the special case when x_i is Weibull distributed with parameter a (see Section 2.3.1), we obtain a model that belongs to both model families, implying a multiplicative relationship between the covariates and the intensity function, as well as between the covariates and the log duration. In this case, (2.39) corresponds to (2.32), hence, $a\varepsilon_i = \epsilon_i^*$ is standard extreme value distributed.

A more general class of models that nests both the PI model and the AFT model is introduced by Ridder (1990) and is called the *Generalized Accelerated Failure Time (GAFT) model*. It is given by

$$\ln g(x_i) = z'_{i-1}\gamma + \varepsilon_i, \tag{2.40}$$

where $g(\cdot)$ is an arbitrary non-decreasing function defined on $[0, \infty)$ and ε_i follows an unknown continuous distribution with p.d.f $f(\cdot)$. The GAFT model nests the PI model if $f(\cdot) = f_{\epsilon^*}$ and nests the AFT model if $g(x) = x$. Ridder (1990) illustrates that the GAFT model can be identified non-parametrically.

2.3.2 Duration Models

Instead of specifying the intensity function, a point process can be alternatively directly described by the process of durations between subsequent points. In order to ensure non-negativity, the most simple approach is to specify the durations in terms of a log-linear regression model,

$$\ln x_i = z'_{i-1}\gamma + \varepsilon_i, \tag{2.41}$$

where ε_i is some i.i.d. error term. Such a model is easily estimated by OLS. As discussed in Section 2.3.1, this type of model belongs to the class of AFT models or – for the special case when ε_i is standard extreme value distributed – also to the class of PI models. However, generally, regression models for log durations are associated with AFT representations.

---

[13] See Kalbfleisch and Prentice (1980).

2.3.3 Count Data Models

An alternative representation of a point process model is obtained by specifying the joint distribution of the number of points in equally spaced intervals of length Δ. Denote N_j^Δ as the number of events in the interval $[j\Delta, (j+1)\Delta)$, i.e.,

$$N_j^\Delta := N((j+1)\Delta) - N(j\Delta), \qquad j = 1, 2, \ldots \qquad (2.42)$$

The specification of a count data models require to aggregate the data which naturally induces a certain loss of information. Denote z_j^Δ as a covariate vector associated with N_j^Δ, then, the most simple count data model is given by the Poisson model by assuming that

$$N_j^\Delta | z_j^\Delta \sim Po(\lambda). \qquad (2.43)$$

More general models are obtained by using the NegBin distribution (see, for example, Cameron and Trivedi, 1998) or the double Poisson distribution introduced by Efron (1986).

2.4 Censoring and Time-Varying Covariates

In this section, we discuss two phenomena which typically occur in traditional duration data (like unemployment data) and also play a role in financial point processes: The effects of censoring and of time-varying covariates. As will be illustrated, both phenomena are easily incorporated in an intensity framework.

2.4.1 Censoring

A typical property of economic duration data is the occurrence of censoring leading to incomplete spells. Therefore, a wide strand of econometric duration literature focusses on the consideration of censoring mechanisms[14]. In the context of financial point processes, censoring occurs if there exist intervals in which the point process cannot be observed directly. This might be, for example, due to non-trading periods, like nights, weekends or holidays. Assume in the following that it is possible to identify whether a point t_i lies within such a censoring interval. Consider, for example, a point process where the points are associated with the occurrence of a cumulative price change of given size.[15] Then, prices move during non-trading periods as well (due to trading on other markets), but can be observed at the earliest at the beginning of the

---

[14] See, for example, Horowitz and Neumann (1987, 1989), the survey by Neumann (1997), Gørgens and Horowitz (1999) or Orbe, Ferreira, and Nunez-Anton (2002).

[15] Such processes will be discussed in more details in Section 3.1.2 of Chapter 3 and in Section 6.5 of Chapter 6.

next trading day. In this case, we only know *that* a price event occurred but we do not know *when* it exactly occurred.[16] Hence, we can identify only the minimum length of the corresponding duration (i.e., the time from the previous point t_{i-1} to the beginning of the censoring interval) and the maximum length of the spell (i.e., the time from the end of the censoring interval to the next point t_{i+1}).

In the following, t_i^l and t_i^u with $t_i^l \leq t_i \leq t_i^u$ are defined as the boundaries of a potential censoring interval around t_i, and c_i is defined as the indicator variable that indicates whether t_i occurs within a censoring interval, i.e.

$$c_i = \begin{cases} 1 & \text{if } t_i \in (t_i^l; t_i^u), \\ 0 & \text{if } t_i = t_i^l = t_i^u . \end{cases}$$

In the case of left-censoring, right-censoring or left-right-censoring, the non-observed duration x_i can be isolated by the boundaries $x_i \in [x_i^l, x_i^u]$ where the lower and upper boundary x_i^l and x_i^u are computed corresponding to

$$x_i \in \begin{cases} [t_i - t_{i-1}^u; t_i - t_{i-1}^l] & \text{if } c_{i-1} = 1, c_i = 0 \quad \text{(left-censoring)} \\ [t_i^l - t_{i-1}; t_i^u - t_{i-1}] & \text{if } c_{i-1} = 0, c_i = 1 \quad \text{(right-censoring)} \\ [t_i^l - t_{i-1}^u; t_i^u - t_{i-1}^l] & \text{if } c_{i-1} = 1, c_i = 1 \quad \text{(left-right-censoring)}. \end{cases}$$
$$(2.44)$$

A common assumption that is easily fulfilled in the context of financial data is the assumption of *independent censoring*. This assumption means that the censoring mechanism is determined exogenously and is not driven by the duration process itself. For a detailed exposition and a discussion of different types of censoring mechanisms, see e.g. Neumann (1997).

Under the assumption of independent censoring, the likelihood can be decomposed into

$$\mathcal{L}(W; \theta | c_1, \ldots, c_n) = \mathcal{L}(W; \theta, c_1, \ldots, c_n) \cdot \mathcal{L}(c_1, \ldots, c_n). \quad (2.45)$$

Hence, the second factor does not depend on the parameters of the model, and thus the parameter vector θ is estimated by maximizing the first factor of (2.45). Therefore, the corresponding log likelihood function is given by

---

[16] However, note that we cannot identify whether more than one price movement occurred during the non-trading period.

$$\ln \mathcal{L}(W, c_1, \ldots, c_n; \theta) = \sum_{i=1}^{n} (1 - c_i)(1 - c_{i-1}) \cdot (-\Lambda(t_{i-1}, t_i) + \ln \lambda(t_i; \mathfrak{F}_{t_i}))$$

$$+ \sum_{i=1}^{n} c_{i-1}(1 - c_i) \cdot \ln \left(S(t_i - t_{i-1}^u) - S(t_i - t_{i-1}^l) \right)$$

$$+ \sum_{i=1}^{n} (1 - c_{i-1})c_i \cdot \ln \left(S(t_i^l - t_{i-1}) - S(t_i^u - t_{i-1}) \right)$$

$$+ \sum_{i=1}^{n} c_{i-1}c_i \cdot \ln \left(S(t_i^l - t_{i-1}^u) - S(t_i^u - t_{i-1}^l) \right). \quad (2.46)$$

The first term in (2.46) is the log likelihood contribution of a non-censored duration while the following terms are the contributions of left-censored, right-censored and left-right-censored observations.

2.4.2 Time-Varying Covariates

In the following, we discuss the treatment of time-varying covariates $z(t)$. Here, the impact of the covariates on the intensity is not constant during the time from t_{i-1} to t_i, but is time-varying. However, the formulation of a continuous-time model for the covariate path is rather difficult, since it requires the identification of the precise timing of events on the continuous path. In order to circumvent these difficulties, one typically builds a model based on a discrete-time framework. Hence, it is common to proceed by assuming that the occurrence of events can occur at discrete points of time only. This leads to a discretization of the intensity concept.

By following the notation in Lancaster (1997), denote $\mathcal{Z}_i(x)$ as the path of $z(t)$ from $z(t_{i-1})$ to $z(t_{i-1} + x)$ and $\mathcal{Z}_i(x_1, x_2)$ as the path of $z(t)$ from $z(t_{i-1}+x_1)$ to $z(t_{i-1}+x_2)$. Furthermore, according to the framework outlined in Section 2.3.1, the observed durations are divided into K intervals and it is assumed that the durations x_i can only take the discrete values $\bar{x}_k$, $k = 1, \ldots, K-1$. Moreover, it is assumed that the discretization is sensitive enough to capture each particular point, i.e., $\min\{x_i\} \geq \bar{x}_1 \; \forall \, i$.

Then, a discrete formulation of the intensity function at the points $t_{i-1}+\bar{x}_k$ is obtained by[17]

$$\bar{\lambda}(t_{i-1} + \bar{x}_k; \mathcal{Z}_i(\bar{x}_k)) = \text{Prob}\left[x_i = \bar{x}_k \,|\, x_i \geq \bar{x}_k; \mathcal{Z}_i(\bar{x}_k)\right] \quad (2.47)$$

and can be computed based on

---

[17] See also Lancaster (1997).

$$\text{Prob}\left[x_i \geq \bar{x}_k, \mathcal{Z}_i(\bar{x}_k)\right] = \prod_{j=1}^{k-1} \left[1 - \bar{\lambda}(t_{i-1} + \bar{x}_j; \mathcal{Z}_i(\bar{x}_j))\right] \qquad (2.48)$$

$$\times \prod_{j=1}^{k-1} \text{Prob}\left[\mathcal{Z}_i(\bar{x}_{j-1}, \bar{x}_j) \,|\, x_i \geq \bar{x}_j, \mathcal{Z}_i(\bar{x}_{j-1})\right]$$

and

$$\text{Prob}[x_i = \bar{x}_k, x_i \geq \bar{x}_k, \mathcal{Z}_i(\bar{x}_k)] = \bar{\lambda}(t_{i-1} + \bar{x}_k; \mathcal{Z}_i(\bar{x}_k)) \qquad (2.49)$$

$$\times \prod_{j=1}^{k-1} \left[1 - \bar{\lambda}(t_{i-1} + \bar{x}_j; \mathcal{Z}_i(\bar{x}_j))\right]$$

$$\times \prod_{j=1}^{k} \text{Prob}\left[\mathcal{Z}_i(\bar{x}_{j-1}, \bar{x}_j) \,|\, x_i \geq \bar{x}_j, \mathcal{Z}_i(\bar{x}_{j-1})\right].$$

The expressions (2.48) and (2.49) simplify in the case of a so-called *exogenous covariate process*, i.e., following the definition in Lancaster (1997), if and only if

$$\text{Prob}\left[\mathcal{Z}_i(x, x + \Delta) \,|\, x_i \geq x + \Delta, \mathcal{Z}_i(x)\right]$$
$$= \text{Prob}\left[\mathcal{Z}_i(x, x + \Delta) \,|\, \mathcal{Z}_i(x)\right], \quad \forall\, x \geq 0, \Delta > 0. \qquad (2.50)$$

Hence, exogeneity of the covariate process means that the information that no further event has been observed until $x + \Delta$ has no predictability for the further path of the covariates from x to $x + \Delta$. In this case, (2.48) and (2.49) simplify to

$$\text{Prob}\left[x_i \geq \bar{x}_k, \mathcal{Z}_i(\bar{x}_k)\right] = \prod_{j=1}^{k-1} \left[1 - \bar{\lambda}(t_{i-1} + \bar{x}_j; \mathcal{Z}_i(\bar{x}_j))\right] \qquad (2.51)$$

and

$$\text{Prob}[x_i = \bar{x}_k, x_i \geq \bar{x}_k, \mathcal{Z}_i(\bar{x}_k)] = \bar{\lambda}(t_{i-1} + \bar{x}_k; \mathcal{Z}_i(\bar{x}_k))$$
$$\times \prod_{j=1}^{k-1} \left[1 - \bar{\lambda}(t_{i-1} + \bar{x}_j; \mathcal{Z}_i(\bar{x}_j))\right]. \qquad (2.52)$$

Thus, only when the covariates are exogenous, (2.48) and (2.49) can be interpreted as conditional probabilities, given the covariate path. However, even when they are valid probabilities they can *never* be interpreted as the values of a conditional survivor function or probability density function of x_i given $\mathcal{Z}_i(\bar{x}_j)$ at the point $\bar{x}_k$. This is due to the fact that $\mathcal{Z}_i(\bar{x}_j)$ is itself a function of t and the conditioning event of the conditional survivor function or conditional p.d.f. changes when the argument itself changes.

Nevertheless, even though it cannot be interpreted as the value of a (discrete) p.d.f., it can be used to draw statistical inference. Hence, the intensity function, given the covariate path, is always defined even when there exists no counterpart that is interpretable as a conditional density or conditional survivor function. These relationships illustrate the importance of the intensity concept in modelling duration data.

2.5 Outlook on Dynamic Extensions

The following section provides a brief overview of different possibilities for dynamic extensions of the concepts discussed in Section 2.3 of this chapter. In this sense, the section serves as an introduction to the particular models that will be discussed in more detail in the Chapters 5-7 of the book.

The implementation of autoregressive structures in point processes can be performed in alternative ways. According to the three possible specifications of a point process, dynamics can be introduced either in the intensity process, the duration process or in the counting process. A priori, it is quite unclear which way should be preferred and whether one specification is superior to another. Ultimately, the particular concepts have to be judged by their ability to result in well specified empirical models that provide a satisfying fit to the data and whose coefficients may be readily economically interpretable. Nonetheless, in the context of a dynamic framework, it is necessary to have a closer look at the fundamental differences between the particular approaches and at their strengths and weaknesses with regard to the specific properties of financial point processes.

The most simple and probably most intuitive way to model autoregressive point processes is to specify an autoregressive process in terms of the durations. A particular class of *autoregressive duration models* has been proposed by Engle and Russell (1997, 1998) and Engle (1996, 2000) and will be considered in more detail in Chapter 5 of this monograph. The major advantage of such a concept is its practicability since standard time series packages for ARMA or GARCH models can more or less directly applied to the duration data. Presumably for this reason, an autoregressive duration approach is the most common type of financial point process model and is widespread in the younger financial econometrics literature (for detailed quotations, see Chapter 5).

Nonetheless, autoregressive duration models also reveal major drawbacks. First, they are not easily extended to a multivariate framework. The problem which has to been resolved is that in a multivariate context the particular processes occur asynchronously, and thus, there exist no joint points that can be used to couple the processes. For this reason, it is extremely difficult to estimate contemporaneous correlations between the particular autoregressive processes. Second, the treatment of censoring mechanisms is rather difficult. In the case of censoring, the exact timing of particular points of the process cannot be observed and can only be approximated by a corresponding interval. Such effects lead to problems in an autoregressive framework because the information about the exact length of a spell is needed for the sequel of the time series. Hence, modified approaches have to be applied in this context. One possible solution is, for example, to build a modified model where non-observed durations are replaced by a function of their corresponding conditional expectations. However, the implementation of such modifications in a time series context is not straightforward in any case (see, for example, the discussion in Gerhard, 2001). Third, in a pure duration framework, it is difficult to account for time-varying covariates. As discussed already in Section 2.4.2, an intensity representation is more useful and convenient.

The key advantage of *autoregressive intensity models* (see Chapter 7) over autoregressive duration approaches is that they allow to model point processes in a continuous-time framework. Thus, while the duration between two points is by definition observed at the particular points t_i themselves, the intensity function is defined at *any* point in time. This feature plays an important role in the modelling of time-varying covariates and in the context of autoregressive multivariate point process models. In an intensity framework, it is possible to couple the particular processes at *each* point of the pooled process, and thus, the problem of asynchronous processes vanishes. *Multivariate autoregressive intensity* models will be discussed in more detail in Chapter 7.

However, as in a duration framework, the consideration of censoring structures is quite difficult. For a censored observation, the occurrence time, and thus the intensity function at the particular time, is not exactly measurable. Thus, when these (unobserved) realizations are needed in the sequel of a dynamic model, approximations have to be used. Therefore, for the modelling of point processes that are subjected to strong censoring mechanisms, alternative approaches are necessary. One valuable alternative is to build a dynamic model based on a function of the survivor function. The survivor function plays an important role in the context of censoring (see Section 2.4.1), since it allows to isolate the non-observed occurrence times in terms of survivor probabilities with respect to the corresponding censoring bounds. Hence, exploiting the relationship between the survivor function and the integrated intensity function (see (2.20)), an alternative dynamic point process model can be built based on the integrated intensity function. Moreover, as illustrated in Chapter 6, *autoregressive integrated intensity models* are valuable approaches for dynamic extensions of semiparametric proportional intensity models. In this context, the discretization approach presented in Section 2.3.1 will be applied to obtain a semiparametric estimation of a non-specified baseline intensity in a dynamic framework.

A further possibility to model autoregressive point processes is to specify an *autoregressive count data* model. The strength of this approach is that these models are based (per assumption) on equi-distant time intervals. For this reason, they are easily extended to a multivariate framework (see Davis, Rydberg, Shephard, and Streett, 2001 or Heinen and Rengifo, 2003). However, the treatment of censoring mechanisms and the inclusion of time-varying covariates is rather difficult in this context. Since the focus on this book lies mainly on duration and intensity models, a detailed considerations of dynamic count data models is beyond its scope.

3

Economic Implications of Financial Durations

Financial point processes are defined on the basis of single events that are related to the trading process on a financial market. Hence, by defining the event of interest in a particular way, we obtain a specific type of financial point process. Obviously, the choice of this event heavily depends on the economical or statistical objective which is being investigated. In this chapter, we present different types of financial point processes and discuss their economic implications.

Section 3.1 introduces alternative financial duration concepts. Here, we consider trade and quote durations that are strongly related to market microstructure issues. Furthermore, we discuss aggregated duration processes, such as price durations, directional change durations and (excess) volume durations that play an important role for volatility and liquidity quantification. Section 3.2 illustrates the role of trade durations in market microstructure theory. Here, the informational content of trade durations and their relationship with the key market microstructure variables, such as the price, the bid-ask spread and the trading volume is examined in more detail. Section 3.3 illustrates the use of price durations in building alternative risk and volatility measures. Such concepts explicitly account for time structures in the price process and provide a worthwhile framework for particular issues in financial risk management. Section 3.4 is concerned with the quantification of liquidity. In this section, the use of volume and price durations in constructing measures for particular dimensions of liquidity is illustrated in more detail.

3.1 Types of Financial Durations

In the following section, we focus on a (multivariate) point process associated with the complete trading process of a particular financial asset over a given time span. By selecting certain points of this process, different types of financial duration processes are generated. The selection of particular points is commonly referred to as the "thinning" of the point process. In general, we can distinguish between two different selection procedures. The most simple way to thin a point process is based on *single* marks of the process. Consider, for example, a mark which indicates whether an event is a trade or a quote. Then, accordingly the (thinned) point process of trades or quotes is generated.

The second type of selection procedure uses *sequences* of marks as selection criterion. In these cases, the decision to select a particular point depends not only on its own marks, but also on the marks of previous and/or following points. Such criteria are typically applied to capture sequences in price or volume processes. In the following subsections we briefly summarize different types of (thinned) financial point processes.

3.1.1 Selection by Single Marks

Trade Durations

Trade durations are the most common type of financial durations and are defined as the time between subsequent transactions. They measure the speed of the market, which is associated with the trading intensity. Since it is natural to associate a trade with the demand for liquidity, a trade duration is related to the intensity for liquidity demand. Refinements of this concept lead to specific *buy* or *sell trade durations*. These are defined as the time between consecutive buys or consecutive sells and measure the trading intensity on the particular sides of the market.

Quote and Limit Order Durations

In a price driven market[1], the quotes posted by a market maker characterize the supply of liquidity. Hence, quote durations, defined as the time between the occurrence of new quotes, are natural measures for the intensity of liquidity supply.

In an order driven market, limit order durations, defined as the time between consecutive arrivals of limit orders, reflect the activity in the limit order book. In studying limit order book dynamics, it is of particular interest to differentiate between different types of limit orders. For example, one can examine limit orders with respect to the side of the market or to their position in the queues of the order book.

---

[1] For a concise overview of different types of market and trading systems see, for example, Bauwens and Giot (2001).

3.1.2 Selection by Sequences of Marks

Price Durations

Price durations are generated by selecting points according to their price information. Let p_i, a_i and b_i be the process of transaction prices, best ask prices and best bid prices, respectively. Define in the following i' with $i' < i$ as the index of the *most recently* selected point of the point process. Then, as proposed by Engle and Russell (1997), a series of price durations is generated by thinning the process according to the following rule:

Retain point i, $i > 1$, if $|p_i - p_{i'}| \geq dp$.

The variable dp gives the size of the underlying cumulative absolute price change and is exogenous. The first point of the thinned point process is selected exogenously and typically corresponds to the first point ($i = 1$) of the original point process. In order to avoid biases caused by the bid-ask bounce (see Roll, 1984), a valuable alternative is to generate price durations not on the basis of transaction prices p_i, but on midquotes as defined by $mq := (a_i + b_i)/2$. As will be discussed in more detail in Section 3.3 of this chapter, price durations are strongly related to alternative risk and volatility measures.

Directional Change Durations

In technical analysis, turning points of fundamental price movements are of particular interest since they are associated with optimal times to buy or to sell. In this context, the time between local extrema (directional change duration), provides important information concerning time structures in the price process. A directional change duration is generated according to the following procedure:

Retain point i, $i > 1$, if

(i) $p_i - p'_i \geq (\leq) \, dp$
     and if there exists a point, indexed by i'' with $i'' > i'$, for which
(ii) $p_i \geq (\leq) \, p_j$ with $j = i + 1, \ldots, i''$-1
(iii) $p_i - p_{i''} \geq (\leq) \, dp$.

In this context, dp gives the *minimum* price difference between consecutive local extreme values.

Volume Durations

Volume durations are defined as the time until a certain aggregated volume is absorbed by the market. A volume duration is generated formally by retaining point i, $i > 1$ if $\sum_{j=i'+1}^{i} vol_j \geq dv$, where dv represents the predetermined amount of the cumulated volume and vol_i the transaction volume associated with trade i. Volume durations capture the volume intensity and, as will be discussed in Section 3.4, they are proxies for the time-varying market liquidity. Accordingly, *buy* or *sell volume durations* characterize the volume intensity on the particular sides of the market.

Excess Volume Durations

A natural indicator for the presence of information on the market is the time it takes to trade a given amount of excess buy or sell volume. The so-called excess volume durations measure the intensity of (one-sided) demand for liquidity and are formally created by retaining point i, $i > 1$, if $|\sum_{j=i'+1}^{i} y_j^b vol_j| \geq dv$, where y_j^b is an indicator variable that takes the value 1 if a trade is buyer-initiated and -1 if a transaction is seller-initiated. The threshold value dv of the excess demand is fixed exogenously and is determined by the risk preference of the liquidity supplier. As will be illustrated in Section 3.4.4, excess volume durations are a valuable tool in evaluating the risk of a liquidity supplier with respect to asymmetric information and inventory problems and can be used for empirical tests of market microstructure hypotheses.

3.2 The Role of Trade Durations in Market Microstructure Theory

3.2.1 Traditional Market Microstructure Approaches

As stated by Madhavan (2000), market microstructure theory is concerned with "the process by which investors' latent demands are ultimately translated into prices and volumes." Thus, central issues in market microstructure theory are related to price formation and price discovery, as well as to information diffusion and dissemination in markets.

Traditional microstructure theory provides two major frameworks to explain price setting behavior: inventory models and information-based models[2]. The first branch of the literature investigates the uncertainty in the order flow and the inventory risk and optimization problem of liquidity suppliers[3]. The latter branch models market dynamics and adjustment processes of prices by using insights from the theory of asymmetric information and adverse selection. This framework is based on the crucial assumption of the existence of differently informed traders. It is assumed that there exist so-called "informed traders", who trade due to private information and "liquidity traders",

---

[2] For a comprehensive overview see, for example, O'Hara (1995), or the survey by Madhavan (2000).

[3] See, for example, Garman (1976), Stoll (1978), Amihud and Mendelson (1980) or Ho and Stoll (1981).

who trade due to exogenous reasons, like, for example, due to portfolio adjustments or liquidity aspects. The assumption of heterogeneous groups of traders provides the basis for a plethora of asymmetric information models.[4] In these approaches, uninformed market participants deduce from the trading process the existence of information in the market. Here, the trading process itself serves as a source of information. Therefore, on the basis of the assumption of diversely informed, heterogeneous traders, information-based market microstructure settings model relationships between the price change, the bid-ask spread, the trading volume and the trading intensity. The main theoretical findings with respect to the key microstructure variables are systemized as follows:

(i) The role of the transaction volume: In the Easley and O'Hara (1987) model, traders are allowed to trade either small or large quantities, but are not allowed to refrain from trading. Thus, in this setting, the time between transactions plays no role. However, it is shown that large quantities indicate the existence of information. Blume, Easley, and O'Hara (1994) investigate the informational role of volume when traders receive information signals of different quality in each period. The authors analyze how the statistical properties of volume relate to the behavior of market prices. They show that traders can infer from the volume about the quality and quantity of information in the market. The crucial result is that the volume provides additional information that cannot be deduced from the price statistics. Barclay and Warner (1993) examine the strategic behavior of market participants and analyze the proportion of cumulative price changes that occur in certain volume categories. Based on an empirical study, they conclude that most of the cumulative price change is due to medium-sized trades. This result is consistent with the hypothesis that informed traders tend to use medium volume sizes.[5]

(ii) The role of the bid-ask spread: A wide range of studies examine the informational role of the bid-ask spread. In the Glosten and Milgrom (1985) model, it is shown that the market maker sets the quoted prices equal to his conditional expectations of the asset value, given the type of the trade (buy or sell) and the possible type of the trader (informed trader or liquidity trader). Thus, in this early information-based model, the market maker determines the spread in such a way that it compensates for the risk due to adverse selection. The higher the probability that he transacts at a loss due to trading with market participants with superior information, the higher the bid-ask spread. The idea of the market maker as a Bayesian learner who updates his beliefs by observing the trading flow is further developed in the setting of Easley and O'Hara (1992). They extend the Glosten and Milgrom (1985)

---

[4] See, for example, Bagehot (1971), Copeland and Galai (1983), Glosten and Milgrom (1985), Kyle (1985), Diamond and Verrecchia (1987), Admati and Pfleiderer (1988) or Easley and O'Hara (1987, 1992) among others.

[5] A paper with a related focus is Kempf and Korn (1999) who empirically analyze the relationship between unexpected net order flow and price changes and find highly nonlinear relationships.

model by explicitly accounting for the role of time. In this approach, the market maker uses no-trade-intervals to infer the existence of new information. Thus, a negative correlation between the lagged durations and the magnitude of the spread is derived.

(iii) The role of the trade duration: Diamond and Verrecchia (1987) propose a rational expectation model with short selling constraints. They assert that, in this setting, the absence of a trade is associated with the occurrence of "bad" news. Thus, the absence of a trade is informative and should be correlated with price volatility. In this framework, time matters only because of the imposed short selling restrictions. Easley and O'Hara (1992), however, analyze the role of time and its relationship to information in a more general model. In this approach, informed traders increase the speed of trading since they want to exploit their informational advantage. Thus, the time between consecutive trades is negatively correlated with the magnitude of information on the market. This result leads to testable hypotheses regarding the impact of trade durations on other market microstructure variables, such as spreads or volumes.

(iv) Relationship between volume and volatility: A string of market microstructure literature is concerned with investigating the relationship between trading volume and price volatility. On the basis of information-based approaches, Easley and O'Hara (1987) and Blume, Easley, and O'Hara (1994) derive a positive correlation between volume and volatility. Alternative approaches rely on the mixture-of-distribution hypothesis introduced by Clark (1973) and further developed by Tauchen and Pitts (1983). In the latter approach it is assumed that an underlying latent stochastic process drives both the returns and the volume leading to a positive correlation between volumes and volatility.[6]

3.2.2 Determinants of Trade Durations

As illustrated in the last subsection, trade durations play an important role in market microstructure theory since they are used as proxies for the existence of information in the market, and thus, are predictors for other market microstructure variables. However, a central question is, whether the trading intensity itself is predictable based on past trading sequences. For two reasons, predictions of the future trading intensity could be interesting.

First, transactions characterize the individual demand for liquidity. Consequently, it is natural to associate trade durations with the intensity for liquidity demand. In this context, a liquidity supplier might be interested in predictions of the expected liquidity demand in order to assess his inventory risk. Such an aspect is strongly related to the issues of traditional inventory-based microstructure theory where uncertainties in the order flow and liquid-

---

[6] For refinements of this idea, see Andersen (1996) or Liesenfeld (1998, 2001). However, note, that these studies focus exclusively on daily data.

ity suppliers' risks due to inventory optimization problems play an important role.

Second, in addition to the question of *why* an agent trades, it is also of particular importance to analyze *when* he will enter the market. Obviously, these decisions are strongly related. For example, it is natural to assume that an informed trader has an incentive to exploit his informational advantage and to enter the market as quickly as possible. The same argument might hold for a technical trader who enters the market as soon as possible whenever he receives a technical trading signal. In contrast, a liquidity trader is more interested in avoiding high transaction costs. He prefers to trade in liquid periods that are associated with low trading costs. Thus, the relationship between market microstructure variables and the expected trading intensity sheds some light on the question of how these observable factors drive the market participant's preference for immediacy.

By referring to the microstructure relationships summarized in the previous subsection, we will derive testable hypotheses with respect to the impact of these variables on the expected trade duration. According to 3.2.1 (i), large trading volumes are associated with the existence of information on the market. On the one hand, a large volume increases the probability of the entry of informed traders to the market. On the other hand, a large volume might be a technical trading signal attracting technical traders as well. However, as already discussed in (i), strategic aspects of informed trading might also matter. If informed traders tend to camouflage their informational advantage and consequently trade medium volume quantities, then one would expect medium trading volumes to have the strongest impact on trading intensity. This leads to a nonlinear relationship between trade durations and the past trading volume. These implications are summarized in the following hypothesis:

H1: Volume and trade durations

(a) Large volumes decrease subsequent trade durations.
(b) The relationship between trading volumes and subsequent trade durations is nonlinear.

Following 3.2.1 (ii), a market maker sets the bid and ask prices in order to compensate for the risk due to adverse selection. Thus, the width of the bid-ask spread reflects his assessment of informed trading on the market. However, on the other hand, a high spread increases the transaction costs, and thus, should decrease the future trading intensity. This implication is tested by analyzing whether the magnitude of the bid-ask spread influences the succeeding trade durations positively:

H2: Bid-ask spreads and trade durations

Bid-ask spreads are positively correlated with subsequent trade durations.

The informational role of trade durations itself has important implications for the autocorrelation structure of the duration process. In the Easley and O'Hara (1992) model, small trade durations are signals for the existence of informed trading. Accordingly, small durations increase the probability of the entry of further informed and technical traders and lead to positively autocorrelated trade durations. Admati and Pfleiderer (1988) have a different explanation for duration clustering. In their equilibrium setting, liquidity traders prefer to minimize their transaction costs. Moreover, they illustrate that it is optimal for informed traders to behave similarly. As a consequence, trading is clustered, and trade durations should be positively autocorrelated. However, duration clustering might also be explained by referring to the mixture-of-distribution approach. Whenever there exists an underlying stochastic process that drives trading volume and volatility simultaneously, it is natural to assume that this drives the timing of transactions as well. Summarizing these implications, we can state the following hypothesis:

H3: Dynamic properties of trade durations

Trade durations are positively autocorrelated.

The last main hypothesis of this section is related to the volatility of the price process. The assumption of the informativeness of past price sequences is based on the noisy rational expectation equilibrium models of Hellwig (1982) and Diamond and Verrecchia (1981). These models analyze rational expectation equilibria in a market where investors learn from past prices. This implies that large price changes indicate the existence of information, and thus, increase informed and technical traders' preferences for immediacy. We test this implication using the following hypothesis:

H4: Absolute price changes and trade durations

Absolute price changes are negatively correlated with subsequent trade durations.

An empirical test of these hypotheses will be performed in Section 5.5.2 of Chapter 5.

3.3 Risk Estimation based on Price Durations

3.3.1 Duration-Based Volatility Measurement

The use of price durations as defined in Section 3.1.2 opens up alternative ways of estimating volatility and price change risks. By definition they account for time structures in the price process, and are of particular interest whenever an investor is able to determine his risk in terms of a certain price movement. On the basis of price durations, it is possible to estimate first passage times in the price process, i.e. the time until a certain price limit is exceeded. Equivalently, they allow for the quantification of the risk for a given price change within a particular time interval. Such analyzes play an important role in several issues of financial economics. For example, in Value-at-Risk (VaR) investigations, it is of interest to determine with which probability a certain VaR will be exceeded within a given time interval. Or alternatively, what is the expected time until the next required portfolio adjustment?

Moreover, price durations are strongly related to the price volatility of a particular stock. In order to clarify this relationship in more detail, it is worthwhile to consider the expected conditional volatility per time unit measured over the next trade duration x_{i+1}, i.e.

$$\sigma^2(t_i) := \mathrm{E}\left[\frac{1}{x_{i+1}}\left(r_{i+1} - \bar{r}_{i+1}\right)^2 \middle| \mathfrak{F}_{t_i}\right], \quad i = 1, \ldots, n, \qquad (3.1)$$

$$\text{with } r_i := \frac{p_i - p_{i-1}}{p_{i-1}},$$

$$\text{and } \bar{r}_i := \mathrm{E}\left[r_i \middle| \mathfrak{F}_{t_{i-1}}\right] := 0.$$

Here, r_i denotes the simple net return with a conditional mean assumed to be zero.[7] Note that (3.1) is quite unprecise with respect to the measure that is used within the integral. This aspect will be considered in more detail below.

In a GARCH framework, the event of interest is not the price at every transaction, but the price observed at certain points in time. Intra-day aggregates are typically used, e.g. on the basis of 5 minute or 10 minute intervals[8]. Define in the following Δ as the length of the aggregation interval and $p(t)$ as the price that is valid at t. Then, a volatility measure based on *equi-distant* time intervals is obtained by

$$\sigma^2_{(r^\Delta)}(j\Delta) := \mathrm{E}_r\left[r^\Delta((j+1)\Delta)^2 \middle| \mathfrak{F}_{j\Delta}\right]\frac{1}{\Delta}, \quad j = 1, \ldots, n^\Delta, \qquad (3.2)$$

$$\text{with } r^\Delta(t) := \frac{p(t) - p(t - \Delta)}{p(t - \Delta)}, \qquad (3.3)$$

where $t = j\Delta$, $j = 1, \ldots, n^\Delta$ are the equi-distant time points associated with Δ and n^Δ denotes the sample size of Δ minute intervals. Thus, $\sigma^2_{(r^\Delta)}(t)$ is not based on the transaction process, but on certain (equi-distant) points in time and is typically estimated by standard GARCH-type models.[9] In general, volatility measurement based on equi-distant time intervals raises the question of an optimal aggregation level. A major problem is that the criteria of optimality for the determination of the aggregation level are not yet clear. Moreover, Aït-Sahalia and Mykland (2003) illustrate that the ignoring of the

---

[7] For the ease of exposition, no log returns are used. However, using log returns would not change the general proceeding.

[8] See e.g. Andersen and Bollerslev (1998a, b) for further references.

[9] In this context, it is common to approximate $p(t)$ by the most recent observed price before t.

effects of sampling discreteness and sampling randomness can lead to strong biases.

Henceforth, an alternative to an equi-distant interval framework is to specify a bivariate process, which accounts for the stochastic nature of both the process of price changes and the process of trade durations. These specifications allow volatility estimates on the basis of a bivariate distribution. By defining the price change variable $d_i := p_i - p_{i-1}$, we have thus

$$\sigma^2_{(d,x)}(t_i) := \mathop{\mathrm{E}}_{d,x} \left[\frac{1}{x_{i+1}} \cdot d^2_{i+1} \,\middle|\, \mathfrak{F}_{t_i} \right] \frac{1}{p_i^2}, \quad i = 1, \dots, n. \tag{3.4}$$

The limitation of such models to the analysis of price changes d_i is standard and not substantial since these models are usually estimated on the basis of transaction data. However, the process of price changes on the transaction level has a peculiar property which needs to be accounted for. Price changes take on only a few different values depending on the tick size and liquidity of the traded asset.[10] Most models concentrating on the bivariate process account explicitly for those market microstructure effects, like Russell and Engle (1998), Gerhard and Pohlmeier (2002), Liesenfeld and Pohlmeier (2003) or Rydberg and Shephard (2003). Ghysels and Jasiak (1998) and Grammig and Wellner (2002) analyze a bivariate process based on a GARCH specification which explicitly accounts for the stochastic nature of the price intensity. Modelling the bivariate distribution of price changes and trade durations on a trade-to-trade level yields valuable information for the analysis of the market microstructure. However, such an analysis is of limited use for risk assessment because of the tediousness of aggregating the underlying models in order to obtain a valid estimator for the overall risk of a series.

A volatility estimator based on price durations is proposed by Gerhard and Hautsch (2002b). They start with the assumption that a decision maker in need of a risk measure is able to express the size of a significant price change, dp. In this framework, the bivariate distribution of r_i and x_i is no longer of interest since r_i is reduced to the ratio of a constant and a conditionally deterministic variable. Define $\{t_i^{dp}\}_{i \in \{1,2,\dots,n^{dp}\}}$ to be the sequence of points of the thinned point process associated with price durations with respect to price change sizes of dp. Moreover, define $x_i^{dp} := t_i^{dp} - t_{i-1}^{dp}$ as the corresponding price duration. Then, we can formulate the conditional volatility per time measured over the spell of a price duration as

---

[10] For an analysis of the effects of neglected discreteness, see Harris (1990), Gottlieb and Kalay (1985) or Ball (1988). For the implications of the tick size on the distributional and dynamical properties of the price process, see e.g. Hautsch and Pohlmeier (2002).

$$\sigma^2_{(x^{dp})}(t_i^{dp}) := \operatorname*{E}_{x}\left[\left.\frac{1}{x_{i+1}^{dp}}\right|\mathfrak{F}_{t_i^{dp}}\right]\left(\frac{dp}{p_{t_i^{dp}}}\right)^2 \tag{3.5}$$

$$= \sigma^{*2}_{(x^{dp})}(t_i^{dp}) \cdot \frac{1}{p^2_{t_i^{dp}}}, \quad i = 1,\ldots,n^{dp}, \tag{3.6}$$

where $\sigma^{*2}_{(x^{dp})}(t_i^{dp})$ stands for the conditional price change volatility from which the conditional volatility of returns can easily be recovered according to (3.6). Note that the estimation of $\sigma^2_{(x^{dp})}(t_i^{dp})$ necessitates the estimation of the conditional expectation $\operatorname{E}_x\left[\left.\frac{1}{x_{i+1}^{dp}}\right|\mathfrak{F}_{t_i^{dp}}\right]$. This requires either specifying a stochastic process for $\frac{1}{x_i^{dp}}$ itself or, alternatively, computing the conditional distribution $\frac{1}{x_i^{dp}}$ using a transformation of the conditional distribution of x_i^{dp}. The latter procedure is typically quite cumbersome and requires the use of simulation techniques. For this reason, in Section 6.6 of Chapter 6 we suggest an approximation procedure based on a categorization approach.

Note that (3.6) is only defined at the particular points t_i^{dp} of the underlying thinned point process associated with dp-price durations. Thus, it does not provide a continuous-time volatility estimation. Alternatively, an instantaneous volatility is defined as[11]

$$\tilde{\sigma}^2(t) := \lim_{\Delta\downarrow 0}\frac{1}{\Delta}\operatorname{E}\left[\left.\left(\frac{p(t+\Delta)-p(t)}{p(t)}\right)^2\right|\mathfrak{F}_t\right]. \tag{3.7}$$

Equation (3.7) can be formulated in terms of the intensity function associated with the process of dp-price changes $\{t_i^{dp}\}_{i\in\{1,2,\ldots,n^{dp}\}}$. Thus,

$$\tilde{\sigma}^2_{(x^{dp})}(t) := \lim_{\Delta\downarrow 0}\frac{1}{\Delta}\operatorname{Prob}\left[|p(t+\Delta)-p(t)|\geq dp\,|\mathfrak{F}_t\right]\cdot\left[\frac{dp}{p(t)}\right]^2 \tag{3.8}$$

$$= \lim_{\Delta\downarrow 0}\frac{1}{\Delta}\operatorname{Prob}\left[(N^{dp}(t+\Delta)-N^{dp}(t))>0\,|\mathfrak{F}_t\right]\cdot\left[\frac{dp}{p(t)}\right]^2$$

$$= \lambda^{dp}(t;\mathfrak{F}_t)\cdot\left[\frac{dp}{p(t)}\right]^2,$$

where $N^{dp}(t)$ denotes the counting process associated with cumulated absolute dp-price changes and $\lambda^{dp}(t;\mathfrak{F}_t)$ the corresponding dp-price change intensity function. Expression (3.8) gives the expectation of the conditional dp-price change volatility per time in the next instant and thus, allows for a continuous picture of volatility over time.

Applications of the volatility estimators given by (3.6) and (3.8) to analyze intraday and interday volatility are given in the Chapters 6 and 7, respectively.

---

[11] See Engle and Russell (1998).

3.3.2 Economic Implications of Directional Change Durations

Note that price durations are defined based on absolute cumulative price changes, and thus, provide no implications with respect to the *direction* of price changes. A slightly different concept is based on directional change durations (as defined in Section 3.1.2). These durations are defined as the time until the next directional change of the price process. In difference to price durations, they provide no direct information with respect to the *exact* size of price changes between subsequent turning points[12]. Thus, they are of limited use for VaR analyzes. However, directional change durations play an important role in technical analysis since the corresponding turning points are optimal points for buy and sell decisions of intraday traders. Obviously, price change durations are strongly related to price durations and can be used to construct modified volatility measures that incorporate not only the magnitude but also the direction of price changes.

3.4 Liquidity Measurement

3.4.1 The Liquidity Concept

Liquidity has been recognized as an important determinant of the efficient working of a market. Following the conventional definition of liquidity, an asset is considered as liquid if it can be traded quickly, in large quantities and with little impact on the price.[13] According to this concept, the measurement of liquidity requires to account for three dimensions of the transaction process: time, volume and price. Kyle (1985) defines liquidity in terms of the tightness indicated by the bid-ask spread, the depth corresponding to the amount of one sided volume that can be absorbed by the market without inducing a revision of the bid and ask quotes and resiliency, i.e. the time in which the market returns to its equilibrium.

The multi-dimensionality of the liquidity concept is also reflected in theoretical and empirical literature, where several strings can be divided: A wide range of the literature is related to the bid-ask spread as a measure of liquidity,[14] the price impact of volumes[15] and the analysis of market depth[16].

---

[12] Note that they are defined only based on a *minimum* price change between two local extrema.

[13] See, for example, Keynes (1930), Demsetz (1968), Black (1971) or Glosten and Harris (1988).

[14] See, for example, Conroy, Harris, and Benet (1990), Greene and Smart (1999), Bessembinder (2000) or Elyasiani, Hauser, and Lauterbach (2000).

[15] See, for example, Chan and Lakonishok (1995), Keim and Madhavan (1996) or Fleming and Remolona (1999).

[16] See, for example, Glosten (1994), Biais, Hillion, and Spatt (1995), Bangia, Diebold, Schuermann, and Stroughair (2002) or Giot and Grammig (2002).

In the next subsections, we illustrate how to use volume and price durations to quantify different aspects of liquidity. Section 3.4.2 discusses the application of volume durations to capture the time and volume dimension of liquidity. Sections 3.4.3 and 3.4.4 deal with concepts that allow for the measurement of market depth using price and excess volume durations.

3.4.2 Volume Durations and Liquidity

Natural liquidity measures are based on volume durations that capture the time and volume dimensions of the intraday trading process. Even though volume durations do not account for the price impact, they provide a reasonable measure of time costs of liquidity (see Gouriéroux, Jasiak, and LeFol, 1999). Consider, for example, a trader who wants to execute a large order, but wants to avoid the costs of immediacy induced by a high bid-ask-spread. Then, he has the option of splitting his order and distributing the volume over time.[17] Such a trader is interested in the time he has to wait until the execution of the complete order. Then, the expected volume duration will allow him to quantify the (time) costs of liquidity.

By defining volume durations not only based on the amount of volume shares, but also on the type of the corresponding transactions, it is possible to capture different components of the trading process. Hence, buy (sell) volume durations might be interpreted as the waiting time until a corresponding (unlimited) market order is executed. In this sense, forecasts of volume durations are associated with predictions of the absorptive capacities of the market, especially for the particular market sides. Alternatively, by the measurement of the time until a given volume on both market sides is traded, one obtains a liquidity measure that also accounts for the balance between the market sides. In this case, a market period is defined as liquid if unlimited market orders are executed quickly on both sides of the market. For a deeper analysis of such different types of volume durations, see Hautsch (2001).

3.4.3 The VNET Measure

Clearly, the neglect of the price impact of volumes is a restriction of the volume duration concept. However, in empirical research the estimation of the expected price impact for a given trading volume is an important task. In electronic trading systems, the price impact is determined by the market reaction curve, which gives the hypothetical transaction price to be expected for a certain buy or sell order and thus, combines tightness with market depth. The stronger the price impact of a hypothetical volume, the lower the depth of the market.

In a market maker market, the price impact of a volume is determined by the (closed) order book of the market maker, and thus the corresponding

---

[17] Such behavior might also be reasonable due to strategic reasons.

posted bid/ask price. The larger the volume a trader wants to buy or sell, the larger the spread posted by the market maker in order to account for his adverse selection risk and his inventory costs. Thus, the investor has to bear liquidity costs that arise from the difference between the market price and the ask (bid) quote placed by the market maker. Since the order book of the market maker is unobservable, it is difficult to quantify the price impact of a hypothetical volume.

A natural way to estimate the *realized* market depth is to relate the net trading volume to the corresponding price change over a fixed interval of time. However, as discussed in Engle and Lange (2001), using a too small fixed time interval can lead to measurement problems because the excess demand or the corresponding price change often can be zero. In contrast, using longer intervals reduces the ability of the measure to capture short-run dynamics that are of particular interest when the market is very active. For this reason, Engle and Lange (2001) propose the VNET measure, which measures the log net directional (buy or sell) volume over a price duration. Using a price duration as an underlying time interval avoids the aforementioned problems and links market volatility to market depth. Consider the sequence of points associated with dp-price changes, $\{t_i^{dp}\}$. Then, VNET is computed as

$$
VNET_i := \ln \left| \sum_{j=N(t_{i-1}^{dp})}^{N(t_i^{dp})} y_j^b v_j \right|, \tag{3.9}
$$

where y_i^b is an indicator variable that takes on the value 1 if a trade is buyer-initiated and -1 if a transaction is seller-initiated and $N(t)$ denotes the counting function associated with the transaction process. Hence, VNET measures the excess volume that can be traded before prices exceed a given threshold and therefore can be interpreted as the intensity in which excess demand flows into the market. An application of this concept in analyzing the depth of the market is found in Engle and Lange (2001).

3.4.4 Measuring (Il)liquidity Risks using Excess Volume Durations

According to traditional information-based market microstructure models, the presence of information on the market is associated with fast trading and large one-sided volumes confronting liquidity suppliers with risks due to inventory problems and adverse selection. Hautsch (2003) proposes to use excess volume durations as defined in Section 3.1.2 as natural indicators for the presence of information on the market. Excess volume durations measure the intensity of the (one-sided) demand for liquidity. Thus, they allow us to account for the time and volume dimension of information-based trading. In this framework, small durations indicate a high demand of one-sided volume per time and reflect a high (il)liquidity risk. In contrast, long excess volume durations indicate either thin trading (long trade durations), or heavy, balanced trading.

However, neither of the latter two cases confront the liquidity supplier with high liquidity risks. In this sense, excess volume durations serve as natural indicators for informed trading inducing with adverse selection and inventory risks.

Furthermore, similar to the VNET measure proposed by Engle and Lange (2001), the absolute price change measured over the corresponding duration period provides a measure of the realized market depth. The main difference between both concepts is the underlying time interval over which the price impact is measured. In the framework of excess volume durations, market depth is related to the intensity of the demand for (one-sided) liquidity and thus to liquidity supplier's inventory and adverse selection risk. Hence, it allows to analyze price adjustments in situations which are characterized by large imbalances between the buy and sell trading flow.

An interesting issue to investigate is whether market microstructure variables that reflect liquidity providers' assessments of information on the market can be used to predict the future intensity of excess volumes. In the following we derive several hypotheses from information-based market microstructure models which will be tested in Chapter 5. The following two hypotheses are related to price and quote changes measured over excess volume durations. The absolute price change associated with a given excess volume is a proxy for the realized market depth and is linked to the slope of the market reaction curve. This slope is determined by liquidity suppliers' expectations of future price movements and thus, by the information on the market. Hence, the magnitude of the price impact for a given excess volume should be the higher, the higher the amount of information in the market, and thus, the higher the expected excess demand. Therefore, given the amount of excess volume and the length of the underlying duration period, the market maker adjusts his posted quotes the stronger, the stronger his expectations regarding the future excess demand intensity, and thus his assessment of risk. This is formulated in hypothesis H5:

H5: Price impact and the expected excess demand

> The absolute price change associated with a given excess volume is negatively correlated with the expected excess volume duration.

According to market microstructure theory, the market maker's expectations concerning inventory and adverse selection risk are reflected in the quoted bid and ask price. In order to compensate for adverse selection risks, the bid-ask spread should be the wider, the higher the probability for information based trading:

H6: The impact of the bid-ask spread

> The length of an excess volume duration is negatively correlated with (i) the width of the spread posted at the beginning of the spell and with (ii) an increase of the spread during the previous duration.

The next hypothesis covers possible asymmetric behavior of market participants. It is a well known result that "bad" news lead to stronger price and volatility responses than "good" news (see, for example, Black, 1976, Christie, 1982, French, Schwert, and Stambaugh, 1987 or, recently, Hautsch and Hess, 2002). According to this phenomenon, the excess demand intensity should be stronger for "bad" news than for "good" news:

H7: Asymmetric reactions of market participants

> The (signed) price change associated with a given excess volume is positively correlated with the length of the subsequent excess volume duration.

A further interesting question is whether the time in which a given excess volume is traded, plays an important role. I.e., does trading a certain excess volume in a short period results in stronger price reactions than trading the same amount over a longer time interval? An obvious argument is that a market maker has better chances of inferring from the trading flow when he is confronted with a more and thus uniform demand than in periods where he faces large single orders entering the market. In order to explore the relationship between the price impact of a given excess volume and the length of the *contemporaneous* time spell, we state hypothesis H8 as following:

H8: The price impact and the contemporaneous excess volume duration

> The price impact of excess volumes is the higher, the longer the length of the underlying volume duration. Correspondingly, a higher than expected excess demand intensity decreases the price impact.

These hypotheses will be empirically tested by applying the framework of ACD models in Section 5.5.3 of Chapter 5.

Statistical Properties of Financial Durations

In this chapter, we present the transaction databases that are used for several empirical applications in this book. We discuss the data preparation and illustrate the statistical properties of alternative types of financial duration series based on various assets and exchanges. Section 4.1 deals with peculiar problems which have to be taken into account when transaction data sets are prepared. Section 4.2 describes the databases considered in this book, consisting of data from the New York Stock Exchange (NYSE), the German Stock Exchange, the London International Financial Futures Exchange (LIFFE), the EUREX trading platform in Frankfurt, as well as from the Australian Stock Exchange (ASX). Sections 4.3 through 4.5 illustrate the main statistical properties of different types of financial durations. During this book we focus mainly on highly liquid stocks, which are of particular interest in most empirical studies. For this reason, we restrict our descriptive analysis in the following sections to blue chips traded at the individual exchanges.

4.1 Data Preparation Issues

4.1.1 Matching Trades and Quotes

A typical problem occurs when trades and quotes are recorded in separate trade and quote databases, like, for example, in the Trade and Quote (TAQ) database released by the NYSE. In this case, it is not directly identifiable whether a quote which has been posted some seconds before a transaction was already valid at the corresponding trade. In order to reduce the problem of potential mismatching, typically the "five-seconds rule" proposed by Lee and Ready (1991) is applied. According to this rule, a particular trade is associated with the quote posted at least five seconds before the corresponding transaction. Lee and Ready (1991) illustrate that this rule leads to the lowest rates of mismatching.

4.1.2 Treatment of Split-Transactions

A further problem emerges from so-called "split-transactions". Split-trans-actions arise when an order on one side of the market is matched against several orders on the opposite side. Typically, such observations occur in electronic trading systems when the volume of an order exceeds the capacities of the first level of the opposing queue of the limit order book. In these cases, the orders are automatically matched against several opposing order book entries. Often, the recorded time between the particular "sub-transactions" is extremely small[1] and the corresponding transaction prices are equal or show an increasing (or respectively, decreasing) sequence. In most studies, zero durations are simply aggregated to one single transaction. However, as argued by Veredas, Rodriguez-Poo, and Espasa (2002), the occurrence of such observations might also be due to the fact that the limit orders of many traders are set for being executed at round prices, and thus, trades executed at the same time do not necessarily belong to the same trader.[2] In these cases, a simple aggregation of zero durations would lead to mismatching. For this reason, we use an algorithm to consolidate split-transactions as proposed by Grammig and Wellner (2002). According to this algorithm, a trade is identified as a split-transaction when the durations between the sub-transactions are smaller than one second, and the sequence of the prices (associated with a split-transaction on the bid (ask) side of the order book) are non-increasing (non-decreasing). For simplicity, the time stamp and corresponding price of the split-transaction is determined by the corresponding last sub-transaction, while the volume of the particular sub-trades is aggregated. An alternative and slightly more precise method would be to treat the corresponding trade duration as right-censored and to compute the price as the (volume weighted) average of the prices of the sub-transactions. However, note that such a proceeding would lead to a disappearance of the discreteness of the price process.

4.1.3 Identification of Buyer- and Seller-Initiated Trades

Typically, it is not possible to identify whether a trade is seller- or buyer-initiated.[3] For this reason, the initiation of trades is indirectly inferred from the price and quote process. The most commonly used methods of inferring the trade direction are the tick test, the quote method, as well as hybrid methods which combine both methods (see e.g. Finucane, 2000). The tick test uses

---

[1] Often it is a matter of measurement accuracy that determines whether sub-transactions have exactly the same time stamp or differ only by hundredths of a second.

[2] Based on limit order book data from the ASX, it is possible to identify the particular brokers executing a transaction. In fact, it is shown that sometimes different transactions are executed within one second.

[3] An exception is the limit order book database from the ASX, where each trade is directly characterized as a buy or a sell.

previous trades to infer the trade direction. According to this method, a trade is classified as a buy (sell) if the current trade occurs at a higher (lower) price than the previous trade. If the price change between the transactions is zero, the trade classification is based on the last price that differs from the current price. The quote method is based on the comparison of the transaction price and the midquote. Whenever the price is above (below) the midquote, the trade is classified as a buy (sell). Here, we use a combination of both methods as proposed by Lee and Ready (1991), where the quote method is used to classify all transactions that do not occur at the midquote, and the tick test is applied to all remaining cases.

4.2 Transaction Databases and Data Preparation

In the following, we present the different databases used in this book and explain the corresponding data preparation. To avoid data peculiarities induced by special opening or closing (auction) procedures, we generally remove all observations outside of the regular trading hours. Furthermore, all non-valid trades are deleted. Trades which cannot be identified as particular sub-trades of a split-transaction are not deleted and remain as zero durations in the sample.[4]

4.2.1 NYSE Trading

Trading at the New York Stock Exchange (NYSE) is based on a so-called hybrid system, i.e., the trading mechanism combines a market maker system with an order book system. For each stock, one market maker (specialist) has to manage the trading and quote process and has to guarantee the provision of liquidity, when necessary, by taking the other side of the market. He manages the order book, posts bid and ask prices and is responsible for the matching of the buy and sell side of the market. The NYSE releases the so-called "Trade and Quote (TAQ) database" which contains detailed information on the intraday trade and quote process. The TAQ database consists of two parts: the trade database and the quote database. The trade database contains transaction prices, trading volumes, the exact time stamp (to the second) and attribute information concerning the validity of the transaction. The quote database consists of time stamped (best) bid and ask quotes as posted by the market maker, the volume for which the particular quote is valid (market depth), as well as additional information on the validity of the quotes. Because the NYSE features a hybrid trading mechanism, the quotes reported in the quote database can be quotes that are posted by the specialist, limit orders from market participants posted in the limit order book, or

---

[4] Nonetheless, in the empirical applications in Chapters 5, 6 and 7, these observations are deleted.

limit orders submitted by traders in the trading crowd. For more details, see Hasbrouck, Sofianos, and Sosebees (1993) or Bauwens and Giot (2001).

For the applications in this book, we extract data from the January-May TAQ (2001) CD-ROM. Regular trading at the NYSE starts at 9:30 and ends at 16:00. The matching of the trade and quote database is performed following the "five-seconds rule" as described in Section 4.1.1, while the Lee and Ready (1991) rule (see Section 4.1.3) is applied to identify buy and sell trades. Note that all trades that occurred before any quote has been posted at the beginning of a trading day, are not identifiable. These trades are removed whenever a buy-sell classification is required (as in the case of generating buy or sell volume durations). The identification of split-transactions is complicated by the fact that the NYSE trading is a hybrid system. Here, split-transactions can occur due to order-book mechanisms as described in Section 4.1.2 or because several traders (nearly) simultaneously buy (or sell) at the same price. Since these particular mechanisms are not clearly identifiable, we use a slightly modified consolidation procedure that identifies only those trades as parts of a split-transaction for which the durations between the corresponding sub-transactions are smaller than one second and the prices are *equal*. Moreover, whenever two quotes are posted within one second, the corresponding first quote is removed from the sample.

4.2.2 XETRA Trading

The German Exchange Electronic Trading (XETRA) is an electronic screen-based system consisting of an open order book that is observable for all market participants. XETRA is a double continuous auction system with an opening and closing call auction at the beginning and at the end of the trading day, respectively, and a mid-day call auction. During the normal trading period, trading is based on an automatic order matching procedure. Limit orders enter the queues of the order book according to strict price-time priority. In difference to other electronic trading systems, like the Paris Bourse or the ASX (see Section 4.2.5), in the XETRA system, market orders are fully executed until complete fill. Hence, entering a market order leads to a matching of the complete volume, however, induces the risk due to the possible incurred price impact. XETRA trading is completely anonymous and does not reveal the identity of the traders.

We use a data set released by the "Deutsche Finanzdatenbank" (DFDB), Karlsruhe, which contains time stamped transaction prices and volumes from January to December 1999. During this sample period, it is necessary to account for a change in the trading hours. Before the 09/20/99, trading starts at 8:30 GMT and ends at 17:00 GMT. On 09/20/99, the opening hours changed to 9:00 GMT while closing hours changed to 17:30 GMT. Split-transactions are identified according to the rule described in Section 4.1.2. Note that this data set consists only on the resulting market orders and contains no information about the order book. For this reason, no best bid or best ask prices are available.

4.2.3 Frankfurt Floor Trading

In addition to the XETRA system, German stock trading is organized on eight trading floors[5]. The most influential and liquid floor trading is located at the Frankfurt Stock Exchange. The floor trading is characterized by a dominant role of the market maker who manages the closed order book, matches the incoming orders with the entries of the book and is responsible for the price discovery. By posting corresponding bid and ask prices, he guarantees market liquidity on both sides of the market. As with XETRA trading, here, we use a data set released by the DFDB containing time stamped prices and volumes from January to December 1999. The trading hours correspond to the trading hours in the XETRA system. Split-transactions are consolidated as they are in the TAQ data: by identifying trades as sub-transactions only if the particular prices are equal. As for the XETRA data, no spreads are available.

4.2.4 Bund Future Trading at EUREX and LIFFE

The Bund future is one of the most actively traded future contracts in Europe and is a notional 6% German government bond of EURO 100,000 with a face value which matured in 8.5 to 10.5 years at contract expiration.[6] There are four contract maturities per year: March, June, September and December. Prices are denoted in basis points of the face value. One tick is equivalent to a contract value of EURO 10 (during the observation period it was DEM 25). The contract is traded in almost identical design at the electronic trading at EUREX and the floor trading at LIFFE.

The Bund future trading at EUREX (former "Deutsche Terminbörse", DTB), Frankfurt, is based on an electronic screen-based trading system. It is a continuous auction system with automatic electronic order matching similar to the XETRA system (see Section 4.2.2). The data used in this book is released by the EUREX and consists of time stamped prices and volumes. Since the data set contains no order book information, no spreads are available. The data preparation is performed according to the rules described for the XETRA trading. The descriptive statistics presented in the Sections 4.3 through 4.5 are based on one contract from 12/02/96 to 03/06/97.

The floor trading system at LIFFE is a dealer driven system with open outcry. In difference to the floor trading at the Frankfurt Stock Exchange, an official order book does not exist. The particular market makers post bid and ask quotes that are only valid as long "breath is warm" (see Franke and

---

[5] In Berlin, Bremen, Düsseldorf, Frankfurt, Hamburg, Hannover, München and Stuttgart.

[6] During the chosen observation period, 1994-1997, the bond was of DEM 250,000.

Hess, 2000). The transactions are recorded by price reporters who enter the transaction prices in the computer system and thus, guarantee immediate publication. One exception is the transaction volume which is published with short delay and is reported quite inaccurately.[7] For this reason, the single transaction volumes are of limited value. In this book, we use time stamped prices, volumes, and the corresponding bid and ask prices extracted from a CD released by LIFFE[8]. The empirical study in Section 6.6 of Chapter 6 uses data on 11 contracts corresponding to 816 trading days between 04/05/94 and 06/30/97. The descriptive statistics presented in the following section are based on one contract from 03/03/97 to 06/05/97. During the sample period used, open outcry trading took place between 7:00 and 16:15 GMT. The consolidation of split-transactions is conducted as it is for the TAQ and the Frankfurt floor data.

4.2.5 ASX Trading

The Australian Stock Exchange (ASX) is a continuous double auction electronic market. The continuous auction trading period is preceded and followed by an opening call auction. Normal trading takes place continuously between 10:09 and 16:00 Sydney time on Monday to Friday. Limit orders are queued in the buy and sell queues according to a strict time-price priority order. Any buy (sell) order entered that has a price that is greater (less) than existing queued sell (buy) orders, will be executed immediately. The order will be automatically matched to the extent of the volume that is available at the specified limit price. Then, the executed order results in a trade and will be deleted from the queues of the limit order book. Orders that are executed immediately are *market orders*. Orders entered with a price that does not overlap the opposite order queue remain in the book as *limit orders*. Entered orders that are partially executed are a *combination* of a market order for the immediately executed volume and a limit order for the remaining volume. All orders and trades are always visible to the public. Order prices are always visible, however orders may be entered with an undisclosed (hidden) volume if the total value of the order exceeds AUD 200,000. Although undisclosed volume orders are permitted, sufficient information is available to unambiguously reconstruct transactions. The identity of the broker who entered an order is not public information, but is available to all other brokers. Thus, this database allows for an exact identification of split-transactions. Hence, corresponding multiple trade records are easily recovered and aggregated into a single trade record.

The ASX data sets used in this book contain time stamped prices, volumes and identification attributes (inclusive the buy/sell identification) of all trades

---

[7] In particular, the price reporters mainly use approximative values in order to record whether a volume is low-sized, medium-sized or large-sized.

[8] Hence, in this context no matching of separate trade and quote databases are required.

and limit orders during July 2002. Thus, this database allows for a complete reconstruction of the limit order book. In this context, data from the opening and closing call auctions periods are not utilized and all crossings and off market trades are removed. The reconstruction of the limit order book requires to account explicitly for deletions and changes of queued orders. Modifying the order volume downwards does not affect order priority. In contrast, modifying the order volume upwards automatically creates a new order at the same price as the original order with the increase in volume as the volume of the newly created order. Note that this avoids loss of priority on the existing order volume. A modification of the order price so that the price overlaps the opposite order queue will cause immediate or partial execution of the order according to the rules described in Section 4.2.2. Modifying the price otherwise causes the order to move to the lowest time priority within the new price level.

4.3 Statistical Properties of Trade, Limit Order and Quote Durations

Table 1 shows descriptive statistics of trade and quote durations for NYSE trading, XETRA trading, Frankfurt floor trading, as well as Bund future trading at EUREX and LIFFE. We focus on the AOL, IBM and GE stock traded at the NYSE, and the Allianz, BASF and Henkel stock traded at the German Stock Exchange. The NYSE blue chips are traded extremely frequently, with trade and quote durations averaging between 5 and 12 seconds. These stocks belong to the most actively traded stocks in the world. The Allianz and BASF stock are among the most liquid blue chips in the German Stock Index (DAX). However, average trade durations of approximately 50 seconds for XETRA trading and 170 seconds for Frankfurt floor trading illustrate that the German Stock Exchange is clearly less liquid than the NYSE. The Henkel stock belongs to the less liquid DAX stocks, with average durations of about 140 seconds in the XETRA system. For the Bund future trading, we observe that during the corresponding sample period, the liquidity at LIFFE was nearly as high as at EUREX[9].

Table 2 presents the same descriptive statistics of trade and limit order durations for three liquid stocks traded at the ASX. Here, we focus on the BHP, NAB and MIM stock. BHP and NAB are among the most actively traded stocks at the ASX, with average trade durations of between 17 and 23 seconds. MIM is clearly less liquid with average trade durations of 134 seconds.

We observe lower limit order durations than trade durations (with exception of the NAB stock). This indicates a higher intensity in the limit order

---

[9] However, in the meantime, this relationship has changed dramatically in favor of the EUREX.

Table 1: Descriptive statistics of trade and quote durations (number of observations, mean, standard deviation, minimum, maximum, quantiles and Ljung-Box $(\chi^2(20))$ statistic). Based on trading at NYSE, XETRA, Frankfurt floor, EUREX and LIFFE.

	Obs	Mean	S.D.	Min	Max	0.05q	0.25q	0.50q	0.75q	0.95q	LB(20)
\multicolumn{12}{c}{trade durations NYSE (01/02/01 to 05/31/01)}
AOL	214030	11.33	14.22	0	817	1	3	6	14	38	12427
IBM	340238	7.13	6.79	0	205	1	3	5	9	20	43807
GE	258182	8.31	8.87	0	160	1	3	5	10	25	9201
\multicolumn{12}{c}{quote durations NYSE (01/02/01 to 05/31/01)}
AOL	191727	12.03	17.75	0	629	0	2	6	14	45	19827
IBM	414049	5.88	7.11	1	941	1	2	4	7	18	26413
GE	298614	7.56	9.69	0	212	0	2	4	9	26	28348
\multicolumn{12}{c}{buy durations NYSE (01/02/01 to 05/31/01)}
AOL	111720	21.65	35.06	0	817	1	4	9	24	86	2937
IBM	187318	12.95	16.87	0	386	1	4	7	15	44	11278
GE	139489	15.36	21.77	0	600	2	4	8	18	54	1727
\multicolumn{12}{c}{sell durations NYSE (01/02/01 to 05/31/01)}
AOL	102040	23.66	40.09	0	949	1	4	9	26	97	2567
IBM	152680	15.87	22.00	0	543	1	4	8	19	57	5603
GE	118475	18.06	26.45	0	621	2	4	9	21	66	1221
\multicolumn{12}{c}{trade durations XETRA (01/04/99 to 12/30/99)}
Allianz	170282	45.54	67.10	0	1469	2	7	20	56	174	37220
BASF	151448	51.22	72.50	0	1807	2	8	25	65	188	43596
Henkel	28166	141.06	207.68	0	4219	3	17	65	180	533	3782
\multicolumn{12}{c}{trade durations Frankfurt floor (01/04/99 to 12/30/99)}
Allianz	43156	168.31	199.54	1	4308	9	45	106	217	538	39195
BASF	41372	175.84	213.95	2	4586	10	43	106	225	583	28163
Henkel	32060	521.84	548.15	1	7856	41	163	356	687	1556	8111
\multicolumn{12}{c}{Bund future EUREX (12/02/96 to 03/06/97)}
B-Future	198827	10.79	21.38	0	886	1	2	4	11	40	187041
\multicolumn{12}{c}{Bund future LIFFE (03/03/97 to 06/05/97)}
B-Future	160230	12.64	18.11	0	1081	1	4	8	15	38	202231

Overnight spells are ignored. Descriptive statistics in seconds.

process than in the market order process. Comparable effects are observed for trading on the NYSE. Here, the liquidity supply associated with the arrival of new quotes is more active than the liquidity demand. Focussing on distributional aspects of the durations, we observe in general overdispersion, i.e., the standard deviation exceeds the mean, which is quite typical for trade (and quote) durations. Therefore, evidence against an exponential distribution is found.[10] Overdispersion effects are also reflected in the distributional shape of trade durations.

---

[10] Of course, this has to be tested statistically. However, for ease of exposition, during this section we refrain from more precise statistical inference.

Table 2: Descriptive statistics of trade and limit order durations (number of observations, mean, standard deviation, minimum, maximum, quantiles and Ljung-Box ($\chi^2(20)$) statistic). Based on ASX trading. Sample period 07/01/02 to 07/31/02.

	Obs	Mean	S.D.	Min	Max	0.05q	0.25q	0.50q	0.75q	0.95q	LB(20)
					trade durations						
BHP	27108	17.91	32.32	0	2107	1	3	8	21	65	9600
NAB	21490	22.65	40.72	0	2451	1	4	10	26	84	8401
MIM	3617	134.41	207.05	0	2953	3	18	63	162	503	976
					limit order durations						
BHP	27447	17.68	30.83	0	2078	0	3	9	21	62	9202
NAB	20809	23.38	40.64	0	2481	1	4	11	28	81	8687
MIM	5891	82.41	122.34	0	2342	2	13	40	102	299	2241
					buy durations						
BHP	16640	29.17	49.05	0	2157	1	5	14	35	105	6564
NAB	11851	41.03	73.07	0	2451	1	7	18	48	145	5693
MIM	1664	287.24	548.11	0	6040	3	26	97	305	1155	261
					sell durations						
BHP	10468	46.36	84.06	0	2130	1	7	20	52	173	4676
NAB	9639	50.46	83.60	0	2457	2	9	23	58	187	4184
MIM	1953	246.81	395.10	0	6203	4	33	111	287	956	341

Overnight spells are ignored. Descriptive statistics in seconds.

Figure 1 shows kernel density plots of trade durations for the Allianz, BASF and Henkel stock traded at XETRA and on the Frankfurt floor. The density depicts a strong right-skewed shape, indicating a high occurrence of relatively low durations and a strongly declining proportion of longer durations[11]. A slightly lower dispersion is observed for trade durations based on floor trading compared to trade durations based on electronic trading. This finding is also reflected in a lower ratio between the standard deviation and mean.

The Ljung-Box (LB) statistics in Tables 1 and 2 formally test the null hypothesis that the first 20 autocorrelations are zero and are $\chi^2(20)$ distributed with a critical value of 31.41 at the 5% significance level. Based on these statistics, the null hypothesis of no autocorrelation is easily rejected for all types of trade and quote durations.

Figures 2 through 5 show the autocorrelation functions (ACF) of trade, quote and limit order durations. In general, we observe highly significant autocorrelations. Interestingly, these patterns differ between the particular markets and trading systems. Trade durations on the NYSE exhibit relatively low first order autocorrelations between 0.05 and 0.1 (Figure 2). However, the duration processes are very persistent. This property is indicated by

---

[11] Since the corresponding kernel density plots associated with the other stocks are quite similar, they are not shown here.

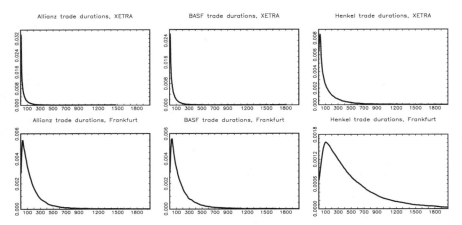

Figure 1: Kernel density plots (Epanechnikov kernel with optimal bandwidth) of trade durations for Allianz, BASF and Henkel. Based on XETRA and Frankfurt floor trading.

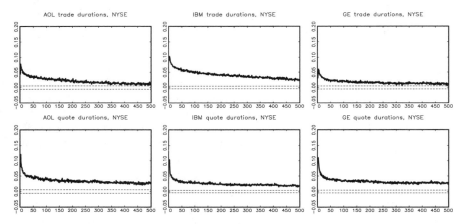

Figure 2: Autocorrelation functions of trade and quote durations for AOL, IBM and GE. Based on NYSE trading. Dotted lines: approx. 99% confidence interval. The x-axis denotes the lags in terms of durations.

autocorrelation functions that decay with a slow, hyperbolic rate which is quite typical for long memory processes. A similar pattern is observed for XETRA and Frankfurt floor trading (Figure 3). Interestingly, the floor trading in Frankfurt yields trade durations that clearly have a higher low order autocorrelation (about 0.2) than the XETRA trading (between 0.05 and 0.1).

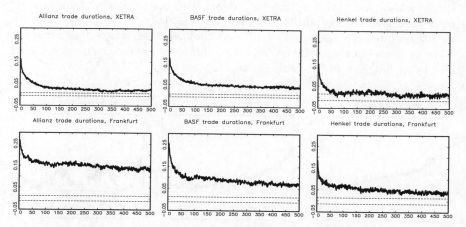

Figure 3: Autocorrelation functions of trade durations for Allianz, BASF and Henkel. Based on XETRA and Frankfurt floor trading. Dotted lines: approx. 99% confidence interval. The x-axis denotes the lags in terms of durations.

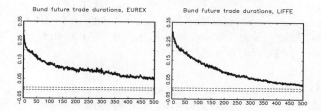

Figure 4: Autocorrelation functions of trade durations for the Bund future. Based on EUREX and LIFFE trading. Dotted lines: approx. 99% confidence interval. The x-axis denotes the lags in terms of durations.

Thus, slight evidence is provided for the fact that floor trading seems to induce higher autocorrelations in the trading intensity. Distinctly different autocorrelation functions are observed for Bund future trade durations (Figure 4). On the one hand, they reveal explicitly higher low order autocorrelations and thus, a stronger clustering of trading activity. On the other hand, the autocorrelation functions decline with a higher rate of decay. Hence, these processes appear to be less persistent than the other duration series. Again, we observe differences between electronic and floor trading. This confirms the results obtained for the Allianz and BASF trading on the German Stock Exchange. Once more, the trading intensity on the floor seems to be more strongly clustered than in the computerized system. A possible reason could be the higher anonymity of traders in electronic trading systems. Franke and Hess (2000) find empirical evidence for an influence of the type of the trading system on the information diffusion within the market. By comparing electronic trading with floor trading, the authors argue that especially in periods of high information intensity, floor trading allows for higher information dissemination. Thus, it turns out that higher information diffusion seems to cause also higher serial dependencies in the resulting trading intensity.

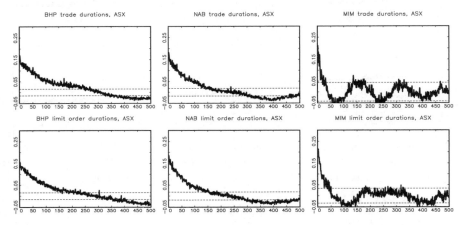

Figure 5: Autocorrelation functions of trade and limit order durations for BHP, NAB and MIM. Based on ASX trading. Dotted lines: approx. 99% confidence interval. The x-axis denotes the lags in terms of durations.

The ACF depicted for ASX trade durations (Figure 5) reveals a quite interesting pattern that differs from the autocorrelation shapes obtained for the other trade duration series. While for low order autocorrelations the shape is similar to those for the other time series, the ASX trade duration processes reveal a clearly lower persistence. The ACF shows a higher rate of decay and even becomes significantly negative up to approximately 200 lags. Hence, the intensity of the trading process is not clustered over a complete trading day (or even longer), but seems to reverse after a certain time period. Such a time series structure is observed for none of the other series and thus seems to be caused by this particular trading system. For the MIM stock, we observe a significant seasonality pattern caused by the fact that higher lags cover a time span lasting over more than one trading day.

Summarizing these findings, we can conclude that the form of the trading system seems to have a strong impact on the dynamics of the resulting trading process. Obviously, the strength and persistence of serial dependencies in the trading intensity mainly differ between the individual exchanges and less between the different assets traded in the same trading system.

Figures 6 through 9 show the intraday seasonality patterns of trade, quote and limit order durations based on cubic spline regressions. For NYSE

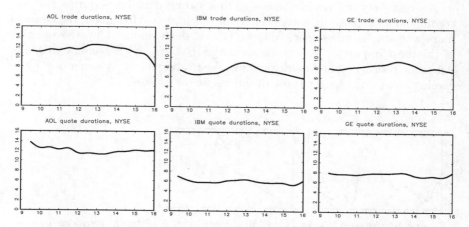

Figure 6: Cubic spline function (30 minute nodes) of trade and quote durations for AOL, IBM and GE. Based on NYSE trading. The x-axis denotes the local calendar time.

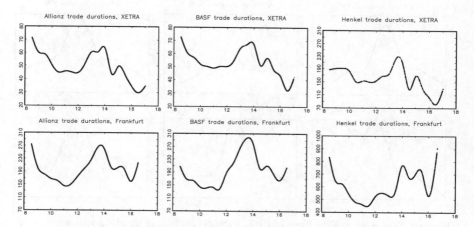

Figure 7: Cubic spline function (30 minute nodes) of trade durations for Allianz, BASF and Henkel. Based on XETRA and Frankfurt floor trading. The x-axis denotes the local calendar time.

trading (Figure 6), we only find weak seasonal variations during the trading day. Slightly longer trade durations are observed around noon indicating the existence of well known "lunchtime effects". Interestingly, such effects are not found for quote durations. Instead, the intensity of quote renewals is nearly constant during a trading day. Clearly more pronounced patterns are shown for XETRA and Frankfurt floor trading (Figure 7). Here, the market starts with a relatively low trading intensity that strongly increases during the first trading hours and significantly declines at lunch time. After noon, the German market again becomes more active. This is likely induced by the opening of the most important American exchanges (CBOT, NYSE and NASDAQ). After the processing of American news, the trading intensity declines slightly coming along with longer trade durations before closure.

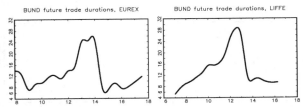

Figure 8: Cubic spline function (30 minute nodes) of trade durations for the Bund future. Based on EUREX and LIFFE trading. The x-axis denotes the local calendar time.

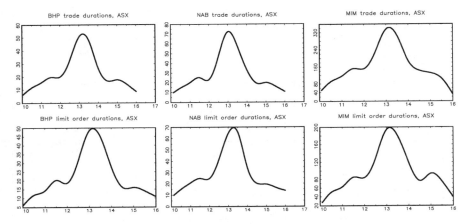

Figure 9: Cubic spline function (30 minute nodes) of trade and limit order durations for BHP, NAB and MIM. Based on ASX trading. The x-axis denotes the local calendar time.

Interestingly, the intraday seasonality pattern of the Bund future trading (Figure 8) presents a different picture. In contrast to trading at the German Stock Exchange, we observe a very active trading period after the opening of the market followed by a clear lunchtime effect. However, in the afternoon, a sharp drop of the seasonality function is observed. This is much more pronounced than for the stock market. Thus, the Bund future market seems to react more strongly to American news than the stock market. For ASX trading (Figure

9) we observe a seasonality pattern that is very similar to Bund future trading
and is again clearly dominated by a strong lunchtime dip around noon.

4.4 Statistical Properties of Price Durations

Table 3 shows descriptive statistics of price durations based on particular
stocks traded at the different markets. We choose stock-specific values for
the size of the underlying cumulative price change dp. This leads to three
aggregation levels that are approximately comparable. For the IBM and the
BHP stock, as well as the LIFFE Bund future the price durations are generated based on midquotes. For all other stocks, the price durations are based

Table 3: Descriptive statistics for price durations (number of observations, mean,
standard deviation, minimum, maximum, quantiles and Ljung-Box ($\chi^2(20)$)
statistic). Based on trading at NYSE, XETRA, Frankfurt floor, EUREX, LIFFE
and ASX.

	Obs	Mean	S.D.	Min	Max	0.05q	0.25q	0.50q	0.75q	0.95q	LB(20)
IBM price durations, NYSE (01/02/01 to 05/31/01)											
$dp = 0.10$	24510	98	130	1	3517	8	26	57	120	326	13098
$dp = 0.25$	6899	346	498	1	9694	25	84	192	402	1175	2388
$dp = 0.50$	2042	1109	1522	3	14689	59	256	602	1321	3954	480
BASF price durations, XETRA (01/04/99 to 12/30/99)											
$dp = 0.05$	35380	216	308	1	7804	6	40	113	268	776	9912
$dp = 0.10$	14698	515	815	1	17721	11	82	242	608	1933	2747
$dp = 0.25$	2773	2378	3232	1	29329	53	405	1165	2930	8996	306
BASF price durations, Frankfurt floor (01/04/99 to 12/30/99)											
$dp = 0.05$	12787	554	721	6	15942	33	137	321	687	1843	1580
$dp = 0.10$	7488	918	1215	8	16427	45	210	506	1125	3218	739
$dp = 0.25$	1939	2899	3408	11	24500	163	691	1628	3802	9965	204
Bund future price durations, EUREX (12/02/96 to 03/06/97)											
$dp = 2.00$	18450	115	216	1	9140	3	18	51	127	429	14488
$dp = 4.00$	4300	483	843	1	13339	9	73	213	551	1759	1456
$dp = 6.00$	1812	1095	1666	1	15632	28	195	552	1257	3759	448
Bund future price durations, LIFFE (03/03/97 to 06/05/97)											
$dp = 2.00$	17607	114	191	1	4134	8	26	59	127	397	15848
$dp = 4.00$	3890	513	883	1	14531	27	102	238	560	1855	1022
$dp = 6.00$	1592	1235	2158	3	24795	51	219	552	1302	4591	97
BHP price durations, ASX (07/01/02 to 07/31/02)											
$dp = 0.05$	8115	59	141	1	3863	2	6	18	58	236	1861
$dp = 0.15$	1468	326	641	1	8007	4	32	122	338	1235	293
$dp = 0.25$	513	914	1417	1	9907	19	124	360	995	3890	109

on the corresponding transaction prices since no midquotes are available.[12] The lowest aggregation level is associated with price durations that include on average 10 underlying transactions. The highest aggregated price durations include approximately 500 transactions and typically last over about 2 hours. For all price duration series, the standard deviation exceeds the mean. Therefore, we observe a clear overdispersion that is more pronounced than for trade durations and thus again evidence against an exponential distribution.

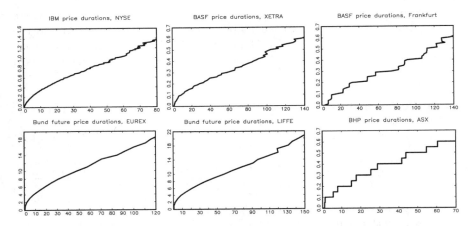

Figure 10: Relationship between the size of cumulative absolute price changes dp (y-axis) and the average length of the resulting price duration (in minutes) (x-axis). Based on IBM (NYSE trading), BASF (XETRA and Frankfurt floor trading), the Bund future (EUREX and LIFFE trading) and BHP (ASX trading).

Figure 10 shows the relationship between the size of the underlying cumulative absolute price change dp and the average length of the resulting price duration. The particular graphs depict slightly concave functions having quite similar shapes. Note that the stepwise functions for Frankfurt floor trading and ASX trading are due to the existence of minimum tick sizes of EURO 0.05 for BASF and of AUD 0.05 for BHP. The relationship between dp and the length of the resulting price duration can be expressed in the form of a simple power function, $\bar{x}_{(dp)} = \omega \cdot dp^p$, where $\bar{x}_{(dp)}$ denotes the mean price duration with respect to dp. Such a power function is well-known in describing the relationship between the absolute size of returns and the time over which the return is measured (see, for example Dacorogna, Gencay, Müller, Olsen, and Pictet, 2001). By analyzing exchange rate returns, Dacorogna et. al. (2001) find quite stable power exponents providing evidence for the existence of a power law. Here, we find highly significant coefficients for p with values 0.74

---

[12] Note that for the Bund future, prices changes are given in terms of ticks. See also Section 4.2.4.

(IBM), 0.89 (BASF, XETRA), 1.01 (BASF, Frankfurt), 0.56 (Bund future, EUREX), 0.60 (Bund future, LIFFE) and 0.64 (BHP). Thus, based on the analyzed data, we find a relatively high variation in the estimated power exponents yielding no support for the hypothesis of a "universal" power law.

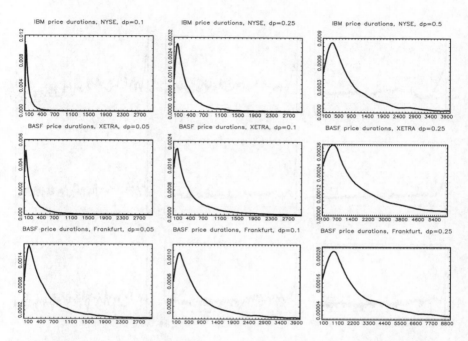

Figure 11: Kernel density plots (Epanechnikov kernel with optimal bandwidth) of price durations for IBM (NYSE trading) and BASF (XETRA and Frankfurt floor trading).

However, note that an extensive study of such a relationship is beyond the scope of this analysis.

Figure 11 shows the kernel density plots of the IBM and BASF price duration series based on the different aggregation levels. Not surprisingly, they are quite similar to those of the trade durations, revealing a strong right-skewed shape. Figures 12 and 13 show the autocorrelation functions of the individual price duration series. Note that for the higher aggregation levels, we compute the ACF based on fewer lags because of smaller sample sizes. In most cases, the highest first order autocorrelations are observed for the lowest or the middle aggregation level while for the highest aggregation levels weaker serial dependencies are obtained. Thus, no monotonic relationship between the magnitude of the first order autocorrelation and the underlying aggregation level is observed. Moreover, the autocorrelation functions reveal clear seasonality patterns. This is due to the fact that the higher lags of the ACF are associated with time periods that date back to the previous trading days. In general, the autocorrelation functions decline slowly, however, for higher lags, they become insignificant and tend toward zero. Moreover, the persistence seems to decline for higher aggregation levels.

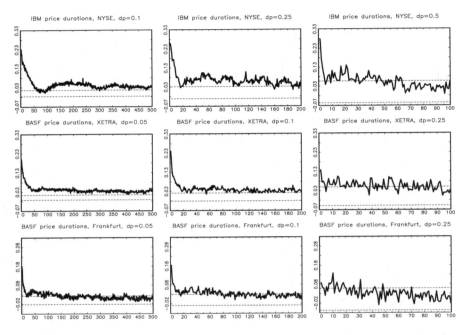

Figure 12: Autocorrelation functions of price durations for IBM (NYSE trading) and BASF (XETRA and Frankfurt floor trading). Dotted lines: approx. 99% confidence interval. The x-axis denotes the lags in terms of durations.

Figure 14 presents intraday seasonality plots of price durations. As illustrated in Chapter 3, price durations are inversely related to volatility. Thus, these patterns can be interpreted as inverse volatility patterns. It turns out that the intraday seasonality pattern of price durations is quite different from the daily shape of trade durations. Especially for NYSE, XETRA and Frankfurt floor trading, clear differences between deterministic intraday patterns of the trade intensity and the price intensity are observed. In general, the seasonal shape of price durations is strongly governed by significant lunchtime effects. Hence, a high volatility is observed after the opening and before the closing of the market, while dropping sharply around noon. Nevertheless, for the European markets, we find a further dip around 14:00 GMT which is probably caused by the arrival of American trading news. Note that in this

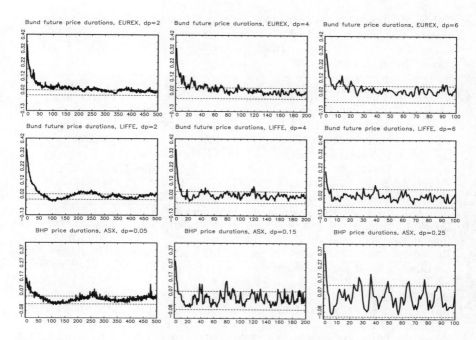

Figure 13: Autocorrelation functions of price durations for the Bund future (EUREX and LIFFE trading) and BHP (ASX trading). Dotted lines: approx. 99% confidence interval. The x-axis denotes the lags in terms of durations.

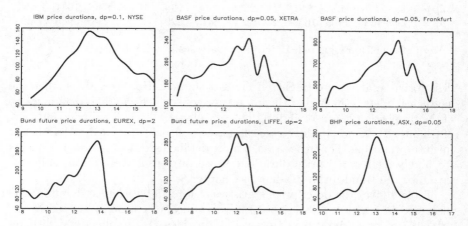

Figure 14: Cubic spline function (30 minute nodes) of price durations for IBM ($dp = 0.1$), BASF ($dp = 0.05$), the Bund future ($dp = 2$) and BHP ($dp = 0.05$). The x-axis denotes the local calendar time.

Table 4: Descriptive statistics of directional change durations (number of observations, mean, standard deviation, minimum, maximum, quantiles and Ljung-Box ($\chi^2(20)$) statistic). Based on trading at NYSE, XETRA, Frankfurt floor, EUREX, LIFFE and ASX.

	Obs	Mean	S.D.	Min	Max	0.05q	0.25q	0.50q	0.75q	0.95q	LB(20)
IBM directional change durations, NYSE (01/02/01 to 05/31/01)											
$dp = 0.10$	13946	172	190	1	3154	15	54	114	223	523	3387
$dp = 0.25$	4870	488	589	1	7889	48	147	304	595	1556	922
$dp = 0.50$	1682	1329	1626	3	17043	105	370	765	1638	4565	314
BASF directional change durations, XETRA (01/04/99 to 12/30/99)											
$dp = 0.10$	13254	569	785	1	12024	10	93	295	727	2058	2696
$dp = 0.20$	3839	1855	2400	1	25574	33	352	956	2422	6705	514
$dp = 0.40$	854	5918	5613	3	28214	312	1637	4230	8477	17276	33
BASF directional change durations, Frankfurt floor (01/04/99 to 12/30/99)											
$dp = 0.10$	5100	1318	1670	8	19021	69	301	749	1639	4467	540
$dp = 0.20$	2432	2506	3098	9	26920	122	533	1405	3180	8765	381
$dp = 0.40$	539	7002	5968	115	28453	636	2298	5018	10547	19364	24
Bund future directional change durations, EUREX (12/02/96 to 03/06/97)											
$dp = 3.00$	6591	317	503	1	9268	6	60	158	374	1118	4105
$dp = 5.00$	2307	884	1293	1	19898	33	179	474	1052	2999	781
$dp = 7.00$	1168	1636	2055	1	19898	67	385	915	2029	6088	291
Bund future directional change durations, LIFFE (03/03/97 to 06/05/97)											
$dp = 3.00$	6077	330	475	1	7560	28	93	186	380	1087	2403
$dp = 5.00$	2208	882	1369	2	19317	75	211	455	1011	2886	233
$dp = 7.00$	1034	1812	2586	17	25199	116	419	934	2078	6663	67
BHP directional change durations, ASX (07/01/02 to 07/31/02)											
$dp = 0.05$	5685	85	190	1	4023	2	7	26	83	342	1065
$dp = 0.15$	1140	422	840	1	9636	4	40	151	456	1641	155
$dp = 0.25$	428	1105	1636	2	12914	47	229	542	1256	3986	42

Overnight spells are ignored. Descriptive statistics in seconds.

descriptive analysis, overnight effects are ignored. However, the higher the aggregation level, the higher the importance of such effects due to an increase of the proportion of censored durations (see also the discussion in Section 2.4.1 of Chapter 2). Hence, in these cases, ignoring overnight spells can induce heavy misinterpretations. For this reason, a more detailed study of seasonality effects for highly aggregated price durations based on censored duration models is given for the Bund future trading in Chapter 6.

Table 4 gives the descriptive statistics of *directional change durations*. As discussed in Chapter 3, Section 3.1.2, this concept is strongly related to the principle of price durations. However it provides also implications with respect to the direction of price changes. Similar to the price durations case, we use three different aggregation levels to describe price movements over different time horizons. Note that the underlying aggregation level dp defines the *minimum* price change between two consecutive local extrema. Therefore, the resulting waiting times are significantly higher than for corresponding price durations using the same aggregation level. For example, for the IBM stock, the mean directional change duration for an aggregation level of $dp = 0.01$ is approximately 3 minutes and is approximately 6 minutes for the corresponding price duration.

In general, the statistical properties of directional change durations are quite similar to those of price durations. Again, in most cases, we observe overdispersion. However, the ratio between standard deviation and mean is lower than for price durations and in some cases reveals underdispersion for highly aggregated directional change durations in XETRA and Frankfurt floor trading (see Table 4). The density kernel plots are quite similar to those of price durations and trade durations. For this reason, we refrain from showing them in this context. The autocorrelation functions (Figures 15 and 16) depict highly significant first order autocorrelations with values between approximately 0.15 and 0.23 (for the lowest and middle aggregation levels). Thus, a clear predictability of the next directional change duration using lagged observations is shown. Hence, even for high

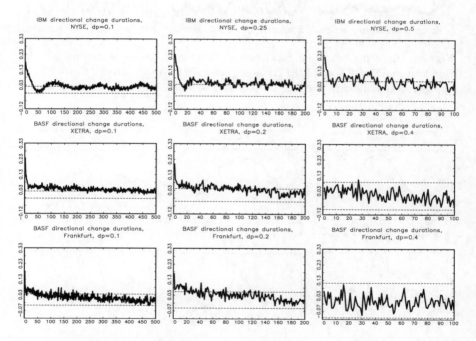

Figure 15: Autocorrelation functions of directional change durations for IBM (NYSE trading) and BASF (XETRA and Frankfurt floor trading). The dotted lines show the approx. 99% confidence interval. The x-axis denotes the lags in terms of durations.

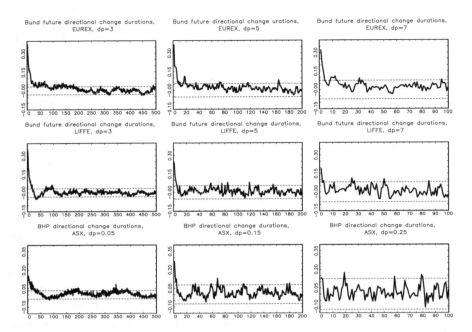

Figure 16: Autocorrelation functions of directional change durations for the Bund future (EUREX and LIFFE trading) and BHP (ASX trading). Dotted lines: approx. 99% confidence interval. The x-axis denotes the lags in terms of durations.

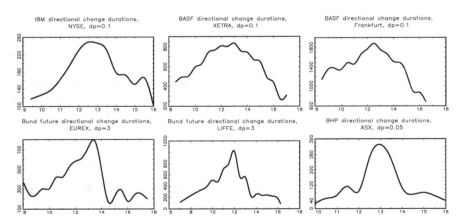

Figure 17: Cubic spline function (30 minute nodes) of directional change durations for IBM ($dp = 0.1$), BASF ($dp = 0.1$), the Bund future ($dp = 3$) and BHP ($dp = 0.05$). The x-axis denotes the local calendar time.

aggregation levels, evidence for clustering effects of directional change durations is provided. Thus, the smaller the time interval between a local maximum (minimum) and the subsequent minimum (maximum), the smaller the following spell until the next maximum (minimum). However, the overall serial dependence declines and becomes insignificant for higher aggregation levels, exhibiting an ACF that converges relatively fast to zero. Especially for higher aggregation levels, the autocorrelations are insignificant for nearly all lags due to the low sample range. As for the price durations series, the autocorrelation functions are subjected to significant seasonality structures. The intraday seasonality pattern (Figure 17) is quite similar to the seasonality function of price durations. Once again, it depicts high activity after the opening and before the closure of the market and is interrupted by strong lunchtime effects.

4.5 Statistical Properties of (Excess) Volume Durations

Table 5 presents the descriptive statistics of volume durations with respect to three different aggregation levels associated with small, medium and large cumulated volumes.[13] The properties of volume durations differ from trade durations and price durations with respect to their distributional and dynamical properties. While the distribution of trade durations and price durations reveal a clear overdispersion, volume durations exhibit underdispersion, i.e.,

Table 5: Descriptive statistics of volume durations (number of observations, mean, standard deviation, minimum, maximum, quantiles and Ljung-Box ($\chi^2(20)$) statistic). Based on trading at NYSE, XETRA, Frankfurt floor, EUREX and ASX.

	Obs	Mean	S.D.	Min	Max	0.05q	0.25q	0.50q	0.75q	0.95q	LB(20)
IBM volume durations, NYSE (01/02/01 to 05/31/01)											
$dv = 50000$	14001	172	125	3	1209	40	87	139	221	421	54421
$dv = 200000$	3680	649	408	31	3370	184	357	546	838	1460	7824
$dv = 500000$	1433	1620	937	189	5741	485	931	1386	2112	3466	2400
BASF volume durations, XETRA (01/04/99 to 12/30/99)											
$dv = 20000$	16535	452	405	1	5344	42	167	339	618	1234	24447
$dv = 50000$	7074	1026	799	3	8992	136	440	823	1404	2575	12020
$dv = 100000$	3549	1966	1392	28	9736	309	919	1644	2698	4665	6027
BASF volume durations, Frankfurt (01/04/99 to 12/30/99)											
$dv = 5000$	9216	752	680	4	8675	54	263	568	1034	2067	6934
$dv = 10000$	5259	1277	1020	12	14559	151	569	1020	1735	3224	5402
$dv = 20000$	2790	2294	1664	34	20362	437	1082	1904	3113	5486	4037

Overnight spells are ignored. Descriptive statistics in seconds.

---

[13] Note that for the LIFFE Bund future trading no volume durations are computed since the volume is measured quite inaccurately (see also Section 4.2.4).

Table 5 cont'd:

	Obs	Mean	S.D.	Min	Max	0.05q	0.25q	0.50q	0.75q	0.95q	LB(20)
Bund future volume durations, EUREX (12/02/96 to 03/06/97)											
$dv = 500$	10085	209	267	1	6088	24	68	134	251	637	27589
$dv = 2000$	2610	772	806	27	12994	121	307	555	971	2112	3308
$dv = 4000$	1287	1507	1407	66	18997	275	675	1133	1873	3958	941
BHP volume durations, ASX (07/01/02 to 07/31/02)											
$dv = 100000$	1660	290	336	1	4177	33	102	201	343	829	1010
$dv = 200000$	870	550	567	8	5436	108	235	399	625	1570	491
$dv = 400000$	436	1083	976	48	6987	271	530	795	1216	3264	182

Overnight spells are ignored. Descriptive statistics in seconds.

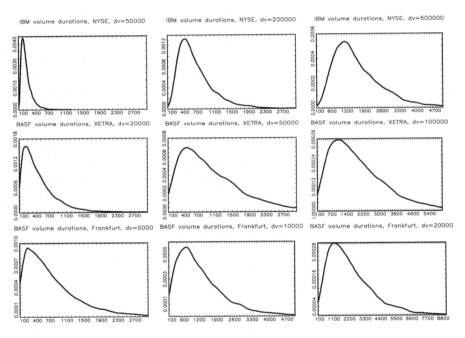

Figure 18: Kernel density plots (Epanechnikov kernel with optimal bandwidth) of volume durations for IBM, traded at the NYSE, and BASF, traded at XETRA and on the Frankfurt floor.

the standard deviation is lower than the mean. Thus, the distribution of volume durations is clearly less dispersed than for price durations, having a lower fraction of extreme long or extreme small durations leading to a more symmetric density function (see Figure 18). Focussing on the dynamic properties

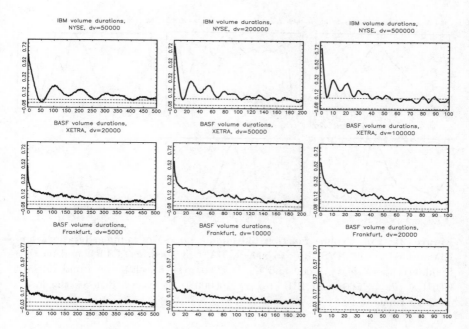

Figure 19: Autocorrelation functions of volume durations for IBM (NYSE trading) and BASF (XETRA and Frankfurt floor trading). Dotted lines: approx. 99% confidence interval. The x-axis denotes the lags in terms of durations.

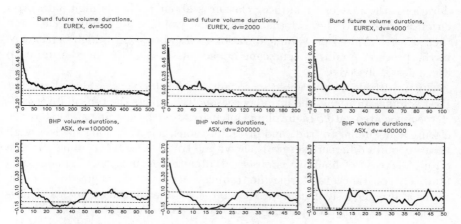

Figure 20: Autocorrelation functions of volume durations for the Bund future trading (EUREX and LIFFE trading) and BHP (ASX trading). Dotted lines: approx. 99% confidence interval. The x-axis denotes the lags in terms of durations.

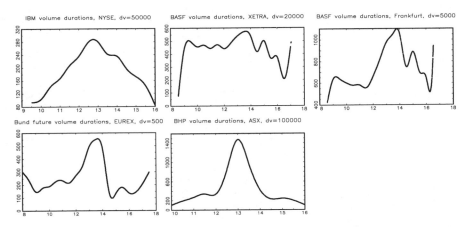

Figure 21: Cubic spline function (30 minute nodes) of volume durations for IBM traded at the NYSE ($dv = 50000$), BASF, traded at XETRA and on the Frankfurt floor ($dv = 20000$ and $dv = 5000$, respectively), the Bund future, EUREX ($dv = 500$) and BHP ($dv = 100000$). The x-axis denotes the local calendar time.

of volume durations, the Ljung-Box statistics in Table 5 indicate a stronger serial dependence than for trade or price durations. Figures 19 and 20 show the autocorrelation functions of the particular types of volume durations. For all assets, the ACF starts from a quite high level. For example, for IBM volume durations, the first order autocorrelation is about 0.7. This high autocorrelation is caused by the fact that transaction volumes are strongly clustered themselves.[14] Moreover, volume duration series also reveal a high persistence in the process. Again, a significant impact of seasonalities on the duration dynamics is observed.

Figure 21 shows the cubic seasonality spline for volume durations. For IBM volume durations, the intraday seasonality depicts the typical inverse U-shape associated with a high volume intensity after the opening and before the closure and, similar to trade and price durations, relatively long spells around noon. A quite different picture is shown for BASF XETRA trading. Here, a strong decrease of liquidity is observed after the opening that remains at a relatively constant level during the trading day except of a slight dip at noon. In the afternoon, a significant increase in the volume intensity is shown. This is probably caused by the opening of the American market. Interestingly, on the Frankfurt floor, the lunchtime effect is clearly more pronounced than in the XETRA system. Similar patterns are observed for the Bund future trading at EUREX and the BHP trading at ASX. Here, the seasonality pattern is

---

[14] However, note that these effects are *not* caused by split-transactions since such effects already have been taken into account. See Section 4.1.

dominated by clear lunchtime effects leading to the well known inverse U-shape of intraday volume intensity.

Table 6: Descriptive statistics of excess volume durations (number of observations, mean, standard deviation, minimum, maximum, quantiles and Ljung-Box $(\chi^2(20))$ statistic). Based on NYSE trading and ASX trading.

	Obs	Mean	S.D.	Min	Max	0.05q	0.25q	0.50q	0.75q	0.95q	LB(20)
AOL excess volume durations, NYSE (01/02/01 to 05/31/01)											
$dv = 50000$	10260	239	306	1	4110	17	64	137	290	793	4926
$dv = 100000$	4055	564	664	4	5831	41	143	330	708	1919	1333
$dv = 200000$	1314	1243	1129	14	6316	112	373	875	1763	3625	326
IBM excess volume durations, NYSE (01/02/01 to 05/31/01)											
$dv = 25000$	10442	227	312	1	4104	14	50	116	272	815	6543
$dv = 50000$	4138	495	582	3	5139	32	118	279	644	1709	1937
$dv = 100000$	1622	859	771	3	4637	74	283	617	1201	2422	642
BHP excess volume durations, ASX (07/01/02 to 07/31/02)											
$dv = 25000$	2897	166	280	1	4189	4	29	86	193	560	821
$dv = 50000$	1211	394	563	1	6106	16	89	223	469	1261	193
$dv = 100000$	444	1056	1390	1	9755	37	266	649	1273	3169	45
NAB excess volume durations, ASX (07/01/02 to 07/31/02)											
$dv = 5000$	4944	98	192	1	4100	2	12	40	107	361	1709
$dv = 10000$	2276	212	406	1	7546	5	34	97	234	747	400
$dv = 20000$	899	527	849	2	7995	19	111	256	571	1926	135

Overnight effects are ignored. Descriptive statistics in seconds.

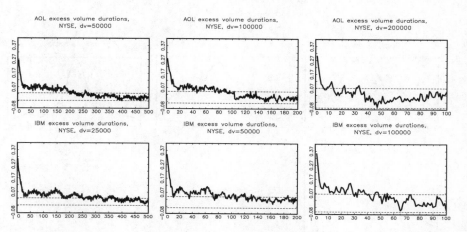

Figure 22: Autocorrelation functions of excess volume durations for AOL and IBM (NYSE trading). Dotted lines: approx. 99% confidence interval. The x-axis denotes the lags in terms of durations.

Table 6 presents the descriptive statistics of *excess volume durations* based on the AOL and IBM stock traded at the NYSE and the BHP and NAB stock, traded at the ASX. We use three aggregation levels which are roughly comparable to the aggregations used for volume durations. With respect to the distributional properties, no clear evidence for underdispersed or overdispersed distributions is found. However, for most of the analyzed excess volume

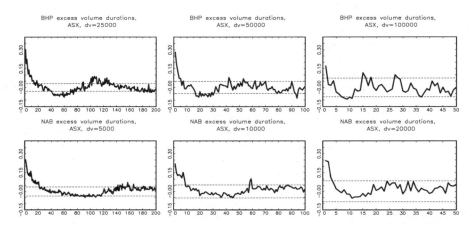

Figure 23: Autocorrelation functions of excess volume durations for BHP and NAB (ASX trading). Dotted lines: approx. 99% confidence interval. The x-axis denotes the lags in terms of durations.

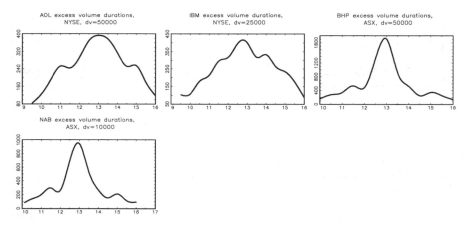

Figure 24: Cubic spline function (30 minute nodes) of excess volume durations for AOL and IBM (NYSE trading), as well as for BHP and NAB (ASX trading). The x-axis denotes the local calendar time.

duration series, the standard deviation exceeds the mean, reflecting overdispersion. Since the kernel density plots are quite similar to those of pure volume durations, they are not shown here. The autocorrelation functions (Figures 22 and 23), as well as the Ljung-Box statistics in Table 6 indicate strong serial dependencies, with relatively high first order autocorrelations of approximately 0.25. Hence, periods of high excess demand intensities are clustered. However, the serial dependence is not as strong as it is for volume durations or price durations. The ACF declines abruptly and tends toward insignificant values for higher lags. The intraday seasonality pattern (Figure 24) is quite similar to that for volume durations and also displays an inverted U-shape.

4.6 Summarizing the Statistical Findings

Drawing conclusions from the previous sections, we can summarize the following facts with respect to the statistical properties of financial durations:

(i) Trade durations, quote (limit order) durations and price durations are overdispersed while volume durations are underdispersed. Excess volume durations yield mixed evidence. In most cases, clear evidence against an exponential distribution is found.

(ii) Trade durations show the lowest autocorrelations, however, they reveal the strongest persistence. Especially for highly liquid assets, the trading intensity shows characteristics of long memory processes with an ACF that declines with a slow, hyperbolic rate.

Price and volume durations are more strongly autocorrelated than trade durations, however, these processes seem to be less persistent. The highest first order autocorrelation is observed for time series of volume durations, indicating a strong clustering of the volume intensity. The lowest persistence is observed for directional change durations and excess volume durations. Nonetheless, highly significant autocorrelations of lower orders are found.

(iii) The dynamic properties of financial durations seem to be influenced by the trading form and the institutional settings. Slight evidence is provided that floor trading is characterized by a stronger clustering of the trading intensity than electronic trading.

(vi) All types of financial durations are subjected to strong intraday seasonality patterns, which is also reflected in the autocorrelation functions. In particular, the weakest seasonality effects are found for trade and quote durations. For the particular aggregates (price durations, as well as volume durations), inverted U-shape seasonality patterns are observed, providing evidence for strong lunchtime effects.

5

Autoregressive Conditional Duration Models

This chapter deals with dynamic models for financial *duration* processes. As already discussed in Chapter 2, duration approaches are the most common way to model point processes since they are easy to estimate and allow for straightforward computations of forecasts. In Section 5.1, we discuss autoregressive models for log durations as a natural starting point. In Section 5.2, we present the basic form of the autoregressive conditional duration (ACD) model proposed by Engle and Russell (1997, 1998). Because it is the most common type of autoregressive duration model and is extensively considered in recent econometrics literature, we discuss the theoretical properties and estimation issues in more detail. Section 5.3 deals with extensions of the basic ACD model. In this section, we discuss generalizations with respect to the functional form of the basic specification. Section 5.4 is devoted to specification tests for the ACD model. Here, we focus on (integrated) conditional moment tests as a valuable framework to test the conditional mean restriction implied by the ACD model. Finally, in Section 5.5 we illustrate several applications of the ACD model. The first application deals with the evaluation of different ACD specifications based on trade and price durations by using the testing framework considered in Section 5.4. A further objective is to empirically test the market microstructure hypotheses derived in Chapter 3. Moreover, we apply the ACD model to quantify illiquidity risks on the basis of excess volume durations.

5.1 ARMA Models for (Log-)Durations

A natural starting point for an autoregressive duration model is to specify an (autoregression) model for log durations. Since log durations are not subjected to non-negativity restrictions, traditional time series models are easily applicable. Hence, a simple ARMA model for log durations is given by

$$\ln x_i = \omega + \sum_{j=1}^{P} \alpha_j \ln x_{i-j} + \sum_{j=1}^{Q} \beta_j \tilde{\varepsilon}_{i-j} + \tilde{\varepsilon}_i, \quad i = 1, \ldots, n, \qquad (5.1)$$

where $\tilde{\varepsilon}_i$ is a white noise random variable. As illustrated in Chapter 2, (auto-)regression models for (log-)durations belong to the class of AFT models. Thus, covariates, including in this context past durations, accelerate or decelerate the time to failure. A quasi maximum likelihood (QML) estimator for $\theta = (\omega, \alpha, \beta)$ is obtained by estimating the model under the normality assumption for $\tilde{\varepsilon}_i$, implying a conditionally log normal distribution for x_i. Based on QML estimates, the empirical distribution of the residuals $\hat{\tilde{\varepsilon}}_i$ yields a nonparametric estimate of the underlying distribution, and thus, the baseline hazard.

More sophisticated specifications for log durations are obtained by ARMA-GARCH type specifications. For instance, the conditional mean function of log durations can be specified according to (5.1), while the conditional (log) duration volatility, h_i^v, follows a standard GARCH process. Thus,

$$\tilde{\varepsilon}_i = \sqrt{h_i^v} u_i, \quad u_i \sim N(0, 1)$$

$$h_i^v = \omega^v + \sum_{j=1}^{P^v} \alpha_j^v \tilde{\varepsilon}_{i-j}^2 + \sum_{j=1}^{Q^v} \beta_j^v h_{i-j}^v, \qquad (5.2)$$

where $.^v$ indexes the corresponding volatility parameters. For trade durations or volume durations, the duration volatility admits an economically reasonable interpretation as liquidity risk (see, for example, Ghysels, Gouriéroux, and Jasiak, 1998). By exploiting to the asymptotic properties of the QML estimator of the Gaussian GARCH model (see Bollerslev and Wooldridge, 1992)[1], the autoregressive parameters of (5.1) and (5.2) are estimated consistently.

Note that the separability of the conditional mean and the conditional variance of log durations is implied by the normality assumption. However, such a separation of the two first moments is not straightforward for plain durations. In general, distributions defined on a positive support typically imply a strict relationship between the first moment and higher order moments and do not allow to disentangle the conditional mean and variance function. For example, under the exponential distribution, all higher order moments directly depend on the first moment. Thus, a parameterization of the conditional mean implies per (distributional) assumption also a parameterization of higher order conditional moments. Ghysels, Gouriéroux, and Jasiak (1998) argue that such distributional assumptions are too restrictive and are not flexible enough for a modelling of the duration dynamics. For this reason they propose a dynamic two factor model.[2]

---

[1] For more details, see Section 5.2.2, where the asymptotic properties of the GARCH QML estimator are carried over to ACD models.

[2] For more details, see Section 5.3.5.

However, researchers are often not interested in modelling (and forecasting) of log durations but of plain durations. Moreover, Dufour and Engle (2000a) illustrate that the forecast performance of auto-regressions in log-durations perform rather poorly compared to more sophisticated ACD specifications as presented in the following subsection. An alternative specification is given by an ARMA model for plain durations where the innovations follow a distribution defined on positive support. Hence,

$$x_i = \omega + \sum_{j=1}^{P} \alpha_j x_{i-j} + \sum_{j=1}^{Q} \beta_j \tilde{\varepsilon}_{i-j} + \tilde{\varepsilon}_i, \qquad (5.3)$$

where $\omega > 0$, $\alpha_j \geq 0$, $\beta_j \geq 0$. A QML estimator for $\theta = (\omega, \alpha, \beta)$ is obtained by assuming a standard exponential distribution for $\tilde{\varepsilon}_i$. Then it is proposed to maximize the quasi maximum likelihood function

$$\ln \mathcal{L}_{QML}(W; \theta) = - \sum_{i=1}^{n} \tilde{\varepsilon}_i = - \sum_{i=1}^{n} \left[x_i - \omega - \sum_{j=1}^{P} \alpha_j x_{i-j} - \sum_{j=1}^{Q} \beta_j \tilde{\varepsilon}_{i-j} \right],$$
$$(5.4)$$

which is the true log likelihood if the p.d.f. of $\tilde{\varepsilon}_i$ were the exponential density. Under correct specification of the conditional mean function, we obtain consistent estimates for the parameter vector θ.

A drawback of this approach is that in this case the marginal distribution of the resulting duration process is obviously not exponential. Thus, in difference to Gaussian ARMA models, the relationship between the conditional distribution and the marginal distribution of the durations is not obvious. Lawrence and Lewis (1980) propose an exponential ARMA (EARMA) model which is based on i.i.d. exponential innovations and leads to an exponential marginal distribution. This result is achieved by specifying a linear autoregressive model for a stationary variable that is based on a probabilistic choice between different linear combinations of independent exponentially distributed random variables.

5.2 The ACD Model

The most popular autoregressive duration approach is proposed by Engle (1996, 2000) and Engle and Russell (1997, 1998). The basic idea of the autoregressive conditional duration (ACD) model is a dynamic parameterization of the conditional mean function

$$\Psi_i := \Psi_i(\theta) = \mathrm{E}[x_i | \mathfrak{F}_{t_{i-1}}; \theta], \qquad (5.5)$$

where θ denotes a $M \times 1$ parameter vector. It is assumed that the standardized durations

$$\varepsilon_i = \frac{x_i}{\Psi_i}$$

follow an i.i.d. process defined on positive support with $\mathrm{E}[\varepsilon_i] = 1$. Obviously, the ACD model can be regarded as a GARCH model for duration data. Different types of ACD models can be divided either by the choice of the functional form used for the conditional mean function Ψ_i or by the choice of the distribution for ε_i.

In the following subsection we present the theoretical properties of the basic ACD specification. In Section 5.2.2, we illustrate that the QML properties of GARCH models can be carried over to the ACD framework. Section 5.2.3 deals with distributional extensions of the ACD model that are estimated by ML. Section 5.2.4 considers the treatment of seasonalities and the inclusion of explanatory variables in the ACD framework.

5.2.1 The Basic ACD Framework

The basic ACD specification is based on a linear parameterization of the conditional mean function

$$\Psi_i = \omega + \sum_{j=1}^{P} \alpha_j x_{i-j} + \sum_{j=1}^{Q} \beta_j \Psi_{i-j}, \qquad (5.6)$$

where $\omega > 0$, $\alpha \geq 0$, $\beta \geq 0$. It can be rewritten in terms of an intensity representation

$$\lambda(t; \mathfrak{F}_t) = \tilde{\lambda}_\varepsilon \left(\frac{x(t)}{\Psi_{\tilde{N}(t)+1}} \right) \frac{1}{\Psi_{\tilde{N}(t)+1}}, \qquad (5.7)$$

where $\tilde{\lambda}_\varepsilon(s)$ denotes the hazard function of the ACD residual ε_i. It is easy to see that the ACD model belongs to the class of AFT models since past dynamics influence the rate of failure time. Changes of the intensity function during a spell are only induced by the hazard shape of ε_i. Hence, new information enters the model only at the particular points t_i. However, (5.7) could be used as the starting point for generalized specifications by directly parameterizing the intensity function and allowing for news arrival within a spell.[3] Clearly, such specifications require to switch completely from a duration framework to an intensity framework, which is discussed in more detail in Chapter 7. However, note that the basic idea of the ACD model is to (dynamically) parameterize the conditional duration mean rather than the intensity function itself. Thus, the complete dynamic structure, as well as the influence of covariates is captured by the function Ψ_i which can per construction only updated at the points t_i.

---

[3] Related specifications which extend the ACD model to allow for time-varying covariates have been proposed by Lunde (2000) and Hamilton and Jorda (2002).

The conditional mean of the ACD model is given by definition as $\mathrm{E}[x_i|\mathfrak{F}_{t_{i-1}}] = \Psi_i$, whereas the unconditional mean and the conditional variance are

$$\mathrm{E}[x_i] = \mathrm{E}[\Psi_i] \cdot \mathrm{E}[\varepsilon_i] = \frac{\omega}{1 - \sum_{j=1}^{P} \alpha_j - \sum_{j=1}^{Q} \beta_j} \tag{5.8}$$

$$\mathrm{Var}[x_i|\mathfrak{F}_{t_{i-1}}] = \Psi_i^2 \cdot \mathrm{Var}[\varepsilon_i]. \tag{5.9}$$

The derivation of the unconditional variance requires the computation of $\mathrm{E}[\Psi_i^2]$ which is quite cumbersome in the case of an ACD(P, Q) model. Hence, for ease of exposition, we illustrate the unconditional variance for the ACD(1,1) model. In this case, $\mathrm{E}[x_i^2]$ is computed by

$$\begin{aligned} \mathrm{E}[x_i^2] &= \mathrm{E}[\Psi_i^2] \cdot \mathrm{E}[\varepsilon_i^2] \\ &= \frac{\omega^2 \mathrm{E}[\varepsilon_i^2]}{1 - \beta^2 - 2\alpha\beta - \alpha^2 \mathrm{E}[\varepsilon_i^2]} + \frac{(2\omega\alpha + 2\omega\beta)\omega \mathrm{E}[\varepsilon_i^2]}{(1 - \beta^2 - 2\alpha\beta - \alpha^2 \mathrm{E}[\varepsilon_i^2])(1 - \alpha - \beta)} \end{aligned} \tag{5.10}$$

and thus, the unconditional variance is given by

$$\mathrm{Var}[x_i] = \mathrm{E}[x_i]^2 \cdot \left[\frac{\mathrm{E}[\varepsilon_i^2](1 - \alpha^2 - \beta^2 - 2\alpha\beta) - (1 - \beta^2 - 2\alpha\beta - \alpha^2 \mathrm{E}[\varepsilon_i^2])}{1 - \beta^2 - 2\alpha\beta - \alpha^2 \mathrm{E}[\varepsilon_i^2]} \right]. \tag{5.11}$$

Hence, in the special case of an exponential distribution, we obtain $\mathrm{E}[\varepsilon_i^2] = 2$, and therefore, $\mathrm{Var}[x_i]$ is given by

$$\mathrm{Var}[x_i] = \mathrm{E}[x_i]^2 \cdot \left[\frac{1 - \beta^2 - 2\alpha\beta}{1 - \beta^2 - 2\alpha\beta - 2\alpha^2} \right]. \tag{5.12}$$

It is easy to see that $\mathrm{Var}[x_i] > \mathrm{E}[x_i]^2$, thus, the ACD model implies excess dispersion, i.e., the unconditional standard deviation exceeds the unconditional mean. This property might be regarded as the counterpart to the "overkurtosis property" of the Gaussian GARCH model.

By introducing the martingale difference $\eta_i := x_i - \Psi_i$, the ACD(P,Q) model can be written in terms of an ARMA(max(P,Q),Q) model for plain durations

$$x_i = \omega + \sum_{j=1}^{max(P,Q)} (\alpha_j + \beta_j)x_{i-j} - \sum_{j=1}^{Q} \beta_j \eta_{i-j} + \eta_i. \tag{5.13}$$

Based on the ARMA-representation of the ACD model, the first order ACF is easily derived as

$$\rho_1 := \mathrm{Cov}(x_i, x_{i-1}) = \frac{\alpha_1(1 - \beta_1^2 - \alpha_1\beta_1)}{1 - \beta_1^2 - 2\alpha_1\beta_1}, \tag{5.14}$$

while the Yule-Walker equations are given by

$$\text{Cov}(x_i, x_{i-h}) = \sum_{j=1}^{max(P,Q)} (\alpha_j + \beta_j) \text{Cov}(x_j, x_{j-h}).$$

The covariance stationarity conditions of the ACD model are similar to the covariance stationarity conditions of the GARCH model and are ensured by

$$\sum_{j=1}^{P} \alpha_j + \sum_{j=1}^{Q} \beta_j < 1. \tag{5.15}$$

5.2.2 QML Estimation of the ACD Model

A natural choice for the distribution of ε_i is the exponential distribution. As discussed in Chapter 2, the exponential distribution is the central distribution for stochastic processes defined on positive support and can be seen as the counterpart to the normal distribution for random variables defined on the complete support. Therefore, the specification of an Exponential-ACD (EACD) model is a natural starting point. Even though the assumption of an exponential distribution is quite restrictive for many applications[4], it has the major advantage that it leads to a QML estimator for the ACD parameters. Then, the quasi log likelihood function is given by

$$\ln \mathcal{L}_{QML}(W; \theta) = \sum_{i=1}^{n} l_i(\theta) = - \sum_{i=1}^{n} \left[\ln \Psi_i + \frac{x_i}{\Psi_i} \right], \tag{5.16}$$

where $l_i(\theta)$ denotes the log likelihood contribution of the i-th observation. The score and the Hessian are given by

$$\frac{\partial \ln \mathcal{L}_{QML}(W; \theta)}{\partial \theta} = \sum_{i=1}^{n} s_i(\theta) = - \sum_{i=1}^{n} \frac{\partial \Psi_i}{\partial \theta} \cdot \frac{1}{\Psi_i} \left[\frac{x_i}{\Psi_i} - 1 \right] \tag{5.17}$$

$$\frac{\partial^2 \ln \mathcal{L}_{QML}(W; \theta)}{\partial \theta \partial \theta'} = \sum_{i=1}^{n} h_i(\theta) \tag{5.18}$$

$$= \sum_{i=1}^{n} \left\{ \frac{\partial}{\partial \theta'} \left[\frac{1}{\Psi_i} \frac{\partial \Psi_i}{\partial \theta} \right] \left(\frac{x_i}{\Psi_i} - 1 \right) - \frac{1}{\Psi_i} \frac{\partial \Psi_i}{\partial \theta} \frac{\partial \Psi_i}{\partial \theta'} \frac{x_i}{\Psi_i^2} \right\},$$

where $s_i(\theta)$ is a $M \times 1$ vector denoting the i-th contribution to the score matrix and $h_i(\theta)$ is a $M \times M$ matrix denoting the i-th contribution to the Hessian matrix. Under correct specification of the model, i.e., $\Psi_i = \Psi_{i,0}$, where $\Psi_{i,0} := \Psi_i(\theta_0) = \mathrm{E}\left[x_i | \mathfrak{F}_{t_{i-1}}; \theta_0 \right]$ denotes the "true" conditional mean function, it follows that $\varepsilon_i = x_i/\Psi_i$ is stochastically independent of Ψ_i and has

---

[4] See also the descriptive statistics in Chapter 4.

an expectation of one. Hence, the score $s_i(\theta)$ is a martingale difference with respect to the information set $\mathfrak{F}_{t_{i-1}}$ and

$$\mathrm{E}\left[\left.\frac{\partial^2 \ln \mathcal{L}_{QML}(W;\theta_0)}{\partial\theta\partial\theta'}\right|\mathfrak{F}_{t_{i-1}}\right] = \sum_{i=1}^{n}\tilde{h}_i(\theta_0) = -\sum_{i=1}^{n}\mathrm{E}\left[\left.\frac{1}{\Psi_{i,0}^2}\frac{\partial\Psi_{i,0}}{\partial\theta}\frac{\partial\Psi_{i,0}}{\partial\theta'}\right|\mathfrak{F}_{t_{i-1}}\right],$$

(5.19)

where $\tilde{h}_i(\theta) := \mathrm{E}\left[h_i(\theta)|\mathfrak{F}_{t_{i-1}}\right]$. Thus, the correct specification of the conditional mean function is an essential prerequisite to establish the QML property of the EACD estimator. Engle (2000) illustrates that the results of Bollerslev and Wooldridge (1992) can be directly applied to the EACD model. These results are summarized in the following theorem:

Theorem 5.1 (Theorem 2.1 of Bollerslev and Wooldridge, 1992, Theorem 1 of Engle, 2000).
Assume the following regularity conditions:

(i) *Θ is a compact parameter space and has nonempty interior; Θ is a subset of $\mathbb{R}^M$.*

(ii) *For some $\theta_0 \in int\,\Theta$, $\mathrm{E}\left[x_i|\mathfrak{F}_{t_{i-1}};\theta_0\right] = \Psi_i(\theta_0) := \Psi_{i,0}$.*

(iii) (a) *$\Psi_i(\theta) := \Psi_i$ is measurable for all $\theta \in \Theta$ and is twice continuously differentiable on int Θ for all x_i;*
(b) *Ψ_i is positive with probability one for all $\theta \in \Theta$.*

(iv) (a) *θ_0 is the identifiable unique maximizer of $n^{-1}\sum_{i=1}^{n}\mathrm{E}[l_i(\theta) - l_i(\theta_0)]$;*
(b) *$\{l_i(\theta) - l_i(\theta_0)\}$ satisfies the UWLLN $\forall\, i = 1, 2, \ldots, n$.*

(v) (a) *$\{h_i(\theta_0)\}$ and $\{\tilde{h}_i(\theta_0)\}$ satisfy the WLLN;*
(b) *$\{h_i(\theta) - h_i(\theta_0)\}$ satisfies the UWLLN;*
(c) *$A^\circ := n^{-1}\sum_{i=1}^{n}\mathrm{E}[\tilde{h}_i(\theta_0)]$ is uniformly positive definite.*

(vi) (a) *$\{s_i(\theta_0)s_i(\theta_0)'\}$ satisfies the WLLN;*
(b) *$B^\circ := n^{-1}\sum_{i=1}^{n}\mathrm{E}[s_i(\theta_0)s_i(\theta_0)']$ is uniformly positive definite;*
(c) *$B^{\circ-1/2}n^{-1/2}\sum_{i=1}^{n}s_i(\theta_0)s_i(\theta_0)'\xrightarrow{d}N(0,I_M)$.*

(vii) (a) *$\{\tilde{h}_i(\theta) - \tilde{h}_i(\theta_0)\}$ satisfies the UWLLN;*
(b) *$\{s_i(\theta)s_i(\theta)' - s_i(\theta_0)s_i(\theta_0)'\}$ satisfies the UWLLN.*

Then,

$$\left[A^{\circ-1}B^\circ A^{\circ-1}\right]^{-1/2}\sqrt{n}(\hat{\theta} - \theta_0) \xrightarrow{d} N(0,I_M).$$

Furthermore,

$$\hat{A}^\circ - A^\circ \xrightarrow{p} 0 \qquad and \qquad \hat{B}^\circ - B^\circ \xrightarrow{p} 0,$$

where

$$\hat{A}^\circ = \frac{1}{n}\sum_{i=1}^{n}\tilde{h}_i(\hat{\theta}) \qquad and \qquad \hat{B}^\circ = \frac{1}{n}\sum_{i=1}^{n}s_i(\hat{\theta})s_i(\hat{\theta})'.$$

Proof: See Bollerslev and Wooldridge (1992), p. 167. □

This theorem illustrates that the maximization of the quasi log likelihood function (5.16) leads to a consistent estimate of θ without specifying the density function of the disturbances. As pointed out by Bollerslev and Wooldridge (1992), the matrix $A^{\circ -1} B^{\circ} A^{\circ -1}$ is a consistent estimator of the White (1982) robust asymptotic variance covariance of $\sqrt{n}(\hat{\theta} - \theta_0)$. A variance covariance estimator that is robust not only against distributional misspecification but also against dynamic misspecification in the ACD errors is obtained by following Newey and West (1987) and estimating $\hat{B}^{\circ}$ by

$$\hat{B}^{\circ} = \hat{\Gamma}_0 + \sum_{j=1}^{J} \left(1 - \frac{j}{J+1} \right) (\hat{\Gamma}_j + \hat{\Gamma}_j'), \tag{5.20}$$

where

$$\hat{\Gamma}_j = n^{-1} \sum_{i=j+1}^{n} s_i(\hat{\theta}) s_i(\hat{\theta})' \tag{5.21}$$

and J denotes the exogenously given truncation lag order. Note that the assumptions of Bollerslev and Wooldridge (1992) are quite strong since they require asymptotic normality of the score vector and uniform weak convergence of the likelihood and its second derivative. Moreover, empirical studies provide evidence (see also Chapter 4) that typical high-frequency duration processes (like for example trade duration processes) are highly persistent and nearly "integrated". In the integrated case, the unconditional mean of Ψ_i is not finite leading to non-normal limiting distributions of the estimator. In order to cope with these problems, Lee and Hansen (1994) establish the asymptotic properties for the IGARCH(1,1) QML estimator under quite weaker assumptions than in Bollerslev and Wooldridge (1992). Engle and Russell (1998) illustrate that these results are easily carried over to the EACD(1,1) case. The main results of Lee and Hansen (1994) are summarized as follows:

Theorem 5.2 (Theorems 1 and 3 of Lee and Hansen, 1994, Corollary of Engle and Russell, 1998). *Assume the following regularity conditions:*

(i) $\theta_0 = (\omega_0, \alpha_0, \beta_0) \in int\ \Theta$.
(ii) $\mathrm{E}\left[x_i \middle| \mathfrak{F}_{t_{i-1}}; \theta_0 \right] := \Psi_i(\theta_0) = \Psi_{i,0} = \omega_0 + \alpha_0 x_{i-1} + \beta_0 \Psi_{i-1}$.
(iii) $\varepsilon_i = x_i / \Psi_{i,0}$ *is strictly stationary and ergodic.*
(iv) ε_i *is non-degenerate.*
(v) $\mathrm{E}\left[\varepsilon_i^2 \middle| \mathfrak{F}_{t_{i-1}} \right] < \infty$ *a.s.*
(vi) $\sup_i \mathrm{E}\left[\ln \beta_0 + \alpha_0 \varepsilon_i \middle| \mathfrak{F}_{t_{i-1}} \right] < 0$ *a.s.*
(vii) $\ln \mathcal{L}(W; \theta) = \sum_{i=1}^{n} l_i(\theta) = -\sum_{i=1}^{n} \left[\ln \Psi_i + \frac{x_i}{\Psi_i} \right]$,
    where $\Psi_i = \omega + \alpha x_{i-1} + \beta \Psi_{i-1}$.
(viii) θ_0 *is the identifiable unique maximizer of* $n^{-1} \sum_{i=1}^{n} \mathrm{E}[l_i(\theta) - l_i(\theta_0)]$.

Then,

$$\left[A^{\circ-1}B^\circ A^{\circ-1}\right]^{-1/2}\sqrt{n}(\hat\theta-\theta_0)\overset{d}{\to}N(0,I_M)$$

and

$$\hat A^\circ - A^\circ \overset{p}{\to} 0 \qquad and \qquad \hat B^\circ - B^\circ \overset{p}{\to} 0.$$

Proof: See Lee and Hansen (1994), p. 47ff. □

Note that these results are also valid for integrated duration processes, i.e. for the case where $\alpha + \beta = 1$. Moreover, the standardized durations are not necessarily assumed to follow an i.i.d. process, however, it is required that they are strictly stationary and ergodic. This property generalizes the results of Bollerslev and Wooldridge (1992) to a broader class of models, including, for example, the class of so-called semiparametric ACD models introduced by Drost and Werker (2001). However, note that these results are based on the linear EACD(1,1) model and cannot necessarily carried over to more general cases, like (nonlinear) EACD(P,Q) models. In any case, a crucial necessity for the QML estimation of the ACD model is the validity of the conditional mean restriction, i.e., the correct specification of the conditional mean function Ψ_i. This assumption will be explored in more detail in Section 5.4.

As illustrated by Engle and Russell (1998), an important implication of the strong analogy between the Gaussian GARCH model and the Exponential ACD model is that the ACD model can be estimated by GARCH software. In particular, the ACD parameter vector θ can be estimated by taking $\sqrt{x_i}$ as the dependent variable in a GARCH regression where the conditional mean is set to zero. Therefore, given the QML properties of the EACD model, the parameter estimates $\hat\theta$ are consistent, but inefficient.

However, most empirical studies[5] show that the assumption of an exponential distribution for the standardized durations does not hold. Thus, QML parameter estimates are biased in finite samples. Grammig and Maurer (2000) analyze the performance of different ACD specifications based on Monte Carlo studies and show that the QML estimation of these models may perform poorly in finite samples, even in quite large samples such as 15,000 observations. For this reason, they propose to specify the ACD model based on more general distributions. Bauwens, Giot, Grammig, and Veredas (2000) investigate the predictive performance of ACD models by using density forecasts. They illustrate that the predictions can be significantly improved by allowing for more flexible distributions. In these cases, the ACD model is not estimated by QML but by standard ML. This issue will be considered in the next subsection.

5.2.3 Distributional Issues and ML Estimation of the ACD Model

Clearly, the ACD model can be specified based on every distribution defined on a positive support. In duration literature, a standard way to obtain more

---

[5] See also in Chapter 4 or in Section 5.5 of this chapter.

flexible distributions is to use mixture models. In this framework, a specific parametric family of distributions is mixed with respect to a heterogeneity variable leading to a mixture distribution[6]. The most common mixture model is obtained by multiplying the integrated hazard rate by a random heterogeneity term. In most applications, a gamma distributed random variable is used leading to a mixture model which is analytically tractable and allows for the derivation of simple results.

Below, we give a small classification over the most common types of mixture models leading to an extremely flexible family of distributions. Consider a Weibull distribution which can be written in the form

$$\frac{x}{\lambda} = \frac{u}{a}, \quad a > 0, \tag{5.22}$$

where λ and a are the scale and shape parameter of the distribution, respectively, and u is a random variable which follows an unit exponential distribution. Note that in this case, $(x/\lambda)^a$ is the integrated hazard function. A gamma mixture of Weibull distributions is obtained by multiplying the integrated hazard by a random heterogeneity term following a gamma distribution[7] $v \sim \mathcal{G}(\eta, \eta)$ with mean $\mathrm{E}[v] = 1$ and variance $\mathrm{Var}[v] = \eta^{-1}$. Then, the resulting mixture model follows a Burr distribution with parameters λ, a and η.

More flexible models are obtained by a generalization of the underlying Weibull model and by assuming that $u \sim \mathcal{G}(m, m)$ leading to a duration model based on the generalized gamma distribution (see Appendix A). The generalized gamma family of density functions nests the Weibull family when $m = 1$ and the (two-parameter) gamma distribution when $a = 1$. Both types of distributions have been already successfully applied in the ACD framework leading to the Burr ACD model (Grammig and Maurer, 2000) and the generalized gamma ACD model (Lunde, 2000).

Note that both models belong to different distribution families and are not nested. An extension of the generalized gamma ACD model which also nests one member of the Burr family is based on a gamma mixture of the generalized gamma distribution leading to the generalized F distribution (see Kalbfleisch and Prentice, 1980 or Lancaster, 1997). The generalized F distribution is obtained by assuming the generalized gamma model as basic duration model and multiplying the integrated hazard by a gamma variate $v \sim \mathcal{G}(\eta, \eta)$. Then, the marginal density function of x is given by

$$f(x) = \frac{a x^{am-1} [\eta + (x/\lambda)^a]^{(-\eta-m)} \eta^\eta}{\lambda^{am} \mathcal{B}(m, \eta)}, \tag{5.23}$$

where $\mathcal{B}(\cdot)$ describes the complete Beta function with $\mathcal{B}(m, \eta) = \frac{\Gamma(m)\Gamma(\eta)}{\Gamma(m+\eta)}$. [8] The moments of the generalized F distribution are given by

---

[6] For an overview of mixture distributions, see e.g. Lancaster (1997).

[7] See Appendix A.

[8] See also Appendix A.

$$\mathrm{E}[x^s] = \lambda\eta^{1/a}\frac{\Gamma(m+s/a)\Gamma(\eta-s/a)}{\Gamma(m)\Gamma(\eta)}, \quad s < a\eta. \qquad (5.24)$$

Hence, the generalized F ACD model is based on three parameters a, m and η, and thus, nests the generalized gamma ACD model for $\eta \to \infty$, the Weibull ACD model for $m = 1, \eta \to \infty$ and the log-logistic ACD model for $m = \eta = 1$. Figures 25-27 show the hazard functions implied by the generalized F distribution based on different parameter combinations.

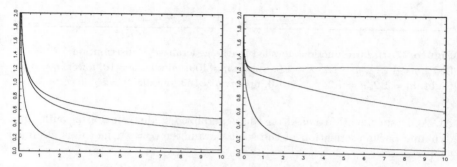

Figure 25: Hazard functions implied by the generalized F distribution.
Left: $m = 0.8$, $a = 0.8$, $\lambda = 1.0$, upper: $\eta = 100$, middle: $\eta = 10$, lower: $\eta = 0.5$.
Right: $m = 0.8$, $a = 1.1$, $\lambda = 1.0$, upper: $\eta = 100$, middle: $\eta = 10$, lower: $\eta = 0.5$.

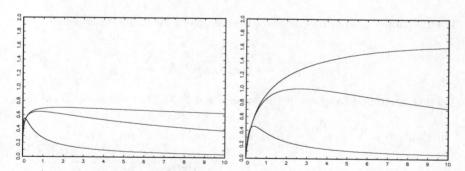

Figure 26: Hazard functions implied by the generalized F distribution.
Left: $m = 1.4$, $a = 0.9$, $\lambda = 1.0$, upper: $\eta = 100$, middle: $\eta = 10$, lower: $\eta = 0.5$.
Right: $m = 1.4$, $a = 1.2$, $\lambda = 1.0$, upper: $\eta = 100$, middle: $\eta = 10$, lower: $\eta = 0.5$.

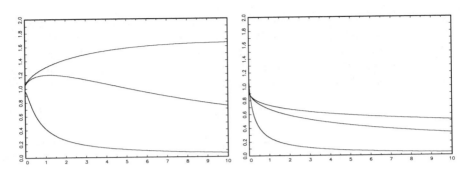

Figure 27: Hazard functions implied by the generalized F distribution.
Left: $m = 1.2^{-1}$, $a = 1.2$, $\lambda = 1.0$, upper: $\eta = 100$, middle: $\eta = 10$, lower: $\eta = 0.5$.
Right: $m = 1.2$, $a = 1.2^{-1}$, $\lambda = 1.0$, upper: $\eta = 100$, middle: $\eta = 10$, lower: $\eta = 0.5$.

An ACD specification based on the generalized F distribution is built on a dynamic parameterization of the scale parameter $\phi := \lambda$, i.e., it is assumed that

$$\Psi_i = \phi_i \eta^{1/a} \frac{\Gamma(m + 1/a)\Gamma(\eta - 1/a)}{\Gamma(m)\Gamma(\eta)} = \phi_i \zeta(a, m, \eta) \qquad (5.25)$$

with

$$\zeta(a, m, \eta) := \frac{\eta^{1/a}\Gamma(m + 1/a)\Gamma(\eta - 1/a)}{\Gamma(m)\Gamma(\eta)}. \qquad (5.26)$$

Then, ϕ_i follows an autoregressive process, given by

$$\phi_i = \frac{\omega}{\zeta(a, m, \eta)} + \sum_{j=1}^{P} \frac{\alpha_j}{\zeta(a, m, \eta)} x_{i-j} + \sum_{j=1}^{Q} \beta_j \phi_{i-j}. \qquad (5.27)$$

The log likelihood function of the generalized F ACD model is obtained by

$$\ln \mathcal{L}(W; \theta) = \sum_{i=1}^{n} \ln \frac{\Gamma(m + \eta)}{\Gamma(m)\Gamma(\eta)} + \log a - am \log \phi_i + (am - 1) \log x_i \qquad (5.28)$$
$$- (\eta + m) \log \left[\eta + (\phi_i^{-1} x_i)^a \right] + \eta \ln(\eta).$$

5.2.4 Seasonalities and Explanatory Variables

As illustrated in Chapter 4, financial duration processes are typically subjected to strong seasonality effects. One common solution in the ACD framework is to generate seasonally adjusted series by partialling out the time-of-day effects. In this context, the durations are decomposed into a deterministic and stochastic component. Engle and Russell (1998) assume that deterministic seasonality effects act multiplicatively, thus

$$x_i = \breve{x}_i s(t_{i-1}),\qquad(5.29)$$

where $\breve{x}_i$ denotes the seasonal adjusted duration and $s(t_i)$ the seasonality component at t_i. Then, the conditional mean is given by

$$\Psi_i = \mathrm{E}\left[\breve{x}_i|\,\mathfrak{F}_{t_{i-1}}\right]s(t_{i-1}) = \breve{\Psi}_i s(t_{i-1}).\qquad(5.30)$$

The deterministic seasonality function $s(t)$ can be specified in different ways. The most common specification is based on a linear or cubic spline function. A valuable alternative is to use the flexible Fourier series approximation proposed by Andersen and Bollerslev (1998b) based on the work of Gallant (1981). Assuming a polynomial of degree Q, the non-stochastic seasonal trend term is of the form

$$s(t) = s(\delta^s, \bar{t}, Q) = \delta^s \cdot \bar{t} + \sum_{j=1}^{Q}\left(\delta_{c,j}^s \cos(\bar{t} \cdot 2\pi j) + \delta_{s,j}^s \sin(\bar{t} \cdot 2\pi j)\right),\quad(5.31)$$

where δ^s, $\delta_{c,j}^s$, and $\delta_{s,j}^s$ are the seasonal coefficients to be estimated and $\bar{t} \in [0,1]$ is a normalized intraday time trend defined as the number of seconds from opening of the exchange until t divided by the length of the trading day. The seasonality function can be jointly estimated with the ACD parameters. However, often it is common to apply a two-step estimation approach. In this context, in the first step durations are seasonally filtered and in the second step parameters are estimated on the basis of the deseasonalized dependent variable.[9]

Explanatory variables can be included in two different ways. The first possibility is to include them directly in the conditional mean function, thus

$$\Psi_i = \omega + \sum_{j=1}^{P}\alpha_j x_{i-j} + \sum_{j=1}^{Q}\beta_j \Psi_{i-j} + z_{i-1}'\gamma,\qquad \text{or}\qquad(5.32)$$

$$(\Psi_i - z_{i-1}'\gamma) = \omega + \sum_{j=1}^{P}\alpha_j x_{i-j} + \sum_{j=1}^{Q}\beta_j(\Psi_{i-j} - z_{i-1-j}'\gamma).\qquad(5.33)$$

Note that (5.32) and (5.33) do not necessarily ensure nonnegativity of the implied process. Alternatively, covariates can also enter the regression function in terms of some non-negative function, like, for example, in exponential form. However, in most applications (see also Section 5.5.3) non-negativity is satisfied without explicit restrictions. Note that (5.32) implies a dynamic inclusion of explanatory variables, i.e., in this specification the covariates enter

---

[9] An alternative procedure is discussed by Veredas, Rodriguez-Poo, and Espasa (2002). They propose a semiparametric estimator where the seasonal components are jointly estimated non-parametrically with the parameters of the ACD model.

the ACD specification in an infinite lag structure (see, e.g. Hendry, 1995). In contrast, (5.33) implies a static inclusion of additional regressors.

Alternatively, explanatory variables might be included as an additional scaling function, i.e., in this context, $\breve{x}_i$ is defined as the duration standardized by seasonality and covariate effects. Thus

$$\breve{x}_i := \frac{x_i}{s(t_{i-1})g(z'_{i-1}\gamma)} \qquad (5.34)$$

and Ψ_i is given by

$$\Psi_i = \breve{\Psi}_i s(t_{i-1})g(z'_{i-1}\gamma), \qquad (5.35)$$

where $g(\cdot)$ denotes some non-negative mapping function.

5.3 Extensions of the ACD Framework

This section is concerned with generalizations of the basic ACD model. Section 5.3.1 discusses so-called augmented ACD models that extend the basic linear specification, (5.6), in several directions. As illustrated in Section 5.3.2, these approaches belong to the class of generalized polynomial random coefficient autoregressive models for which theoretical properties can be derived following the results of Carrasco and Chen (2002). In Section 5.3.3, we review extensions of the ACD framework that can be summarized as regime-switching ACD parameterizations. Section 5.3.4 discusses briefly long memory ACD specifications while Section 5.3.5 considers further extensions that have been proposed in recent literature.

5.3.1 Augmented ACD Models

In this section, we consider extensions in two directions. On the one hand, we discuss ACD models that allow for additive as well as multiplicative stochastic components, i.e. specifications, where lagged innovations enter the conditional mean function additively and/or multiplicatively. On the other hand, we illustrate parameterizations that allow not only for linear but for more flexible news impact curves. Several empirical studies[10] illustrate that a linear news impact function is too restrictive to capture the adjustment process of the conditional mean to recent durations. Moreover, in Section 5.5.1 of this Chapter, it will be shown that more flexible parameterizations are essential to ensure the validity of the conditional mean restriction implied by the ACD model. For simplicity of exposition, the following considerations are restricted to models with a lag order of $P = Q = 1$.

---

[10] See, for example, Dufour and Engle (2000a), Zhang, Russell, and Tsay (2001) or Fernandes and Grammig (2001).

Additive ACD (AACD) Model

An alternative specification that has not been considered yet in the literature is based on an *additive* innovation component and is given by

$$\Psi_i = \omega + \alpha\varepsilon_{i-1} + \beta\Psi_{i-1}. \tag{5.36}$$

This specification implies a linear news impact curve with a slope given by α. Note that the basic ACD specification, (5.6), implies a slope of $\alpha\Psi_{i-1}$. Hence, while in the basic ACD model the lagged innovation enters Ψ_i multiplicatively, here, it enters Ψ_i additively without any interaction with Ψ_{i-1}.

Additive and Multiplicative ACD (AMACD) Model

A more general specification that encompasses specifications (5.6) and (5.36) is based on an *additive* and *multiplicative* innovation component, i.e.,

$$\Psi_i = \omega + (\alpha\Psi_{i-1} + \nu)\varepsilon_{i-1} + \beta\Psi_{i-1}. \tag{5.37}$$

This specification implies a news impact curve with a slope given by $\alpha\Psi_{i-1}+\nu$. Thus, the lagged innovation enters the conditional mean function additively, as well as multiplicatively. In this sense, the (so called) AMACD model is more flexible and nests the linear ACD model for $\nu = 0$ and the AACD model for $\alpha = 0$.

Logarithmic ACD (LACD) Model

Bauwens and Giot (2000) and Lunde (2000) propose a logarithmic ACD model[11] that ensures the non-negativity of durations without any parameter restrictions and is obtained by

$$\begin{aligned} \ln\Psi_i &= \omega + \alpha\ln\varepsilon_{i-1} + \beta\ln\Psi_{i-1} \\ &= \omega + \alpha\ln x_{i-1} + (\beta - \alpha)\ln\Psi_{i-1} \qquad (\text{LACD}_1). \end{aligned} \tag{5.38}$$

Because of the logarithmic transformation, this model implies a concave news impact curve, i.e., the news impact difference between small innovations ("negative" surprises, $\varepsilon_i < 1$) is larger than between large innovations ("positive" surprises, $\varepsilon_i > 1$).

Bauwens and Giot (2000) also propose an alternative parameterization given by

$$\begin{aligned} \ln\Psi_i &= \omega + \alpha\varepsilon_{i-1} + \beta\ln\Psi_{i-1} \\ &= \omega + \alpha(x_{i-1}/\Psi_{i-1}) + \beta\ln\Psi_{i-1} \qquad (\text{LACD}_2), \end{aligned} \tag{5.39}$$

which implies a convex news impact curve.

---

[11] In some studies, this model is also called "Nelson type" ACD model since it resembles the EGARCH specification proposed by Nelson (1991).

Box-Cox ACD (BACD) Model

Hautsch (2001) suggests an additive ACD model based on power transformations of Ψ_i and ε_i:

$$\Psi_i^{\delta_1} = \omega + \alpha\varepsilon_{i-1}^{\delta_2} + \beta\Psi_{i-1}^{\delta_1}, \qquad (5.40)$$

where $\delta_1, \delta_2 > 0$. It is easy to see that this model can be written in terms of Box-Cox transformations. Thus,

$$\frac{\Psi_i^{\delta_1} - 1}{\delta_1} = \tilde{\omega} + \tilde{\alpha}(\varepsilon_{i-1}^{\delta_2} - 1)/\delta_2 + \beta\frac{\Psi_{i-1}^{\delta_1} - 1}{\delta_1}, \qquad (5.41)$$

where

$$\tilde{\omega} = \frac{\omega + \alpha + \beta - 1}{\delta_1} \qquad \text{and} \qquad \tilde{\alpha} = \frac{\alpha\delta_2}{\delta_1}.$$

This specification allows for concave, convex, as well as linear news impact functions. It nests the AACD model for $\delta_1 = \delta_2 = 1$, the LACD_1 model for $\delta_1 \to 0$, $\delta_2 \to 0$ and the LACD_2 model for $\delta_1 \to 0$, $\delta_2 = 1$. For $\delta_1 \to 0$, it coincides with a Box-Cox ACD specification proposed by Dufour and Engle (2000a).

EXponential ACD (EXACD) Model

Alternative parameterizations are obtained by the assumption of piece-wise linear news impact functions. Dufour and Engle (2000a) introduce the so called EXponential[12] ACD model that captures features of the EGARCH specification proposed by Nelson (1991). This model allows for a linear news impact function that is kinked at $\varepsilon_{i-1} = 1$:

$$\ln \Psi_i = \omega + \alpha\varepsilon_{i-1} + c|\varepsilon_{i-1} - 1| + \beta\ln\Psi_{i-1}. \qquad (5.42)$$

Thus, for durations shorter than the conditional mean ($\varepsilon_{i-1} < 1$), the news impact curve has a slope $\alpha - c$ and an intercept $\omega + c$, while for durations longer than the conditional mean ($\varepsilon_{i-1} > 1$), slope and intercept are $\alpha + c$ and $\omega - c$, respectively.

Augmented Box-Cox ACD (ABACD) Model

While the EXACD model allows for news impact curves that are kinked at $\varepsilon_{i-1} = 1$ only, a valuable generalization is to parameterize also the position of the kink. Using the parameterization for modelling asymmetric GARCH

---

[12] They use this notation to prevent confusion with the Exponential ACD model (EACD) based on an exponential distribution.

processes introduced by Hentschel (1995), we obtain a specification that we call augmented Box-Cox ACD model:

$$\Psi_i^{\delta_1} = \omega + \alpha(|\varepsilon_{i-1} - b| + c(\varepsilon_{i-1} - b))^{\delta_2} + \beta\Psi_{i-1}^{\delta_1}. \tag{5.43}$$

In this specification, the parameter b gives the position of the kink while δ_2 determines the shape of the piecewise functions around the kink. For $\delta_2 > 1$, the shape is convex and for $\delta_2 < 1$, it is concave. It nests the BACD model for $b = 0$ and $c = 0$. Note that this model does not encompass the basic ACD model since the ABACD specification is based on an additive stochastic component.

Even though this specification of the news impact function allows for more flexibility, it has one major drawback since the parameter restriction $|c| <= 1$ has to be imposed in order to circumvent complex values whenever $\delta_2 \neq 1$. Note that this restriction is binding in the case where the model has to be fitted to data that imply an upward kinked concave news impact function. Such a pattern is quite typical for financial durations (see also the empirical results in Section 5.5.1) and is only possible for $c < -1$ (and $\alpha < 0$). Hence, in such a case, a restricted version of the BACD model must be estimated. In particular, for cases when c converges to the boundary, either δ_2 or, alternatively, $|c|$ has to be fixed to 1. Setting $\delta_2 = 1$ implies a piecewise linear news impact function that is kinked at b. In that case, the model is reformulated as

$$\frac{\Psi_i^{\delta_1} - 1}{\delta_1} = \tilde{\omega} + \tilde{\alpha}\varepsilon_{i-1} + \tilde{c}|\varepsilon_{i-1} - b| + \beta\frac{\Psi_{i-1}^{\delta_1} - 1}{\delta_1}, \tag{5.44}$$

where

$$\tilde{\omega} = \frac{\omega - \alpha c b + \beta - 1}{\delta_1}, \qquad \tilde{\alpha} = \frac{\alpha c}{\delta_1}, \qquad \text{and} \qquad \tilde{c} = \frac{\alpha}{\delta_1}.$$

Such a specification nests the EXACD model for $\delta_1 \to 0$ and $b = 1$. Alternatively, by setting $c = 1$, the specification becomes

$$\Psi_i^{\delta_1} = \omega + \tilde{\alpha}(\varepsilon_{i-1} - b)^{\delta_2} \mathbb{1}_{\{\varepsilon_{i-1} \geq b\}} + \beta\Psi_{i-1}^{\delta_1},$$

where $\tilde{\alpha} = 2^{\delta_2}\alpha$. Thus, the news impact function is zero for $\varepsilon_{i-1} \leq b$ and follows a concave (convex) function for $\delta_2 < 1$ ($\delta_2 > 1$). Correspondingly, setting $c = -1$ leads to

$$\Psi_i^{\delta_1} = \omega + \tilde{\alpha}(b - \varepsilon_{i-1})^{\delta_2} \mathbb{1}_{\{\varepsilon_{i-1} \leq b\}} + \beta\Psi_{i-1}^{\delta_1}.$$

Note that the latter two restricted ABACD specifications do not nest the EXACD model but the BACD model for $b = 0$ ($b \to \infty$) whenever c is set to 1 (-1).

Hentschel ACD (HACD) Model

An alternative nonlinear ACD model is proposed by Fernandes and Grammig (2001). They introduce a specification that is the direct counterpart to the augmented GARCH process proposed by Hentschel (1995). Though Fernandes and Grammig call it augmented ACD model, here, we call it H(entschel)-ACD model in order to avoid confusion with other augmented ACD processes considered in this section. The HACD model is given by

$$\Psi_i^{\delta_1} = \omega + \alpha \Psi_{i-1}^{\delta_1}(|\varepsilon_{i-1} - b| + c(\varepsilon_{i-1} - b))^{\delta_2} + \beta \Psi_{i-1}^{\delta_1}. \tag{5.45}$$

This specification is quite similar to the ABACD model. However, the main difference is that the HACD model is based on a *multiplicative* stochastic component (since $\Psi_{i-1}^{\delta_1}$ acts multiplicatively with a function of ε_{i-1}) while the ABACD model is based on an *additive* stochastic component. Therefore, the HACD model imposes the same parameter restriction for c as in the ABACD case. Since the HACD model includes a multiplicative stochastic component, it nests the basic linear ACD model for $\delta_1 = \delta_2 = 1$, $b = c = 0$. Furthermore, it coincides with a special case of the ABACD model for $\delta_1 \to 0$. Therefore, it nests also the LACD$_1$ model for $\delta_1 \to 0$, $\delta_2 \to 0$, $b = c = 0$ and the LACD$_2$ model for $\delta_1 \to 0$, $\delta_2 = 1$, $b = c = 0$. Moreover, it coincides with the Box-Cox specification introduced by Dufour and Engle (2000a) for $\delta_1 \to 0$, $b = c = 0$ and with the EXACD model for $\delta_1 \to 0$, $\delta_2 = b = 1$. However, in general, it does not encompass the AMACD, BACD and ABACD models since it is based on a multiplicative stochastic component.

Augmented ACD (AGACD) Model

An encompassing model that nests all specifications outlined above is given by

$$\Psi_i^{\delta_1} = \omega + \alpha \Psi_{i-1}^{\delta_1}(|\varepsilon_{i-1} - b| \tag{5.46}$$
$$+ c(\varepsilon_{i-1} - b))^{\delta_2} + \nu(|\varepsilon_{i-1} - b| + c(\varepsilon_{i-1} - b))^{\delta_2} + \beta \Psi_{i-1}^{\delta_1}.$$

We call this specification augmented ACD model since it combines the HACD model with the ABACD specification. Note that this ACD model allows for both additive and multiplicative stochastic coefficients, and therefore implies an additive as well as a multiplicative impact of past shocks on the conditional mean function. It encompasses all specifications nested by the HACD model, as well as all other models based on additive stochastic components. In particular, it nests the AMACD model for $\delta_1 = \delta_2 = 1$, $b = c = 0$, the BACD model for $\alpha = b = c = 0$ and the ABACD model for $\alpha = 0$. The AGACD, the HACD and the ABACD models coincide for $\delta_1 \to 0$. Therefore, the parameters α and ν are only separately identifiable if $\delta_1 > 0$. A further generalization would be to specify different news impact parameters b, c and δ_2 for the additive and the stochastic component. However, in this case the estimation of the model becomes tricky due to numerical complexity.

Nonparametric News Impact ACD (NPACD) Model

A further type of ACD specification is obtained by modelling the news response in a non-parametrically way. In the spirit of Engle and Ng (1993), we consider an ACD model based on a nonparametric news impact curve which is parameterized as a linear spline function with nodes at given break points of ε_{i-1}. In particular, the range of ε_{i-1} is divided into K intervals where K^- (K^+) denotes the number of intervals in the range $\varepsilon_{i-1} < 1$ ($\varepsilon_{i-1} > 1$) with $K = K^- + K^+$. We denote the breakpoints by $\{\bar{\varepsilon}_{K^-}, \ldots, \bar{\varepsilon}_{-1}, \bar{\varepsilon}_0, \bar{\varepsilon}_1, \ldots, \bar{\varepsilon}_{K^+}\}$. Then, the NPACD model is given by

$$\Psi_i = \omega + \sum_{k=0}^{K^+} \alpha_k^+ \mathbb{1}_{\{\varepsilon_{i-1} \geq \bar{\varepsilon}_k\}}(\varepsilon_{i-1} - \bar{\varepsilon}_k) \qquad (5.47)$$

$$+ \sum_{k=0}^{K^-} \alpha_k^- \mathbb{1}_{\{\varepsilon_{i-1} < \bar{\varepsilon}_k\}}(\varepsilon_{i-1} - \bar{\varepsilon}_k) + \beta\Psi_{i-1},$$

where α_j^+ and α_j^- denote the coefficients associated with the piecewise linear spline. Alternatively, the model can also be specified in terms of a logarithmic transformation of Ψ_i. In this form, the model is more easy to estimate since it does not require any non-negativity restrictions. Note that the particular intervals must not be equally sized, nor do we need the same number of intervals on each side of $\bar{\varepsilon}_0$. As pointed out by Engle and Ng (1993), a slow increase in K as a function of the sample size should asymptotically give a consistent estimate of the news impact curve. This specification allows for extremely flexible (nonlinear) news responses, but does not necessarily nest the other (parametric) ACD specifications.

5.3.2 Theoretical Properties of Augmented ACD Models

As suggested by Fernandes and Grammig (2001), a valuable possibility to classify the particular augmented ACD specifications is to express them in the form of a generalized polynomial random coefficient autoregressive model as introduced by Carrasco and Chen (2002). Therefore, we define a class of augmented ACD models using the form

$$\vartheta(\Psi_i) = \mathcal{A}(\varepsilon_i)\vartheta(\Psi_{i-1}) + \mathcal{C}(\varepsilon_i), \qquad (5.48)$$

where $\vartheta(\cdot)$ is a continuous function and ε_i is assumed to be an i.i.d. random variable. Based on (5.48), we can classify all ACD specifications discussed above by using a corresponding parameterization for $\vartheta(\cdot)$, $\mathcal{A}(\cdot)$ and $\mathcal{C}(\cdot)$. This is shown in Table 7. The theoretical properties for this class of models are derived by Fernandes and Grammig (2001). In particular, they refer to the general results of Carrasco and Chen (2002) and Mokkadem (1990), which provide sufficient conditions ensuring β-mixing, strict stationarity and the

Table 7: Classification of ACD models

Typ	$\vartheta(\Psi_i)$	$\mathcal{A}(\varepsilon_i)$	$\mathcal{C}(\varepsilon_i)$
		Linear and logarithmic ACD models	
ACD	Ψ_i	$\alpha\varepsilon_{i-1} + \beta$	ω
AACD	Ψ_i	β	$\omega + \alpha\varepsilon_{i-1}$
LACD$_1$	$\ln\Psi_i$	β	$\omega + \alpha\ln\varepsilon_{i-1}$
LACD$_2$	$\ln\Psi_i$	β	$\omega + \alpha\varepsilon_{i-1}$
		Nonlinear ACD models	
BACD	$\Psi_i^{\delta_1}$	β	$\omega + \alpha\varepsilon_{i-1}^{\delta_2}$
EXACD	$\ln\Psi_i$	β	$\omega + \alpha\varepsilon_{i-1} + c|\varepsilon_{i-1} - 1|$
ABACD	$\Psi_i^{\delta_1}$	β	$\omega + \alpha(|\varepsilon_{i-1} - b| + c(\varepsilon_{i-1} - b))^{\delta_2}$
HACD	$\Psi_i^{\delta_1}$	$\alpha(|\varepsilon_{i-1} - b| + \beta$ $+ c(\varepsilon_{i-1} - b))^{\delta_2}$	ω
		Augmented ACD models	
AMACD	Ψ_i	$\alpha\varepsilon_{i-1} + \beta$	$\omega + \nu\varepsilon_{i-1}$
AGACD	$\Psi_i^{\delta_1}$	$\alpha(|\varepsilon_{i-1} - b| + \beta$ $+ c(\varepsilon_{i-1} - b))^{\delta_2}$	$\omega + \nu(|\varepsilon_{i-1} - b| + c(\varepsilon_{i-1} - b))^{\delta_2}$
		Nonparametric ACD models	
NPACD	Ψ_i or $\ln\Psi_i$	β	$\omega + \sum_{k=0}^{K^+} \alpha_k^+ \mathbb{1}_{\{\varepsilon_{i-1} \geq \bar\varepsilon_k\}}(\varepsilon_{i-1} - \bar\varepsilon_k)$ $+ \sum_{k=0}^{K^-} \alpha_k^- \mathbb{1}_{\{\varepsilon_{i-1} < \bar\varepsilon_k\}}(\varepsilon_{i-1} - \bar\varepsilon_k)$

existence of higher order moments for the class of generalized polynomial random coefficient autoregressive models. We reproduce the main findings in the following proposition:

Proposition 5.3 (Proposition 1 of Fernandes and Grammig, 2001, Proposition 2 and 4 of Carrasco and Chen, 2000). *Assume that the durations x_i, $i = 1, \ldots, n$, follow the process $x_i = \Psi_i\varepsilon_i$, where Ψ_i is given by one of the processes in Table 7. Assume that ε_i is an i.i.d. random variable that is independent of Ψ_i and presume that the marginal probability distribution of ε_i is absolutely continuous with respect to the Lebesgue measure on $(0, \infty)$. Let $\varrho(\mathcal{A}(\cdot))$ be the largest eigenvalue in absolute value of $\mathcal{A}(\cdot)$. Moreover, assume that*

(i) $\mathcal{A}(\cdot)$ and $\mathcal{C}(\cdot)$ are polynomial functions that are measurable with respect to the sigma field generated by ε_i.

(ii) $\varrho(\mathcal{A}(0)) < 1$.

(iii) $\mathrm{E}[\mathcal{A}(\varepsilon_i)]^s < 1$ and $\mathrm{E}[\mathcal{C}(\varepsilon_i)]^s < \infty$ for some integers $s \geq 1$.

Then, the process $\{\Psi_i\}$ is Markov geometrically ergodic and $\mathrm{E}[\Psi_i]^s < \infty$. If the processes $\{\Psi_i\}_{i=1}^n$ and $\{x_i\}_{i=1}^n$ are initialized from their ergodic distribution, then they are strictly stationary and β-mixing with exponential decay.

Proof: See Carrasco and Chen (2002), Proposition 2 and 4. $\square$

Therefore, the establishment of stationarity conditions for the particular ACD specifications requires imposing restrictions on the functions $\mathcal{A}(\cdot)$ and $\mathcal{C}(\cdot)$.[13]

5.3.3 Regime-Switching ACD Models

In the following subsection, we summarize several further functional extensions of the ACD model. All models have in common that they allow for regime-dependence of the conditional mean function. The regimes are determined either based on observable variables, like in threshold ACD or competing risks ACD models, or based on unobservable factors, as in the Markov switching or the stochastic conditional duration model.

Threshold ACD (TACD) Model

Zhang, Russell, and Tsay (2001) introduce a threshold ACD model which allows the expected duration to depend nonlinearly on past information variables. The TACD model can be seen as a generalization of the threshold GARCH models introduced by Rabemananjara and Zakoian (1993) and Zakoian (1994). By using the categorization introduced in Chapter 2, a K-regime TACD(P,Q) model is given by

$$\begin{cases} x_i = \Psi_i \varepsilon_i^{(k)} \\ \\ \Psi_i = \omega^{(k)} + \sum_{j=1}^{P} \alpha_j^{(k)} x_{i-j} + \sum_{j=1}^{Q} \beta_j^{(k)} \Psi_{i-j}, \end{cases} \qquad \text{if } x_{i-1} \in (\bar{x}_{k-1}, \bar{x}_k] \tag{5.49}$$

where $\omega^{(k)} > 0$, $\alpha_j^{(k)} \geq 0$ and $\beta_j^{(k)} \geq 0$ are regime-switching ACD parameters. The error term $\varepsilon_i^{(k)}$ is assumed to follow some distribution $f_\varepsilon(s; \theta_\varepsilon^{(k)})$ depending on regime-switching distribution parameters $\theta_\varepsilon^{(k)}$. Correspondingly to the basic ACD model, the model can be rewritten as

$$x_i = \omega^{(k)} + \sum_{j=1}^{\max(P,Q)} (\alpha_j^{(k)} + \beta_j^{(k)}) x_{i-j} - \sum_{j=1}^{Q} \beta_j^{(k)} \eta_{i-j} + \eta_i. \tag{5.50}$$

Hence, the TACD model corresponds to a threshold ARMA($K; \max(P,Q), Q$) process[14]. Obviously, the TACD model is strongly related to the NPACD model. However, the main difference is that in the NPACD model only the impact of lagged innovations is regime-dependent while in the TACD model

---

[13] A detailed analysis of the theoretical properties of the specific class of LACD models is given by Bauwens, Galli, and Giot (2003).

[14] For more details concerning threshold autoregressive (TAR) models, see, for example, Tong (1990).

all autoregressive and distributional parameters are allowed to be regime-switching. Hence, in this sense, the TACD model is more general and allows for various forms of nonlinear dynamics. However, the derivation of theoretical properties is not straightforward in this framework. Zhang, Russell, and Tsay (2001) derive conditions for geometric ergodicity and the existence of moments for the TACD(1,1) case. Moreover, in the NPACD model, the threshold values associated with the particular regimes are given exogenously while in the TACD model they are endogenous and are estimated together with the other parameters. The estimation has to be performed by using a grid search algorithm over the thresholds $\bar{x}_k$ and maximizing the conditional likelihood for each combination of the grid thresholds. Clearly, for higher numbers of K, this procedure becomes computational quite cumbersome. However, Zhang, Russell, and Tsay (2001) estimate a three-regime TACD(1,1) model for IBM trade durations and find strong evidence for the existence of different regimes and nonlinearities in the news response. Moreover, during the analyzed time period (November 1990 to January 1991), they identify several structural breaks in trade durations which can be matched with real economic events.

Smooth Transition ACD (STACD) Models

Note that in the TACD model, as well as in the NPACD model, the transition from one state to another state is not smooth but follows a jump process. Thus, a further alternative would be to allow for a smooth transition which is driven by some transition function. Smooth transition autoregressive (STAR) models have been considered, for example, by Granger and Teräsvirta (1993), Teräsvirta (1994) and Terärsvirta (1998) for the conditional mean of financial return processes and by Hagerud (1997), Gonzalez-Rivera (1998) and Lundbergh and Teräsvirta (1999) for conditional variance processes. By adopting this idea in the ACD framework, a STACD(P, Q) model is given by

$$\Psi_i = \omega + \sum_{j=1}^{P} \left[\alpha_{1j} + \alpha_{2j}\varsigma(\varepsilon_{i-j})\right] \varepsilon_{i-j} + \sum_{j=1}^{Q} \beta_j \Psi_{i-j}, \tag{5.51}$$

where $\varsigma(\cdot)$ denotes a transition function which has the form

$$\varsigma(\varepsilon_i) = 1 - \exp(-\nu\varepsilon_{i-1}), \quad \nu \geq 0. \tag{5.52}$$

Here, the transition between the two states is smooth and is determined by the function

$$\begin{cases} \varsigma(\varepsilon_i) = 1 & \text{if } \varepsilon_i \to \infty, \\ \varsigma(\varepsilon_i) = 0 & \text{if } \varepsilon_i = 0. \end{cases} \tag{5.53}$$

For $\nu = 0$, the STACD model coincides with the AACD model with innovation parameter α_{11}. Clearly, this structure can be extended in several directions, such as proposed, for example, by Lundbergh and Teräsvirta (1999) or Lundbergh, Teräsvirta, and van Dijk (2003).

The Markov Switching ACD (MSACD) Model

Hujer, Vuletic, and Kokot (2002) propose a Markov switching ACD model where the conditional mean function depends on an unobservable stochastic process, which follows a Markov chain. Define in the following R_i^* as a discrete valued stochastic process which indicates the state of the unobservable Markov chain and takes the values $1, 2, \ldots, R$. Then, the MSACD model is given by

$$x_i = \Psi_i \varepsilon_i, \quad \varepsilon_i \sim f_\varepsilon(s) \tag{5.54}$$

$$\Psi_i = \sum_{r=1}^{R} \text{Prob}\left[R_i^* = r \,\middle|\, \mathfrak{F}_{t_{i-1}}\right] \cdot \Psi_i^{(r)}, \tag{5.55}$$

$$f_\varepsilon(s) = \sum_{r=1}^{R} \text{Prob}\left[R_i^* = r \,\middle|\, \mathfrak{F}_{t_{i-1}}\right] \cdot f_\varepsilon(s; \theta_\varepsilon^{(r)}), \tag{5.56}$$

where $\Psi_i^{(r)}$ denotes the regime-specific conditional mean function

$$\Psi_i^{(r)} = \omega^{(r)} + \sum_{j=1}^{P} \alpha_j^{(r)} x_{i-j} + \sum_{j=1}^{Q} \beta_j^{(r)} \Psi_{i-j}^{(r)} \tag{5.57}$$

and $f_\varepsilon(s; \theta_\varepsilon^{(r)})$ is the regime-dependent conditional density for ε_i. The Markov chain is characterized by a transition matrix P^* with elements $p_{km}^* = \text{Prob}\left[R_i^* = k \,\middle|\, R_{i-1}^* = m\right]$. Hence, the marginal density of x_i is given by

$$f(x_i|\mathfrak{F}_{t_{i-1}}) = \sum_{r=1}^{R} \text{Prob}\left[R_i^* = r \,\middle|\, \mathfrak{F}_{t_{i-1}}\right] \cdot f(x_i|R_i^* = r; \mathfrak{F}_{t_{i-1}}). \tag{5.58}$$

Hujer, Vuletic, and Kokot (2002) estimate the MSACD model based on the Expectation-Maximization (EM) algorithm proposed by Dempster, Laird, and Rubin (1977). They show that the MSACD outperforms the basic (linear) ACD model and leads to a better description of the underlying duration process. In this sense, they confirm the results of Zhang, Russell, and Tsay (2001) that nonlinear ACD specifications are more appropriate to model financial duration processes. Moreover, Hujer, Vuletic and Kokot illustrate that the unobservable regime variable can be interpreted in the light of market microstructure theory and allows for more sophisticated tests of corresponding theoretical hypotheses.

The Stochastic Conditional Duration (SCD) Model

Note that in the MSACD model the latent variable follows a Markov process, i.e., the state of the process depends only on the state of the previous observation. An alternative specification has been proposed by Bauwens and

Veredas (1999)[15]. They assume that the conditional mean function Ψ_i given the history $\mathfrak{F}_{t_{i-1}}$ is not deterministic but is driven by a latent AR(1) process. Hence, the SCD model is given by

$$x_i = \Psi_i \varepsilon_i, \tag{5.59}$$

$$\ln \Psi_i = \omega + \beta \ln \Psi_{i-1} + u_i, \tag{5.60}$$

where ε_i denotes the usual i.i.d. ACD innovation term and u_i is conditionally normally distributed, i.e., $u_i | \mathfrak{F}_{t_{i-1}} \sim$ i.i.d. $N(0, \sigma^2)$ independently of ε_i. Hence, the marginal distribution of x_i is determined by a mixture of a lognormal distribution of Ψ_i and the distribution of ε_i. Economically, the latent factor can be interpreted as information flow (or general state of the market) that cannot be observed directly but drives the duration process. In this sense, the SCD model is the counterpart to the stochastic volatility model introduced by Taylor (1982). Bauwens and Veredas (1999) analyze the theoretical properties of the SCD model and illustrate that the SCD model, even with restrictive distributional assumptions for ε_i, is distributional quite flexible and allows for a wide range of different (marginal) hazard functions of the durations. In a comparison study with the Log-ACD model they find a clear outperformance of the SCD model. The model is estimated by simulated maximum likelihood (see e.g. Section 7.3.5 in Chapter 7) or, as proposed by Bauwens and Veredas (1999), based on QML by applying the Kalman filter.[16]

ACD-GARCH and Competing Risks ACD Models

An alternative way to introduce regime-switching structures in the ACD framework is to allow the ACD parameters to depend on states which are determined by other factors, like, for example the price or the volume process. These models play an important role in studying the interdependencies and causalities between durations, prices and volumes. An important string of the literature focusses on the relationship between price change volatility and the time between consecutive trades. Engle (2000) addresses this issue by proposing an ACD-GARCH model which allows for a simultaneous modelling of the trade duration process and the volatility process of trade-to-trade price changes. He models the trade durations based on an ACD model while the trade-to-trade volatility dynamics are captured by a GARCH model, conditionally on the contemporaneous trade duration. Ghysels and Jasiak (1998) propose a GARCH specification for irregularly spaced time intervals. By applying the results on temporal aggregation properties of GARCH processes established by Drost and Nijman (1993) and Drost and Werker (1996), they

---

[15] See also Meddahi, Renault, and Werker (1998) who introduced a similar specification.

[16] A further alternative to estimate the SCD model has been proposed by Strickland, Forbes, and Martin (2003) based on Monte Carlo Markov Chain (MCMC) techniques.

specify a random coefficient GARCH model where the coefficients are driven by an ACD model for the duration between consecutive trades. Grammig and Wellner (2002) extend this approach by allowing for feedback effects from the volatility process back to the duration process. This approach enables to study the interdependencies between the volatility and the duration process.

Hasbrouck (1991) and Dufour and Engle (2000b) propose a vector autoregressive (VAR) system to model the causality relationships between trade durations, prices and volumes. This approach has been extended in several directions by Manganelli (2002), Spierdijk, Nijman, and van Soest (2002) and Spierdijk (2002). A simultaneous modelling of the price and duration process by explicitly accounting for the discreteness of price changes has been proposed by Gerhard and Pohlmeier (2002). They model the trade durations using a Log-ACD specification while the process of discrete price changes is captured by a dynamic ordered response model introduced by Gerhard (2001). By analyzing Bund future transaction data, they find empirical evidence for a positive simultaneous relationship between the size of price changes and the length of the underlying time interval between two transactions. Renault and Werker (2003) study the simultaneous relationship between price volatility and trade durations in a continuous-time framework. They confirm the results of Gerhard and Pohlmeier (2002) and illustrate that a considerable proportion of intraday price volatility is caused by duration dynamics.

Russell and Engle (1998) propose the autoregressive conditional multinomial (ACM) model which combines an ACD model for the trade duration process with a dynamic multinomial model for the process of discrete price changes.[17] A related model has been introduced by Bauwens and Giot (2003) who propose a two-state competing risks ACD specification to model the duration process in dependence of the direction of the contemporaneous midquote change. They specify two competing ACD models (for upwards midquote movements versus downwards midquote movements) with regime-switching parameters that are determined by the direction of lagged price movements. Then, at each observation the end of only one of the two spells can be observed (depending on whether one observes an increase or decrease of the midquote) while the other spell is censored. Engle and Lunde (2003) propose a similar competing risks model to study the interdependencies between the trade and quote arrival process. However, they impose an asymmetric treatment of both processes since they assume that the trade process is the "driving process" which is completely observed while the quote process is subjected to censoring mechanisms.

---

[17] A decomposition of the price process into a process for the size of price changes and a process for the direction of price changes is proposed by Rydberg and Shephard (2003). Liesenfeld and Pohlmeier (2003) specify an extension that combines the approaches of Russell and Engle (1998) and Rydberg and Shephard (2003).

5.3.4 Long Memory ACD Models

As illustrated in Chapter 4, financial duration series often exhibit a strong persistence and are close to the "integrated case". A typical indication for the existence of long range dependencies is an autocorrelation function which displays no exponential decay but a slow, hyperbolic rate of decay. While the existence of long memory patterns in return and volatility series have been already explored in much detail[18], only a few approaches pay attention to such effects in financial duration series. Engle (2000) applies a two-component model as proposed by Ding and Granger (1996), given by

$$\Psi_i = w\Psi_{1,i} + (1-w)\Psi_{2,i}, \tag{5.61}$$

$$\Psi_{1,i} = \alpha_1 x_{i-1} + (1-\alpha_1)\Psi_{1,i-1}, \tag{5.62}$$

$$\Psi_{2,i} = \omega + \alpha_2 x_{i-1} + \beta_2 \Psi_{2,i-1}. \tag{5.63}$$

Hence, Ψ_i consists of the weighted sum of two components, $\Psi_{1,i}$ and $\Psi_{2,i}$ with weights w and $1-w$, respectively. Equation (5.62) is an integrated ACD(1,1) specification capturing long-term movements, while (5.63) is a standard ACD(1,1) specification that captures short-term fluctuations in financial durations. As illustrated by Ding and Granger (1996), the resulting two-component process is covariance stationary, but allows for a slower decay of the ACF compared to the corresponding standard model. Engle (2000) shows that the two component ACD model improves the goodness-of-fit and captures the duration dynamics in a better way than the basic ACD model. Ding and Granger (1996) analyze the limiting case of a multi-component model by increasing the number of components towards infinity. They show that such a model implies autocorrelation functions that reflect the typical characteristics of long memory processes. Moreover, it is shown that the resulting model is closely related to the fractionally integrated GARCH (FIGARCH) model proposed by Baillie, Bollerslev, and Mikkelsen (1996). Following this string of the literature, Jasiak (1998) proposes a fractionally integrated ACD (FIACD) model which is the counterpart to the FIGARCH model. The FIACD model is based on a fractionally integrated process for Ψ_i and is given by

$$[1 - \beta(L)]\Psi_i = \omega + [1 - \beta(L) - \phi(L)(1-L)^d]x_i, \tag{5.64}$$

where $\phi(L) := 1 - \alpha(L) - \beta(L)$, $\alpha(L) := \sum_{j=1}^{P} \alpha_j L^j$ and $\beta(L) := \sum_{j=1}^{Q} \beta_j L^j$ denote polynomials in terms of the lag operator L. The fractional integration operator $(1-L)^d$ (with $0 < d < 1$) is given by

$$(1-L)^d = \sum_{j=0}^{\infty} \binom{d}{j} (-1)^j L^j = \sum_{j=0}^{\infty} \frac{\Gamma(j-d)}{\Gamma(-d)\Gamma(j+1)} L^j, \tag{5.65}$$

where $\Gamma(\cdot)$ denotes the gamma function. The FIACD model is not covariance stationary and implies infinite first and second unconditional moments of x_i.

---

[18] For an overview, see Beran (1994) or Baillie (1996).

Nevertheless, following the results of Bougerol and Picard (1992), it can be shown that it is strictly stationary and ergodic. By applying the FIACD model to trade durations from the Alcatel stock traded at the Paris Stock Exchange, Jasiak finds empirical evidence for the existence of long memory patterns.

5.3.5 Further Extensions

Semiparametric ACD Models

The assumption of independently, identically distributed innovations ε_i is a standard assumption in this class of time series models. However, Drost and Werker (2001) argue that the i.i.d. assumption for the standardized durations could be too restrictive to describe financial duration processes accurately. They relax the assumption of independent innovations to semiparametric alternatives, like Markov innovations and martingale innovations, and call it semiparametric ACD model. The authors illustrate how to compute the semiparametric efficiency bounds for the ACD parameters by adopting the concept of the efficient score-function[19]. By applying this concept to ACD models based on different types of innovation processes, they illustrate that even small dependencies in the innovations can induce sizeable efficiency gains of the efficient semiparametric procedure over the QML procedure.

The Stochastic Volatility Duration Model

As already discussed in Section 5.1, one drawback of autoregressive duration models (like EARMA or ACD models) is that they do not allow for separate dynamic parameterizations of higher order moments. This is due to the fact that typical duration distributions, like the exponential distribution, imply a direct relationship between the first moment and higher order moments. To cope with this problem, Ghysels, Gouriéroux, and Jasiak (1998) propose a two factor model which allows one to estimate separate dynamics for the conditional variance (duration volatility).

Consider an i.i.d. exponential model with gamma heterogeneity, $x_i = u_i/\lambda_i$, where u_i is an i.i.d. standard exponential variate and the hazard λ_i is assumed to depend on some heterogeneity component v_i. Thus,

$$\lambda_i = av_i, \quad v_i \sim \mathcal{G}(\eta, \eta) \tag{5.66}$$

with v_i assumed to be independent of u_i. Ghysels, Gouriéroux and Jasiak consider the equation $x_i = u_i/(av_i)$ as a two factor formulation and rewrite this expression in terms of Gaussian factors. Hence, x_i is expressed as

$$x_i = \frac{\mathcal{G}^{-1}(1, \Phi(m_1))}{a\mathcal{G}^{-1}(\eta, \Phi(m_2))} = \frac{H(1, m_1)}{aH(\eta, m_2)}, \tag{5.67}$$

---

[19] See Drost, Klaassen, and Werker (1997).

where m_1 and m_2 are i.i.d. standard normal random variables, $\Phi(\cdot)$ is the c.d.f. of the standard normal distribution and $\mathcal{G}^{-1}(m, \alpha)$ is the α-quantile function of the $\mathcal{G}(m, m)$ distribution. The function $H(1, m_1)$ can be solved as $H(1, m_1) = -\log(1 - \Phi(m_1))$ while $H(\eta, m_2)$ allows for no simple analytical solution and has to be approximated numerically (for more details, see Ghysels, Gouriéroux, and Jasiak, 1998).

This (static) representation of an exponential duration model with gamma heterogeneity as the function of two independent Gaussian random variables is used as starting point for a dynamic specification of the stochastic volatility duration (SVD) model. Ghysels, Gouiéroux and Jasiak propose to specify the process $m_i = (m_{1,i}\, m_{2,i})'$ in terms of a VAR representation, where the marginal distribution is constrained to be a $N(0, I)$ distribution, where I denotes the identity matrix. This restriction ensures that the marginal distribution of the durations belongs to the class of exponential distributions with gamma heterogeneity. Thus,

$$m_i = \sum_{j=1}^{P} \Lambda_j m_{i-j} + \varepsilon_i, \tag{5.68}$$

where Λ_j denotes the matrix of autoregressive VAR parameters and ε_i is a Gaussian white noise random variable with variance covariance matrix $\Sigma(\Lambda)$ which ensures that $\text{Var}(m_i) = I$.

The SVD model belongs to the class of nonlinear dynamic factor models for which the likelihood typically is difficult to evaluate. Therefore, Ghysels, Gouiéroux and Jasiak propose a two step procedure. In the first step, one exploits the property that the marginal distribution of x_i is a Pareto distribution depending on the parameters a and η (see Appendix A). Thus, a and η can be estimated by (Q)ML. In the second step, the autoregressive parameters Λ_j are estimated by using the methods of simulated moments (for more details, see Ghysels, Gouriéroux, and Jasiak, 1998).

Ghysels, Gouiéroux and Jasiak apply the SVD model to analyze trade durations from the Alcatel stock traded at the Paris Stock Exchange and find empirical evidence for strong dynamics in both factors. However, it has to be noted that the higher dynamic flexibility of the SVD model comes along with strong distributional assumptions. In particular, Bauwens, Giot, Grammig, and Veredas (2000) show that the assumption of a Pareto distribution for the marginal distribution of x_i is inappropriate for specific types of financial durations (particularly volume durations) which makes the estimation of the Pareto parameters a and η extremely cumbersome. Moreover, they find a relatively poor prediction performance of the SVD model compared to alternative ACD specifications.

5.4 Testing the ACD Model

In this section, we discuss several procedures to test the ACD model. Section 5.4.1 illustrates briefly simple residual checks. Section 5.4.2 reviews methods for density forecast evaluations as valuable tools to investigate the goodness-of-fit. Such tests have been successfully applied to ACD models by Bauwens, Giot, Grammig, and Veredas (2000) and Dufour and Engle (2000a). In Sections 5.4.3 through 5.4.5 we focus explicitly on specification tests to test for the form of the conditional mean function. As illustrated in Section 5.2.2, the correct specification of the conditional mean function Ψ_i, i.e., the validity of the conditional mean restriction, $\Psi_i = \Psi_{i,0}$, is an essential prerequisite for the QML properties of the ACD model. To test this moment restriction, we consider several types of specification tests. As proposed by Engle and Russell (1994), a simple possibility is to specify Lagrange Multiplier (LM) tests which have optimal power against local alternatives.[20] In Section 5.4.3, we use specific LM tests against sign bias alternatives in the news impact function. Such tests allow to test whether the functional form of the conditional mean function is appropriate to account for (possible nonlinear) news response effects. More general tests against violations of the conditional mean restriction are discussed in Section 5.4.4. Here, we discuss conditional moment (CM) tests as introduced by Newey (1985). These tests are consistent against a finite number of possible alternatives since they rely on a finite number of conditional moment restrictions. In Section 5.4.5 we illustrate the use of integrated conditional moment (ICM) tests proposed by Bierens (1982 and 1990). By employing an infinite number of conditional moments, this test possesses the property of consistency against all possible alternatives, and thus, is a generalization of the CM test. In Section 5.4.6, we present a Monte Carlo study where we analyze the size and the power of different forms of LM, CM and ICM tests on the basis of various types of augmented ACD models as data generating processes. Applications of the proposed testing framework will be illustrated in Section 5.5.

5.4.1 Simple Residual Checks

One obvious way to evaluate the goodness-of-fit of the ACD model is to investigate the dynamical and distributional properties of the residuals

$$\hat{\varepsilon}_i = x_i/\hat{\Psi}_i, \ i = 1, \dots, n \tag{5.69}$$

implied by the model. Under correct model specification, the series must be i.i.d. Hence, Ljung-Box or Box-Pierce statistics based on the centered ACD residuals can be used to analyze whether the specification is able to account for the inter-temporal dependence in the duration process.

---

[20] Engle and Russell (1994) apply these tests to test against neglected market microstructure variables.

Furthermore, the residual series should have a mean of one and a distribution which should correspond to the specified distribution of the ACD errors. Graphical checks of the residual series can be performed based on QQ-plots. Alternatively, moment conditions implied by the particular distributions might be investigated to evaluate the goodness-of-fit. In the case of an exponential distribution, a simple moment condition implies the equality of the mean and the standard deviation. Engle and Russell (1998) propose a test for no excess dispersion based on the statistic $\sqrt{n}((\hat{\sigma}_{\hat{\varepsilon}}^2 - 1)/\sigma_\varepsilon)$, where $\hat{\sigma}_{\hat{\varepsilon}}^2$ is the sample variance of $\hat{\varepsilon}_i$ and σ_ε is the standard deviation of $(\varepsilon_i - 1)^2$. Under the exponential null hypothesis, $\hat{\sigma}_{\hat{\varepsilon}}^2$ should be equal to one and σ_ε^2 should be equal to $\sqrt{8}$. Engle and Russell show that under the null, this test statistic has a limiting standard normal distribution.

5.4.2 Density Forecast Evaluations

Another way to evaluate the goodness-of-fit is to evaluate the in-sample density forecasts implied by the model. Diebold, Gunther, and Tay (1998) propose an evaluation method based on the probability integral transform[21]

$$q_i = \int_{-\infty}^{x_i} f(s)ds. \tag{5.70}$$

They show that under the null hypothesis, i.e., correct model specification, the distribution of the q_i series is i.i.d. uniform. Hence, testing the q_i series against the uniform distribution allows to evaluate the performance of the density forecasts[22]. Therefore, a goodness-of-fit test is performed by categorizing the probability integral transforms q_i and computing a χ^2-statistic based on the frequencies of the particular categories

$$\chi^2 = \sum_{k=1}^{K} \frac{(n_k - n\hat{p}_k^*)^2}{n\hat{p}_k^*},$$

where K denotes the number of categories, n_k the number of observations in category k and $\hat{p}_k^*$ the estimated probability to observe a realization of q_i in category k.

More sophisticated tests against distributional misspecification are proposed by Fernandes and Grammig (2000) based on the work of Aït-Sahalia (1996). They suggest a nonparametric testing procedure that is based on the distance between the estimated parametric density function and its nonparametric estimate. This test is very general, since it tests for correctness of the complete (conditional) density. However, note that such general tests

---

[21] See Rosenblatt (1952).

[22] For more details, see e.g. Bauwens, Giot, Grammig, and Veredas (2000) or Dufour and Engle (2000a), who applied this concept to the comparison of alternative financial duration models.

are not appropriate for explicitly testing the conditional mean restriction of ACD models. Clearly, these tests have power against misspecifications of the conditional mean function, but they do not allow to identify whether a possible rejection is due to a violation of distributional assumptions caused by misspecifications of higher order conditional moments or due to a violation of the conditional mean restriction. However, especially in the context of QML estimation of the ACD model, one is mainly interested in the validity of the conditional mean restriction but not necessarily in the correct specification of the complete density. Procedures that explicitly test this particular conditional moment restriction are discussed in the following subsections.

5.4.3 Lagrange Multiplier Tests

In econometric literature, the Lagrange Multiplier (LM) test has proven to be a useful diagnostic tool for detecting model misspecifications.[23] To perform the LM test, it is necessary to specify a general model which encompasses the model under the null. Assume that the ACD specification under the null hypothesis H_0 is a special case of a more general model of the form

$$x_i = \Psi_i \varepsilon_i$$
$$\Psi_i = \Psi_{0,i} + \theta_a' z_{ai}, \tag{5.71}$$

where $\Psi_{0,i}$ denotes the conditional mean function under the null depending on a $M^0 \times 1$ parameter vector θ_0 and z_{ai} and θ_a denote the vectors of M^a missing variables and additional parameters, respectively. Thus, we can test for correct specification of the null model by testing the parameter restriction $H_0 : \theta_a = 0$.

In the following we restrict our consideration to the case of QML estimation of the ACD model, i.e., we maximize the quasi log likelihood function as given in (5.16). Denote $\hat{\theta}^0$ as the QML estimate under the H_0. Then, the standard form of the LM test is obtained by

$$\Upsilon_{LM} = \iota' s(\hat{\theta}^0) \mathcal{I}(\hat{\theta}^0)^{-1} s(\hat{\theta}^0)' \iota \overset{a}{\sim} \chi^2(M^a), \tag{5.72}$$

where ι denotes a $n \times 1$ vector of ones and $s(\hat{\theta}^0) = (s_1(\hat{\theta}^0)', \ldots, s_n(\hat{\theta}^0)')$ is the $n \times (M^0 + M^a)$ matrix whose (i,j)-th element is the (i,j)-th contribution to the score evaluated under the null, $\partial l_i(\hat{\theta}^0)/\partial \theta_j$, and θ_j is the j-th element of θ. Moreover, $\mathcal{I}(\hat{\theta}^0)$ denotes the information matrix evaluated at $\hat{\theta}^0$. When $\mathcal{I}(\hat{\theta}^0)$ is estimated based on the outer product of gradients (OPG), then Υ_{LM} can be computed as n times the coefficient of determination of the linear regression of ι on $s(\hat{\theta}^0)$.

Engle and Ng (1993) propose specifying z_{ai} in terms of sign bias and size bias variables $\mathbb{1}_{\{\hat{\varepsilon}_{i-1}<1\}}$, $\mathbb{1}_{\{\hat{\varepsilon}_{i-1}<1\}}\hat{\varepsilon}_{i-1}$ and $\mathbb{1}_{\{\hat{\varepsilon}_{i-1}\geq 1\}}\hat{\varepsilon}_{i-1}$. This type of LM

---

[23] See for example, Breusch (1978), Breusch and Pagan (1979, 1980), Godfrey (1978) or the overview in Engle (1984).

test allows one to investigate whether there are systematic patterns in the ACD residual series. In such a case, the chosen ACD specification is obviously not appropriate for capturing the news impact function implied by the data. Notice that a LM test based on sign bias and size bias variables does not allow one to test particular types of ACD models against each other (as, for instance, a Log ACD model against a Box-Cox ACD model). However, it can be interpreted as a test against more general alternatives (in terms of the news impact function) that do not necessarily belong to one of the ACD specifications discussed in the previous subsections.

In addition to a test against joint significance of θ_a, tests of the single null hypotheses $H_0 : \theta_{a,j} = 0$, where $\theta_{a,j}$ is associated with single sign bias variables, provide information with respect to negative and positive sign bias. Such diagnostic information can be used to examine in which way the misspecified model has to be improved.

An alternative way to detect nonlinear dependencies between the ACD residuals and the past information set is proposed by Engle and Russell (1998). The main idea is to divide the residuals into bins which range from 0 to ∞ and to regress $\{\hat{\varepsilon}_i\}$ on indicators which indicate whether the previous duration has been in one of the categories. If the ACD residuals are i.i.d., then we should find no predictability in this regression and thus the coefficient of determination should be zero. In case of the rejection of the null hypothesis, the potentially significant coefficients of indicators associated with single bins provide information with respect to the source of misspecification.

5.4.4 Conditional Moment Tests

The LM test discussed in the previous subsection has optimal power against local (parametric) alternatives and is a special case of the CM test. The main idea behind the conditional moment (CM) test is to test the validity of conditional moment restrictions implied by the data which hold when the model is correctly specified. In the ACD framework, a natural moment condition is obtained by the conditional mean restriction. The null hypothesis of a correct functional form of the ACD model can be formulated in terms of the ACD innovation ε_i,

$$H_0: \quad \mathrm{E}\left[\varepsilon_i - 1|\, \mathfrak{F}_{t_{i-1}}\right] = \mathrm{E}\left[x_i/\Psi_i - 1|\, \mathfrak{F}_{t_{i-1}}\right] = 0, \qquad (5.73)$$

or, alternatively, in terms of the martingale difference η_i, i.e.,

$$H_0: \quad \mathrm{E}\left[\eta_i|\, \mathfrak{F}_{t_{i-1}}\right] = \mathrm{E}\left[x_i - \Psi_i|\, \mathfrak{F}_{t_{i-1}}\right] = 0. \qquad (5.74)$$

Asymptotically, this distinction should make no difference, however, in finite samples it can be an issue. Correspondingly, the alternative hypotheses are formulated as

$$H_1: \quad \mathrm{E}\left[x_i/\Psi_i - 1|\, \mathfrak{F}_{t_{i-1}}\right] \neq 0,$$

or

$$H_1: \quad \mathrm{E}\left[x_i - \Psi_i \mid \mathfrak{F}_{t_{i-1}}\right] \neq 0,$$

respectively.

Newey (1985) proposes building a CM test based on J *unconditional* moment restrictions

$$m_j(x_i; \theta) = r(x_i; \theta) w_j(x_{i-1}, x_{i-2}, \dots; \theta_0), \quad j = 1, \dots, J, \qquad (5.75)$$

where $w_j(\cdot)$ is a weighting function associated with the j-th unconditional moment restriction and is based on the duration (or also covariate) history. Thus, $r(x_i; \theta)$ denotes the *conditional* moment restriction associated with the ith observation. Under the null, $w_j(\cdot)$ is orthogonal to $r(x_i; \theta)$, and thus, $\mathrm{E}[m_j(\cdot)] = 0$. As shown by Newey (1985), an asymptotically χ^2 distributed test statistic is built on the vector of unconditional moment restrictions and the vector of scores. Define $m_j(x; \hat{\theta}^0)$ as the $n \times 1$ vector associated with the j-th unconditional moment restriction for all n observations computed based on $\hat{\theta}^0$. Furthermore, $m(x; \hat{\theta}^0) = (m_1(x; \hat{\theta}^0) \dots m_J(x; \hat{\theta}^0))$ denotes the $n \times J$ matrix of all J unconditional moments and, correspondingly, $s(\hat{\theta}^0)$ denotes the $n \times M^0$ score matrix evaluated at $\hat{\theta}^0$. Moreover, define $g(x; \hat{\theta}^0) = (m(x; \hat{\theta}^0) \, s(\hat{\theta}^0))$ as the $n \times (J + M^0)$ matrix that includes $m(x; \hat{\theta}^0)$ and $s(\hat{\theta}^0)$. Then, the CM test statistic is given by

$$\Upsilon_{CM} = \iota' g(x; \hat{\theta}^0)[g(x; \hat{\theta}^0)' g(x; \hat{\theta}^0)]^{-1} g(x; \hat{\theta}^0)' \iota \overset{a}{\sim} \chi^2(M^a), \qquad (5.76)$$

and can be computed as the coefficient of determination of the linear regression of ι on $g(x; \hat{\theta}^0)$.

According to the null hypotheses (5.73) or (5.74), a straightforward choice of the conditional moment restriction is

$$r(x_i; \theta) = \varepsilon_i - 1 = x_i/\Psi_i - 1 \qquad \text{or} \qquad (5.77)$$
$$r(x_i; \theta) = \eta_i = x_i - \Psi_i. \qquad (5.78)$$

Valuable choices for the weighting functions $w_j(\cdot)$ are lagged sign bias variables (as discussed in Section 5.4.3) and/or functionals (e.g. moments) of past durations.

A well known result is that the power of the CM test depends heavily on the choice of the weighting functions. Newey (1985) illustrates how to obtain an optimal conditional moment test with maximal local power. It is shown that the LM test corresponds to an optimal CM test in the case of a particular local alternative. However, since the CM test is based on a finite number of conditional moment restrictions, it cannot be consistent against *all* possible alternatives.

5.4.5 Integrated Conditional Moment Tests

Bierens (1990) illustrates that any CM test of functional form can be converted into a χ^2 test that possesses the property of consistency against all possible alternatives. While Bierens assumes that the data are i.i.d., de Jong (1996) extends the "Bierens test" to the case of serial dependent data. He suggests a tractable solution to the key problem that time series usually are functions of an infinite number of random variables. In the following, we assume that the duration process is strictly stationary and obeys the concept of ν-stability. Moreover it is supposed that $E|\varepsilon_i - 1| < \infty$.

The main idea behind a consistent conditional moment test is based on the following lemma:

Lemma 5.4 (Lemma 1, Bierens, 1990). *Let ϱ be a random variable satisfying the condition $E[|\varrho|] < \infty$, and let u be a bounded random variable in $\mathbb{R}$ with $\mathrm{Prob}[E(\varrho|u) = 0] < 1$. Then, the set $\mathfrak{S} = \{t_1 \in \mathbb{R} : E[\varrho \exp(t_1 u)] = 0\}$ is countable, and thus, has Lebesgue measure zero.*

Proof: See Bierens (1990), p. 1456. $\square$

The main idea behind the proof of this lemma is to show that $E[\varrho \exp(t_1 u)] \neq 0$ in a neighborhood of $t_1 = t_1'$ where t_1' is such that $E[\varrho \exp(t_1' u)] = 0$. However, to achieve this result, the exponential function is not essential. Stinchombe and White (1992) illustrate that one can use any function whose coefficients of series expansion around zero are all unequal to zero[24]. Thus, $\{t_1 \in \mathbb{R} : \varrho \exp(t_1 u)\}$ can be interpreted as an infinite set of (unconditional) moment restrictions associated with the random variable ϱ and the conditioning variable u. These moment conditions are indexed by the parameter t_1. Therefore, the principle of Bierens' consistent conditional moment test is to employ a class of weighting functions which are indexed by a continuous nuisance parameter t_1. Since this nuisance parameter is integrated out, the resulting test is called integrated conditional moment (ICM) test.

Note that in the time series case, a variable typically depends upon an infinite number of conditioning variables. Thus, Lemma 5.4 has to be applied to the case where ϱ depends on an infinite sequence of conditioning variables u. de Jong (1996) extends the "Bierens test" to the time series case by mimicking the idea behind Lemma 5.4 for a $(d \times 1)$ vector of nuisance parameters $t = (t_1, \ldots, t_d)$. The variable d is associated with the number of lags that are taken into account and is given by $d = \min(i - 1, L)$, where L denotes some predetermined integer number. Clearly, a conditional moment test based on $d = L < n$ does not allow us to *consistently* test the hypotheses H_0 and H_1, and thus, $d = i - 1$ should be preferred. However, de Jong points out that if $n < L < \infty$, the resulting test will not be different from the corresponding test that would result from choosing $L = \infty$.

---

[24] For more details, see Stinchombe and White (1992) or Bierens and Ploberger (1997).

In order to construct the ICM test statistic, assume that the model is misspecified, i.e., $\text{Prob}[\text{E}\,[r(x_i;\theta_0)|\,x_{i-1},x_{i-2},\ldots] = 0] < 1$ and replace the conditioning information by $\xi(x_{i-1}),\xi(x_{i-2}),\ldots$, where $\xi(\cdot)$ is a measurable one-to-one mapping from $\mathbb{R}$ to a compact bounded subset of $\mathbb{R}$. Then, the set

$$\mathfrak{S} = \left\{ \mathsf{t} \in \mathbb{R}^d : \text{E}\left[r(x_i;\theta_0)\exp\left(\sum_{j=1}^{d} \mathsf{t}_j\xi(x_{i-j}) \right) \right] = 0 \right\} \tag{5.79}$$

has Lebesgue measure zero. de Jong (1996) suggests a consistent CM test based on the (1×1) unconditional sample moment function

$$\hat{M}_n(\mathsf{t}) = n^{-1/2} \sum_{i=1}^{n} r(x_i;\hat{\theta}^0)\exp\left(\sum_{j=1}^{d} \mathsf{t}_j\xi(x_{i-j}) \right). \tag{5.80}$$

The moment restriction (5.80) corresponds to an extension of the unconditional moment function proposed by Bierens (1990) who assumes i.i.d. data and a fixed number of exogenous conditional variables. $r(\cdot)$ can be chosen as given by (5.77) or (5.78). However, as suggested by Bierens (1982), an alternative is to combine $r(\cdot)$ in addition with some weighting function $w(x_{i-1},x_{i-2},\ldots;\hat{\theta}^0)$. Hence, an alternative ICM moment function is

$$\hat{M}_n(\mathsf{t}) = n^{-1/2} \sum_{i=1}^{n} r(x_i;\hat{\theta}^0)w(x_{i-1},x_{i-2},\ldots;\hat{\theta}^0)\exp\left(\sum_{j=1}^{d} \mathsf{t}_j\xi(x_{i-j}) \right). \tag{5.81}$$

The use of a weighting function $w(\cdot)$ in the conditional moment restriction is irrelevant for consistency, but, it can improve the finite sample properties of the test. Note that under the alternative hypothesis H_1, $\underset{n\to\infty}{\text{plim}}\ \hat{M}_n(\mathsf{t}) \neq 0$ for all t except in a set with Lebesgue measure zero. Hence, by choosing a vector $\mathsf{t} \notin \mathfrak{S}$, a consistent CM test is obtained. However, the crucial point is that $\mathfrak{S}$ depends on the distribution of the data, and thus, it is impossible to choose a fixed vector t for which the test is consistent. A solution to this problem is to achieve test consistency by maximizing a function of $\hat{M}_n(\mathsf{t})$ over a compact subset $\varXi$ of $\mathbb{R}^d$. The main idea is that a vector t which *maximizes* a test statistic based on $\hat{M}_n(\mathsf{t})$ cannot belong to the set $\mathfrak{S}$.

The use of a functional of $\sup_{\mathsf{t}\in\varXi} |\hat{M}_n(\mathsf{t})|$ as test statistic requires to consider the convergence of probability measures on metric spaces. In order to derive the asymptotic properties of an ICM test based on (5.80) for the case when $d = L = \infty$, de Jong suggests to introduce a space for infinite sequences $\{\mathsf{t}_1, \mathsf{t}_2, \ldots\}$

$$\varXi = \{\mathsf{t} : \ \mathfrak{a}_j \le \mathsf{t}_j \le \mathfrak{b}_j\ \forall j\,;\ \mathsf{t}_j \in \mathbb{R}\}, \tag{5.82}$$

where $\mathfrak{a}_j < \mathfrak{b}_j$ and $|\mathfrak{a}_j|$, $|\mathfrak{b}_j| \leq \mathfrak{B}j^{-2}$ for some constant $\mathfrak{B}$. Furthermore, by using the metric

$$||\mathfrak{t} - \mathfrak{t}'|| = \left(\sum_{j=1}^{\infty} j^2 |\mathfrak{t}_j - \mathfrak{t}'_j| \right)^{1/2}, \tag{5.83}$$

de Jong proves that $(\Xi, ||\cdot||)$ is a compact metric space (see de Jong, 1996, Lemma1). Moreover, he illustrates that the relationship found in Lemma 5.4 also holds for the time series case. This is stated in the following theorem which is central to ensure the test consistency when $L = \infty$ and is an extension of Lemma 5.4 to the case of infinite-dimensional vectors $\mathfrak{t}$:

Theorem 5.5 (Theorem 2, de Jong, 1996). *Let the compact metric space Ξ and the metric $||\cdot||$ be defined as above and assume that the model is misspecified, i.e., $\mathrm{Prob}[\mathrm{E}\,[r(x_i; \theta_0)|\,x_{i-1}, x_{i-2}, \ldots] = 0] < 1$. Then, if*

$$\mathrm{E}\left[r(x_i; \theta_0) \exp\left(\sum_{j=1}^{\infty} \mathfrak{t}_j \xi(x_{i-j}) \right) \right] = 0, \tag{5.84}$$

for some $\mathfrak{t} \in \Xi$, and if Ξ' is an arbitrary neighborhood of Ξ, we can find a $\mathfrak{t}'$ in Ξ' such that

$$\mathrm{E}\left[r(x_i; \theta_0) \exp\left(\sum_{j=1}^{\infty} \mathfrak{t}'_j \xi(x_{i-j}) \right) \right] \neq 0. \tag{5.85}$$

Proof: See de Jong (1996), p. 18. $\square$

However, according to the discussion of Lemma 5.4, the existence of a countable number of vectors $\mathfrak{t} \in \Xi$ implies that under the alternative hypothesis H_1, $\underset{n \to \infty}{\mathrm{plim}} \, \hat{M}_n(\mathfrak{t}) = 0$, which prevents us from obtaining a consistent test. However, test consistency is achieved by maximizing the test statistic over the compact space Ξ. de Jong proves that under a set of regularity conditions[25], under H_0, for each $\mathfrak{t} \in \Xi$, the process $\hat{W}_n(\mathfrak{t}) = \hat{M}_n(\mathfrak{t})/(\hat{s}^2(\mathfrak{t}))^{1/2}$ converges in distribution to a $N(0, 1)$ distribution, where $\hat{s}^2(\mathfrak{t})$ denotes the estimated asymptotic variance of $\hat{M}_n(\mathfrak{t})$ which can be estimated based on the gradients of $r(x_i; \hat{\theta}^0)$.[26] Moreover, it is shown that under H_1, there exists some function $\vartheta(\cdot)$, such that

$$\left| n^{-1/2} \sup_{\mathfrak{t} \in \Xi} |\hat{W}_n(\mathfrak{t})| - \sup_{\mathfrak{t} \in \Xi} |\vartheta(\mathfrak{t})| \right| = o(1) \tag{5.86}$$

---

[25] These regularity conditions mainly rule out extreme behavior of the asymptotic variance of $M_n(\mathfrak{t})$ and ensure that the random functionals under consideration are properly defined random variables. For a discussion of these regularity conditions, see Bierens (1990).

[26] For more details, see de Jong (1996).

with $\sup_{t \in \Xi} |\vartheta(t)| > 0$ (see Theorem 4 in de Jong, 1996). Therefore, the test is consistent.

However, the limiting distribution of the test statistic $\sup_{t \in \Xi} |\hat{W}_n(t)|$ is case-dependent which prevents the use of generally applicable critical values. de Jong introduces a simulation procedure based on a conditional Monte Carlo approach. In particular, he shows that under the null, the moment restriction (5.80) has the same asymptotic finite-dimensional distribution as

$$\hat{M}_n^*(t) = n^{-1/2} \sum_{i=1}^n \varpi_i r(x_i; \hat{\theta}^0) \exp\left(\sum_{j=1}^d t_j \xi(x_{i-j})\right) \tag{5.87}$$

point-wise in t, where ϖ_i are bounded i.i.d. random variables independent of x_i and $r(x_i; \hat{\theta}^0)$ with $E[\varpi_i^2] = 1$. Thus, the distribution of $\hat{M}_n(t)$ can be approximated based on the simulation of n-tuples of ϖ_i. de Jong proves that the critical regions obtained by the simulation of $\hat{M}_n^*(t)$ are asymptotically valid (see de Jong, 1996, Theorem 5).

As stated above, a consistent ICM test rests on the statistic $\sup_{t \in \Xi} |\hat{W}_n(t)|$. The calculation of this test statistic is quite cumbersome since it requires the maximization over a parameter space of dimension $n - 1$. For this reason, de Jong suggests finding another continuous functional of $\hat{M}_n^*(t)$ that possesses the same consistency property but is more easily calculated. In particular, he proposes to use the functional

$$\Upsilon_{ICM} = n^{-1} \int_{\Xi} \hat{M}_n(t)^2 f_1(t_1) dt_1 \ldots f_j(t_j) dt_j \ldots, \tag{5.88}$$

where the integrations run over an infinite number of t_j. According to (5.82), each t_j is integrated over the subset of $\mathbb{R}$, such that $a_j \leq |t_j| \leq b_j$. $f_j(t)$ denote a sequence of density functions that integrate to one over the particular subsets. de Jong (1996) shows that, under H_1, Υ_{ICM} is asymptotically equivalent to

$$\int_{\Xi} \vartheta(t)^2 f_1(t_1) dt_1 \ldots f_j(t_j) dt_j \ldots > 0 \tag{5.89}$$

which ensures test consistency. Since $\hat{M}_n(t)$ can be written as a double summation and the integrals can be calculated one at a time, we obtain a functional, that is much easier to calculate than $\sup_{t \in \Xi} |\hat{M}_n(t)|$. By choosing a uniform distribution for $f_j(t_j)$, i.e., $f_j(s) = (b_j - a_j)^{-1} \mathbb{1}_{\{a_j \leq s \leq b_j\}}$, the ICM test statistic Υ_{ICM} results in

$$\Upsilon_{ICM} = n^{-1} \sum_{i=1}^n \sum_{j=1}^n r(x_i; \hat{\theta}^0) r(x_j; \hat{\theta}^0) \prod_{s=1}^d \left\{ \frac{1}{b_j - a_j} \left[\xi(x_{i-s}) + \xi(x_{j-s})\right]^{-1} \right.$$

$$\tag{5.90}$$

$$\times \left[\exp(b_j(\xi(x_{i-s}) + \xi(x_{j-s}))) - \exp(a_j(\xi(x_{i-s}) + \xi(x_{j-s})))\right] \Big\}.$$

Summarizing, an application of the ICM testing procedure to test the ACD model requires to perform the following steps:

(i) *(Q)ML estimation of the ACD model:* Estimate the particular ACD model by QML and calculate the conditional moment restriction $r(x_i; \hat{\theta}^0)$.

(ii) *Choice of a_j and b_j:* Choose values for $\mathfrak{a}_j$ and $\mathfrak{b}_j$, that define the parameter space Ξ. de Jong (1996) suggests using $\mathfrak{a}_j = \mathfrak{A}j^{-2}$ and $\mathfrak{b}_j = \mathfrak{B}j^{-2}$ where the values $\mathfrak{A}$ and $\mathfrak{B}$ $(0 \leq \mathfrak{A} < \mathfrak{B})$ can be chosen arbitrarily. Asymptotically, the choice of $\mathfrak{A}$ and $\mathfrak{B}$ should have no impact on the power of the test. However, in finite samples, it may have an influence. While there is no general rule determining the range covered by $\mathfrak{A}$ and $\mathfrak{B}$, the Monte Carlo simulations of de Jong (1996) and also the Monte Carlo experiments made in this study (see Section 5.4.6) suggest choosing a small range, for example, $\mathfrak{A} = 0$ and $\mathfrak{B} = 0.5$.

(iii) *Choice of $\xi(\cdot)$:* According to Lemma 5.4, the function $\xi(\cdot)$ must be a one-to-one mapping from $\mathbb{R}$ to a compact bounded subset of $\mathbb{R}$. Asymptotically, the specific choice of the function $\xi(\cdot)$ is irrelevant. However, Bierens (1990) proposes using $\xi(x) = \arctan(x)$. In this study, we use $\xi(x) = arctan(0.01 \cdot x) \cdot 100$, which is a bounded function as well, but has the advantage of being linear in the relevant region which improves the small sample properties of the test.

(iv) *Choice of d:* Choose the number of lags taken into account. Note that in the case of dependent data, the test consistency is only ensured by accounting for all feasible lags, $d = i - 1$. Then, the dimension of the parameter space under consideration grows with the sample size. An alternative which does not require as much computer time, would be to choose a fixed value $d = L < n$. In this case however, the test does not allow us to test the moment condition for an infinite number of conditioning variables *consistently*.

(v) *Simulation of n-tuples of ϖ_i:* Simulate M^{ICM} n-tuples of (bounded) i.i.d. random variables $\varpi_{i,m}$, $i = 1,\dots,n$ for $m = 1,\dots,M^{ICM}$ with $\mathrm{E}[\varpi_{i,m}^2] = 1$. Following de Jong (1996), we generate ϖ_i such that $\mathrm{E}[\varpi_i = 1] = \mathrm{E}[\varpi_i = -1] = 0.5$.

(vi) *Computation of the test statistic and simulation of the critical values:*
   • Compute the test statistic Υ_{ICM} according to (5.90).
   • For each n-tuple of $\varpi_{\cdot,m}$, compute the simulated test statistic

$$\tilde{\Upsilon}_{m,ICM} = n^{-1} \sum_{i=1}^{n} \sum_{j=1}^{n} (\varpi_{i,m} r(x_i; \hat{\theta}^0))(\varpi_{j,m} r(x_j; \hat{\theta}^0)) \tag{5.91}$$

$$\times \prod_{s=1}^{d} \left\{ \frac{1}{\mathfrak{b}_j - \mathfrak{a}_j} [\xi(x_{i-s}) + \xi(x_{j-s})]^{-1} \right.$$
$$\left. \times [\exp(\mathfrak{b}_j(\xi(x_{i-s}) + \xi(x_{j-s}))) - \exp(\mathfrak{a}_j(\xi(x_{i-s}) + \xi(x_{j-s})))] \right\}.$$

(vii) *Computation of simulated p-values:* Since the critical region of the test has the form $(C, \infty]$, we compute the simulated p-value of the ICM test as

$$pv_{ICM} = \frac{1}{M^{ICM}} \sum_{m=1}^{M^{ICM}} \mathbb{1}_{\{\tilde{\Upsilon}_{m,ICM} \leq \Upsilon_{ICM}\}}. \tag{5.92}$$

5.4.6 Monte Carlo Evidence

In order to gain deeper insights into the size and the power of the proposed diagnostic tests, we conduct an extensive Monte Carlo study. We draw samples of size $n = 500$ which is quite small for typical high-frequency financial data. Therefore, the study allows us to evaluate the small sample properties of the particular tests. Each Monte Carlo experiment is repeated $n^{MC} = 1000$ times. For each replication, a (linear) ACD(1,1) model is estimated. In order to check the size and the power of the tests, we use 11 data generating processes based on different augmented ACD specifications. In particular, we use parameterizations of AMACD, LACD$_1$, BACD, EXACD, HACD and NPACD models. We choose a relatively high value for the persistence parameter ($\beta = 0.7$) while the corresponding function of innovation parameters associated with the various (nonlinear) news impact functions is approximately 0.1 for $\varepsilon_{i-1} = 1$. Therefore, the chosen parameterizations are quite realistic for typical financial duration processes.

Table 8 shows the parameter settings of the particular data generating processes. Figures 28 and 29 graphically illustrate the difference between the particular data generating processes and the corresponding estimated linear ACD specification in terms of the news impact function. In particular, we plot the news response function implied by the various data generating processes together with the (linear) news impact function computed based on the average estimated ACD(1,1) parameters over all replications.

In experiment (1), the data generating process corresponds to an ACD(1,1) model. This experiment allows us to check the *size* of the tests. Thus, the small difference between the corresponding linear news impact functions of

Table 8: Underlying data generating processes in the Monte-Carlo study. Categories for the NPACD model: $0.1, 0.2, \ldots, 2.0$ with $\bar{\varepsilon}_0 = 1.0$.

	(1) ACD	(2) AMACD	(3) AMACD	(4) LACD$_1$	(5) BACD	(6) BACD
ω	0.200	0.050	0.000	0.055	0.165	0.212
α_1	0.100	0.150	0.050	0.100	0.100	0.100
β_1	0.700	0.700	0.700	0.700	0.700	0.700
δ_1					0.800	0.500
δ_2					1.500	0.500
ν		0.100	0.250			

	(7) EXACD	(8) EXACD	(9) HACD	(10) NPACD	(11) NPACD
ω	-0.090	0.010	0.380	-0.047	-0.086
α_1	0.200	0.200	0.400		
α_2					
β_1	0.700	0.700	0.700	0.700	0.700
δ_1			1.000		
δ_2			0.400		
b			1.000		
c	-0.150	-0.300	-0.800		
α_1					
α_1^+				-0.060	0.020
α_2^+				0.050	0.020
α_3^+				0.050	0.020
α_4^+				0.100	0.020
α_5^+				0.100	0.020
α_6^+				0.150	0.020
α_7^+				0.200	0.020
α_8^+				-0.200	0.020
α_9^+				-0.270	0.020
α_{10}^+				-0.050	0.020
α_1^-				-0.020	0.000
α_2^-				-0.020	0.000
α_3^-				0.020	0.000
α_4^-				0.030	0.000
α_5^-				0.100	0.000
α_6^-				0.100	0.000
α_7^-				0.150	0.100
α_8^-				0.150	0.100
α_9^-				0.200	0.100
α_{10}^-				0.120	0.100

the "true" model and the estimated model in Picture (1) of Figure 28 reveals a slight small sample bias. Experiments (2)-(11) are associated with an examination of the *power* of the diagnostic tests to detect different types of model misspecifications. In particular, experiments (2) and (3) are based on a data generating process assuming an AMACD specification.[27] Experiments (4)-(6) assume concave and convex news impact curves based on $LACD_1$ and BACD processes, while the specifications (7) and (8) are based on upward and downward kinked news impact functions (EXACD model). Experiment (9) assumes a highly nonlinear news response function based on a HACD process, and experiments (10) and (11) are based on a NPACD process implying a non-monotonic news impact curve.

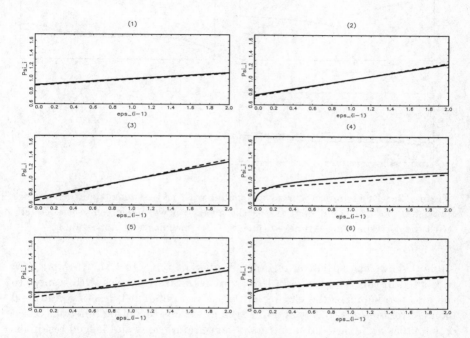

Figure 28: News impact curves associated with the assumed data generating processes (1)-(6) (solid line) versus the average news impact curve associated with the corresponding estimated (linear) ACD specification (dotted line).

---

[27] Note that the news impact curves are plotted for a fixed value of Ψ_{i-1}. Thus, the difference between a multiplicative and an additive ACD model cannot be fully revealed by the corresponding pictures.

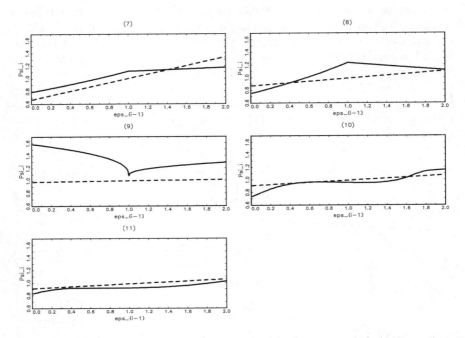

Figure 29: News impact curves associated with the assumed data generating processes (7)-(11) (solid line) versus the average news impact curve associated with the corresponding estimated (linear) ACD specification (dotted line).

Table 9 shows the specifications of the different types of LM, CM and ICM tests used in this study. The LM tests use sign bias variables with respect to the first two lags. For the CM tests, we use two different types of conditional moment restrictions ((5.77) and (5.78)) and various weighting functions based on sign bias variables and moments of lagged durations and lagged residuals. Correspondingly, we evaluate various specifications of ICM tests using different conditional moment restrictions with and without weighting functions based on the first and second moments of lagged durations. In this context, a further issue is to gain more insights into the properties of the ICM test statistic in dependence of the parameters $\mathfrak{A}$, $\mathfrak{B}$ and L. As discussed in Section 5.4.5, consistency of the ICM test is only ensured for the space $\varXi$ growing in dimension to infinity with the sample size. However, for the case $d = i - 1$ and for large samples, the calculation of the ICM test statistic and of the simulated critical values is computational quite intensive.

Table 9: Parameterizations of the specification tests used in the Monte-Carlo studies and in the empirical applications.

	LM tests and F-test against nonlinearities in the news response			
LM_1	$z_{ai}^1 = \big(\mathbb{1}_{\{\hat{\varepsilon}_{i-1}<1\}},\ \mathbb{1}_{\{\hat{\varepsilon}_{i-1}<1\}}\hat{\varepsilon}_{i-1},\ \mathbb{1}_{\{\hat{\varepsilon}_{i-1}\geq1\}}\hat{\varepsilon}_{i-1}\big)'$			
LM_2	$z_{ai}^2 = \big(z_{ai}^{1\prime},\ \mathbb{1}_{\{\hat{\varepsilon}_{i-2}<1\}},\ \mathbb{1}_{\{\hat{\varepsilon}_{i-2}<1\}}\hat{\varepsilon}_{i-2},\ \mathbb{1}_{\{\hat{\varepsilon}_{i-2}\geq1\}}\hat{\varepsilon}_{i-2},\ \big)'$			
NL_1	bins for $\hat{\varepsilon}_{i-1}:\ (0,0.2),\ldots,(1.6,1.8),(2.0,2.5),\ldots,(3.5,4.0)$			
NL_2	bins for $\hat{\varepsilon}_{i-1}$ and $\hat{\varepsilon}_{i-2}:\ (0,0.2),\ldots,(1.6,1.8),(2.0,2.5),\ldots,(3.5,4.0)$			
	CM tests			
$CM.^1$	$r(\cdot)=\hat{\varepsilon}_i-1$			
$CM.^2$	$r(\cdot)=x_i-\hat{\Psi}_i$			
CM_1	$w_1(\cdot)=\big(x_{i-1},\ x_{i-1}^2,\ x_{i-1}^3,\ \hat{\varepsilon}_{i-1},\ \hat{\varepsilon}_{i-1}^2,\ \hat{\varepsilon}_{i-1}^3\big)'$			
CM_2	$w_2(\cdot)=\big(w_1(\cdot)',\ x_{i-2},\ x_{i-2}^2,\ x_{i-2}^3,\ \hat{\varepsilon}_{i-2}\,\hat{\varepsilon}_{i-2}^2,\ \hat{\varepsilon}_{i-2}^3\big)'$			
CM_3	$w_3(\cdot)=\big(x_{i-1},\ z_{ai}^{1\prime}\big)'$			
CM_4	$w_4(\cdot)=\big(x_{i-1},\ z_{ai}^{1\prime},\ x_{i-2},\ z_{ai}^{2\prime}\big)'$			
CM_5	$w_5(\cdot)=(x_{i-1},x_{i-2},\ldots,x_{i-10})'$			
	ICM tests $(M^{ICM}=100)$			
ICM_1	$r(\cdot)=x_i-\hat{\Psi}_i$	$w(\cdot)=1$	$\mathfrak{A}=0$	$\mathfrak{B}=1$ $L=1$
ICM_2	$r(\cdot)=x_i-\hat{\Psi}_i$	$w(\cdot)=\big(x_{i-1},\ x_{i-1}^2\big)'$	$\mathfrak{A}=0$	$\mathfrak{B}=0.1$ $L=1$
ICM_3	$r(\cdot)=x_i-\hat{\Psi}_i$	$w(\cdot)=\big(x_{i-1},\ x_{i-1}^2\big)'$	$\mathfrak{A}=0$	$\mathfrak{B}=0.5$ $L=1$
ICM_4	$r(\cdot)=\hat{\varepsilon}_i-1$	$w(\cdot)=\big(x_{i-1},\ x_{i-1}^2\big)'$	$\mathfrak{A}=0$	$\mathfrak{B}=0.1$ $L=1$
ICM_5	$r(\cdot)=x_i-\hat{\Psi}_i$	$w(\cdot)=1$	$\mathfrak{A}=0$	$\mathfrak{B}=1$ $L=5$
ICM_6	$r(\cdot)=x_i-\hat{\Psi}_i$	$w(\cdot)=\big(x_{i-1},\ x_{i-1}^2\big)'$	$\mathfrak{A}=0$	$\mathfrak{B}=0.1$ $L=50$
ICM_7	$r(\cdot)=\hat{\varepsilon}_i-1$	$w(\cdot)=\big(x_{i-1},\ x_{i-1}^2\big)'$	$\mathfrak{A}=0$	$\mathfrak{B}=0.1$ $L=50$

Therefore, we investigate the impact of the choice of $d=L$ on the power of the test. We compare the corresponding ICM test statistics computed based on predetermined lag values $L=1$, $L=5$ and $L=50$. Moreover, we examine the influence of the choice of the parameters $\mathfrak{A}$ and $\mathfrak{B}$ defining the space Ξ. The crucial point is that we cannot derive a generally applicable rule to determine the parameters $\mathfrak{A}$ and $\mathfrak{B}$. Therefore, we choose the parameter values similar to de Jong (1996), i.e., we fix $\mathfrak{A}$ to zero and choose different values of $\mathfrak{B}$ varying between 0.1 and 1.

Tables 10 and 11 show the actual rejection rates of the particular test statistics based on the 5% and 10% level. The first three rows display the average values of the ACD(1,1) parameters estimated for each replication and are associated with the dotted linear news impact functions depicted in the plots of Figures 28 and 29. Note that experiment (1) evaluates the *size* of the tests since the estimated model is in line with the assumed data generating process. It is shown that the estimated ACD(1,1) parameters indicate a slight small sample bias. For most of the tests, we find slight over-rejections, i.e., the size exceeds the nominal values. The best results are found for the

Table 10: Actual rejection frequencies of the applied diagnostic tests (see Table 9) based on the data generating processes (1)-(6) (see Table 8). Size of simulated samples: $n = 500$. Number of replications: $n^{MC} = 1000$. Estimated model: ACD(1,1).

	(1) ACD		(2) AMACD		(3) AMACD		(4) LACD		(5) BCACD		(6) BCACD	
	5%	10%	5%	10%	5%	10%	5%	10%	5%	10%	5%	10%
Median of estimated ACD(1,1) parameters												
ω	0.22	0.22	0.13	0.13	0.18	0.18	0.31	0.31	0.26	0.26	0.32	0.32
α	0.09	0.09	0.26	0.26	0.33	0.33	0.11	0.11	0.23	0.23	0.09	0.09
β	0.67	0.67	0.60	0.60	0.47	0.47	0.55	0.55	0.49	0.49	0.57	0.57
Rejection frequencies of LM tests and F-tests												
LM_1	0.07	0.13	0.05	0.11	0.13	0.22	0.16	0.25	0.15	0.23	0.07	0.13
LM_2	0.08	0.14	0.09	0.15	0.23	0.34	0.20	0.30	0.15	0.26	0.07	0.12
NL_1	0.05	0.10	0.05	0.09	0.06	0.11	0.09	0.18	0.09	0.16	0.06	0.11
NL_2	0.05	0.10	0.05	0.10	0.05	0.09	0.09	0.17	0.08	0.15	0.06	0.11
Rejection frequencies of CM tests												
CM_1^1	0.05	0.11	0.11	0.19	0.32	0.45	0.16	0.26	0.13	0.24	0.08	0.15
CM_1^2	0.05	0.11	0.12	0.21	0.33	0.47	0.15	0.24	0.13	0.22	0.08	0.14
CM_2^1	0.06	0.13	0.13	0.23	0.28	0.43	0.19	0.32	0.14	0.25	0.08	0.16
CM_2^2	0.06	0.12	0.11	0.22	0.26	0.40	0.18	0.30	0.13	0.24	0.07	0.16
CM_3^1	0.06	0.13	0.13	0.21	0.36	0.48	0.18	0.28	0.20	0.30	0.08	0.15
CM_3^2	0.06	0.12	0.14	0.25	0.30	0.46	0.18	0.29	0.18	0.28	0.08	0.16
CM_4^1	0.08	0.14	0.15	0.23	0.35	0.47	0.21	0.33	0.19	0.32	0.09	0.16
CM_4^2	0.07	0.14	0.13	0.22	0.28	0.42	0.19	0.32	0.18	0.31	0.09	0.17
CM_5^1	0.10	0.16	0.14	0.22	0.31	0.44	0.12	0.22	0.12	0.22	0.10	0.18
CM_5^2	0.09	0.16	0.13	0.22	0.30	0.42	0.12	0.22	0.12	0.22	0.10	0.19
Rejection frequencies of ICM tests												
ICM_1	0.04	0.08	0.14	0.23	0.26	0.36	0.13	0.21	0.05	0.11	0.06	0.12
ICM_2	0.05	0.09	0.13	0.22	0.28	0.40	0.14	0.23	0.04	0.10	0.07	0.13
ICM_3	0.06	0.12	0.15	0.24	0.25	0.35	0.17	0.25	0.05	0.12	0.10	0.18
ICM_4	0.04	0.08	0.11	0.20	0.27	0.37	0.13	0.22	0.02	0.06	0.06	0.11
ICM_5	0.04	0.09	0.15	0.25	0.30	0.41	0.13	0.21	0.06	0.13	0.06	0.13
ICM_6	0.05	0.08	0.11	0.18	0.28	0.39	0.14	0.22	0.04	0.09	0.07	0.13
ICM_7	0.03	0.06	0.09	0.16	0.26	0.37	0.13	0.21	0.02	0.06	0.06	0.12

ICM tests which reveal a size rather close to the nominal values of 5% and 10%, respectively. Experiments (2)-(11) allow us to gain insights into the *power* properties of the particular tests. In general, the power of the tests is relatively low. However, it should be noted that the sample size used ($n = 500$) is quite small for typical financial duration series. Moreover, most experiments are based on data generating processes that are quite realistic for typical financial duration series. Thus, the strength of the misspecification is chosen

Table 11: Actual rejection frequencies of the applied diagnostic tests (see Table 9) based on the data generating processes (7)-(11) (see Table 8). Size of simulated samples: $n = 500$. Number of replications: $n^{MC} = 1000$. Estimated model: ACD(1,1).

	(7) EXACD		(8) EXACD		(9) HACD		(10) NPACD		(11) NPACD	
	5%	10%	5%	10%	5%	10%	5%	10%	5%	10%
Median of estimated ACD(1,1) parameters										
ω	0.33	0.33	0.34	0.34	0.33	0.33	0.32	0.32	0.33	0.33
α	0.16	0.16	0.11	0.11	0.01	0.01	0.08	0.08	0.07	0.07
β	0.50	0.50	0.52	0.52	0.64	0.64	0.57	0.57	0.57	0.57
Rejection frequencies of LM tests and F-tests										
LM_1	0.17	0.28	0.67	0.79	0.13	0.20	0.10	0.17	0.15	0.25
LM_2	0.21	0.31	0.74	0.84	0.13	0.22	0.12	0.20	0.21	0.29
NL_1	0.10	0.18	0.35	0.50	0.07	0.12	0.08	0.14	0.06	0.10
NL_2	0.11	0.17	0.36	0.47	0.07	0.12	0.08	0.14	0.08	0.12
Rejection frequencies of CM tests										
CM_1^1	0.16	0.25	0.58	0.70	0.10	0.18	0.10	0.21	0.14	0.25
CM_1^2	0.14	0.24	0.57	0.70	0.10	0.18	0.10	0.21	0.13	0.23
CM_2^1	0.18	0.29	0.65	0.78	0.11	0.20	0.13	0.24	0.18	0.30
CM_2^2	0.16	0.29	0.64	0.76	0.11	0.19	0.14	0.23	0.17	0.28
CM_3^1	0.19	0.29	0.66	0.77	0.11	0.18	0.13	0.20	0.17	0.27
CM_3^2	0.19	0.28	0.66	0.77	0.11	0.19	0.13	0.21	0.16	0.25
CM_4^1	0.22	0.34	0.74	0.84	0.12	0.21	0.14	0.24	0.23	0.34
CM_4^2	0.23	0.33	0.74	0.84	0.11	0.21	0.13	0.22	0.20	0.31
CM_5^1	0.15	0.25	0.36	0.49	0.12	0.19	0.18	0.27	0.26	0.39
CM_5^2	0.14	0.24	0.34	0.47	0.11	0.19	0.17	0.27	0.25	0.37
Rejection frequencies of ICM tests										
ICM_1	0.16	0.25	0.43	0.56	0.04	0.08	0.14	0.22	0.15	0.21
ICM_2	0.17	0.27	0.52	0.62	0.05	0.09	0.13	0.19	0.12	0.17
ICM_3	0.19	0.28	0.50	0.60	0.07	0.11	0.15	0.22	0.15	0.20
ICM_4	0.18	0.26	0.50	0.62	0.06	0.09	0.12	0.18	0.11	0.17
ICM_5	0.16	0.25	0.42	0.55	0.04	0.07	0.14	0.22	0.15	0.21
ICM_6	0.17	0.27	0.52	0.63	0.05	0.09	0.12	0.19	0.13	0.18
ICM_7	0.16	0.25	0.51	0.62	0.05	0.09	0.12	0.19	0.12	0.17

in a realistic way. The lowest rejection frequencies are found for the second Box-Cox type data generating process (experiment (6)). This result is not surprising since this process is associated with a news response function that is relatively close to a linear one (see Figure 28), and thus is well fitted by a linear ACD model. The highest power for the particular tests is found in experiment (8), which is based on an EXACD data generating process. This is due to the fact that the assumed news impact function is kinked upwards (downwards) for values $\hat{\varepsilon}_{i-1} < 1$ ($\hat{\varepsilon}_{i-1} \geq 1$), and thus is strongly misspecified. Interestingly, we also find comparable high rejection frequencies for DGP (3), which is associated with an AMACD process based on additive and multiplicative components. This result indicates that the distinction between additive and multiplicative stochastic components can be an important issue and a source of model misspecification. For all other experiments, we find relatively similar rejection rates. It turns out that especially nonlinearities in the *tails* of the news impact curve imply misspecifications that are more easy detected. This finding is especially illustrative for DGP (9), which implies a news impact function that is highly nonlinear for values of $\hat{\varepsilon}_{i-1}$ around 1, but tends to linearity for very small and very large past innovations. Therefore, even though this news response curve is highly nonlinear, it seems to be more difficult to detect this form of model misspecification.

By comparing the power of the particular tests, it is shown that the CM test generally has the highest power. The fact that particularly the CM test based on weighting function $w_4(\cdot)$ outperforms the other tests in most cases, illustrates that it is crucial to apply a specification test that is sensitive against nonlinearities in the news response function. We find slight differences between the CM tests using the conditional moment restriction (5.77) or (5.78). However, we do not find one specification to be superior. Furthermore, even though the LM test has optimal power against local alternatives, it underperforms the CM test in nearly all cases. Thus, for more general forms of misspecification, the power of the LM test seems to diminish. However, its power is the higher the stronger the nonlinearity of the news impact function. Not surprisingly, the LM test reveals comparably low rejection frequencies for the experiments based on AMACD and BACD data generating processes. The corresponding F-tests against nonlinearities in the news impact curve perform poorly. Even though they are based on a high number of bins and should be sensitive against nonlinearities, they indicate only a quite low power. The ICM test underperforms the CM test and the LM test in most cases. However, its power is satisfying for a sample size of $n = 500$ and for the chosen forms of model misspecifications. It turns out that the power of the test seems to be largely insensitive against the choice of the test parameters since no specification outperforms the other settings in general. Moreover, it shows a comparable high power in cases of model misspecifications that are *not* easily detected by sign bias variables, like, for example, in experiments (2), (3) and (6) based on AMACD and BACD type data generating processes.

Summarizing the Monte Carlo results, we can conclude that the CM test seems to be the most appropriate diagnostic tool to detect functional misspecification. Not surprisingly, the results also illustrate that this result depends on the choice of the weighting function. The same implication holds for the LM test which has only maximal power when the DGP is a local alternative. Hence, as illustrated in this study, a major advantage of the ICM test is that its power seems to be largely robust against the particular choice of the test parameters. Summing up, it turns out that it is reasonable to combine these particular types of diagnostic tests in order to detect different forms of model misspecification.

5.5 Applications of ACD Models

This section illustrates several applications of ACD models to analyze financial duration processes. Section 5.5.1 is devoted to the estimation and evaluation of different types of ACD models on the basis of trade and price durations. Here, several ACD specifications are evaluated by using the testing framework proposed in the previous subsections. In Section 5.5.2, we apply the ACD model to test the market microstructure hypotheses derived in Section 3.2.2 of Chapter 3. In this context, we analyze trade duration series based on different stocks and exchanges. Section 5.5.3 deals with the evaluation of (il)liquidity risks based on excess volume durations. By estimating Box-Cox ACD specifications based on seven different excess volume duration series from NYSE trading, we empirically check the hypotheses formulated in Section 3.4.4 of Chapter 3.

5.5.1 Evaluating ACD Models based on Trade and Price Durations

We estimate and evaluate several ACD specifications based on financial duration data from the NYSE. The sample covers the period from 01/02/01 to 05/31/01 and is extracted from the TAQ database. The basic data preparation is performed as described in Chapter 4, Sections 4.1 and 4.2.

We analyze the four stocks AOL, Coca-Cola, Disney and GE and focus on trade durations and price durations. In order to obtain price durations which are on average of comparable length we use \$0.10 midquote changes for the AOL stock and \$0.05 midquote changes for the other three stocks. In order to obtain comparable sample sizes, the trade durations are constructed using only ten trading days from 03/19/01 to 03/30/01. The particular duration series are seasonally adjusted by applying the two-step procedure as described in Section 5.2.4, where in the first stage a cubic spline function based on 30 minutes nodes is estimated. Descriptive statistics of plain durations, as well as seasonally adjusted durations for the four stocks are given in Table 12.

For each type of financial duration, we estimate seven different ACD specifications, the basic ACD model, the LACD$_2$ model, the BACD model, the EXACD model, the ABACD model, the AGACD model and a logarithmic version of the NPACD model. The particular time series are re-initialized every trading day, i.e., serial dependencies between observations of different trading days are excluded. Since the major focus of this study lies on the analysis of the conditional mean function, the particular specifications are estimated by QML. The lag orders are chosen according to the Bayes Information Criterion (BIC). For all duration series, except for Disney and GE price durations,

Table 12: Descriptive statistics and Ljung-Box statistics of plain and seasonally adjusted trade durations and price durations for the AOL, Coca-Cola, Disney and GE stock traded at the NYSE. Data extracted from the TAQ database, sample period from 01/02/01 to 05/31/01 for price durations and from 03/19/01 to 03/30/01 for trade durations.

| | AOL | | | | Coca-Cola | | | |
| | trade durations | | price durations | | trade durations | | price durations | |
	plain	adj.	plain	adj.	plain	adj.	plain	adj.
Obs	20988	20988	10083	10083	15174	15174	12971	12971
Mean	11.12	1.00	237.93	1.00	15.30	1.00	183.65	1.00
SD	13.63	1.22	302.96	1.14	19.23	1.25	248.43	1.28
LB(20)	851	723	2809	2138	232	138	3166	2318

| | Disney | | | | GE | | | |
| | trade durations | | price durations | | trade durations | | price durations | |
	plain	adj.	plain	adj.	plain	adj.	plain	adj.
Obs	16272	16272	9618	9618	25101	25101	16008	16008
Mean	14.33	1.00	248.00	1.00	8.88	1.00	150.49	0.99
SD	16.98	1.15	341.51	1.24	9.24	1.03	195.00	1.20
LB(20)	548	194	1842	823	875	803	8466	4827

Price durations for AOL stock based on $0.10 midquote changes and for the Coca-Cola, Disney and GE stock based on $0.05 midquote changes. Descriptive statistics of durations in seconds.

we find an ACD(2,1) parameterization as the best specification. For the Disney and GE price durations, we choose an ACD(2,2) specification. Note that for the particular types of nonlinear ACD specifications, the inclusion of a second lag is not straightforward because it also requires a parameterization of the second lag news impact function. Hence, it doubles the corresponding news response parameters which is not especially practicable for the high parameterized ABACD, AGACD and NPACD models. For this reason, we choose a parameterization that allows us, on the one hand, to account for higher order dynamics, but on the other hand, ensures model parsimony. In particular, we model the news impact of the second lag in the same way as for the first lag, i.e., based on the same parameters δ_1, δ_2, c, and double only the autoregressive parameters α, β, c and ν. Then, the AGACD(2,2) model is given by

$$\Psi_i^{\delta_1} = \omega + \sum_{j=1}^{2} \left\{ \alpha_j \Psi_{i-j}^{\delta_1} (|\varepsilon_{i-j} - b| + c_j(\varepsilon_{i-j} - b))^{\delta_2} \right. \tag{5.93}$$

$$\left. + \nu_j (|\varepsilon_{i-j} - b| + c_j(\varepsilon_{i-j} - b))^{\delta_2} \right\} + \sum_{j=1}^{2} \beta_j \Psi_{i-j}^{\delta_1}.$$

Correspondingly, the NPACD(2,2) model is given by

$$\Psi_i = \omega + \sum_{k=0}^{K^+} \alpha_k^+ \mathbb{1}_{\{\varepsilon_{i-1} > \bar{\varepsilon}_k\}} (\varepsilon_{i-1} - \bar{\varepsilon}_k) + \sum_{k=0}^{K^-} \alpha_j^- \mathbb{1}_{\{\varepsilon_{i-1} < \bar{\varepsilon}_k\}} (\varepsilon_{i-1} - \bar{\varepsilon}_j)$$

$$+ \sum_{k=0}^{K^+} (\alpha_j^+ + a^+) \mathbb{1}_{\{\varepsilon_{i-1} \geq \bar{\varepsilon}_k\}} (\varepsilon_{i-1} - \bar{\varepsilon}_j)$$

$$+ \sum_{k=0}^{K^-} (\alpha_j^- + a^-) \mathbb{1}_{\{\varepsilon_{i-1} < \bar{\varepsilon}_k\}} (\varepsilon_{i-1} - \bar{\varepsilon}_j) + \sum_{j=1}^{2} \beta_j \Psi_{i-j}. \tag{5.94}$$

Tables 13 through 20 give the estimation results as well as the diagnostics for the different types of financial durations. Two difficulties have to be considered in this context. First, for nonlinear models involving absolute value functions in the news impact function, the estimation of the Hessian matrix is quite cumbersome because of numerical problems. For this reason, we estimate the asymptotic standard errors by the OPG estimator of the variance covariance matrix. Second, in several specifications, the coefficients δ_1 and c converge to their boundaries. In these cases, the coefficients are set to their boundaries, and the restricted models are re-estimated. In this context, see the discussion in Subsection 5.3.1.

By analyzing the estimation results, the following findings can be summarized. First, for trade durations, the innovation parameters are quite low, while the persistence parameters are close to one. Note that we do not impose explicit non-negativity restrictions on the autoregressive parameters. For this reason, even negative values for α_1 are obtained in several regressions. However, in these cases, they are overcompensated for by positive values for α_2.[28] Nonetheless, based on the estimations we do not find any violations of the non-negativity restriction of Ψ_i. A strong persistence is revealed also for price durations, while the innovation component is clearly higher than for trade durations.

Second, by comparing the goodness-of-fit of the specifications based on the BIC values, we find the best performance for EXACD and BACD models. Especially for price durations, the more simple (linear and logarithmic) models are rejected in favor of the (A)BACD and EXACD model. For trade durations, no clear picture is revealed. For the AOL and the Coca-Cola stock, the AGACD and EXACD model are the best specification, while for the Disney and the GE stock, the basic ACD model leads to the highest BIC. Nonetheless, for all specifications, we observe the strongest increase of the log-likelihood function when the (L)ACD model is extended to a BACD or EXACD model. This result illustrates that for both types of financial durations, it is crucial to account for nonlinear news impact effects. However, the NPACD model leads

---

[28] In this context, it has to be noted that in a nonlinear ACD specification, an upward kinked concave news impact function actually implies a negative value for α; see also Subsection 5.3.1.

Table 13: QML estimates of various types of augmented ACD models for AOL trade durations. Sample perido 03/19/01 to 03/30/01. Standard errors based on OPG estimates (in parantheses). NPACD model estimated based on the category bounds $(0.1, 0.2, 0.5, 1.0, 1.5, 2.0, 3.0)$ with $\bar\varepsilon_0 = 1.0$. Diagnostics: Log Likelihood (LL), Bayes Information Criterion (BIC), mean $(\bar{\bar\varepsilon}_i)$, standard deviation (S.D.) and Ljung-Box statistic with respect to 20 lags (LB) of ACD residuals.

	ACD	LACD	BACD	EXACD	ABACD	AGACD		NPACD
ω	0.009	-0.022	-0.024	-0.024	-0.006	-0.001	ω	-0.014
	(0.001)	(0.001)	(0.007)	(0.002)	(0.010)	(0.012)		(0.004)
α_1	-0.005	-0.003	-0.002	-0.020	0.013	-0.102		
	(0.005)	(0.005)	(0.007)	(0.009)	(0.015)	(0.023)		
α_2	0.030	0.025	0.036	0.038	0.021	0.185		
	(0.005)	(0.005)	(0.011)	(0.009)	(0.013)	(0.037)		
β_1	0.967	0.991	0.990	0.989	0.984	0.935	β_1	0.981
	(0.002)	(0.001)	(0.002)	(0.002)	(0.002)	(0.019)		(0.002)
δ_1			1.585		0.703	1.897	α_1^+	0.030
			(0.350)		(0.332)	(0.662)		(0.010)
δ_2			1.055		0.611	0.647	α_2^+	-0.017
			(0.077)		(0.079)	(0.074)		(0.019)
b					0.569	0.499	α_3^+	0.003
					(0.034)	(0.027)		(0.016)
c_1				0.029	-0.915	1.000**	α_4^+	-0.017
				(0.012)	(0.286)	-**		(0.007)
c_2				-0.022	0.890	0.061	α_1^-	-0.037
				(0.012)	(0.632)	(0.144)		(0.014)
ν_1						0.091	α_2^-	-0.076
						(0.021)		(0.020)
ν_2						-0.077	α_3^-	0.029
						(0.033)		(0.048)
							α_4^-	-0.448
								(0.199)
							a^+	0.090
								(0.025)
							a^-	0.002
								(0.004)
Obs	20988	20988	20988	20988	20988	20988		20988
LL	-20576	-20578	-20572	-20571	-20554	-20540		-20545
BIC	-20596	-20598	-20602	-20601	-20599	-20589		-20605
$\bar{\bar\varepsilon}_i$	1.000	1.001	1.001	1.002	1.001	1.001		1.001
S.D.	1.164	1.164	1.162	1.161	1.154	1.152		1.154
LB	32.179	33.222	31.937	32.915	30.590	29.072		29.640

**: Parameter set to boundary.

Table 14: QML estimates of various types of augmented ACD models for AOL $0.10 price durations. Sample period 01/02/01 to 05/31/01. Standard errors based on OPG estimates (in parantheses). NPACD model estimated based on the category bounds $(0.1, 0.2, 0.5, 1.0, 1.5, 2.0, 3.0)$ with $\bar{\varepsilon}_0 = 1.0$. Diagnostics: Log Likelihood (LL), Bayes Information Criterion (BIC), mean $(\bar{\bar{\varepsilon}}_i)$, standard deviation (S.D.) and Ljung-Box statistic with respect to 20 lags (LB) of ACD residuals.

	ACD	LACD	BACD	EXACD	ABACD	AGACD		NPACD
ω	0.029	-0.080	-0.512	-0.066	0.021		ω	0.036
	(0.005)	(0.005)	(0.215)	(0.006)	(0.009)			(0.014)
α_1	0.129	0.106	0.848	0.190	-0.233			
	(0.013)	(0.011)	(0.340)	(0.014)	(0.035)			
α_2	-0.039	-0.027	-0.304	-0.071	0.130			
	(0.013)	(0.011)	(0.137)	(0.015)	(0.030)			
β_1	0.883	0.967	0.968	0.967	0.968		β_1	0.972
	(0.010)	(0.005)	(0.005)	(0.006)	(0.005)			(0.005)
δ_1			-*		-*		α_1^+	0.035
			-*		-*			(0.026)
δ_2			0.164		1.000**		α_2^+	-0.037
			(0.071)		-**			(0.053)
b					0.697		α_3^+	0.059
					(0.061)			(0.047)
c_1				-0.161	-1.172		α_4^+	-0.027
				(0.020)	(0.069)			(0.024)
c_2				0.084	-0.934		α_1^-	0.245
				(0.020)	(0.090)			(0.028)
ν_1							α_2^-	0.211
								(0.057)
ν_2							α_3^-	0.463
								(0.177)
							α_4^-	0.064
								(0.410)
							a^+	-0.366
								(0.040)
							a^-	-0.008
								(0.009)
Obs	20988	20988	20988	20988	20988			20988
LL	-20576	-20578	-20572	-20571	-20554			-20545
BIC	-20596	-20598	-20602	-20601	-20599			-20605
$\bar{\bar{\varepsilon}}_i$	1.005	1.005	1.005	1.008	1.004			1.006
S.D.	1.051	1.051	1.047	1.047	1.043			1.045
LB	10.056	9.867	13.955	10.623	11.453			12.427

*: Estimation based on a logarithmic specification.
**: Parameter set to boundary.

Table 15: QML estimates of various types of augmented ACD models for Coca-Cola trade durations. Sample period 03/19/01 to 03/30/01. Standard errors based on OPG estimates (in parantheses). NPACD model estimated based on the category bounds $(0.1, 0.2, 0.5, 1.0, 1.5, 2.0, 3.0)$ with $\bar{\varepsilon}_0 = 1.0$. Diagnostics: Log Likelihood (LL), Bayes Information Criterion (BIC), mean ($\bar{\hat{\varepsilon}}_i$), standard deviation (S.D.) and Ljung-Box statistic with respect to 20 lags (LB) of ACD residuals.

	ACD	LACD	BACD	EXACD	ABACD	AGACD		NPACD
ω	0.122	-0.038	0.100	-0.041	0.010		ω	-0.036
	(0.016)	(0.003)	(0.043)	(0.004)	(0.010)			(0.011)
α_1	-0.024	-0.025	-0.026	-0.075	0.066			
	(0.004)	(0.004)	(0.038)	(0.009)	(0.011)			
α_2	0.065	0.063	0.058	0.104	-0.062			
	(0.005)	(0.005)	(0.087)	(0.009)	(0.011)			
β_1	0.837	0.869	0.872	0.868	0.871		β_1	0.899
	(0.019)	(0.017)	(0.018)	(0.018)	(0.018)			(0.014)
δ_1			0.483		-*		α_1^+	0.062
			(0.736)		-*			(0.021)
δ_2			0.672		1.000**		α_2^+	-0.071
			(0.089)		-**			(0.039)
b					1.420		α_3^+	0.031
					(0.203)			(0.031)
c_1				0.080	-0.773		α_4^+	-0.033
				(0.012)	(0.119)			(0.013)
c_2				-0.068	-1.393		α_1^-	-0.124
				(0.013)	(0.187)			(0.020)
ν_1							α_2^-	-0.153
								(0.038)
ν_2							α_3^-	0.245
								(0.112)
							α_4^-	-0.863
								(0.335)
							a^+	0.230
								(0.026)
							a^-	0.009
								(0.004)
Obs	15174	15174	15174	15174	15174			15174
LL	-15059	-15060	-15056	-15047	-15045			-15038
BIC	-15079	-15079	-15085	-15075	-15079			-15095
$\bar{\hat{\varepsilon}}_i$	1.000	1.000	1.000	1.000	1.000			1.000
S.D.	1.245	1.246	1.249	1.245	1.246			1.242
LB	17.146	15.823	16.083	15.254	15.761			17.527

*: Estimation based on a logarithmic specification.
**: Parameter set to boundary.

Table 16: QML estimates of various types of augmented ACD models for Coca-Cola \$0.05 price durations. Sample period 01/02/01 to 05/31/01. Standard errors based on OPG estimates (in parantheses). NPACD model estimated based on the category bounds $(0.1, 0.2, 0.5, 1.0, 1.5, 2.0, 3.0)$ with $\bar{\varepsilon}_0 = 1.0$. Diagnostics: Log Likelihood (LL), Bayes Information Criterion (BIC), mean $(\bar{\hat{\varepsilon}}_i)$, standard deviation (S.D.) and Ljung-Box statistic with respect to 20 lags (LB) of ACD residuals.

	ACD	LACD	BACD	EXACD	ABACD	AGACD		NPACD	
ω	0.012	-0.050	-0.004	-0.036	0.035		ω	0.020	
	(0.002)	(0.003)	(0.027)	(0.003)	(0.007)			(0.009)	
α_1	0.079	0.075	0.037	0.183	-0.170				
	(0.005)	(0.005)	(0.061)	(0.011)	(0.032)				
α_2	-0.020	-0.025	-0.019	-0.113	0.132				
	(0.003)	(0.004)	(0.031)	(0.010)	(0.025)				
β_1	0.930	0.988	0.987	0.990	0.991		β_1	0.989	
	(0.004)	(0.002)	(0.002)	(0.002)	(0.002)			(0.002)	
δ_1			0.081		-*		α_1^+	0.021	
			(0.135)		-*			(0.018)	
δ_2			0.275		1.096		α_2^+	0.010	
			(0.044)		(0.137)			(0.034)	
b					1.016		α_3^+	-0.034	
					(0.103)			(0.029)	
c_1				-0.179	-1.000**		α_4^+	0.006	
				(0.013)	-**			(0.014)	
c_2				0.136	-0.801		α_1^-	0.226	
				(0.013)	(0.035)			(0.020)	
ν_1							α_2^-	0.246	
								(0.038)	
ν_2							α_3^-	0.324	
								(0.101)	
							α_4^-	-0.157	
								(0.199)	
							a^+	-0.394	
								(0.028)	
							a^-	0.005	
								(0.003)	
Obs	12971	12971	12971	12971	12971			12971	
LL	-12072	-12081	-12016	-11996	-11996			-11991	
BIC	-12091	-12100	-12044	-12024	-12029			-12048	
$\bar{\hat{\varepsilon}}_i$	1.008	1.008	1.033	1.005	1.005			1.004	
S.D.	1.208	1.210	1.234	1.196	1.196			1.194	
LB	30.816	31.776	26.809	23.016	24.431			23.217	

*: Estimation based on a logarithmic specification.
**: Parameter set to boundary.

to the overall highest log-likelihood values for all series indicating the best fit to the data.

Third, the estimated Box-Cox parameters $\hat{\delta}_1$ and $\hat{\delta}_2$ are almost always lower than one for price durations, while for trade durations, values greater and less than one are obtained. These results lend opposition to the linear ACD and, in most cases, also oppose the LACD model. Hence, we notice that price duration processes imply concave news impact curves, i.e., the adjustments of the conditional expected mean are stronger in periods of smaller than expected price durations (volatility shocks) than in periods with unexpectedly low price intensities. Corresponding results are found based on the EXACD model because the mostly highly significant negative parameters for c imply upward kinked concave shaped news response curves. For trade durations, the picture is less clear since we obtain evidence for concave, as well as convex news impact curves.

Fourth, in most cases only restricted versions of the augmented ACD models are estimated since either δ_1 tends to zero and/or $|c|$ tends to one. Since most duration series seem to imply a concave news impact function, it is not surprising that the second restriction is binding for nearly all series. In the first case, the model is estimated under the restriction $\delta \to 0$, which is performed by estimating the model based on a logarithmic transformation. Note that this has consequences for the estimates of ω, α and ν since a logarithmic model belongs to the class of Box-Cox models and not to the class of power ACD models, which imply parameter transformations from ω, α and c to $\tilde{\omega}$, $\tilde{\alpha}$ and $\tilde{c}$ as illustrated in Subsection 5.3.1. In these cases, the AGACD and the ABACD models coincide. Thus, this provides evidence against the HACD and in favor of the ABACD model. In the second case, two restricted versions of the model are re-estimated: one specification under the restriction $|c| = 1$ and one model under $\delta_2 = 1$. Then, we choose the specification that leads to the higher log-likelihood value. Based on our estimates, it turns out that in general, neither of the restricted models outperforms the other. In general, we observe that the extension from the BACD/EXACD model to the ABACD model is supported by the data since we find an increase in the log-likelihood function and significant coefficients in most cases. Hence, further flexibility of the news impact function seems to be an important issue. Nevertheless, the additional flexibility of the AGACD model seems not to be required in most cases. However, for AOL and GE trade durations, we observe significant values for ν. This lends support for the need for additive, as well as multiplicative stochastic factors. Especially for the AOL stock, this extra flexibility leads to a strong increase in the log likelihood function and also the BIC value.

Fifth, the NPACD model is estimated using the categorization $\{0.1, 0.2, 0.5, 1.0, 1.5, 2.0, 3.0\}$ with $\bar{\varepsilon}_0 = 1.0$ which allows for more flexibility concerning very small and very large innovations. For most of the thresholds, we obtain significant estimates. However, since these coefficients belong to a spline function, it is not useful to interpret them separately.

Table 17: QML estimates of various types of augmented ACD models for Disney trade durations. Sample period 03/19/01 to 03/30/01. Standard errors based on OPG estimates (in parantheses). NPACD model estimated based on the category bounds $(0.1, 0.2, 0.5, 1.0, 1.5, 2.0, 3.0)$ with $\bar{\varepsilon}_0 = 1.0$. Diagnostics: Log Likelihood (LL), Bayes Information Criterion (BIC), mean $(\bar{\hat{\varepsilon}}_i)$, standard deviation (S.D.) and Ljung-Box statistic with respect to 20 lags (LB) of ACD residuals.

	ACD	LACD	BACD	EXACD	ABACD	AGACD		NPACD
ω	0.034	-0.023	-0.065	-0.023	-0.020	-0.026	ω	0.010
	(0.006)	(0.003)	(0.028)	(0.003)	(0.021)	(0.017)		(0.007)
α_1	0.018	0.017	0.076	0.038	0.050	0.070		
	(0.006)	(0.006)	(0.031)	(0.011)	(0.023)	(0.052)		
α_2	0.007	0.006	0.042	-0.009	0.025	-0.037		
	(0.006)	(0.006)	(0.029)	(0.011)	(0.019)	(0.046)		
β_1	0.941	0.970	0.954	0.968	0.956	0.916	β_1	0.969
	(0.008)	(0.006)	(0.008)	(0.006)	(0.008)	(0.047)		(0.006)
δ_1			3.268		2.920	4.996	α_1^+	0.012
			(0.966)		(1.061)	(2.335)		(0.014)
δ_2			0.881		0.741	0.895	α_2^+	-0.034
			(0.126)		(0.104)	(0.148)		(0.027)
b					0.510	0.362	α_3^+	0.041
					(0.104)	(0.150)		(0.023)
c_1				-0.034	1.000**	1.000**	α_4^+	-0.021
				(0.014)	-**	-**		(0.010)
c_2				0.026	1.000**	1.000**	α_1^-	0.069
				(0.014)	-**	-**		(0.019)
ν_1						-0.003	α_2^-	0.019
						(0.035)		(0.028)
ν_2						0.065	α_3^-	-0.115
						(0.035)		(0.086)
							α_4^-	1.057
								(0.354)
							a^+	-0.082
								(0.031)
							a^-	0.006
								(0.004)
Obs	16272	16272	16272	16272	16272	16272		16272
LL	-16147	-16147	-16141	-16144	-16141	-16140		-16138
BIC	-16166	-16166	-16170	-16173	-16175	-16183		-16197
$\bar{\hat{\varepsilon}}_i$	1.002	1.001	1.001	1.001	1.001	1.001		0.999
S.D.	1.135	1.134	1.134	1.135	1.135	1.133		1.131
LB	20.796	21.195	19.971	21.438	20.126	19.861		20.288

**: Parameter set to boundary.

Table 18: QML estimates of various types of augmented ACD models for Disney \$0.05 price durations. Sample period 01/02/01 to 05/31/01. Standard errors based on OPG estimates (in parantheses). NPACD model estimated based on the category bounds $(0.1, 0.2, 0.5, 1.0, 1.5, 2.0, 3.0)$ with $\bar{\varepsilon}_0 = 1.0$. Diagnostics: Log Likelihood (LL), Bayes Information Criterion (BIC), mean $(\bar{\bar{\varepsilon}}_i)$, standard deviation (S.D.) and Ljung-Box statistic with respect to 20 lags (LB) of ACD residuals.

	ACD	LACD	BACD	EXACD	ABACD	AGACD		NPACD
ω	0.055	-0.045	-0.038	-0.032	-0.029	-0.046	ω	0.040
	(0.010)	(0.008)	(0.018)	(0.005)	(0.015)	(0.025)		(0.011)
α_1	0.170	0.136	0.265	0.210	0.180	-0.125		
	(0.011)	(0.008)	(0.066)	(0.012)	(0.045)	(0.125)		
α_2	-0.089	-0.092	-0.184	-0.147	-0.124	0.056		
	(0.013)	(0.009)	(0.048)	(0.012)	(0.032)	(0.100)		
β_1	0.983	1.291	1.296	1.302	1.286	1.429	β_1	1.245
	(0.099)	(0.085)	(0.069)	(0.070)	(0.071)	(0.165)		(0.071)
β_2	-0.116	-0.323	-0.329	-0.329	-0.321	-0.389	β_2	-0.277
	(0.080)	(0.080)	(0.065)	(0.066)	(0.066)	(0.145)		(0.067)
δ_1			0.518		0.535	0.047	α_1^+	-0.018
			(0.138)		(0.139)	(0.328)		(0.021)
δ_2			0.339		0.383	0.355	α_2^+	0.170
			(0.052)		(0.045)	(0.052)		(0.041)
b					0.020	0.021	α_3^+	-0.112
					(0.011)	(0.008)		(0.035)
c_1				-0.134	1.000**	1.000**	α_4^+	0.039
				(0.016)	-**	-**		(0.015)
c_2				0.092	1.000**	1.000**	α_1^-	0.286
				(0.015)	-**	-**		(0.023)
ν_1						0.142	α_2^-	0.193
						(0.062)		(0.040)
ν_2						-0.068	α_3^-	0.380
						(0.063)		(0.113)
							α_4^-	-0.073
								(0.222)
							a^+	-0.440
								(0.031)
							a^-	-0.036
								(0.006)
Obs	9618	9618	9618	9618	9618	9618		9618
LL	-9039	-9048	-8986	-8999	-8985	-8982		-8986
BIC	-9062	-9071	-9018	-9031	-9021	-9028		-9046
$\bar{\bar{\varepsilon}}_i$	1.004	1.004	1.004	1.003	1.003	1.004		0.998
S.D.	1.195	1.194	1.194	1.195	1.192	1.193		1.186
LB	33.222	29.055	30.176	31.262	30.373	30.403		32.697

**: Parameter set to boundary.

Table 19: QML estimates of various types of augmented ACD models for GE trade durations. Sample period 03/19/01 to 03/30/01. Standard errors based on OPG estimates (in parantheses). NPACD model estimated based on the category bounds $(0.1, 0.2, 0.5, 1.0, 1.5, 2.0, 3.0)$ with $\bar{\varepsilon}_0 = 1.0$. Diagnostics: Log Likelihood (LL), Bayes Information Criterion (BIC), mean $(\bar{\hat{\varepsilon}}_i)$, standard deviation (S.D.) and Ljung-Box statistic with respect to 20 lags (LB) of ACD residuals.

	ACD	LACD	BACD	EXACD	ABACD	AGACD		NPACD
ω	0.036	-0.038	-0.003	-0.037	0.017	0.005	ω	-0.007
	(0.005)	(0.003)	(0.030)	(0.003)	(0.024)	(0.022)		(0.007)
α_1	-0.001	-0.001	-0.000	-0.014	0.016	-0.091		
	(0.006)	(0.006)	(0.006)	(0.010)	(0.018)	(0.037)		
α_2	0.043	0.038	0.042	0.056	0.023	0.162		
	(0.006)	(0.006)	(0.033)	(0.010)	(0.024)	(0.069)		
β_1	0.922	0.965	0.964	0.966	0.960	0.907	β_1	0.962
	(0.007)	(0.005)	(0.005)	(0.005)	(0.005)	(0.044)		(0.005)
δ_1			0.705		0.497	2.078	α_1^+	0.039
			(0.558)		(0.528)	(1.223)		(0.014)
δ_2			0.758		0.595	0.702	α_2^+	-0.021
			(0.107)		(0.078)	(0.098)		(0.027)
b					0.444	0.430	α_3^+	-0.014
					(0.030)	(0.052)		(0.023)
c_1				0.023	-0.956	1.000**	α_4^+	-0.008
				(0.014)	(0.143)	-**		(0.012)
c_2				-0.031	1.000**	0.573	α_1^-	-0.030
				(0.014)	-**	(0.398)		(0.019)
ν_1						0.084	α_2^-	-0.048
						(0.038)		(0.030)
ν_2						-0.046	α_3^-	-0.227
						(0.052)		(0.096)
							α_4^-	0.636
								(1.290)
							a^+	0.097
								(0.031)
							a^-	0.006
								(0.005)
Obs	25101	25101	25101	25101	25101	25101		25101
LL	-24813	-24815	-24811	-24813	-24803	-24800		-24801
BIC	-24834	-24836	-24841	-24843	-24844	-24851		-24862
$\bar{\hat{\varepsilon}}_i$	1.001	1.000	0.999	1.000	1.000	1.000		1.001
S.D.	1.007	1.007	1.006	1.007	1.006	1.006		1.005
LB	38.695	37.281	36.779	37.039	35.826	40.831		36.380

**: Parameter set to boundary.

Table 20: QML estimates of various types of augmented ACD models for GE $0.05 price durations. Sample period 01/02/01 to 05/31/01. Standard errors based on OPG estimates (in parantheses). NPACD model estimated based on the category bounds $(0.1, 0.2, 0.5, 1.0, 1.5, 2.0, 3.0)$ with $\bar{\varepsilon}_0 = 1.0$. Diagnostics: Log Likelihood (LL), Bayes Information Criterion (BIC), mean ($\bar{\hat{\varepsilon}}_i$), standard deviation (S.D.) and Ljung-Box statistic with respect to 20 lags (LB) of ACD residuals.

	ACD	LACD	BACD	EXACD	ABACD	AGACD		NPACD
ω	0.009	-0.030	-0.093	-0.034	-0.016		ω	0.012
	(0.002)	(0.004)	(0.017)	(0.005)	(0.003)			(0.006)
α_1	0.143	0.122	0.372	0.174	-0.389			
	(0.009)	(0.007)	(0.049)	(0.011)	(0.062)			
α_2	-0.094	-0.092	-0.269	-0.128	0.325			
	(0.007)	(0.006)	(0.036)	(0.010)	(0.054)			
β_1	1.227	1.532	1.434	1.433	1.408		β_1	1.391
	(0.073)	(0.057)	(0.063)	(0.067)	(0.066)			(0.068)
β_2	-0.284	-0.538	-0.442	-0.442	-0.417		β_2	-0.400
	(0.066)	(0.056)	(0.062)	(0.066)	(0.065)			(0.067)
δ_1				$-^*$	$-^*$		α_1^+	0.020
				$-^*$	$-^*$			(0.012)
δ_2			0.407		1.000^{**}		α_2^+	0.051
			(0.051)		$-^{**}$			(0.024)
b					0.355		α_3^+	0.002
					(0.032)			(0.020)
c_1				-0.100	-1.235		α_4^+	0.018
				(0.014)	(0.043)			(0.009)
c_2				0.082	-1.179		α_1^-	0.211
				(0.014)	(0.041)			(0.017)
ν_1							α_2^-	0.223
								(0.029)
ν_2							α_3^-	0.120
								(0.072)
α_1^+							α_4^-	0.512
								(0.181)
α_2^+							a^+	-0.377
								(0.028)
α_3^+							a^-	-0.039
								(0.006)
Obs	16008	16008	16008	16008	16008			16008
LL	-14705	-14706	-14670	-14674	-14660			-14662
BIC	-14729	-14730	-14699	-14708	-14699			-14725
$\bar{\hat{\varepsilon}}_i$	1.002	1.001	1.000	1.001	1.000			1.000
S.D.	1.060	1.057	1.059	1.059	1.056			1.057
LB	33.643	23.238	35.240	29.765	32.469			33.801

*: Estimation based on a logarithmic specification.
**: Parameter set to boundary.

The mostly significant parameters a^+ and a^- indicate that it is also useful to account for flexible news impact effects for the second lag. Figures 30 and 31 depict the news impact functions for the particular duration series computed based on the estimates of the NPACD models. The shape of the estimated news impact curves confirm the estimation results of the parametric ACD models. The news response curves for trade durations reveal high nonlinearities, especially for very small innovations. For two of the stocks (AOL and Coca-Cola), we even observe a news response function that implies a downward shape for low values of ε_{i-1}. Hence, for extremely small innovations (with exception of the GE stock), the first order autocorrelation is negative rather than positive. However, only for larger values of ε_{i-1} the ACF is slightly positive. This finding illustrates that small durations induce significantly different adjustments of the expected mean than long durations which has to be taken into account in the econometric modelling.

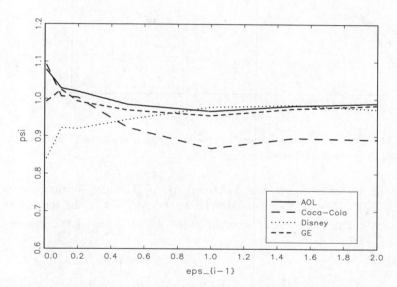

Figure 30: Estimated news impact curves for trade durations of the AOL, Coca-Cola, Disney and GE stock.

This result is in line with the findings of Zhang, Russell, and Tsay (2001) who provide evidence for similar effects based on estimations of the TACD model. Therefore, since such a news impact curve is not in accordance with symmetric ACD models, it is not surprising that the basic linear ACD model, especially for the AOL stock, is misspecified and not sufficient to model trade durations. The news response function for price durations reveals a significantly different shape. The larger (positive) slope of the curve indicates a higher (positive) autocorrelation for price durations. Nonetheless, we notice a nonlinear news impact curve with a strongly increasing pattern for $\varepsilon_{i-1} < 1$ and a nearly flat function for $\varepsilon_{i-1} \geq 1$. Hence, we also observe different adjustment processes for unexpectedly small price durations, i.e., in periods of unexpectedly high volatility.

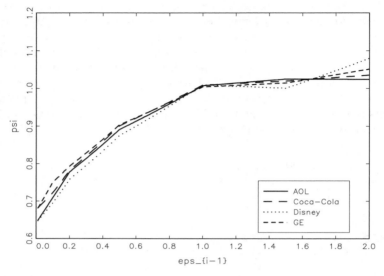

Figure 31: Estimated news impact curves for price durations of the AOL, Coca-Cola, Disney and GE stock.

Sixth, for all processes except the ACD and AGACD model, stationarity is ensured by $\sum_{j=1}^{2} \beta_j < 1$. Correspondingly, in the ACD case, the inequality $\sum_{j=1}^{2} \alpha_j + \beta_j < 1$ has to be satisfied while in the AGACD model, stationarity is ensured by[29]

$$\mathrm{E}\left[\sum_{j=1}^{2} \beta_j + \alpha_j(|\varepsilon_{i-j} - b| + c(\varepsilon_{i-j} - b))^{\delta_2}\right] < 1.$$

Checking these restrictions based on the corresponding parameter estimates, we notice that stationarity is ensured for all considered specifications.[30]

Tables 21 through 24 display the p-values of the applied LM, CM and ICM tests associated with the regressions in Tables 13 through 20. For the LM and CM tests, as well as the F-tests against nonlinearities in the news impact function, we apply the same forms as in the Monte Carlo experiments.

---

[29] See Section 5.3.2.

[30] Since it is cumbersome to verify the stationarity condition for the AGACD model analytically, it is evaluated numerically.

Table 21: P-values of the different types of diagnostic tests (see Table 9) for the estimates based on AOL trade durations and price durations (Tables 13 and 14).

	ACD	LACD	BACD	EXACD	ABACD	AGACD	NPACD
				Trade durations			
LM_1	0.022	0.100	0.053	0.250	0.029	0.085	0.913
SB_1	0.031	0.033	0.040	0.433	0.020	0.028	0.119
PSB_1	0.204	0.037	0.007	0.322	0.010	0.006	0.216
NSB_1	0.060	0.356	0.221	0.438	0.325	0.112	0.204
SB_2	0.235	0.199	0.243	0.113	0.175	0.127	0.094
PSB_2	0.484	0.497	0.419	0.139	0.034	0.080	0.035
NSB_2	0.200	0.172	0.204	0.163	0.144	0.227	0.193
LM_2	0.010	0.016	0.031	0.029	0.013	0.042	0.187
NL_1	0.013	0.011	0.060	0.005	0.026	0.041	0.017
NL_2	0.003	0.001	0.009	0.001	0.008	0.048	0.004
CM_1^1	0.038	0.106	0.168	0.156	0.119	0.446	0.155
CM_1^2	0.030	0.106	0.261	0.173	0.187	0.566	0.211
CM_2^1	0.009	0.012	0.036	0.006	0.041	0.064	0.039
CM_2^2	0.009	0.011	0.052	0.006	0.041	0.060	0.039
CM_3^1	0.025	0.045	0.085	0.015	0.011	0.193	0.208
CM_3^2	0.023	0.066	0.114	0.093	0.039	0.322	0.300
CM_4^1	0.011	0.014	0.060	0.004	0.015	0.105	0.065
CM_4^2	0.006	0.009	0.060	0.007	0.033	0.113	0.052
CM_5^1	0.017	0.004	0.002	0.002	0.003	0.010	0.019
CM_5^2	0.129	0.054	0.013	0.049	0.017	0.058	0.083
ICM	0.504	0.455	0.336	0.287	0.347	0.416	0.346
				Price durations			
LM_1	0.000	0.000	0.093	0.109	0.753		0.343
SB_1	0.000	0.000	0.013	0.062	0.098		0.185
PSB_1	0.000	0.000	0.038	0.058	0.175		0.195
NSB_1	0.081	0.018	0.043	0.118	0.243		0.380
SB_2	0.011	0.001	0.289	0.342	0.287		0.296
PSB_2	0.000	0.000	0.003	0.001	0.000		0.001
NSB_2	0.387	0.193	0.085	0.094	0.093		0.086
LM_2	0.000	0.000	0.006	0.006	0.020		0.013
NL_1	0.000	0.000	0.452	0.047	0.063		0.299
NL_2	0.000	0.000	0.161	0.011	0.007		0.172
CM_1^1	0.000	0.000	0.098	0.052	0.040		0.055
CM_1^2	0.000	0.000	0.394	0.116	0.068		0.076
CM_2^1	0.000	0.000	0.097	0.263	0.225		0.236
CM_2^2	0.000	0.000	0.291	0.481	0.362		0.355
CM_3^1	0.000	0.000	0.044	0.025	0.053		0.017
CM_3^2	0.000	0.000	0.145	0.064	0.082		0.023
CM_4^1	0.000	0.000	0.021	0.026	0.028		0.013
CM_4^2	0.000	0.000	0.146	0.099	0.067		0.027
CM_5^1	0.991	0.805	0.309	0.409	0.485		0.223
CM_5^2	0.841	0.982	0.837	0.767	0.775		0.601
ICM	0.336	0.079	0.396	0.415	0.455		0.445

Table 22: P-values of the different types of diagnostic tests (see Table 9) for the estimates based on Coca-Cola trade durations and price durations (Tables 15 and 16).

	ACD	LACD	BACD	EXACD	ABACD	NPACD
			Trade durations			
LM_1	0.000	0.000	0.003	0.824	0.605	0.004
SB_1	0.000	0.000	0.003	0.106	0.271	0.440
PSB_1	0.225	0.003	0.024	0.115	0.173	0.143
NSB_1	0.061	0.350	0.435	0.253	0.407	0.000
SB_2	0.326	0.434	0.173	0.249	0.255	0.148
PSB_2	0.092	0.108	0.078	0.101	0.100	0.109
NSB_2	0.294	0.328	0.495	0.435	0.424	0.380
LM_2	0.002	0.002	0.008	0.861	0.723	0.009
NL_1	0.005	0.005	0.037	0.815	0.873	0.187
NL_2	0.046	0.046	0.110	0.876	0.921	0.140
CM_1^1	0.000	0.000	0.000	0.101	0.068	0.019
CM_1^2	0.000	0.000	0.000	0.038	0.024	0.036
CM_2^1	0.000	0.000	0.000	0.010	0.012	0.021
CM_2^2	0.000	0.000	0.000	0.003	0.003	0.035
CM_3^1	0.000	0.000	0.001	0.209	0.185	0.005
CM_3^2	0.000	0.000	0.000	0.059	0.026	0.020
CM_4^1	0.001	0.000	0.001	0.046	0.087	0.001
CM_4^2	0.001	0.000	0.000	0.054	0.036	0.010
CM_5^1	0.021	0.034	0.600	0.072	0.095	0.109
CM_5^2	0.005	0.001	0.446	0.019	0.029	0.327
ICM	0.485	0.436	0.376	0.168	0.069	0.405
			Price durations			
LM_1	0.000	0.000	0.001	0.006	0.010	0.343
SB_1	0.000	0.000	0.000	0.496	0.122	0.185
PSB_1	0.000	0.000	0.000	0.147	0.111	0.195
NSB_1	0.008	0.006	0.040	0.000	0.003	0.380
SB_2	0.012	0.007	0.350	0.139	0.225	0.296
PSB_2	0.042	0.071	0.358	0.135	0.119	0.001
NSB_2	0.184	0.147	0.186	0.333	0.211	0.086
LM_2	0.000	0.000	0.001	0.011	0.014	0.013
NL_1	0.000	0.000	0.138	0.480	0.222	0.299
NL_2	0.000	0.000	0.104	0.359	0.119	0.172
CM_1^1	0.000	0.000	0.002	0.024	0.038	0.055
CM_1^2	0.000	0.000	0.003	0.066	0.085	0.076
CM_2^1	0.000	0.000	0.005	0.019	0.023	0.236
CM_2^2	0.000	0.000	0.010	0.040	0.044	0.355
CM_3^1	0.000	0.000	0.003	0.012	0.020	0.017
CM_3^2	0.000	0.000	0.036	0.053	0.084	0.023
CM_4^1	0.000	0.000	0.001	0.004	0.007	0.013
CM_4^2	0.000	0.000	0.014	0.028	0.059	0.027
CM_5^1	0.072	0.042	0.007	0.134	0.111	0.223
CM_5^2	0.477	0.512	0.018	0.395	0.319	0.601
ICM	0.772	0.950	0.594	0.257	0.317	0.445

Since for large samples the computation of the ICM test statistic is quite time consuming, we only use an ICM test specification based on the parameters $\mathfrak{A} = 0$, $\mathfrak{B} = 0.5$, $M^{ICM} = 100$ and $L = 2$. Note that such a test is consistent against misspecifications with respect to the first two lags, but is not consistent for an infinite number of lags.

Tables 21 through 24 display the p-values of the applied LM, CM and ICM tests associated with the regressions in Tables 13 through 20. For the LM and CM tests, as well as the F-tests against nonlinearities in the news impact function, we apply the same forms as in the Monte Carlo experiments. Since for large samples the computation of the ICM test statistic is quite time consuming, we only use an ICM test specification based on the parameters $\mathfrak{A} = 0$, $\mathfrak{B} = 0.5$, $M^{ICM} = 100$ and $L = 2$. Note that such a test is consistent against misspecifications with respect to the first two lags, but is not consistent for an infinite number of lags. For AOL trade durations, we notice that the more simple specifications are rejected on the basis of nearly all diagnostics. In particular, the basic (linear) ACD model seems to be clearly misspecified. The more flexible models, like the BACD, EXACD and ABACD specifications, lead to slightly higher p-values. Nonetheless, especially based on the CM tests, the null hypothesis of a correct specification of the conditional mean is rejected for all models. Better results are obtained for the AGACD and the NPACD model. Thus, accounting for additive *and* multiplicative stochastic factors seems to be essential for this duration series. These findings are in line with the estimation results that also indicate the best goodness-of-fit for these two specifications. The LM and CM test statistics based on weighting functions including functionals of the first lag produce quite low test statistics and thus high p-values. In contrast, model violations are indicated by tests whose weighting functions also account for higher lags. Therefore, the NPACD and AGACD models seem to be appropriate for capturing nonlinearities in the news response of the first lag, but, there are neglected news impact asymmetries in higher lags. For AOL price durations, we observe a misspecification of the ACD and LACD model since the null is clearly rejected based on nearly all test statistics. Significantly better results are found for the BACD, EXACD and AGACD models that allow for a more flexible modelling of the (nonlinear) news impact curve. However, even based on the highly parameterized NPACD model, we still find evidence for a model misspecification. In particular, the low p-values of the CM_3 and CM_4 tests indicate that the specifications do not seem to be flexible enough to capture all nonlinearities in the news response. This leads to a violation of the conditional mean restriction. However, we notice that for both AOL duration series, the ICM test statistics indicate no rejection of the null hypothesis. In these cases, the ICM test does not seem to have enough power to detect these particular forms of model misspecification.

Similar results are found for Coca-Cola trade and price durations. Again, we observe a clear rejection of the null hypothesis for the more simple models.

Table 23: P-values of the different types of diagnostic tests (see Table 9) for the estimates based on Disney trade durations and price durations (Tables 17 and 18).

	ACD	LACD	BACD	EXACD	ABACD	AGACD	NPACD
				Trade durations			
LM_1	0.353	0.328	0.652	0.424	0.643	0.461	0.905
SB_1	0.070	0.069	0.150	0.119	0.169	0.088	0.043
PSB_1	0.165	0.141	0.203	0.093	0.188	0.116	0.102
NSB_1	0.477	0.465	0.465	0.319	0.416	0.397	0.065
SB_2	0.488	0.460	0.282	0.355	0.332	0.322	0.468
PSB_2	0.284	0.274	0.497	0.429	0.478	0.480	0.456
NSB_2	0.343	0.334	0.335	0.252	0.380	0.379	0.347
LM_2	0.649	0.620	0.820	0.778	0.845	0.751	0.991
NL_1	0.302	0.235	0.619	0.692	0.407	0.371	0.749
NL_2	0.556	0.372	0.775	0.871	0.682	0.656	0.871
CM_1^1	0.197	0.240	0.442	0.442	0.450	0.114	0.560
CM_1^2	0.080	0.110	0.379	0.266	0.385	0.080	0.464
CM_2^1	0.267	0.285	0.544	0.369	0.600	0.251	0.507
CM_2^2	0.148	0.172	0.467	0.263	0.513	0.190	0.445
CM_3^1	0.569	0.531	0.790	0.402	0.760	0.672	0.911
CM_3^2	0.604	0.577	0.822	0.507	0.803	0.738	0.900
CM_4^1	0.714	0.735	0.853	0.548	0.902	0.860	0.975
CM_4^2	0.724	0.751	0.905	0.578	0.937	0.917	0.951
CM_5^1	0.679	0.489	0.753	0.576	0.679	0.787	0.776
CM_5^2	0.861	0.818	0.796	0.813	0.732	0.817	0.835
ICM	0.019	0.010	0.009	0.009	0.020	0.030	0.009
				Price durations			
LM_1	0.000	0.000	0.515	0.078	0.352	0.347	0.562
SB_1	0.001	0.000	0.382	0.114	0.368	0.466	0.220
PSB_1	0.000	0.000	0.320	0.017	0.333	0.207	0.236
NSB_1	0.314	0.359	0.245	0.367	0.284	0.311	0.037
SB_2	0.004	0.006	0.089	0.053	0.070	0.132	0.063
PSB_2	0.001	0.001	0.132	0.023	0.156	0.166	0.010
NSB_2	0.019	0.023	0.006	0.002	0.007	0.004	0.000
LM_2	0.000	0.000	0.003	0.000	0.002	0.003	0.000
NL_1	0.000	0.000	0.816	0.211	0.808	0.857	0.676
NL_2	0.000	0.000	0.169	0.005	0.127	0.272	0.147
CM_1^1	0.000	0.000	0.682	0.350	0.540	0.647	0.299
CM_1^2	0.000	0.000	0.611	0.349	0.474	0.613	0.308
CM_2^1	0.000	0.000	0.048	0.002	0.042	0.009	0.003
CM_2^2	0.000	0.000	0.101	0.011	0.085	0.050	0.019
CM_3^1	0.000	0.000	0.694	0.391	0.509	0.569	0.371
CM_3^2	0.000	0.000	0.755	0.441	0.563	0.616	0.170
CM_4^1	0.000	0.000	0.015	0.001	0.010	0.003	0.000
CM_4^2	0.000	0.000	0.052	0.011	0.041	0.023	0.001
CM_5^1	0.001	0.004	0.009	0.004	0.007	0.007	0.002
CM_5^2	0.000	0.000	0.041	0.011	0.031	0.033	0.004
ICM	0.000	0.000	0.633	0.336	0.594	0.614	0.316

Table 24: P-values of the different types of diagnostic tests (see Table 9) for the estimates based on GE trade durations and price durations (Tables 19 and 20).

	ACD	LACD	BACD	EXACD	ABACD	AGACD	NPACD
				Trade durations			
LM_1	0.071	0.297	0.199	0.334	0.224	0.227	0.816
SB_1	0.069	0.083	0.060	0.183	0.335	0.143	0.389
PSB_1	0.244	0.255	0.063	0.189	0.409	0.210	0.366
NSB_1	0.011	0.249	0.243	0.488	0.126	0.295	0.437
SB_2	0.042	0.021	0.121	0.060	0.124	0.268	0.077
PSB_2	0.455	0.377	0.365	0.382	0.438	0.365	0.319
NSB_2	0.003	0.012	0.015	0.029	0.037	0.013	0.025
LM_2	0.022	0.167	0.210	0.291	0.288	0.111	0.596
NL_1	0.051	0.048	0.034	0.089	0.138	0.033	0.092
NL_2	0.045	0.064	0.029	0.065	0.313	0.036	0.187
CM_1^1	0.037	0.063	0.077	0.058	0.097	0.355	0.150
CM_1^2	0.021	0.037	0.046	0.036	0.162	0.250	0.126
CM_2^1	0.005	0.008	0.033	0.006	0.029	0.095	0.038
CM_2^2	0.007	0.011	0.044	0.010	0.056	0.119	0.056
CM_3^1	0.010	0.022	0.007	0.036	0.046	0.153	0.057
CM_3^2	0.007	0.012	0.004	0.018	0.051	0.148	0.027
CM_4^1	0.001	0.002	0.003	0.003	0.016	0.049	0.008
CM_4^2	0.001	0.002	0.004	0.003	0.020	0.048	0.007
CM_5^1	0.000	0.000	0.000	0.000	0.004	0.002	0.001
CM_5^2	0.000	0.000	0.000	0.000	0.003	0.003	0.001
ICM	0.366	0.3560	0.336	0.356	0.317	0.525	0.396
				Price durations			
LM_1	0.000	0.000	0.191	0.197	0.041		0.688
SB_1	0.000	0.000	0.116	0.080	0.009		0.413
PSB_1	0.000	0.000	0.024	0.017	0.002		0.472
NSB_1	0.436	0.251	0.323	0.329	0.096		0.118
SB_2	0.113	0.039	0.115	0.277	0.360		0.103
PSB_2	0.048	0.114	0.451	0.401	0.439		0.192
NSB_2	0.052	0.362	0.003	0.019	0.002		0.004
LM_2	0.000	0.000	0.094	0.113	0.003		0.049
NL_1	0.000	0.000	0.414	0.057	0.403		0.155
NL_2	0.000	0.000	0.065	0.003	0.046		0.005
CM_1^1	0.000	0.000	0.221	0.052	0.524		0.022
CM_1^2	0.001	0.000	0.378	0.101	0.658		0.051
CM_2^1	0.001	0.000	0.144	0.069	0.209		0.029
CM_2^2	0.001	0.000	0.253	0.082	0.273		0.039
CM_3^1	0.000	0.000	0.003	0.003	0.019		0.155
CM_3^2	0.000	0.000	0.014	0.022	0.077		0.244
CM_4^1	0.000	0.000	0.000	0.000	0.002		0.011
CM_4^2	0.000	0.000	0.004	0.002	0.010		0.024
CM_5^1	0.111	0.259	0.025	0.033	0.027		0.010
CM_5^2	0.018	0.493	0.150	0.118	0.228		0.073
ICM	0.901	0.891	0.940	0.960	0.941		0.950

The CM test statistics are particularly high for all specifications, and thus, indicate violations of the conditional mean restriction. The best model fit is observed for the NPACD and ABACD model. Nevertheless, its overall performance is not satisfying since clear model violations are indicated due to size bias effects. Again, the ICM test seems to have not enough power to detect model misspecifications.

Quite a different picture is shown for Disney trade durations. Here, for all models, even for the linear ACD model, the validity of the conditional mean restriction is not rejected. Thus, the news impact curve of this duration series seems to reveal a largely linear pattern that is in accordance with more simple ACD models. This result is confirmed by the plot of the estimated news response curve (Figure 30). It depicts an almost linear pattern with the exception of extremely small innovations. Nonetheless, an interesting finding is that the ICM test statistic displays misspecifications for all estimated specifications. A possible explanation could be that the LM and CM tests do not have enough power to detect the sharp drop in the upper tail of the news impact curve. For Disney price durations, we find similar results as we have found for AOL and Coca-Cola price durations. Again, we notice a clear rejection of the ACD and LACD model, while the more highly parameterized models lead to significantly higher p-values. Once again, the AGACD model is the horse race winner. Nevertheless, as indicated by the low p-values of the corresponding CM and LM tests, the augmented ACD specifications seem unable to fully capture the existing asymmetric news impact effects in the data.

The diagnostic results for GE financial durations are similar to the findings for the AOL and Coca-Cola stock. While the more simple models are largely rejected, the more sophisticated specifications of the conditional mean function yield additional rewards since the test statistics are clearly reduced. The overall best diagnostic results are achieved by the AGACD model for trade durations and by the NPACD model for price durations. Nevertheless, for both types of financial durations, the CM test statistics display rejections of the null hypothesis of correct model misspecification due to unconsidered asymmetries in the news response.

By summarizing the diagnostic results based on the investigated financial duration series, we obtain the following conclusions: First, it seems to be easier to model trade durations than price durations. Second, the overall best performance in terms of a valid conditional mean restriction is obtained by the AGACD model, corresponding to the most flexible parametric ACD specification. Third, nonlinear news response effects exist not only for the first lag, but also for higher order lags which has to be taken into account in the econometric specification. Fourth, LM tests are more conservative than CM tests since they display the lowest test statistics in general. These findings confirm the results of the Monte-Carlo study. Fifth, in most cases, ICM tests seem to have not enough power to detect the particular types of misspecifications. Nevertheless, they provide valuable information concerning more general forms of model violations.

Hence, two major conclusions can be drawn from this study: First, it turns out that there is actually a need to test the validity of the functional form of ACD models since it is shown that more simple models lead to clear violations of the conditional mean restriction. Second, there is a need for ACD models that allow for nonlinear news response functions. The empirical results point out that, in particular, very small innovations that are associated with unexpectedly active trading periods, imply different adjustments of the expected mean, and thus must be treated in a nonlinear way.

5.5.2 Modelling Trade Durations

This subsection deals with empirical tests of the market microstructure hypotheses given in Section 3.2.2 of Chapter 3. We investigate trade duration series based on the AOL and IBM stock traded at the NYSE (sample period 05/07/02 to 05/18/02), the Allianz (All) stock traded at XETRA (X) (sample period 12/01/99 to 12/30/99) and on the Frankfurt floor (F) (sample period 10/01/99 to 12/30/99), Bund future (BF) trading at EUREX (EX) (sample period 01/02/97 to 01/31/97) and LIFFE (L) (sample period 08/01/95 to 08/31/95), as well as the BHP stock traded at the ASX (sample period 07/01/02 to 07/31/02). The particular data sets are sub-samples from the samples presented in Chapter 4. Since their descriptive statistics strongly resembles the statistics illustrated in Chapter 4, we refrain from showing them here. The particular series are seasonally adjusted by running a cubic spline regression based on 30 minutes nodes in the first stage. Furthermore, they are re-initialized at each trading day.

As a starting point, we apply an ARMA-GARCH model for log durations. Such a specification allows us to investigate whether not only the conditional (log-)duration mean but also the conditional (log-)duration volatility is influenced by market microstructure variables. As already discussed in Section 5.1, the trade duration volatility admits an interpretation as liquidity risk. Therefore, the following analysis allows us to investigate whether liquidity risk is predictable based on past sequences of the trading process.

As explanatory variables we use the lagged signed and absolute midquote change d_{i-1} and $|d_{i-1}|$ with $d_{i-1} = mq_{i-1} - mq_{i-2}$, as well as the bid-ask spread $spd_{i-1} := a_{i-1} - b_{i-1}$ associated with t_{i-1}[31]. Moreover, we include dummy variables indicating whether the previous trading volume was greater or equal than its unconditional 25%, 50% or 75% quantile, respectively ($y_{i,1}^v$, $y_{i,2}^v$ and $y_{i,3}^v$).[32] These variables allow us to explore possible nonlinear rela-

---

[31] Note that for the Allianz stock and the EUREX Bund future no spreads are available.

[32] One exception is made for LIFFE Bund future trading. Since in that case, the volume is measured quite inaccurately (see Section 4.2.4 in Chapter 4), we gen-

tionships between the past volume and the current duration (see Hypothesis 1 in Chapter 4).

Table 25 shows ARMA(2,1)-GARCH(2,1)[33] regressions with dynamically included covariates in the conditional mean function, as well as in the conditional variance function. The conditional variance is parameterized in logarithmic form in order to ensure non-negativity. Hence, the estimated model is given by

$$\ln x_i = \omega + \sum_{j=1}^{2} \alpha_j \ln x_{i-j} + \beta_1 \tilde{\varepsilon}_{i-1} + z'_{i-1}\gamma + \tilde{\varepsilon}_i, \quad i = 1, \ldots, n, \qquad (5.95)$$

with

$$\tilde{\varepsilon}_i = \sqrt{h_i^v} u_i, \qquad u_i \sim N(0, 1) \qquad (5.96)$$

$$h_i^v = \exp\left(\omega^v + \sum_{j=1}^{2} \alpha_j^v |\tilde{\varepsilon}_{i-j}| + \beta_1^v \ln h_{i-1}^v + z'_{i-1}\gamma^v\right). \qquad (5.97)$$

The estimates of the autoregressive parameters in the conditional mean function reveal the well known clustering of the trading intensity, and thus strongly support the economic hypothesis H3 (see Section 3.2.2 in Chapter 3). Moreover, $\hat{\alpha}_1$ and $\hat{\beta}_1$ are quite high in absolute terms, however, they nearly compensate each other. This finding indicates a high persistence of the underlying duration process and confirms the results found in the previous subsection. Moreover, for most of the series (with exception of BHP and Frankfurt Allianz trading) we also observe a strong serial dependence in the conditional volatility function. Hence, not only the trading intensity is clustered but also the duration volatility, i.e. the liquidity risk.

Based on the estimated coefficients associated with market microstructure hypotheses variables in the conditional mean function the following findings can be summarized: For lagged absolute price changes, in most cases we obtain a significantly negative coefficient which confirms hypothesis H4. Thus, large absolute price changes indicate the existence of information on the market, and thus increase the speed of trading in future periods. With respect to the impact of lagged *signed* price changes only mixed evidence is found. Interestingly, in five of seven cases we observe significant coefficients associated with the variable d_{i-1}, which indicates the existence of asymmetry effects with respect to negative and positive price changes. Unfortunately, we find contradicting results for the sign of these coefficients, and thus, no clear direction

---

erate only two dummy variables indicating whether the volume was greater or equal to 41 or 100 shares, respectively. These two numbers correspond to the proxies which are used by the price reporters in order to record whether a volume is medium-sized or large-sized.

[33] The selection of the lag order is determined based on BIC.

Table 25: QML-estimates of ARMA(2,1)-GARCH(2,1) models with explanatory variables for log trade durations. Based on the AOL, IBM, Allianz and BHP stock, as well as Bund future trading. QML standard errors in parantheses. Diagnostics: log likelihood function (LL) and Ljung Box statistic of $\hat{\bar{\varepsilon}}_i$ (LB), as well as of $\hat{\bar{\varepsilon}}_i^2$ (LB2).

	AOL	IBM	All (X)	All (F)	BF (EX)	BF (L)	BHP
			Conditional Mean Function				
ω	−0.023	−0.001	−0.004	0.005	0.013	0.002	−0.012
	(0.004)	(0.001)	(0.003)	(0.004)	(0.002)	(0.001)	(0.006)
α_1	0.933	0.996	1.093	0.995	1.025	1.040	1.045
	(0.000)	(0.005)	(0.008)	(0.009)	(0.006)	(0.001)	(0.008)
α_2	0.028	−0.005	−0.098	−0.010	−0.050	−0.057	−0.071
	(0.005)	(0.005)	(0.008)	(0.009)	(0.005)	(0.001)	(0.007)
β_1	0.929	−0.962	−0.962	−0.930	−0.909	−0.926	−0.934
	(0.008)	(0.002)	(0.002)	(0.005)	(0.003)	(0.003)	(0.005)
d_{i-1}	−0.194	−0.040	−0.063	−0.031	0.014	0.008	0.004
	(0.115)	(0.037)	(0.016)	(0.009)	(0.003)	(0.003)	(0.005)
$|d_{i-1}|$	−0.050	−0.182	0.003	0.007	−0.011	−0.036	−0.009
	(0.168)	(0.039)	(0.004)	(0.008)	(0.002)	(0.004)	(0.003)
$y^v_{i-1,1}$	0.003	−0.002	0.003	−0.017	−0.004	−0.017	−0.018
	(0.006)	(0.003)	(0.007)	(0.009)	(0.005)	(0.008)	(0.007)
$y^v_{i-1,2}$	−0.009	0.000	0.001	0.013	−0.024	−0.025	0.005
	(0.007)	(0.003)	(0.007)	(0.009)	(0.005)	(0.021)	(0.007)
$y^v_{i-1,3}$	0.014	0.003	−0.017	−0.036	−0.041	−	−0.024
	(0.006)	(0.002)	(0.006)	(0.008)	(0.005)	−	(0.007)
spd_{i-1}	0.003	0.031	−	−	−	−0.000	0.012
	(0.035)	(0.010)	−	−	−	(0.000)	(0.004)
			Conditional Variance Function				
ω^v	−0.019	−0.013	−0.006	0.159	−0.015	−0.014	0.553
	(0.003)	(0.002)	(0.002)	(0.048)	(0.003)	(0.002)	(0.045)
α^v_1	0.045	−0.003	0.046	0.057	0.052	0.056	0.084
	(0.013)	(0.016)	(0.017)	(0.019)	(0.015)	(0.011)	(0.005)
α^v_2	−0.017	0.022	−0.036	0.000	−0.007	−0.034	0.036
	(0.013)	(0.017)	(0.017)	(0.022)	(0.015)	(0.012)	(0.008)
β^v_1	0.979	0.989	0.991	−0.173	0.968	0.985	0.075
	(0.003)	(0.001)	(0.002)	(0.211)	(0.002)	(0.002)	(0.067)
d_{i-1}	−0.003	−0.032	−0.020	−0.084	0.000	0.004	−0.021
	(0.126)	(0.041)	(0.007)	(0.033)	(0.004)	(0.002)	(0.007)
$|d_{i-1}|$	0.156	0.025	−0.001	−0.034	−0.018	−0.010	0.051
	(0.089)	(0.035)	(0.002)	(0.041)	(0.002)	(0.002)	(0.008)
$y^v_{i-1,1}$	−0.001	−0.005	−0.000	−0.012	−0.003	−0.001	−0.000
	(0.003)	(0.003)	(0.004)	(0.034)	(0.005)	(0.004)	(0.015)
$y^v_{i-1,2}$	−0.006	−0.001	0.004	0.067	−0.010	0.014	−0.001
	(0.004)	(0.003)	(0.004)	(0.035)	(0.005)	(0.013)	(0.014)
$y^v_{i-1,3}$	0.009	0.003	−0.007	0.068	−0.026	−	−0.133
	(0.003)	(0.002)	(0.004)	(0.036)	(0.005)	−	(0.015)
spd_{i-1}	−0.046	−0.003	−	−	−	0.000	−0.050
	(0.019)	(0.009)	−	−	−	(0.000)	(0.017)
obs	22090	33418	16050	14239	37962	42547	27108
LL	−32527	−40059	−26759	−21692	−53171	−57765	−47372
LB(20)	40.686	46.403	24.365	19.116	23.660	26.737	60.621
LB2(20)	27.080	28.977	25.030	55.856	59.634	15.408	17.411

Table 26: QML-estimates of Box-Cox ACD(2,1) models with explanatory variables for trade durations. Based on the AOL, IBM, Allianz and BHP stock, as well as Bund future trading. QML standard errors in parantheses. Diagnostics: log likelihood function (LL), mean ($\bar{\hat{\varepsilon}}_i$), standard deviation (S.D.) and Ljung-Box statistic (LB) of ACD residuals.

	AOL	IBM	All (X)	All (F)	BF (EX)	BF (L)	BHP
ω	−0.024	−0.053	−0.057	−0.075	−0.096	−0.1019	−0.070
	(0.005)	(0.013)	(0.007)	(0.009)	(0.013)	(0.0077)	(0.017)
α_1	0.020	0.077	0.154	0.127	0.247	0.1900	0.231
	(0.007)	(0.015)	(0.023)	(0.015)	(0.028)	(0.0149)	(0.025)
α_2	0.003	−0.014	−0.091	−0.034	−0.084	−0.0670	−0.151
	(0.007)	(0.013)	(0.018)	(0.014)	(0.017)	(0.0102)	(0.023)
β_1	0.987	0.991	0.996	0.985	0.980	0.9844	0.987
	(0.004)	(0.003)	(0.001)	(0.002)	(0.001)	(0.0016)	(0.006)
δ_1	1.052	0.554	0.735	0.734	0.524	0.6712	0.524
	(0.090)	(0.077)	(0.060)	(0.049)	(0.045)	(0.0372)	(0.035)
d_{i-1}	−0.087	−0.055	−0.065	−0.031	0.017	0.0091	−0.003
	(0.097)	(0.041)	(0.015)	(0.009)	(0.004)	(0.0036)	(0.004)
$|d_{i-1}|$	−0.074	−0.198	0.007	0.003	−0.025	−0.0450	−0.013
	(0.125)	(0.067)	(0.005)	(0.008)	(0.003)	(0.0046)	(0.003)
$y^v_{i-1,1}$	0.003	−0.005	0.001	−0.010	−0.004	−0.0146	−0.003
	(0.005)	(0.003)	(0.008)	(0.009)	(0.006)	(0.0094)	(0.011)
$y^v_{i-1,2}$	−0.006	−0.004	0.007	0.011	−0.036	−0.0243	−0.003
	(0.006)	(0.004)	(0.009)	(0.009)	(0.007)	(0.0253)	(0.008)
$y^v_{i-1,3}$	0.007	0.004	−0.020	−0.040	−0.058	−	−0.010
	(0.005)	(0.003)	(0.008)	(0.008)	(0.007)	−	(0.007)
spd_{i-1}	−0.017	0.037	−	−	−	−0.000	0.010
	(0.024)	(0.015)	−	−	−	(0.001)	(0.004)
obs	22090	33418	16050	14239	37962	42547	27108
LL	−21581	−32814	−13309	−13289	−33027	−32986	−26007
$\bar{\hat{\varepsilon}}_i$	1.000	1.000	1.000	1.000	1.000	1.000	1.000
S.D.	1.211	0.891	1.288	1.011	1.175	0.947	1.430
LB(20)	49.836	51.963	23.024	37.035	34.453	46.931	26.523

for the asymmetries can be identified. For the lagged volume, in most regressions a negative impact on the subsequent trade duration is found which confirms hypothesis H1 (a). However, the estimated volume dummies do not reveal any convincing evidence in favor of the hypothesis that medium trading volumes have the strongest impact on the trading intensity, thus, H1 (b) is rejected. With respect to the influence of the bid-ask spread on the subsequent trade duration, a confirmation of hypothesis H2 is found. Particularly, in three of four cases the parameter is significantly positive indicating that traders' preference for immediacy declines when the transaction costs increase.

Concerning the impact of market microstructure variables on the duration volatility, in most cases we observe a negative influence of past signed price changes[34]. Thus, positive (negative) price changes decrease (increase) the liquidity risk which is in line with the well known phenomenon that "bad" news lead to stronger responses in trading activity than "good" news (see also Section 3.4.4 in Chapter 3). With respect to the impact of absolute price changes, no clear results are obtained, thus, liquidity risk seems to be influenced rather by the direction than by the magnitude of past price movements. Also for the impact of the lagged trading volume on the duration volatility no significant effects are observed in most cases. Finally, the estimates of the coefficient related to the bid-ask spread reveal slight evidence for a negative relationship. Hence, the wider the current spread, the lower the subsequent duration volatility.

In order to check the robustness of the results, we re-estimate the models based on a Box-Cox ACD specification (Dufour and Engle, 2000) that ensures the non-negativity restriction and is given by

$$\ln \Psi_i = \omega + \sum_{j=1}^{2} \alpha_j \varepsilon_{i-j}^{\delta_1} + \beta_1 \ln \Psi_{i-1} + z'_{i-1}\gamma. \tag{5.98}$$

Table 26 shows the QML estimation results of Box-Cox ACD(2,1) models for the particular series. In general, the estimates confirm the findings above.

5.5.3 Quantifying (Il)liquidity Risks

A further objective in this section is to apply the ACD framework to examine the dynamics and determinants of the time-varying excess demand intensity and its relationship to liquidity and market depth. Therefore, we test the hypotheses given in Section 3.4.4 of Chapter 3. The empirical analysis consists of two parts: the first part is devoted to the modelling of the time-varying excess demand intensity based on ACD models which are augmented by market microstructure variables. This analysis provides us deeper insights into the risk assessment of liquidity suppliers and allows us to check the hypotheses H5, H6 and H7. In the second part, we examine the relationship between the market depth and the excess demand intensity. In particular, we test hypothesis H8 and explore the behavior of liquidity suppliers in situations which are characterized by information-based trading.

Excess volume durations are constructed based on the AOL, AT&T, Boeing, Disney, IBM, JP Morgan and Philip Morris stock traded at the NYSE. As in the previous subsection, the sample covers the period from 01/02/01 to 05/31/01 and is extracted from the TAQ database. The basic data preparation is performed as described in Chapter 4, Sections 4.1 and 4.2. For each

---

[34] However, they are not significant in all regressions.

asset, we choose two different aggregation levels that ensure a satisfying number of observations per trading day and per stock leading to mean durations between approximately 4 and 10 minutes corresponding to approximately 40 and 130 observations per trading day, respectively. The resulting excess volume duration series are seasonally adjusted by using a cubic spline regression based on 30 minutes nodes in the first stage. Table 27 shows the summary statistics of the resulting excess volume duration series.

Table 27: Descriptive statistics of buy and sell volumes and excess volume durations based on various stocks traded at the NYSE. Database extracted from the 2001 TAQ database, sample period from 01/02/01 to 05/31/01.

	AOL				AT&T			
	buy vol.	sell vol.	$dv =$ 50000	$dv =$ 100000	buy vol.	sell vol.	$dv =$ 50000	$dv =$ 100000
Obs	111720	102040	10260	4055	70357	118844	7134	2932
Mean	6779	6002	233	557	6211	4929	331	740
S.D.	15185	13632	297	697	21430	18611	435	852
LB(20)	-	-	6675	1482	-	-	3362	938

	Boeing				Disney			
	buy vol.	sell vol.	$dv =$ 10000	$dv =$ 20000	buy vol.	sell vol.	$dv =$ 20000	$dv =$ 40000
Obs	81675	78500	10809	4895	74624	90195	8140	3552
Mean	1814	1763	221	479	3091	2533	292	634
S.D.	4750	4735	297	614	8596	7787	394	794
LB(20)	-	-	5021	1350	-	-	2760	1122

	IBM				JP Morgan			
	buy vol.	sell vol.	$dv =$ 25000	$dv =$ 50000	buy vol.	sell vol.	$dv =$ 25000	$dv =$ 50000
Obs	187318	152680	10442	4138	139329	129201	10145	4093
Mean	2464	2106	229	491	2680	2777	235	537
S.D.	4797	4206	320	580	6637	7059	310	628
LB(20)	-	-	7433	2174	-	-	5950	1878

	Philip Morris							
	buy vol.	sell vol.	$dv =$ 25000	$dv =$ 50000				
Obs	84369	91287	9625	4220				
Mean	3747	3475	248	547				
S.D.	9699	9383	326	707				
LB(20)	-	-	4089	1411				

The Determinants of Excess Volume Durations

The econometric framework is based on a Box-Cox ACD model (see Section 5.3.1) that allows for a parsimonious modelling of nonlinearities in the news impact function. Thus, the model is given by

$$((\phi_i \zeta)^{\delta_1} - 1)/\delta_1 = \omega + \sum_{j=1}^{P} \alpha_j \varepsilon_{i-j}^{\delta_2} + \sum_{j=1}^{Q} \beta_j ((\phi_{i-1} \zeta)^{\delta_1} - 1)/\delta_1 + z'_{i-1} \gamma,$$

$$(5.99)$$

where ϕ_i and ζ are defined as in Section 5.3.1. Note that this specification does not necessarily ensure positive values for Ψ_i and x_i. However, our estimates do not reveal any violations of this restriction. In order to account for the specific distributional properties of excess volume durations (see Section 4.5 in Chapter 4), the ACD errors are assumed to follow a generalized F distribution leading to a log likelihood function which is given by (5.28). The choice of the appropriate lag order is performed on the basis of the Bayes Information Criterion (BIC) leading to an ACD specification with a lag order of $P = Q = 1$ as the preferred model. Note that each time series of excess volume durations is re-initialized every trading day.

Table 28 shows the estimated ACD parameters for the different excess volume duration series. The autocorrelation parameters are highly significant and reveal a strong serial dependence in the series of excess demand intensities. Nevertheless, as indicated by the Ljung-Box statistics, with respect to the ACD residuals, nearly all estimated specifications are appropriate for capturing the dynamics of the duration process. In most cases, the Box-Cox parameter δ_1 is found to be between 0 and 0.3. Particularly for the lower aggregation levels, the null hypothesis of a logarithmic ACD specification ($\delta_1 \to 0$) cannot be rejected. In contrast, for the higher aggregation levels, linear specifications ($\delta_1 = 1$), as well as logarithmic specifications are clearly rejected. The estimates of the second Box-Cox parameter δ_2 vary between 0.5 and 0.7 in most regressions yielding evidence for a concave shape of the news impact curve. Economically, such a finding means that the adjustment of traders' expectations is more sensible in periods of unexpectedly high excess demand intensity (i.e. when $\varepsilon_i < 1$) which confirms the results for price durations found in Section 5.5.1. With respect to the distribution parameters a, m and η our results show that the high flexibility of the generalized F distribution is not required in all cases. In most cases, the estimates of η are close to zero indicating that a generalized gamma distribution seems to be sufficient.[35]

---

[35] Note that in three regressions, no convergence of the ML procedure is reached since the parameter η tends to zero and thus, to the boundary of the parameter space. In these cases, the models are re-estimated using a generalized gamma distribution.

Table 28: Estimates of Box-Cox-ACD(1,1) models with explanatory variables for excess volume durations. Based on various stocks traded at the NYSE. Database extracted from the 2001 TAQ database, sample period from 01/02/01 to 05/31/01. QML standard errors in parantheses. Diagnostics: log likelihood function (LL), Bayes Information Criterion (BIC), LR-test for joint significance of all covariates ($\chi^2(6)$-statistic), as well as mean ($\bar{\tilde{\varepsilon}}$), standard deviation (S.D.) and Ljung-Box statistic (LB) of ACD residuals.

	AOL		AT&T		Boeing		Disney
	$dv =$ 50000	$dv =$ 100000	$dv =$ 50000	$dv =$ 100000	$dv =$ 10000	$dv =$ 20000	$dv =$ 20000
ω	1.108	1.392	0.506	0.256	0.237	0.515	0.380
	(0.212)	(0.507)	(0.186)	(0.390)	(0.092)	(0.215)	(0.165)
α	0.212	0.312	0.195	0.125	0.137	0.152	0.109
	(0.031)	(0.087)	(0.031)	(0.051)	(0.022)	(0.038)	(0.022)
β	0.961	0.935	0.961	0.928	0.974	0.943	0.938
	(0.008)	(0.018)	(0.009)	(0.021)	(0.006)	(0.017)	(0.013)
δ_1	0.084	0.3449	0.387	0.512	0.276	0.570	0.074
	(0.171)	(0.129)	(0.143)	(0.136)	(0.153)	(0.172)	(0.172)
δ_2	0.577	0.477	0.641	0.942	0.710	0.728	0.823
	(0.064)	(0.101)	(0.068)	(0.188)	(0.066)	(0.125)	(0.094)
a	1.054	1.140	0.934	0.952	1.238	0.902	1.025
	(0.091)	(0.195)	(0.111)	(0.160)	(0.041)	(0.173)	(0.150)
m	1.700	2.093	1.093	1.299	1.279	0.902	1.009
	(0.205)	(0.527)	(0.167)	(0.291)	(0.067)	(0.215)	(0.190)
η	0.115	0.135	0.142	0.138	–	0.110	0.077
	(0.034)	(0.067)	(0.064)	(0.083)	–	(0.109)	(0.064)
$|dp_{i-1}|$	-0.131	-0.068	-0.165	-0.112	-0.116	-0.130	0.058
	(0.048)	(0.054)	(0.132)	(0.156)	(0.063)	(0.079)	(0.128)
dp_{i-1}	-0.069	0.006	-0.424	-0.358	0.005	-0.018	-0.193
	(0.039)	(0.047)	(0.115)	(0.135)	(0.048)	(0.050)	(0.094)
$\overline{dv}_{i-1}$	-0.115	-0.141	-0.060	-0.036	-0.035	-0.065	-0.049
	(0.019)	(0.044)	(0.017)	(0.034)	(0.009)	(0.021)	(0.016)
$\overline{vol}_{i-1}$	0.001	-0.004	-0.000	-0.020	0.003	-0.009	-0.011
	(0.004)	(0.009)	(0.003)	(0.011)	(0.005)	(0.010)	(0.007)
spd_{i-1}	-0.100	-0.026	-0.145	-0.193	-0.053	-0.060	-0.112
	(0.033)	(0.060)	(0.074)	(0.132)	(0.037)	(0.080)	(0.054)
$dspd_{i-1}$	-0.052	-0.010	0.037	-0.331	-0.053	-0.255	-0.328
	(0.075)	(0.099)	(0.238)	(0.361)	(0.081)	(0.104)	(0.138)
obs	10260	4055	7134	2932	10809	4895	8140
LL	−10323	−4115	−7180	−2974	−9719	−4336	−727
BIC	−8898	−3416	−6217	−2527	−9779	−4396	−7328
$\chi^2(6)$	114.44	55.38	44.53	23.63	48.34	28.32	48.32
$\bar{\tilde{\varepsilon}}$	1.006	1.015	1.006	1.014	1.004	1.009	1.006
S.D.	0.962	1.043	1.050	0.986	1.095	1.067	1.124
LB(20)	15.50	24.96	24.60	24.44	18.37	22.12	20.42

Table 28 cont'd: Estimates of Box-Cox-ACD(1,1) models with explanatory variables for excess volume durations. Based on various stocks traded at the NYSE. Database extracted from the 2001 TAQ database, sample period from 01/02/01 to 05/31/01. QML standard errors in parantheses. Diagnostics: log likelihood function (LL), Bayes Information Criterion (BIC), LR-test for joint significance of all covariates ($\chi^2(6)$-statistic), as well as mean ($\bar{\hat{\epsilon}}$), standard deviation (S.D.) and Ljung-Box statistic (LB) of ACD residuals.

	Disney	IBM		JP Morgan		Philip Morris	
	$dv =$ 40000	$dv =$ 25000	$dv =$ 50000	$dv =$ 25000	$dv =$ 50000	$dv =$ 25000	$dv =$ 50000
ω	-0.039	1.977	2.536	0.843	1.493	0.545	0.856
	(0.454)	(0.356)	(1.014)	(0.198)	(0.408)	(0.147)	(0.314)
α	0.187	0.292	0.339	0.151	0.163	0.137	0.154
	(0.063)	(0.041)	(0.070)	(0.027)	(0.054)	(0.024)	(0.047)
β	0.926	0.945	0.931	0.965	0.927	0.943	0.917
	(0.023)	(0.010)	(0.016)	(0.007)	(0.016)	(0.010)	(0.018)
δ_1	0.013	0.055	0.113	0.057	0.063	0.322	0.354
	(0.091)	(0.093)	(0.123)	(0.110)	(0.099)	(0.101)	(0.084)
δ_2	0.623	0.566	0.545	0.702	0.719	0.746	0.755
	(0.108)	(0.056)	(0.087)	(0.071)	(0.145)	(0.082)	(0.127)
a	0.894	1.615	2.206	0.981	1.001	1.352	0.857
	(0.115)	(0.207)	(0.214)	(0.113)	(0.138)	(0.053)	(0.115)
m	0.937	3.240	5.405	1.158	1.231	1.544	0.856
	(0.155)	(0.617)	(1.007)	(0.173)	(0.224)	(0.098)	(0.146)
η	0.176	0.050	-	0.142	0.086	-	0.187
	(0.078)	(0.027)	-	(0.059)	(0.061)	-	(0.082)
$|dp_{i-1}|$	0.068	-0.025	-0.127	-0.072	-0.140	-0.245	-0.314
	(0.176)	(0.018)	(0.039)	(0.058)	(0.069)	(0.124)	(0.139)
dp_{i-1}	-0.193	0.023	-0.043	-0.101	-0.080	-0.204	-0.161
	(0.095)	(0.018)	(0.027)	(0.045)	(0.049)	(0.082)	(0.095)
dv_{i-1}	-0.015	-0.213	-0.257	-0.091	-0.151	-0.063	-0.091
	(0.041)	(0.035)	(0.093)	(0.019)	(0.037)	(0.014)	(0.028)
$\overline{vol}_{i-1}$	-0.014	0.007	-0.006	0.004	-0.016	-0.008	-0.023
	(0.017)	(0.004)	(0.009)	(0.005)	(0.011)	(0.007)	(0.012)
spd_{i-1}	-0.177	-0.123	-0.094	-0.157	-0.223	-0.254	-0.362
	(0.125)	(0.033)	(0.043)	(0.042)	(0.089)	(0.060)	(0.113)
$dspd_{i-1}$	-0.640	-0.027	-0.055	-0.259	-0.207	-0.615	-0.464
	(0.216)	(0.056)	(0.065)	(0.086)	(0.138)	(0.126)	(0.157)
obs	3552	10442	4138	10145	4093	9625	4220
LL	-3075	-8636	-3382	-8773	-3386	-8464	-3563
BIC	-3132	-8701	-3436	-8837	-3444	-8523	-3621
$\chi^2(6)$	29.961	134.282	77.434	125.772	88.314	186.640	84.601
$\bar{\hat{\epsilon}}$	1.016	1.005	1.013	1.005	1.017	1.008	1.016
S.D.	1.127	1.060	0.999	1.082	1.008	1.107	1.125
LB(20)	19.67	32.11	20.70	36.89	24.42	28.05	21.00

In order to capture past trading activities, the model is augmented by six explanatory variables. By including the absolute midquote change $|dp_{i-1}|$, measured over the previous excess volume spell, we examine whether the magnitude of quote adjustments during the past duration period has some predictability for the future liquidity demand intensity (hypothesis H5). A significantly negative coefficient for five of the seven stocks shows that liquidity suppliers indeed adjust their quotes stronger when they expect high excess demands, and thus high liquidity risks. I.e., the lower the market depth over the previous spell, the smaller the length of the subsequent excess volume duration which confirms hypothesis H5.

Moreover, a highly significant coefficient associated with the signed price change (dp_{i-1}) measured over the previous duration period provides strong evidence in favor of asymmetric behavior of market participants. Surprisingly, the sign of this coefficient is negative disclosing a higher (lower) excess demand intensity after positive (negative) price changes. Obviously, the intensity of excess buy volumes seems to be higher than that of (excess) sell volumes. This is in line with our descriptive statistics, but is clearly in contrast to hypothesis H7.

As a further explanatory variable we include the magnitude of the lagged (excess) volume. In this context, it has to be noticed that a volume duration is measured as the time it takes a certain *minimum* excess volume to trade. Obviously, in the case of a large block trade, this threshold can be clearly exceeded. Since the extent to which the traded volume exceeds the threshold can contain important information for the appraisal of liquidity risk, we include the *effective* excess volume cumulated during the previous spell (dv_{i-1}) as a further regressor. We find a significantly negative impact of this variable in most regressions, yielding additional evidence for a strong clustering of excess volume intensities. Furthermore, we include the average volume per transaction measured over the last duration spell as an additional covariate $(\overline{vol}_{i-1})$. This is due to the fact that small volume durations can be caused by small trade durations and/or higher volumes per trade. However, for most of the series, we only find insignificant coefficients. One explanation for this result could be the correlation between dv_{i-1} and $\overline{vol}_{i-1}$ caused by the fact that large values of dv_{i-1} only occur when the last trade in a duration period is a block trade. Then, the larger the quantity of this final transaction, the larger dv_{i-1} and therefore, the larger the average volume per trade, $\overline{vol}_{i-1}$.

In order to test Hypothesis H6, we include the bid-ask spread posted at the beginning of the corresponding duration period (spr_{i-1}), as well as the change of the spread during the previous spell, $dspr_{i-1} = spr_{i-1} - spr_{i-2}$. Our results confirm hypothesis H6(i) showing that, given the past excess volume intensity, the spread posted by the market maker indicates his assessment of the expected liquidity risk. On the other hand, with respect to the informational content of spread changes during the last spell no clear-cut results are obtained. Only for half of the series, we find a significantly negative impact which supports hypothesis H6(ii). Accordingly, a widening of the bid-ask spread reflects an increase of the expected liquidity risk in the subsequent period.

Finally a likelihood ratio test (see the χ^2-statistics outlined in Table 28) indicates that for all volume duration series, the null hypothesis of no joint significance is clearly rejected. Hence, evidence is found that market microstructure covariates have explanatory power for excess volume intensities.

Measuring Realized Market Depth

As discussed in Chapter 3, the absolute price change over an excess volume spell measures realized market depth and depends on the average slope of the market reaction curve. Table 5.5.3 shows OLS regressions of the absolute price change, $|dp_i|$, measured over an excess volume duration episode on a set of regressors consisting mainly of the variables we have already used in the

Table 29: OLS regressions for the absolute price change measured over an excess volume duration. Dependent variable: $|dp_i|$. Based on various stocks traded at the NYSE. Database extracted from the 2001 TAQ database, sample period from 01/02/01 to 05/31/01. Diagnostics: p-value of F-test for joint significance of explanatory variables (pv_F) and adjusted R-squared. HAC robust standard errors in parantheses.

	AOL		AT&T		Boeing		Disney
	$dv =$ 50000	$dv =$ 100000	$dv =$ 50000	$dv =$ 100000	$dv =$ 10000	$dv =$ 20000	$dv =$ 20000
$|dp_{i-1}|$	0.070	0.077	0.113	0.116	0.077	0.083	0.075
	(0.011)	(0.018)	(0.013)	(0.021)	(0.011)	(0.017)	(0.016)
dv_{i-1}	-0.005	-0.029	-0.001	-0.004	-0.007	-0.015	-0.006
	(0.003)	(0.010)	(0.001)	(0.003)	(0.001)	(0.004)	(0.001)
$\overline{vol}_{i-1}$	-0.005	-0.006	-0.000	-0.000	-0.003	-0.005	-0.000
	(0.001)	(0.003)	(0.000)	(0.001)	(0.000)	(0.002)	(0.000)
spr_{i-1}	0.090	0.177	0.114	0.179	0.062	0.109	0.058
	(0.015)	(0.035)	(0.014)	(0.031)	(0.012)	(0.025)	(0.013)
$dspr_{i-1}$	-0.033	-0.041	-0.036	-0.066	-0.013	-0.051	-0.018
	(0.014)	(0.027)	(0.016)	(0.036)	(0.011)	(0.019)	(0.014)
$\hat{\varepsilon}_i$	0.028	0.023	0.007	0.007	0.027	0.026	0.010
	(0.003)	(0.007)	(0.001)	(0.003)	(0.002)	(0.004)	(0.001)
x_i	0.012	0.027	0.009	0.016	0.003	0.009	0.009
	(0.003)	(0.006)	(0.001)	(0.003)	(0.002)	(0.004)	(0.001)
$const$	0.137	0.472	0.034	0.067	0.118	0.249	0.082
	(0.033)	(0.115)	(0.012)	(0.035)	(0.016)	(0.046)	(0.013)
obs	10259	4045	7133	2931	10808	4894	8139
pv_F	0.000	0.000	0.000	0.000	0.000	0.000	0.000
$\bar{R}^2$	0.188	0.001	0.147	0.165	0.147	0.147	0.114

Table 29 cont'd: OLS regressions for the absolute price change measured over an excess volume duration. Dependent variable: $|dp_i|$. Based on various stocks traded at the NYSE. Database extracted from the 2001 TAQ database, sample period from 01/02/01 to 05/31/01. Diagnostics: p-value of F-test for joint significance of explanatory variables (pv_F) and adjusted R-squared. HAC robust standard errors in parantheses.

	Disney	IBM		JP Morgan		Philip Morris	
	$dv =$ 40000	$dv =$ 25000	$dv =$ 50000	$dv =$ 25000	$dv =$ 50000	$dv =$ 25000	$dv =$ 50000
$|dp_{i-1}|$	0.119	-0.000	0.097	0.062	0.046	0.083	0.085
	(0.030)	(0.000)	(0.018)	(0.014)	(0.020)	(0.011)	(0.017)
dv_{i-1}	-0.005	-0.391	-0.066	-0.012	-0.010	-0.001	-0.012
	(0.004)	(0.379)	(0.026)	(0.002)	(0.008)	(0.001)	(0.005)
$\overline{vol}_{i-1}$	0.000	0.162	-0.008	-0.004	-0.005	-0.002	-0.000
	(0.002)	(0.182)	(0.005)	(0.001)	(0.003)	(0.000)	(0.001)
spr_{i-1}	0.197	-0.234	0.121	0.086	0.184	0.029	0.032
	(0.045)	(0.362)	(0.065)	(0.031)	(0.057)	(0.011)	(0.024)
$dspr_{i-1}$	-0.183	0.184	-0.042	-0.039	-0.064	-0.012	-0.011
	(0.055)	(0.230)	(0.065)	(0.018)	(0.034)	(0.011)	(0.020)
$\hat{\varepsilon}_i$	0.007	-0.001	0.073	0.025	0.025	0.017	0.015
	(0.003)	(0.054)	(0.010)	(0.002)	(0.006)	(0.001)	(0.003)
x_i	0.020	0.064	0.022	0.007	0.025	0.001	0.008
	(0.003)	(0.057)	(0.009)	(0.002)	(0.006)	(0.001)	(0.002)
$const$	0.076	2.873	0.917	0.190	0.212	0.060	0.182
	(0.046)	(2.506)	(0.282)	(0.026)	(0.094)	(0.017)	(0.053)
obs	3551	10441	4137	10144	4092	9624	4219
pv_F	0.000	0.000	0.000	0.000	0.000	0.000	0.000
$\bar{R}^2$	0.188	0.001	0.147	0.165	0.147	0.147	0.114

previous subsection. Moreover, we include the contemporaneous excess volume duration, x_i, as well as the ACD residual $\hat{\varepsilon}_i = x_i/\hat{\Psi}_i$. Clearly, these variables are not weakly exogenous for $|dp_i|$. Nevertheless, one might argue that these regressors can be considered as being under the control of a market participant who influences the demand for liquidity, and thus the length of the underlying volume duration spell by his trading behavior. This specification enables us to explore the determinants of the price impact given the length of the underlying spell. Such a proceeding is associated with the view of a trader who wants to place his volume in an optimal way and is interested in the relationship between the price impact and the time in which the volume is absorbed by the market. We find a confirmation of hypothesis H8 since the mostly highly significant coefficients associated with x_i indicate a positive correlation between the length of the time interval in which a certain excess volume is traded and the corresponding absolute price reaction. Furthermore, it is shown that unexpectedly large (small) duration episodes, as measured by $\hat{\varepsilon}_i$, increase (decrease) the price impact in addition. Hence, the market depth is lower in periods of unexpectedly high excess demand intensity. These results seem to be counter-intuitive since one would expect that traders who trade large quantities quickly would have to bear additional "costs for immediacy". However, it has to be taken into account that this empirical study does not consider the trading of block transactions. Such trades are typically executed on the upstairs market and are based on special agreements between the specialist and the trader (see, for example, Keim and Madhavan 1996) confronting the investor with additional transaction costs induced by higher bid-ask spreads. By explicitly focussing on floor trading at the NYSE, such effects are not captured and an opposite effect seems to prevail. Hence, confirming hypothesis H8, liquidity suppliers do not seem to draw inferences from single large transactions, but from a more permanent one-sided trading flow. Accordingly, the higher the continuity of the one-sided trading flow, the higher the probability of existing information on the market. Obviously,

Table 30: OLS regressions for the absolute price change measured over an excess volume duration. Dependent variable: $|dp_i|$. Based on various stocks traded at the NYSE. Database extracted from the 2001 TAQ database, sample period from 01/02/01 to 05/31/01. Diagnostics: F-test for joint significance of explanatory variables (pv_F) and adjusted R-squared. HAC robust standard errors in parantheses.

	AOL		AT&T		Boeing		Disney
	$dv=$ 50000	$dv=$ 100000	$dv=$ 50000	$dv=$ 100000	$dv=$ 10000	$dv=$ 20000	$dv=$ 20000
$|dp_{i-1}|$	0.062	0.074	0.104	0.119	0.073	0.079	0.066
	(0.012)	(0.018)	(0.015)	(0.023)	(0.012)	(0.018)	(0.017)
dv_{i-1}	-0.007	-0.027	-0.001	-0.006	-0.008	-0.013	-0.007
	(0.003)	(0.010)	(0.001)	(0.003)	(0.002)	(0.005)	(0.001)
$\overline{vol}_{i-1}$	-0.006	-0.007	-0.001	0.000	-0.003	-0.008	0.000
	(0.001)	(0.003)	(0.000)	(0.001)	(0.001)	(0.002)	(0.000)
spr_{i-1}	0.089	0.176	0.130	0.198	0.067	0.107	0.063
	(0.016)	(0.037)	(0.015)	(0.035)	(0.013)	(0.026)	(0.014)
$dspr_{i-1}$	-0.031	-0.039	-0.051	-0.051	-0.015	-0.049	-0.020
	(0.015)	(0.028)	(0.017)	(0.038)	(0.011)	(0.020)	(0.015)
$\hat{\Psi}_i$	0.019	0.032	0.013	0.021	0.007	0.001	0.017
	(0.002)	(0.007)	(0.001)	(0.004)	(0.002)	(0.005)	(0.001)
$const$	0.187	0.485	0.041	0.096	0.160	0.291	0.092
	(0.036)	(0.120)	(0.014)	(0.037)	(0.017)	(0.052)	(0.015)
obs	10259	4054	7133	2931	10808	4894	8139
pv_F	0.000	0.0000	0.000	0.000	0.000	0.000	0.000
$\bar{R}^2$	0.026	0.0309	0.043	0.046	0.021	0.023	0.032

Table 30 cont'd: OLS regressions for the absolute price change measured over an excess volume duration. Dependent variable: $|dp_i|$. Based on various stocks traded at the NYSE. Database extracted from the 2001 TAQ database, sample period from 01/02/01 to 05/31/01. Diagnostics: F-test for joint significance of explanatory variables (pv_F) and adjusted R-squared. HAC robust standard errors in parantheses.

	Disney	IBM		JP Morgan		Philip Morris	
	$dv =$ 40000	$dv =$ 25000	$dv =$ 50000	$dv =$ 25000	$dv =$ 50000	$dv =$ 25000	$dv =$ 50000
$|dp_{i-1}|$	0.112	-0.000	0.083	0.053	0.043	0.072	0.075
	(0.032)	(0.000)	(0.019)	(0.015)	(0.021)	(0.012)	(0.018)
dv_{i-1}	-0.008	-0.394	-0.069	-0.010	-0.009	-0.002	-0.011
	(0.005)	(0.382)	(0.027)	(0.002)	(0.009)	(0.002)	(0.005)
$\overline{vol}_{i-1}$	0.000	0.195	-0.016	-0.006	-0.011	-0.002	-0.002
	(0.002)	(0.218)	(0.006)	(0.001)	(0.003)	(0.000)	(0.001)
spr_{i-1}	0.197	-0.190	0.122	0.093	0.191	0.035	0.032
	(0.046)	(0.321)	(0.067)	(0.032)	(0.058)	(0.012)	(0.026)
$dspr_{i-1}$	-0.187	0.158	-0.039	-0.046	-0.073	-0.017	-0.015
	(0.055)	(0.204)	(0.066)	(0.019)	(0.035)	(0.011)	(0.021)
$\hat{\Psi}_i$	0.022	0.172	0.012	0.009	0.017	0.004	0.005
	(0.004)	(0.164)	(0.011)	(0.002)	(0.007)	(0.001)	(0.003)
$const$	0.122	2.517	1.100	0.214	0.287	0.083	0.216
	(0.049)	(2.071)	(0.304)	(0.028)	(0.104)	(0.019)	(0.055)
obs	3551	10441	4137	10144	4092	9624	4219
pv_F	0.000	0.000	0.000	0.0000	0.000	0.000	0.000
$\bar{R}^2$	0.066	0.001	0.022	0.0222	0.026	0.015	0.013

single large trades are associated with individual liquidity traders, while a more continuous one-sided liquidity demand seems to be attributed to the existence of information.

In order to check the robustness of our estimation results, we re-estimate the model by omitting the contemporaneous variables (Table 30). Here, only the expected excess volume duration, Ψ_i, is included in addition to the other market microstructure regressors. Note that this variable is undoubtedly weakly exogenous since it is based on the information set prevailing at $i-1$. We observe that the estimated coefficients of the particular market microstructure variables change only slightly. Thus, the conditional price impact (given the contemporaneous duration spell) is influenced in the same way as the unconditional price impact. We find a significantly positive coefficient for the absolute midquote change measured over the previous spell. Thus, price reactions related to excess volumes are strongly clustered over time. Furthermore, past excess volumes (per trade) reveal significantly negative coefficients. Hence, the less transactions occur during an excess volume spell (leading to a higher volume per trade), the lower the price reaction in the following period. This result confirms the hypothesis that a more continuous (one-sided) liquidity demand is associated with the existence of information (hypothesis H8). With respect to the spread measured at the beginning of the spell, a significantly positive coefficient is found in most cases. Therefore, the larger the bid-ask spread, the higher the probability of informed trading and thus, the higher the expected market depth.

6

Semiparametric Dynamic Proportional Intensity Models

In this chapter, we discuss dynamic extensions of the proportional intensity (PI) model. This alternative class of models can be interpreted as the direct counterpart to the class of dynamic accelerated failure time (AFT) models considered in the previous chapter. As discussed in Chapter 2, a PI model can be estimated in different ways. One possibility is to adopt a fully parametric approach leading to a complete parameterization of the intensity, including also the baseline intensity function $\lambda_0(t)$. Such a model is consistently estimated by ML given that the chosen parameterization is correct. A further possibility is to refer to the results of Cox (1975) and to consistently estimate the parameter vector γ either by a partial likelihood approach or in a semiparametric way. In this framework, the model requires no specification of $\lambda_0(t)$ but of $\Lambda_0(t_{i-1}, t_i)$ while $\lambda_0(t)$ is estimated semiparametrically or non-parametrically.[1]

In this chapter, we refer to the latter type of models and consider a dynamic extension of *semiparametric* PI models. This class of models is introduced by Gerhard and Hautsch (2002a) and is called *autoregressive conditional proportional intensity (ACPI) model*. The basic idea is to built a dynamic process directly on the *integrated* intensity function $\Lambda(t_{i-1}, t_i)$. It is illustrated that the categorization approach proposed by Han and Hausman (1990) is a valuable starting point for the specification of a dynamic integrated intensity model. This framework leads to a model that allows for a consistent estimation of the model parameters without requiring explicit parametric forms of the baseline intensity. Moreover, as in the non-dynamic case, discrete points of the baseline survivor function can be estimated simultaneously with the dynamic parameters.

A further strength of a dynamic model in terms of the integrated intensity becomes apparent in the context of censoring structures. As discussed in Section 2.4 of Chapter 2, censoring mechanisms occur due to non-trading periods,

---

[1] See, for example, Han and Hausman (1990), Meyer (1990), Horowitz and Neumann (1987) or Horowitz (1996, 1999).

when the exact end of a spell and correspondingly, the beginning of the next spell, cannot be observed directly. Because of the relationship between the integrated intensity and the conditional survivor function (see Chapter 2), an autoregressive model for the integrated intensity is a natural way to account for censoring mechanisms in a dynamic framework.

In Section 6.1, we discuss general difficulties when specifying dynamics in semiparametric PI models and motivate the idea behind the ACPI model. Section 6.2 introduces the semiparametric ACPI model and illustrates the ML estimation. Theoretical properties are discussed in Section 6.3. Here, we focus on the derivation of the theoretical ACF, as well as on the effects of the discretization approach on the estimation quality of the model. Section 6.4 considers extensions of the basic ACPI model, where we discuss regime-switching dynamics and baseline intensities, as well as the consideration of unobserved heterogeneity and censoring mechanisms. Diagnostic tests for the model are given in Section 6.5. Section 6.6 illustrates the application of the ACPI approach for modelling price change volatilities based on censored price durations.

6.1 Dynamic Integrated Intensity Processes

Recall the definition of the standard PI model, as given in (2.24)

$$\lambda(t; z_{\breve{N}(t)}) = \lambda_0(t) \exp(-z'_{\breve{N}(t)} \gamma).$$

Note that we exclude the case of time-varying covariates. In principle, the model can be extended to account for such effects, however, since this would complicate the following exposition considerably, it is beyond the scope of this chapter. We formulate the PI model in terms of a log linear representation of the baseline intensity function, as illustrated in Chapter 2, (2.30),

$$\ln \Lambda_0(t_{i-1}, t_i) = \phi(\mathfrak{F}_{t_{i-1}}; \theta) + \epsilon_i^*, \tag{6.1}$$

where $\phi(\mathfrak{F}_{t_{i-1}}; \theta)$ denotes a possibly nonlinear function depending of past durations, marks and a parameter vector θ. The error ϵ_i^* follows per construction an i.i.d. standard extreme value distribution. Hence, we obtain a regression model based on a standard extreme value distributed error term as a natural starting point for a dynamic extension.

In order to simplify the following considerations, define

$$x_i^* := \ln \Lambda_0(t_{i-1}, t_i) \tag{6.2}$$

as a transformation of the duration x_i. In the case of a non-specified baseline intensity $\lambda_0(t)$, the transformation from x_i to x_i^* is unknown, and thus, x_i^* is interpreted as a latent variable. Then, the PI model can be represented in more general form as a latent variable model

$$x_i^* = \phi(\mathfrak{F}_{t_{i-1}}; \theta) + \epsilon_i^*. \tag{6.3}$$

Using the terminology of Cox (1981), there are two ways to incorporate a dynamic in a latent variable model. On the one hand, there are *observation driven* models, which are characterized by a conditional mean function which is measurable with respect to some *observable* information set. On the other hand, there exist *parameter driven* models whose conditional mean function is measurable with respect to some *unobservable* information set $\mathfrak{F}_{t_i}^* = \sigma(x_i^*, x_{i-1}^*, \dots, x_1^*, z_i, z_{i-1}, \dots, z_1)$. At first glance, the distinction seems point-less since (6.2) is a one-to-one mapping from x_i to x_i^* and thus, the information sets coincide. However, the importance of the distinction will become clear once both approaches have been outlined in greater detail.

An observation driven dynamic model of x_i^* can be depicted as in (6.3). The estimation of this type of model uses the partial likelihood procedure proposed by Cox (1975), which has been shown by Oakes and Cui (1994) to be available, even in the dynamic case. Therefore, the two-step estimation of the parameters θ and the baseline intensity $\lambda_0(t)$ is still possible. A simple example would be to specify $\phi(\cdot)$ in terms of lagged durations (see e.g. Hautsch, 1999). However, it turns out, that the dynamic properties of such models are non-trivial and that in most applications AR type structures are not sufficient to capture the persistence in financial duration processes. The inclusion of a MA term is not easy in the given context and requires to build the dynamic on variables which are unobservable leading directly to a parameter driven model.

Hence, an alternative is to specify the model dynamics directly in terms of the log integrated baseline intensity leading to a *parameter driven* dynamic PI model of the form

$$x_i^* = \phi(\mathfrak{F}_{t_{i-1}}^*; \theta) + \epsilon_i^*. \tag{6.4}$$

In this context two main problems have to be resolved. First, since this specification involves dynamics in terms of the log integrated intensity, the partial likelihood approach proposed by Cox (1975) is not available. This becomes obvious, if one considers an AR(1) process for x_i^* as a special case of (6.4),

$$x_i^* = \alpha x_{i-1}^* + \epsilon_i^* = \alpha \ln \Lambda_0(t_{i-2}, t_{i-1}) + \epsilon_i^*. \tag{6.5}$$

This is simply because x_i^* is a function of the baseline intensity $\lambda_0(t)$ and the latter is left unspecified. Hence, a separate estimation of α and $\lambda_0(t)$ is not possible. Second, a further problem is that x_i^* itself is not observable directly. In general, dynamics attached to a latent variable lead to a maximum likelihood estimator involving a n-fold integral, as has been amply discussed in the literature, see e.g. the surveys by Hajivassiliou and Ruud (1994) and Manrique and Shephard (1998).

To circumvent these problems, Gerhard and Hautsch (2002a) propose a model which is based on the categorization framework introduced by Han and Hausman (1990). As it has been discussed already in Section 2.3.1 of

Chapter 2, this approach is a valuable alternative to the partial likelihood approach and allows for a simultaneous estimation of discrete points of the baseline survivor function and of the parameter vector. A further advantage of a categorization approach is that x_i^* can be observed through a threshold function. This property is exploited by specifying a model based on an observation driven dynamic which does not necessitate the use of extensive simulation methods but allows for a straightforward ML estimation. As it will be demonstrated in more detail in the next subsection, the ACPI model embodies characteristics of observation driven specifications as well as parameter driven models.

6.2 The Semiparametric ACPI Model

By adopting the discretization approach as discussed in Section 2.3.1 of Chapter 2, we recall the definition of μ_k^* as the value of the latent variable x_i^* at the boundary $\bar{x}_k$, (2.33),

$$\mu_k^* := \ln \Lambda_0(t_{i-1}, t_{i-1} + \bar{x}_k), \qquad k = 1, \ldots, K-1.$$

Define in the following

$$x_i^d := k \cdot \mathbb{1}_{\{\bar{x}_{k-1} < x_i \leq \bar{x}_k\}} \tag{6.6}$$

as an ordered integer variable indicating the observed category. Moreover, let $\mathfrak{F}_{t_i}^d := \sigma(x_i^d, x_{i-1}^d, \ldots, x_1^d, z_i, z_{i-1}, \ldots, z_1)$ be the information set generated by the *categorized* durations. This is a standard approach in the analysis of grouped durations, see e.g. Thompson (1977), Prentice and Gloeckler (1978), Meyer (1990), Kiefer (1988), Han and Hausman (1990), Sueyoshi (1995) or Romeo (1999). Clearly, the discretization approach implies a certain loss of information, whereby we have for the information generated by the discretized durations $\mathfrak{F}_{t_i}^d \subset \mathfrak{F}_{t_i}$.

As mentioned above, a further obstacle is that the specification of dynamics based on a latent variable leads to a likelihood involving a n-fold integral. However, the discretization approach allows to circumvent this problem by specifying an observation driven dynamic as proposed by Gerhard (2001). The main idea is to specify the dynamic on the basis of conditional expectations of the error ϵ_i^*,

$$e_i := \mathrm{E}\left[\epsilon_i^* \mid \mathfrak{F}_{t_i}^d\right]. \tag{6.7}$$

Note that e_i relates to the concept of generalized errors, see Gouriéroux, Monfort, Renault, and Trognon (1987), or Bayesian errors, see Albert and Chib (1995).

Then, the ACPI(P,Q) model is given by

$$x_i^* = \phi_i + \epsilon_i^*, \tag{6.8}$$

where $\phi_i := \phi(\mathfrak{F}_{t_{i-1}}; \theta)$ is defined through a recursion, conditioned on an initial ϕ_0,

$$\phi_i = \sum_{j=1}^{P} \alpha_j (\phi_{i-j} + e_{i-j}) + \sum_{j=1}^{Q} \beta_j e_{i-j}. \tag{6.9}$$

Note that the computation of ϕ_i allows us to use a conditioning approach which exploits the observation of ϕ_{i-1}, and thus prevents us from computing cumbersome high-dimensional integrals. Therefore, the semiparametric ACPI model is built on an ARMA structure based on the conditional expectation of the latent variable given the *observable* categorized durations. This specification poses a dynamic extension of the approach of Han and Hausman (1990) since the autoregressive structure is built on values of a transformation of $\lambda_0(t)$ that is assumed to be constant during the particular categories. The covariance stationarity conditions for the ACPI(P,Q) model correspond to the well known covariance stationarity conditions of a standard ARMA model.

The semiparametric ACPI model can be rewritten in terms of the intensity as

$$\lambda(t; \mathfrak{F}^d_{t_{\breve{N}(t)}}) = \lambda_0(t) \exp(-\phi_{\breve{N}(t)}). \tag{6.10}$$

In general, the dynamic structure of the semiparametric ACPI model, given by (6.8) and (6.9), is not difficult to interpret since it is based on a recursive updating structure. In order to provide a deeper understanding of the principle of the model, we consider for illustration a special case of the ACPI model when $\lambda_0(t)$ is assumed to be known. In this case, the model dynamics are easier to understand and the model can be compared with alternative dynamic duration models. When $\lambda_0(t)$ is known, the transformation from x_i to x_i^*, is replaced by a measurable one-to-one function, so that $\mathfrak{F}^d_{t_i} = \mathfrak{F}^*_{t_i} = \mathfrak{F}_{t_i}$. Therefore, $e_i = \epsilon_i^*$ and $\phi_i = \mathrm{E}[x_i^*|\mathfrak{F}^*_{t_{i-1}}] - \mathrm{E}[\epsilon_i^*]$, and thus, the semiparametric ACPI model collapses to a standard ARMA process of the log integrated intensity. Assume in the following a Weibull specification for the baseline intensity, i.e. $\lambda_0(t) = ax(t)^{a-1}$. Then, we obtain a Weibull ACPI(1,1) (WACPI(1,1)) model of the form

$$a \ln x_i = \phi_i + \epsilon_i^*, \tag{6.11}$$

where $\phi_i = \alpha(\phi_{i-1} + \epsilon_{i-1}^*) + \beta \epsilon_{i-1}^*$. Therefore, the WACPI(1,1) model can be written as an ARMA model for log durations based on a standard extreme value distribution, i.e.

$$\ln x_i = \alpha \ln x_{i-1} + \frac{\beta}{a} \epsilon_{i-1}^* + \frac{\epsilon_i^*}{a}. \tag{6.12}$$

However, in the general case where $\lambda_0(t)$ is unknown, the ACPI model does *not* exactly correspond to an ARMA model for log durations. In this case, the parameters α_j and β_j can be only interpreted as approximations to the parameters $\tilde{\alpha}_j$ and $\tilde{\beta}_j$ in the ARMA model of the form

$$x_i^* = \sum_{j=1}^{P} \tilde{\alpha}_j x_{i-1}^* + \sum_{j=1}^{Q} \tilde{\beta}_j \epsilon_{i-j}^* + \epsilon_i^*. \tag{6.13}$$

Clearly, this approximation is the better the finer the chosen categorization (see also Section 6.3.2). In the limiting case for $K \rightarrow \infty$, $\lambda_0(t)$ is quasi-observable, and thus the ACPI model converges to (6.13) with parameters $\alpha_j = \tilde{\alpha}_j$ and $\beta_j = \tilde{\beta}_j$.

An important advantage of the proposed dynamic is that it allows for a straightforward computation of the likelihood function without requiring the use of simulation methods. Thus, a computationally simple maximum likelihood estimator of the dynamic parameters α and β and the parameters of the baseline intensity approximation μ^* is directly available.

The computation of the log likelihood function requires to compute the generalized errors e_i by

$$e_i := \mathrm{E}\left[\epsilon_i^* \,|\, \mathfrak{F}_{t_i}^d\right] = \mathrm{E}\left[\epsilon_i^* \,|\, x_i^d, \phi_i\right]$$

$$= \begin{cases} \frac{\kappa(-\infty, \nu_{i,1})}{1 - S_{\epsilon^*}(\nu_{i,1})} & \text{if } \bar{x}_i = 1, \\ \frac{\kappa(\nu_{i,k-1}, \nu_{i,k})}{S_{\epsilon^*}(\nu_{i,k-1}) - S_{\epsilon^*}(\nu_{i,k})} & \text{if } \bar{x}_i \in \{2, \ldots, K-1\}, \\ \frac{\kappa(\nu_{i,K-1}, \infty)}{S_{\epsilon^*}(\nu_{i,K-1})} & \text{if } \bar{x}_i = K, \end{cases} \tag{6.14}$$

where $\nu_{i,k} := \mu_k^* - \phi_i$, $\kappa(s_1, s_2) := \int_{s_1}^{s_2} u f_{\epsilon^*}(u) du$ and $f_{\epsilon^*}(\cdot)$ and $S_{\epsilon^*}(\cdot)$ denote the p.d.f. and survivor function of the standard extreme value distribution, respectively.[2]

Since the observation driven dynamic enables us to use the standard prediction error decomposition (see e.g. Harvey, 1990), the likelihood is evaluated in a straightforward iterative fashion: The function ϕ_i is initialized with its unconditional expectation $\phi_0 := \mathrm{E}[\phi_i]$. Then, based on the recursion, (6.9), as well as the definition of the generalized errors, (6.14), the likelihood contribution of observation i given the observation rule (6.6) is computed as

$$\mathrm{Prob}\left[x_i^d = k \,\Big|\, \mathfrak{F}_{t_{i-1}}^d\right] = \begin{cases} 1 - S_{\epsilon^*}(\mu_1^* - \phi_i) & \text{if } x_i^d = 1, \\ S_{\epsilon^*}(\mu_1^* - \phi_i) - S_{\epsilon^*}(\mu_2^* - \phi_i) & \text{if } x_i^d = 2, \\ \vdots & \\ S_{\epsilon^*}(\mu_{K-1}^* - \phi_i) & \text{if } x_i^d = K. \end{cases} \tag{6.15}$$

Hence, the log likelihood function is given by

$$\ln \mathcal{L}(W; \theta) = \sum_{i=1}^{n} \sum_{k=1}^{K} \mathbb{1}_{\{x_i^d = k\}} \ln \mathrm{Prob}\left[x_i^d = k \,\Big|\, \mathfrak{F}_{t_{i-1}}^d\right]. \tag{6.16}$$

As illustrated in Section 2.3 of Chapter 2, the baseline survivor function and (discrete) baseline intensity is computed based on (2.34) through (2.36).

---

[2] For an extended discussion of generalized errors in the context of non-dynamic models, see Gouriéroux, Monfort, Renault, and Trognon (1987).

6.3 Properties of the Semiparametric ACPI Model

6.3.1 Autocorrelation Structure

Since the dynamics of the model are incorporated in a *latent* structure, an important question is how the autoregressive parameters can be interpreted in terms of the *observable* durations. This issue will be considered in more details in the following subsection.

The main difficulty is that no closed form expression for the generalized residuals and the p.d.f of the latent variable x_i^* can be given, so that one needs to resort to numerical methods to evaluate the model's ACF. For this reason, we conduct a simulation study. In the following we simulate ACPI processes, (6.8) and (6.9), based on different categorizations. The categories are exogenously given and are associated with predetermined quantiles of the unconditional distribution of x_i^*.[3] Then, we compute the resulting empirical ACF of the ACPI model, (6.8)-(6.9), as well as of the ARMA process in (6.13) for $\alpha = \tilde{\alpha}$ and $\beta = \tilde{\beta}$. As discussed in Section 6.2, this process corresponds to the limiting case of the ACPI process when $K \to \infty$, i.e., when $\lambda_0(t)$ is known. Hence, it can be interpreted in some sense as benchmark process. Moreover, we compute the ACF of the resulting *observable* durations implied by WACPI models with shape parameter $a = 0.5$.[4]

Figures 32 and 33 show the particular autocorrelation patterns for ACPI(1,0) dynamics with parameters $\alpha = 0.5$ and $\alpha = 0.9$, respectively, based on $n = 100,000$ drawings. Each process is simulated under three different groupings for the durations. The first grouping is based on only two categories, in particular below and above the 0.5-quantile of x_i^*. Correspondingly, the other categorizations are based on the 0.25-, 0.5- and 0.75-quantiles, as well as

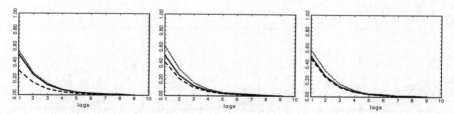

Figure 32: Simulated autocorrelation functions of x_i^* and x_i based on an ACPI(1,0) model with $\alpha = 0.5$. Categorizations based on quantiles of x_i^*. Left: 0.5-quantile, middle: 0.25-, 0.5-, 0.75-quantiles, right: 0.1-, 0.2-,..., 0.9-quantiles. Solid line: ACF of x_i^* based on (6.13). Broken line: ACF of x_i^*, based on (6.8)-(6.9). Dotted line: ACF of x_i.

---

[3] These quantiles are computed based on a simulation study as well.

[4] We also simulated ACPI processes based on more sophisticated functional forms for $\lambda_0(t)$, however, did not find substantially different results.

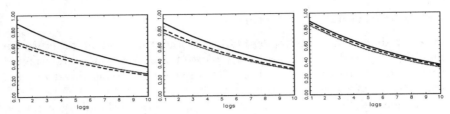

Figure 33: Simulated autocorrelation functions of x_i^* and x_i based on an ACPI(1,0) model with $\alpha = 0.9$. Categorizations based on quantiles of x_i^*. Left: 0.5-quantile, middle: 0.25-, 0.5-, 0.75-quantiles, right: 0.1-, 0.2-,..., 0.9-quantiles. Solid line: ACF of x_i^* based on (6.13). Broken line: ACF of x_i^*, based on (6.8)-(6.9). Dotted line: ACF of x_i.

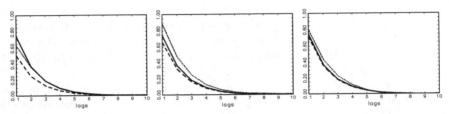

Figure 34: Simulated autocorrelation functions of x_i^* and x_i based on an ACPI(1,1) model with $\alpha = 0.5$ and $\beta = 0.7$. Categorizations based on quantiles of x_i^*. Left: 0.5-quantile, middle: 0.25-, 0.5-, 0.75-quantiles, right: 0.1-, 0.2-,..., 0.9-quantiles. Solid line: ACF of x_i^* based on (6.13). Broken line: ACF of x_i^*, based on (6.8)-(6.9). Dotted line: ACF of x_i.

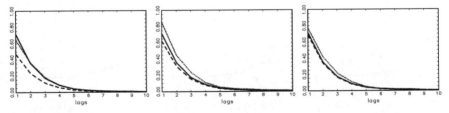

Figure 35: Simulated autocorrelation functions of x_i^* and x_i based on an ACPI(1,1) model with $\alpha = 0.5$ and $\beta = 0.5$. Categorizations based on quantiles of x_i^*. Left: 0.5-quantile, middle: 0.25-, 0.5-, 0.75-quantiles, right: 0.1-, 0.2-,..., 0.9-quantiles. Solid line: ACF of x_i^* based on (6.13). Broken line: ACF of x_i^*, based on (6.8)-(6.9). Dotted line: ACF of x_i.

the 0.1-, 0.2-,...,0.9-quantiles. The former categorization being the worst possible approximation of the true baseline intensity $\lambda_0(t)$ one could possible think of in the context of the given model and the latter being a more realistic case of using a moderate number of thresholds. We observe a close relationship between the different autocorrelation functions. However, this relationship depends rather on the chosen categorization than on the strength of the serial dependence. While for the two-category-grouping clear differences in the autocorrelation functions are observed, quite similar shapes are revealed for the finer categorizations. It is nicely illustrated that the ACF implied by the ACPI model converges towards the ACF of the pure ARMA model for the log integrated intensity, (6.13), when the categorization becomes finer. Moreover, we also observe a quite close relationship between the ACF of the latent variable x_i^* and the observable variable x_i.

These results are confirmed by the Based on ACPI(1,1) processes (Figures 34 and 35). Thus, we can conclude that for a sufficiently fine categorization, the autocorrelation functions of the latent and the observable processes are relatively similar. Hence, it turns out that the ACF implied by the estimated ARMA coefficients of an ACPI model is a good proxy for the ACF of a pure ARMA process for $\ln \Lambda_0(t_{i-1}, t_i)$ and for the ACF of the observed durations.

6.3.2 Evaluating the Estimation Quality

In the previous subsection, we illustrated a quite close relationship between the ACPI dynamic and the corresponding dynamic in the log integrated intensity, (6.13). However, the goal of this subsection is to evaluate the small sample bias incurred by the discretization approach when the model is estimated. Therefore, we perform a Monte Carlo study for different sample sizes and two different categorizations based on $K = 2$ and $K = 10$ categories. The categories are again chosen in accordance with the 0.5-quantile, as well as 0.1-,...,0.9-quantile of x_i^*. The two-category-model is replicated for two sample sizes $n = 50$ and $n = 500$. The model with more thresholds is only estimated for a small sample size $n = 50$. This set-up allows us to compare the improvement achieved by increasing

Table 31: Results of a Monte Carlo study based on semiparametric ACPI(P,Q) models with K categories and n observations.

P	Q	K	n	bias	MSE	MAE
1	0	2	50	-0.006	0.029	0.118
1	0	10	50	0.004	0.009	0.073
1	0	2	500	0.005	0.002	0.037
0	1	2	50	-0.007	0.074	0.202
0	1	10	50	-0.005	0.018	0.093
0	1	2	500	0.015	0.020	0.095

the number of observations versus the benefit of a better approximation of the baseline intensity. Parameter estimations are based on the semiparametric ACPI(1,0) and ACPI(0,1) model. Since here, the focus is on the bias of the dynamic parameters, we fix the threshold parameters to their true values. A range of parameter values for α and β are covered in the simulations, concisely, $\alpha, \beta \in \mathcal{Q} = \{-0.9, -0.8, \ldots, 0.8, 0.9\}$ providing $n_i^{MC} = 1,000$, $i \in \mathcal{Q}$, replications for each value. The errors ϵ_i^* are drawn from the standard extreme value distribution as in the assumed DGP.

Overall results for all $n^{MC} = 19,000$ replications are reported in Table 31. It gives descriptive statistics of the difference between the true parameters and the corresponding estimates, $\alpha^{(i)} - \hat{\alpha}^{(i)}$, and $\beta^{(i)} - \hat{\beta}^{(i)}$, for $i = 1, \ldots, n^{MC}$. Although we aggregate over all parameter values, the small sample properties match the expectation build from asymptotic theory, i.e., the variance decreases over an increasing sample size. The results indicate that even a moderately sized sample of 50 observations is quite sufficient to obtain reasonable results. In particular, for the ACPI(1,0) model the asymptotic properties seem to hold quite nicely. The performance of the ACPI(0,1) model seems to be worse than of the corresponding ACPI(1,0) models. To gain more insight into the consequences, the discretization grid of the durations bears for the estimation, the results of the Monte Carlo experiment are scrutinized with respect to the parameters of the model, α and β. Simulation results for each of the 19 considered values of the true parameter in the DGP are illustrated in the Box plots reported in Figures 36 through 38.

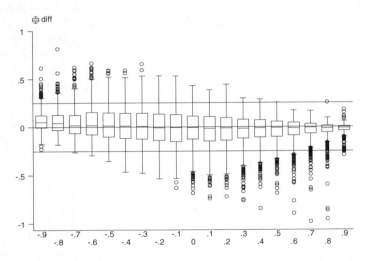

Figure 36: ACPI(1,0), $K = 2$, $n = 50$: Box plots of $\alpha^{(i)} - \hat{\alpha}^{(i)}$ for 19 values of the parameter $\alpha^{(i)}$ in a Monte Carlo study.

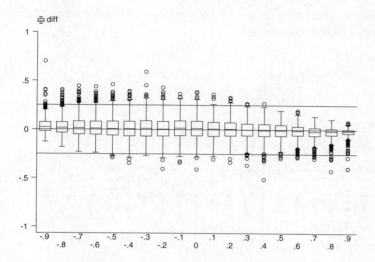

Figure 37: ACPI(1,0), $K = 10$, $n = 50$: Box plots of $\alpha^{(i)} - \hat{\alpha}^{(i)}$ for 19 values of the parameter $\alpha^{(i)}$ in a Monte Carlo study.

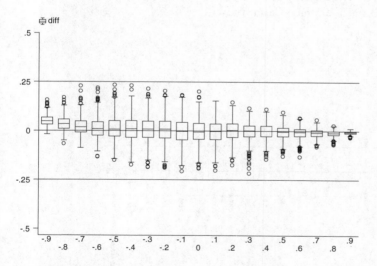

Figure 38: ACPI(1,0), $K = 2$, $n = 500$: Box plots of $\alpha^{(i)} - \hat{\alpha}^{(i)}$ for 19 values of the parameter $\alpha^{(i)}$ in a Monte Carlo study.

Overall, the results are quite encouraging and indicate that the quite considerable bias incurred for an ACPI(1,0) based on $K = 2$ categories is reduced considerably once a higher parameterized model based on $K = 10$ categories is employed. Furthermore, for a reasonable sample size ($n = 500$) even for the two-category-model the performance of the estimator is quite encouraging over all parameter values considered.

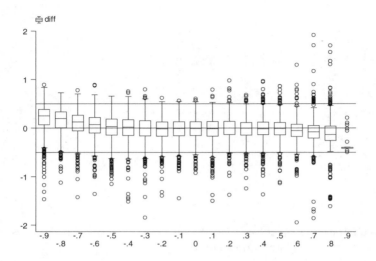

Figure 39: ACPI(0,1), $K = 2$, $n = 50$: Box plots of $\beta^{(i)} - \hat{\beta}^{(i)}$ for 19 values of the parameter $\beta^{(i)}$ in a Monte Carlo study.

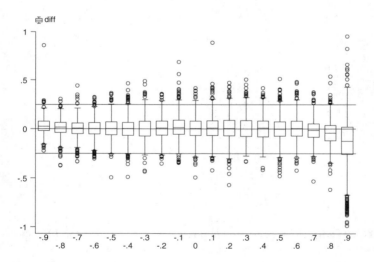

Figure 40: ACPI(0,1), $K = 10$, $n = 50$: Box plots of $\beta^{(i)} - \hat{\beta}^{(i)}$ for 19 values of the parameter $\beta^{(i)}$ in a Monte Carlo study.

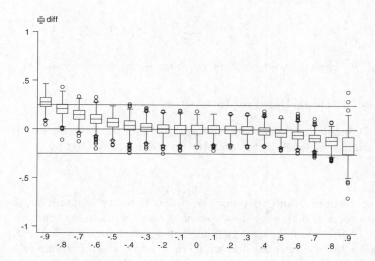

Figure 41: ACPI(0,1), $K = 2$, $n = 500$: Box plots of $\beta^{(i)} - \hat{\beta}^{(i)}$ for 19 values of the parameter $\beta^{(i)}$ in a Monte Carlo study.

Figures 39-41 give the corresponding results for ACPI(0,1) models. Although, qualitatively similar, it is evident from the study that the ACPI(0,1) model performs worse than the corresponding ACPI(1,0) model. After an increase in the number of categories from $K = 2$ to $K = 10$, the approximation reaches about the quality of the ACPI(1,0) process with $K = 2$ categories, except for the parameter value $\beta = 0.9$. The reason for this can be found in the differing ACF of an AR(1) and a MA(1) process. The relatively bad performance of the ACPI(0,1) process for parameters β with a large absolute value is due to the flattening out of the ACF towards the limits of the invertible region.[5]

6.4 Extensions of the ACPI Model

6.4.1 Regime-Switching Dynamics

One valuable extension of the basic ACPI model is to allow for nonlinear news response functions by the inclusion of regime-switching autoregressive parameters. In the given framework, natural regimes are determined by the chosen categorization, and thus the thresholds μ_k^* of the latent variable. Hence, a generalization of the ACPI model[6] to a R-regime-switching ACPI specification is given by

---

[5] See also the similar results for Gaussian models in Gerhard (2001).

[6] For ease of exposition, we consider the case $P = Q = 1$.

$$\phi_i = \sum_{r=1}^{R} \alpha^{(r)} \mathbb{1}_{\{x_{i-1}^d = r\}} (\phi_{i-1} + e_{i-1}) + \sum_{r=1}^{R} \beta^{(r)} \mathbb{1}_{\{x_{i-1}^d = r\}} e_{i-1}, \qquad (6.17)$$

where $\alpha^{(r)}$ and $\beta^{(r)}$ are regime-dependent autoregressive parameters.

6.4.2 Regime-Switching Baseline Intensities

The standard PI model underlies the assumption that, for any two sets of explanatory variables z_1 and z_2, the intensity functions are related by

$$\lambda(t; z_1) \propto \lambda(t; z_2).$$

To obtain more flexibility and to relax the proportionality assumption we stratify the data set and define regime-dependent baseline intensity functions $\lambda_{0,r}(t)$, $r = 1, \ldots, R$. Assume a state defining integer variable $R_i = 1, \ldots, R$, which is at least weakly exogenous and which determines the functional relationship between x_i^* and the baseline intensity $\lambda_{0,r}(t)$. Then, the transformation from x_i to x_i^* is state-dependent and is given by

$$x_i^* = \sum_{r=1}^{R} \mathbb{1}_{\{R_i = r\}} \ln \Lambda_{0,r}(t_{i-1}, t_i), \qquad (6.18)$$

where $\Lambda_{0,r}(t_{i-1}, t_i)$ denotes the regime-dependent integrated baseline intensity. The assumption of a model with R distinct baseline intensities translates to R sets of distinct threshold parameters $\mu_{k,r}^*$, where $k = 1, \ldots, K-1$ and $r = 1, \ldots, R$. Correspondingly, we obtain R baseline survivor functions evaluated at the $K-1$ thresholds[7]

$$S_{0,r}(\bar{x}_k) = \exp(-\exp(\mu_{k,r}^*)), \quad k = 1, \ldots, K-1, \quad r = 1, \ldots, R. \qquad (6.19)$$

Therefore, the generalized residuals are state-dependent as well, i.e.,

$$e_i := \mathrm{E}\left[\epsilon_i^* \mid \mathfrak{F}_{t_i}^d\right] = \sum_{r=1}^{R} \mathbb{1}_{\{R_i = r\}} e_{i,r},$$

where $e_{i,r}$ is computed according to (6.14) based on the corresponding set of threshold parameters $\mu_{k,r}^*$. Hence, a semiparametric ACPI model with state-dependent baseline intensities is defined as

$$\lambda(t; \mathfrak{F}_{t_{\check{N}(t)}}^d) = \sum_{r=1}^{R} \lambda_{0,r}(t) \mathbb{1}_{\{R_{\check{N}(t)} = r\}} \exp(-\phi_{\check{N}(t)}). \qquad (6.20)$$

---

[7] Note that it is also possible to use different categorizations for the duration x_i within each state, thus, the number of thresholds estimated for each state can differ.

The calculation of the log likelihood is based on the procedure proposed in Section 6.2, therefore, we obtain the log likelihood function by

$$\ln \mathcal{L}(W;\theta) = \sum_{i=1}^{n}\sum_{k=1}^{K}\sum_{r=1}^{R} \mathbb{1}_{\{x_i^d=k\}} \mathbb{1}_{\{R_i=r\}} \ln \text{Prob}\left[x_i^d = k \,\Big|\, R_i = r; \mathfrak{F}_{t_{i-1}}^d\right].$$

(6.21)

6.4.3 Censoring

A major advantage of the ACPI model is that it is quite obvious to account for censoring structures. Define in the following $x_i^{d,l}$ and $x_i^{d,u}$ as the discretized counterparts to the duration boundaries x_i^l and x_i^u as given in Section 2.4.1 of Chapter 2. Hence, $x_i^d \in [x_i^{d,l}, x_i^{d,u}]$, where $x_i^{d,l}$ and $x_i^{d,u}$ are computed corresponding to (2.44) by accounting for the observation rule, (6.6). Then, in the case of a censored observation i, the corresponding log likelihood contribution in (6.21) is replaced by

$$\text{Prob}\left[x_i^l \leq x_i^d \leq x_i^u \,\Big|\, \mathfrak{F}_{t_{i-1}}^d, c_{i-1}, c_i, c_{i+1}\right] = S_{\epsilon^*}(\mu^{*l} - \phi_i) - S_{\epsilon^*}(\mu^{*u} - \phi_i),$$

(6.22)

where

$$\begin{cases} \mu^{*l} = \mu_k^* & \text{if} \quad x_i^{d,l} = k+1 \\ \mu^{*u} = \mu_k^* & \text{if} \quad x_i^{d,u} = k. \end{cases}$$

Moreover, the derivation of the generalized error needs to be slightly modified. Hence, in the case of censoring, the conditional expectation of ϵ_i^* is computed as

$$\begin{aligned} e_i &= \text{E}\left[\epsilon_i^* \,\big|\, x_i^d, \phi_i, c_{i-1}, c_i, c_{i+1}\right] = \text{E}\left[\epsilon_i^* \,\big|\, x_i^{d,l}, x_i^{d,u}, \phi_i\right] \\ &= \frac{\kappa(\nu_i^l, \nu_i^u)}{S_{\epsilon^*}(\nu_i^u) - S_{\epsilon^*}(\nu_i^l)}, \end{aligned}$$

(6.23)

where $\nu_i^l := \mu^{*l} - \phi_i$ and $\nu_i^u := \mu^{*u} - \phi_i$.

6.4.4 Unobserved Heterogeneity

A further advantage of the ACPI model is that it is readily extended to allow for unobservable heterogeneity. In duration literature, it is well known that ignoring unobserved heterogeneity can lead to biased estimates of the baseline intensity function.[8] Following Han and Hausman (1990), we specify unobserved heterogeneity effects by a random variable which enters the intensity

---

[8] See e.g. Lancaster (1979) or Heckmann and Singer (1984) among others.

function multiplicatively leading to a mixed ACPI model. From an econometric point of view, accounting for unobserved heterogeneity can be interpreted
as an additional degree of freedom. Furthermore, Lancaster (1997) illustrates
that the inclusion of a heterogeneity variable can capture errors in the variables. In financial duration data, unobservable heterogeneity effects can be
associated with different kinds of traders or different states of the market.

The standard procedure to account for unobserved heterogeneity in the
semiparametric ACPI model is to introduce an i.i.d. random variable $v_{\check{N}(t)}$ in
the specification (6.10) to obtain

$$\lambda(t; \mathfrak{F}^d_{t_{\check{N}(t)}}, v_{\check{N}(t)}) = \lambda_0(t) \cdot v_{\check{N}(t)} \cdot \exp(-\phi_{\check{N}(t)}). \tag{6.24}$$

We assume for the random variable $v_{\check{N}(t)}$ a Gamma distribution with mean
one and variance η^{-1}, which is standard for this type of mixture models, see
e.g. Lancaster (1997) or Section 5.2.3 in Chapter 5. The survivor function of
the compounded model is obtained by integrating out $v_{\check{N}(t)}$

$$S(x_i; \mathfrak{F}^d_{t_{i-1}}) = \left[1 + \eta^{-1} \exp(-\phi_i) \Lambda_0(t_{i-1}, t_i)\right]^{-\eta}. \tag{6.25}$$

Note that this is identical to the survivor function of a BurrII(η) distribution
under appropriate parameterization (see Appendix A).

The latter gives rise to an analogue model based on the discretization
approach outlined in Section 6.2. By augmenting the log linear model of
the integrated baseline intensity by the compounder, we obtain an extended
ACPI(P,Q) model based on the modified latent process

$$x_i^* = \ln(\eta) + \phi_i + \epsilon_i^*, \tag{6.26}$$

where the error term ϵ_i^* follows in this case a BurrII(η) distribution with
density function

$$f_{\epsilon^*}(s) = \frac{\eta \exp(s)}{[1 + \exp(s)]^{\eta+1}}. \tag{6.27}$$

It is easily shown that

$$\lim_{\eta \to \infty} \left[1 + \eta^{-1} \phi_i \Lambda_0(t_{i-1}, t_i)\right]^{-\eta} = \exp(-\Lambda_0(t_{i-1}, t_i) \phi_i),$$

i.e., for $\eta^{-1} = \text{Var}(v_i) \to 0$, the BurrII($\eta$) distribution converges to the standard extreme value distribution. Hence, if no unobservable heterogeneity ef-
fects exist, the model coincides with the basic ACPI model.

The estimation procedure is similar to the procedure described in Section 6.2. The difference is just that the model is now based on a BurrII(η)
distribution. Apart from an obvious adjustment to the generalized errors, the
relationship between the estimated thresholds and the estimation of the distribution function of the error term, as given in (2.34) for the standard extreme
value distribution, is slightly modified to

$$S_0(\bar{x}_k) = \frac{1}{[1 + \exp(\mu_k^* - \ln(\eta))]^{\eta}}, \quad k = 1, \ldots, K-1. \tag{6.28}$$

6.5 Testing the ACPI Model

An obvious way to test for correct specification of the ACPI model is to evaluate the properties of the series of the estimated log integrated intensity $\hat{\epsilon}_i^* = \ln \hat{\Lambda}(t_{i-1}, t_i)$ which should be i.i.d. standard extreme value or BurrII(η) distributed, respectively. However, the difficulty is that we cannot estimate ϵ_i^* but only its conditional expectation $\hat{e}_i = \hat{E}[\epsilon_i^* | \mathfrak{F}_{t_i}^d]$. Thus, the ACPI model has to be evaluated by comparing the distributional and dynamical properties of $\hat{e}_i$ with their theoretical counterparts.

The theoretical mean of e_i is straightforwardly computed as

$$E[e_i] = E[E[\epsilon_i^* | x_i^d, \phi_i]] = E[\epsilon_i^*]. \tag{6.29}$$

However, the computation of higher order moments of e_i is a difficult task. The reason is that the categorization is based on x_i^*, and thus the category boundaries for ϵ_i^*, $\nu_{i,k} = \mu_i^* - \phi_i$, are time-varying and depend itself on lags of e_i. Therefore, the derivation of theoretical moments can only be performed on the basis of the *estimated* model dynamics, and thus, they are of limited value for powerful diagnostic checks of the model. Hence, only upper limits for the moments in the limiting case $K \to \infty$ can be given. In this case, $e_i = \epsilon_i^*$, and thus, the moments of e_i correspond to the moments of the standard extreme value or BurrII(η) distribution, respectively.

The *dynamic* properties of the $\hat{e}_i$ series is evaluated based on a test for serial dependence as proposed by Gouriéroux, Monfort, and Trognon (1985). The test is based on the direct relationship between the score of the observable and the latent model. By accounting for unobserved heterogeneity (see Section 6.4.4), the latent model is written as

$$x_i^* = \ln(\eta) + \phi_i + u_i, \tag{6.30}$$

$$u_i = \tilde{\alpha}_j u_{i-j} + \epsilon_i^*, \tag{6.31}$$

where ϵ_i^* is i.i.d. BurrII(η) distributed and j denotes the tested lag. Then, the null hypothesis is $H_0 : \tilde{\alpha}_j = 0$. The test is based on the score of the observable model. Along the lines of the work of Gouriéroux, Monfort, and Trognon (1985), the observable score is equal to the conditional expectation of the latent score, given the observable categorized variable, i. e.,

$$\frac{\partial \ln \mathcal{L}(W; \theta)}{\partial \theta} = E\left[\frac{\partial \ln \mathcal{L}^*(L^*; \theta)}{\partial \theta} \bigg| \mathfrak{F}_{t_i}^d \right], \tag{6.32}$$

where $\ln \mathcal{L}^*(\cdot)$ denotes the log likelihood function of the *latent* model and L^* denotes the $n \times 1$ vector of the *latent* realizations x_i^*. Under the assumption of a BurrII(η) distribution for ϵ_i^*, the log likelihood function of the latent model is given by

$$\ln \mathcal{L}^*(L^*;\theta) = \sum_{i=j+1}^{n} \ln f_{\epsilon^*}(u_i - \tilde{\alpha}_j u_{i-j})$$

$$= \sum_{i=j+1}^{n} \left[\ln(\eta) + \tilde{\alpha}_j u_{i-j} - u_i - (\eta + 1)\ln\left[1 + \exp(\tilde{\alpha}_j u_{i-j} - u_i)\right]\right].$$

Under the null, the score with respect to $\tilde{\alpha}_j$, $s(\tilde{\alpha}_j)$ is given by

$$s(\tilde{\alpha}_j) = \mathrm{E}\left[\left.\frac{\partial \ln \mathcal{L}^*(L^*;\theta)}{\partial \tilde{\alpha}_j}\right| \mathfrak{F}_{t_i}^d\right] \qquad (6.33)$$

$$= \sum_{i=j+1}^{n} \mathrm{E}\left[\left.e_{i-j}^*\right| \mathfrak{F}_{t_i}^d\right]\left[1 - (\eta + 1)\mathrm{E}\left[\left.\frac{\exp(\epsilon_i^*)}{1 + \exp(\epsilon_i^*)}\right| \mathfrak{F}_{t_i}^d\right]\right]$$

$$= \sum_{i=j+1}^{n} e_{i-j}\left[1 - (\eta + 1)\tilde{e}_i\right],$$

where

$$\tilde{e}_i := \mathrm{E}\left[\left.\frac{\exp(\epsilon_i^*)}{1 + \exp(\epsilon_i^*)}\right| \mathfrak{F}_{t_i}^d\right]. \qquad (6.34)$$

Hence,

$$s(\hat{\tilde{\alpha}}_j) = \sum_{i=j+1}^{n} \hat{e}_{i-j}\left[1 - (\hat{\eta} + 1)\hat{\tilde{e}}_i\right]. \qquad (6.35)$$

Under the null, the expectation of $\hat{\tilde{e}}_i$ is given by $\mathrm{E}[\hat{\tilde{e}}_i] = \frac{1}{\eta+1}$, and thus, $\mathrm{E}[s(\hat{\tilde{\alpha}}_j)] = 0$. Exploiting the asymptotic normality of the score, i.e.,

$$\frac{1}{\sqrt{n}}s(\tilde{\alpha}_j) \xrightarrow{d} N\left(0, \plim \frac{1}{n}\sum_{i=j+1}^{n} e_{i-j}^2\left[1 - (\eta + 1)\tilde{e}_i\right]^2\right),$$

a χ^2-statistic for the null hypothesis $H_0 : \tilde{\alpha}_j = 0$ is obtained by

$$\Upsilon^{(j)} = \frac{\left[\sum_{i=j+1}^{n} \hat{e}_{i-j}\left[1 - (\hat{\eta} + 1)\hat{\tilde{e}}_i\right]\right]^2}{\sum_{i=j+1}^{n} \hat{e}_{i-j}^2\left[1 - (\hat{\eta} + 1)\hat{\tilde{e}}_i\right]^2} \overset{a}{\sim} \chi^2(1). \qquad (6.36)$$

Correspondingly, for the standard extreme value case, the test modifies to

$$\Upsilon^{(j)} = \frac{\left[\sum_{i=j+1}^{n} \hat{e}_{i-j}\left(\hat{\tilde{e}}_i - 1\right)\right]^2}{\sum_{i=j+1}^{n} \hat{e}_{i-j}^2\left[\hat{\tilde{e}}_i - 1\right]^2} \overset{a}{\sim} \chi^2(1) \qquad (6.37)$$

with

$$\tilde{e}_i := \mathrm{E}\left[\exp(\epsilon_i^*)|\, \mathfrak{F}_{t_i}^d\right].\tag{6.38}$$

6.6 Estimating Volatility Using the ACPI Model

In this section, we apply the ACPI model for volatility estimation on the basis of price durations. We use aggregation levels that are associated with large cumulated price movements that often last over several trading days, and thus cause censoring effects. Hence, the appropriate consideration of censoring mechanisms in a dynamic framework is an important issue. Therefore, the ACPI model is a suitable approach for this analysis. We reconsider the study of Gerhard and Hautsch (2002b) on the basis of Bund future transaction data. However, while Gerhard and Hautsch apply a non-dynamic PI model for categorized durations (see Chapter 2), we perform the analysis in a dynamic framework.

6.6.1 The Data and the Generation of Price Events

The study uses LIFFE Bund future transaction data on 11 contracts and 816 trading days between 04/05/94 and 06/30/97 with a total of about $n \approx 2 \cdot 10^6$ transactions. The dataset is prepared as it is described in Section 4.2.4 in Chapter 4. We generate price durations x_i^{dp} as described in Section 3.1.2 of Chapter 3 using the aggregation level $dp \in \{10, 15, 20\}$ ticks, corresponding to 7491, 3560, and 2166 observations.

Investigating volatility patterns based on price changes of different sizes yields indications concerning the size of fluctuations of the price process. Assume a price process, which fluctuates frequently around a certain price level causing, for example, a high number of 10 tick price events, but nearly no 20 tick price events. Such a pattern leads to a low 20 tick volatility in combination with a high 10 tick volatility. In contrast, consider a price process causing the same frequency of 10 tick price events, however, coming along with high amplitudes. In such a case, the 10 tick volatility reveals the same pattern as in the example above, but the 20 tick volatility is higher. Thus, the comparison of duration-based volatilities associated with absolute price changes of different sizes yields deeper insights into the frequencies and amplitudes of price movements.

As already discussed in Chapter 2, price change events can also be caused by news occurring during non-trading periods inducing censoring structures. Since the resulting price event is observable at the earliest at the beginning of the next trading day, it is impossible to identify whether the price movement is caused by overnight news or by recent information. In order to overcome this problem, we consider the first price event occurring within the first 15 minutes of a trading day as censored, i.e., this price change is assumed to be driven by events occurring during the non-trading period. In contrast, price events observed after the first 15 minutes of a trading day are identified to be driven by recent information. For these observations, the duration is measured exactly as the time since the last observation of the previous trading day.

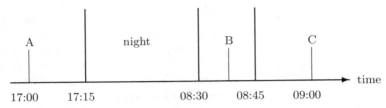

Figure 42: Identification of censored durations in the Bund future trading.

Figure 42 illustrates the identification rule graphically. It shows the arrival times of three transactions occurring on two subsequent trading days. According to our identification rule, the spell between A and B is assumed to be censored and has a minimum length of 15 minutes and a maximum length of 15

Table 32: Distribution of categorized price durations based on 10, 15 and 20 tick price changes. Based on BUND futures trading at LIFFE. Means and standard deviations (only for non-censored durations) in minutes. Censored observations are stated by their corresponding lower boundaries.

	10 Ticks				15 Ticks			
Categories	Non-Censored			Cens.	Non-Censored			Cens.
	Obs	Mean	Std.Dv.	Obs	Obs	Mean	Std.Dv.	Obs
[0', 5']	670	2.37	1.48	68	125	2.35	1.57	21
(5', 10']	618	7.46	1.45	52	117	7.76	1.43	19
(10', 20']	1071	14.72	2.84	96	237	14.60	2.87	45
(20', 40']	1345	29.06	5.77	139	399	29.81	5.66	79
(40', 1h]	762	49.42	5.75	85	336	49.11	5.78	52
(1h, 2h]	974	84.46	16.81	79	489	43.92	16.64	85
(2h, 3h]	375	145.91	17.05	32	251	84.90	17.78	36
(3h, 4h]	214	207.81	17.09	5	146	146.65	17.23	15
(4h, 6h]	249	293.42	36.50	12	240	204.80	35.66	23
(6h, 8h]	89	407.12	33.21	21	138	294.75	34.35	30
(8h, 24h]	311	1089.20	177.29	11	339	406.28	222.35	52
(24h, 36h]	59	1621.92	168.52	0	83	1110.03	148.70	0
(36h, 48h]	36	2708.46	174.35	0	40	2731.32	160.93	0
(48h, ∞)	118	5050.50	2397.13	0	163	4841.94	2931.29	0
Total	6891	321.62	912.70	600	3103	539.95	1309.20	457

Table 32 cont'd:

Categories	20 Ticks Non-Censored			Cens.
	Obs	Mean	Std.Dv.	Obs
[0', 5']	34	1.98	1.48	7
(5', 10']	26	1.64	5.21	4
(10', 20']	84	14.98	2.92	23
(20', 40']	149	31.09	5.79	35
(40', 1h]	148	50.00	5.90	27
(1h, 2h]	243	87.33	17.46	55
(2h, 3h]	134	149.25	16.55	27
(3h, 4h]	85	209.46	17.03	13
(4h, 6h]	158	293.29	34.62	15
(6h, 8h]	130	406.71	37.31	30
(8h, 24h]	294	1145.20	218.72	79
(24h, 36h]	116	1627.54	155.37	3
(36h, 48h]	51	2694.04	170.45	0
(48h, ∞)	196	4875.33	2809.73	0
Total	1848	968.74	1739.66	318

hours and 30 minutes. In contrast, the duration episode between A and C can be measured exactly as 16 hours.

Based on the sample used, we obtain 600, 457, and 318 censored durations for the 10, 15 and 20 tick aggregation level. As dependent variable we use the categorized price duration. We choose a categorization which ensures satisfactory frequencies in each category. For this reason, the categorization differs for the particular aggregation levels. Table 32 shows the distribution of the categorized durations based on the finest grouping which is used for the 10 tick level.

In order to account for seasonalities, a Fourier series approximation $s(t) = s(\delta^s, \bar{t}, Q)$ of order $Q = 6$ is employed (see (5.31) in Chapter 5). This approach allows for a direct but parsimonious assessment of the volatility patterns based on different frequencies, like intraday seasonalities in contrast to time-to-maturity seasonalities. The Fourier series approximation for intraday seasonalities is based on the normalized intraday time $\bar{t} \in [0, 1]$ given by

$$\bar{t} = \frac{\text{seconds since 7:30}}{\text{seconds between 7:30 and 16:15}}, \quad (6.39)$$

while for time-to-maturity seasonalities

$$\bar{t} = \frac{\text{days to maturity}}{150} \quad (6.40)$$

is used. To account for observations with values of more than 150 days to maturity we include a dummy variable. Note that the seasonality variables are included statically, i.e., the ACPI model is given by

$$x_i^* = \phi_i + s(t_{i-1}) + \epsilon_i^* \tag{6.41}$$

$$\phi_i = \sum_{j=1}^{P} \alpha_j (\phi_{i-j} + e_{i-j}) + \sum_{j=1}^{Q} \beta_j e_{i-j}.$$

By accounting for the fact that the durations are categorized, we reformulate the conditional price change volatility as

$$\sigma_{(x^{dp})}^{*2}(t_i^{dp}) = (dp)^2 \sum_{k=1}^{K-1} \text{Prob}\left[x_{i+1}^{d,dp} = k \,\middle|\, \mathfrak{F}_{t_i}^d; s(t_i)\right] \tag{6.42}$$

$$\cdot \, \text{E}\left[\frac{1}{x_{i+1}^{d,dp}} \,\middle|\, x_{i+1}^{d,dp} = k; \mathfrak{F}_{t_i}^d, s(t_i)\right],$$

where $x_i^{d,dp}$ denotes the categorized dp-price duration. Moreover, we assume that in the context of grouped durations the approximation

$$\text{E}\left[\frac{1}{x_{i+1}^{d,dp}} \,\middle|\, x_{i+1}^{d,dp} = k; \mathfrak{F}_{t_i}^d, s(t_i)\right] \approx \text{E}\left[\frac{1}{x_{i+1}^{d,dp}} \,\middle|\, x_{i+1}^{d,dp} = k\right]$$

holds. Then, we obtain an obvious sample estimator for the second factor in the weighted sum of (6.42). Note that this does not exclude the conditioning information, but merely expresses the fact that all information contained in the regressors enters exclusively the first factor $\text{Prob}\left[x_{i+1}^{d,dp} = k \,\middle|\, \mathfrak{F}_{t_i}^d; s(t_i)\right]$ for which an estimator is available by adopting (6.15).

6.6.2 Empirical Findings

Estimation Results

For each aggregation level we estimate three specifications: one model without any seasonality variables (Panel A), one regression with only intraday seasonalities (Panel B) and one specification including all seasonality variables (Panel C).

Tables 33 through 35 give the estimation results for the different ACPI specifications. The choice of the lag order for the autoregressive parameters follows according to the BIC which indicates the ACPI(1,2) as the best specification in nearly all cases[9]. For all three different price durations, it is shown that the goodness of fit is significantly improved by the inclusion of seasonality variables. As indicated by the BIC values, the explanatory power of the time-to-maturity seasonalities increases for larger price changes while the explanatory

---

[9] However, for a matter of convenience we keep this specification for all models.

Table 33: ML estimates of ACPI(1,2) models for Bund future price durations using 10 tick price changes. Standard errors computed based on OPG estimates. Diagnostics: Log Likelihood (LL), Bayes Information Criterion (BIC), mean ($\bar{\tilde{e}}_i$) and standard deviation (S.D.) of ACPI residuals.

	A		B		C	
	est.	S.E.	est.	S.E.	est.	S.E.
Thresholds						
μ_1^* ($\bar{x}_1 = 5'$)	−3.641	0.087	−3.277	0.101	−4.324	0.161
ν_2^* ($\bar{x}_2 = 10'$)	−2.904	0.085	−2.540	0.099	−3.573	0.160
ν_3^* ($\bar{x}_3 = 20'$)	−2.150	0.084	−1.784	0.097	−2.808	0.159
ν_4^* ($\bar{x}_4 = 40'$)	−1.468	0.083	−1.098	0.096	−2.118	0.157
ν_5^* ($\bar{x}_5 = 1h$)	−1.116	0.082	−0.736	0.095	−1.755	0.157
ν_6^* ($\bar{x}_6 = 2h$)	−0.633	0.081	−0.218	0.093	−1.231	0.156
ν_7^* ($\bar{x}_7 = 3h$)	−0.414	0.081	0.026	0.093	−0.979	0.156
ν_8^* ($\bar{x}_8 = 4h$)	−0.272	0.080	0.184	0.093	−0.813	0.156
ν_9^* ($\bar{x}_9 = 6h$)	−0.076	0.080	0.399	0.093	−0.583	0.156
ν_{10}^* ($\bar{x}_{10} = 8h$)	0.009	0.080	0.495	0.095	−0.478	0.157
ν_{11}^* ($\bar{x}_{11} = 24h$)	0.471	0.081	0.988	0.094	0.113	0.155
ν_{12}^* ($\bar{x}_{12} = 36h$)	0.590	0.081	1.108	0.095	0.264	0.155
ν_{13}^* ($\bar{x}_{13} = 48h$)	0.679	0.082	1.196	0.094	0.381	0.155
Intraday seasonalities						
δ_1^s			0.909	0.065	0.992	0.085
$\delta_{s,1}^s$			0.571	0.031	0.605	0.036
$\delta_{s,2}^s$			0.008	0.016	0.013	0.026
$\delta_{s,3}^s$			0.012	0.021	0.006	0.024
$\delta_{s,4}^s$			0.036	0.021	0.031	0.022
$\delta_{s,5}^s$			−0.043	0.021	−0.065	0.022
$\delta_{s,6}^s$			−0.021	0.018	−0.010	0.020
$\delta_{c,1}^s$			0.032	0.023	0.067	0.026
$\delta_{c,2}^s$			0.122	0.021	0.151	0.023
$\delta_{c,3}^s$			0.014	0.021	0.033	0.021
$\delta_{c,4}^s$			0.079	0.021	0.083	0.022
$\delta_{c,5}^s$			0.080	0.020	0.081	0.021
$\delta_{c,6}^s$			−0.031	0.018	−0.030	0.019
Seasonalities over the future's maturity						
$\mathbb{1}_{>150}$					0.261	0.246
δ_1^{*s}					−2.004	0.237
$\delta_{s,1}^{*s}$					−0.110	0.086
$\delta_{s,2}^{*s}$					0.207	0.057
$\delta_{s,3}^{*s}$					−0.048	0.044
$\delta_{s,4}^{*s}$					0.050	0.037
$\delta_{s,5}^{*s}$					0.051	0.034
$\delta_{s,6}^{*s}$					−0.031	0.029
$\delta_{c,1}^{*s}$					0.631	0.049
$\delta_{c,2}^{*s}$					0.010	0.042
$\delta_{c,3}^{*s}$					−0.023	0.037
$\delta_{c,4}^{*s}$					0.028	0.035
$\delta_{c,5}^{*s}$					−0.018	0.031
$\delta_{c,6}^{*s}$					0.011	0.028

Table 33 cont'd:

	A		B		C	
	est.	S.E.	est.	S.E.	est.	S.E.
Dynamic parameters						
α_1	0.923	0.005	0.927	0.005	0.920	0.007
β_1	−0.692	0.009	−0.706	0.009	−0.723	0.011
β_2	−0.072	0.007	−0.066	0.009	−0.075	0.009
Diagnostics						
Obs	7490		7490		7490	
LL	-15142		-14891		-14718	
BIC	-15213		-15020		-14909	
$\bar{\hat{e}}_i$	-0.554		-0.546		-0.559	
S.D.	1.153		1.150		1.161	

power of included intraday patterns declines slightly. Therefore, the influence of time-to-maturity seasonality effects is stronger on higher aggregation levels, while the strength of intraday patterns weakens. Moreover, we observe that the significance of the seasonality variables decreases on higher aggregation levels which is obviously also caused by a smaller sample size. Nevertheless, in

Table 34: ML estimates of ACPI(1,2) models for Bund future price durations using 15 tick price changes. Standard errors computed based on OPG estimates. Diagnostics: Log Likelihood (LL), Bayes Information Criterion (BIC), mean ($\bar{\hat{e}}_i$) and standard deviation (S.D.) of ACPI residuals.

	A		B		C	
	est.	S.E.	est.	S.E.	est.	S.E.
Thresholds						
ν_2^* ($\bar{x}_2 = 10'$)	−3.990	0.182	−3.770	0.208	−4.422	0.246
ν_3^* ($\bar{x}_3 = 20'$)	−3.255	0.179	−3.042	0.205	−3.673	0.243
ν_4^* ($\bar{x}_4 = 40'$)	−2.543	0.178	−2.332	0.203	−2.949	0.241
ν_5^* ($\bar{x}_5 = 1h$)	−2.117	0.177	−1.900	0.202	−2.514	0.241
ν_6^* ($\bar{x}_6 = 2h$)	−1.595	0.175	−1.354	0.200	−1.963	0.239
ν_7^* ($\bar{x}_7 = 3h$)	−1.344	0.175	−1.080	0.200	−1.685	0.239
ν_8^* ($\bar{x}_8 = 4h$)	−1.198	0.174	−0.919	0.199	−1.520	0.238
ν_9^* ($\bar{x}_9 = 6h$)	−0.942	0.173	−0.639	0.198	−1.232	0.237
ν_{10}^* ($\bar{x}_{10} = 8h$)	−0.777	0.172	−0.456	0.198	−1.042	0.237
ν_{11}^* ($\bar{x}_{11} = 24h$)	−0.178	0.169	0.200	0.194	−0.329	0.235
ν_{12}^* ($\bar{x}_{12} = 36h$)	−0.009	0.168	0.376	0.194	−0.127	0.235
ν_{13}^* ($\bar{x}_{13} = 48h$)	0.084	0.168	0.473	0.193	−0.014	0.234

6.6 Estimating Volatility Using the ACPI Model   183

Table 34 cont'd:

Intraday seasonalities				
δ_1^s	0.814	0.141	0.853	0.153
$\delta_{s,1}^s$	0.461	0.054	0.456	0.056
$\delta_{s,2}^s$	−0.068	0.038	−0.066	0.040
$\delta_{s,3}^s$	−0.020	0.034	−0.036	0.037
$\delta_{s,4}^s$	0.023	0.033	0.029	0.035
$\delta_{s,5}^s$	−0.057	0.031	−0.072	0.033
$\delta_{s,6}^s$	−0.034	0.029	−0.022	0.031
$\delta_{c,1}^s$	0.191	0.035	0.228	0.035
$\delta_{c,2}^s$	0.169	0.031	0.189	0.032
$\delta_{c,3}^s$	−0.007	0.029	0.002	0.030
$\delta_{c,4}^s$	0.042	0.030	0.042	0.031
$\delta_{c,5}^s$	0.068	0.029	0.063	0.031
$\delta_{c,6}^s$	−0.044	0.027	−0.042	0.028
Seasonalities over the future's maturity				
$\mathbb{1}_{>150}$			0.418	0.271
δ_1^{*s}			−1.455	0.276
$\delta_{s,1}^{*s}$			−0.038	0.104
$\delta_{s,2}^{*s}$			0.233	0.063
$\delta_{s,3}^{*s}$			−0.048	0.049
$\delta_{s,4}^{*s}$			0.080	0.040
$\delta_{s,5}^{*s}$			0.070	0.034
$\delta_{s,6}^{*s}$			0.026	0.032
$\delta_{c,1}^{*s}$			0.508	0.055
$\delta_{c,2}^{*s}$			0.008	0.044
$\delta_{c,3}^{*s}$			−0.046	0.038
$\delta_{c,4}^{*s}$			0.018	0.036
$\delta_{c,5}^{*s}$			−0.055	0.035
$\delta_{c,6}^{*s}$			0.014	0.030

	A		B		C	
	est.	S.E.	est.	S.E.	est.	S.E.
Dynamic parameters						
α_1	0.962	0.006	0.961	0.006	0.978	0.005
β_1	−0.788	0.014	−0.779	0.013	−0.850	0.014
β_2	−0.084	0.013	−0.081	0.013	−0.085	0.013
Diagnostics						
Obs	3559		3559		3559	
LL	-7480		-7331		-7208	
BIC	-7541		-7445		-7380	
$\bar{\bar{e}}_i$	-0.561		-0.561		-0.557	
S.D.	1.166		1.159		1.169	

Table 35: ML estimates of ACPI(1,2) models for Bund future price durations using 20 tick price changes. Standard errors computed based on OPG estimates. Diagnostics: Log Likelihood (LL), Bayes Information Criterion (BIC), mean ($\bar{\hat{e}}_i$) and standard deviation (S.D.) of ACPI residuals.

	A		B		C	
	est.	S.E.	est.	S.E.	est.	S.E.
Thresholds						
ν_1^* ($\bar{x}_1 = 20'$)	-3.833	0.200	-3.514	0.211	-4.183	0.273
ν_2^* ($\bar{x}_2 = 40'$)	-3.074	0.194	-2.760	0.204	-3.414	0.267
ν_3^* ($\bar{x}_3 = 1h$)	-2.608	0.192	-2.295	0.202	-2.940	0.266
ν_4^* ($\bar{x}_4 = 2h$)	-2.049	0.189	-1.730	0.199	-2.367	0.264
ν_5^* ($\bar{x}_5 = 3h$)	-1.795	0.188	-1.468	0.198	-2.103	0.263
ν_6^* ($\bar{x}_6 = 4h$)	-1.644	0.188	-1.312	0.197	-1.944	0.263
ν_7^* ($\bar{x}_7 = 6h$)	-1.376	0.186	-1.031	0.196	-1.658	0.262
ν_8^* ($\bar{x}_8 = 8h$)	-1.151	0.185	-0.792	0.195	-1.414	0.262
ν_9^* ($\bar{x}_9 = 24h$)	-0.496	0.182	-0.102	0.193	-0.691	0.259
ν_{10}^* ($\bar{x}_{10} = 36h$)	-0.238	0.181	0.162	0.193	-0.398	0.259
ν_{11}^* ($\bar{x}_{11} = 48h$)	-0.106	0.181	0.298	0.191	-0.242	0.257
Intraday seasonalities						
δ_1^s			0.681	0.169	0.622	0.175
$\delta_{s,1}^s$			0.246	0.065	0.192	0.069
$\delta_{s,2}^s$			-0.052	0.047	-0.050	0.048
$\delta_{s,3}^s$			-0.057	0.042	-0.081	0.043
$\delta_{s,4}^s$			0.007	0.042	-0.001	0.044
$\delta_{s,5}^s$			-0.034	0.039	-0.048	0.041
$\delta_{s,6}^s$			-0.025	0.039	-0.041	0.040
$\delta_{c,1}^s$			0.176	0.042	0.209	0.042
$\delta_{c,2}^s$			0.071	0.037	0.105	0.038
$\delta_{c,3}^s$			-0.037	0.038	-0.035	0.039
$\delta_{c,4}^s$			0.053	0.036	0.038	0.036
$\delta_{c,5}^s$			-0.005	0.036	-0.033	0.038
$\delta_{c,6}^s$			-0.003	0.034	0.008	0.036
Seasonalities over the future's maturity						
$\mathbb{1}_{>150}$					0.437	0.325
δ_1^{*s}					-1.298	0.353
$\delta_{s,1}^{*s}$					0.016	0.128
$\delta_{s,2}^{*s}$					0.230	0.077
$\delta_{s,3}^{*s}$					-0.057	0.059
$\delta_{s,4}^{*s}$					0.045	0.051
$\delta_{s,5}^{*s}$					0.046	0.043
$\delta_{s,6}^{*s}$					0.016	0.039
$\delta_{c,1}^{*s}$					0.455	0.063
$\delta_{c,2}^{*s}$					-0.034	0.050
$\delta_{c,3}^{*s}$					-0.063	0.048
$\delta_{c,4}^{*s}$					0.015	0.045
$\delta_{c,5}^{*s}$					-0.042	0.043
$\delta_{c,6}^{*s}$					0.055	0.038

Table 35 cont'd:

	A		B		C	
	est.	S.E.	est.	S.E.	est.	S.E.
	Dynamic parameters					
α_1	0.943	0.009	0.942	0.010	0.970	0.007
β_1	−0.827	0.020	−0.801	0.021	−0.887	0.021
β_2	−0.004	0.018	−0.027	0.019	−0.033	0.020
	Diagnostics					
Obs	2165		2165		2165	
LL	-4483		-4430		-4346	
BIC	-4536		-4534		-4504	
$\bar{\bar{e}}_i$	-0.566		-0.565		-0.561	
S.D.	1.177		1.171		1.179	

Table 36: Test statistics for a test on serial correlation. Based on the regressions in Tables 33 through 35. $\Upsilon^{(j)}$ denotes the $\chi^2(1)$ test statistic as derived in (6.37).

	$dp = 10$			$dp = 15$			$dp = 20$		
	A	B	C	A	B	C	A	B	C
	Thresholds								
$\Upsilon^{(1)}$	0.37	0.42	0.18	0.04	0.06	0.02	0.01	0.03	0.00
$\Upsilon^{(2)}$	9.76***	7.12***	7.90***	1.74	1.85	3.23*	0.79	1.25	1.44
$\Upsilon^{(3)}$	0.77	0.44	0.05	0.49	1.20	1.73	0.40	0.40	0.49
$\Upsilon^{(4)}$	0.99	1.29	0.75	0.97	0.22	0.00	0.11	0.08	0.00
$\Upsilon^{(5)}$	2.65	2.45	2.74*	0.27	2.15	2.36	0.12	0.00	0.39
$\Upsilon^{(6)}$	0.01	0.04	0.03	0.28	0.09	0.00	0.05	0.01	0.03
$\Upsilon^{(7)}$	0.97	0.51	1.61	0.27	0.01	0.06	0.08	0.02	0.10
$\Upsilon^{(8)}$	0.30	0.47	1.05	0.41	0.51	0.73	0.15	0.11	0.00
$\Upsilon^{(9)}$	0.07	0.29	0.00	1.59	1.47	1.87	1.24	1.10	1.06
$\Upsilon^{(10)}$	1.87	1.42	2.36	0.46	0.10	0.21	0.51	0.18	0.37
$\Upsilon^{(11)}$	1.13	0.89	1.42	0.00	0.10	0.76	0.57	0.65	0.68
$\Upsilon^{(12)}$	1.29	0.25	0.40	0.47	0.15	0.35	1.22	1.41	1.56
$\Upsilon^{(13)}$	0.00	0.00	0.04	0.40	0.50	0.16	1.41	0.98	1.17
$\Upsilon^{(14)}$	0.97	0.79	0.12	0.00	0.00	0.01	0.16	0.08	0.33
$\Upsilon^{(15)}$	0.29	0.69	1.03	0.93	0.67	0.40	0.76	0.66	1.05

***, **, *: significance at the 1%, 5% or 10% level.

all regressions the seasonality variables are jointly significant. Concerning the autoregressive parameters we observe highly significant coefficients, indicating strong serial dependencies. The need of two MA terms illustrates the existence of a strong persistence in the data. Hence, even on the highest aggregation level, price durations are strongly clustered.

In order to test for misspecification due to serial dependence in the error terms, we apply the test on serial correlation as illustrated in Section 6.5. Table 36 shows the test statistics for all regressions in the Tables 33 through 35. Based on nearly all test statistics, the null hypothesis of is clearly rejected, thus, it turns out that the ACPI model does a good job in capturing the serial dependence in the data.

In order to check for unobservable heterogeneity effects, we re-estimate the fully parameterized models (specification C) using a gamma compounded ACPI model (Table 37). It is shown that the heterogeneity variance increases with the aggregation level. In particular, for 20 tick price changes we find for the heterogeneity variance a value of 1.156, thus, this specification is close to an ordered logit model.[10] This feature coincides with the results of Hautsch (1999), who obtained only very small heterogeneity effects by analyzing trade durations.

Time-To-Maturity Seasonalities

In order to model time-to-maturity seasonalities, we include a set of flexible Fourier form regressors capturing the effects due to the decreasing time to maturity over the life of a BUND future. Note that a reasonable interpretation of the seasonalities requires to look at all regressors within the series approximation.[11] For this reason, the computation of the resulting price change volatility $\sigma^{*2}_{(x^{dp})}(t_i^{dp})$ is performed based on all coefficients of the Fourier form.

Figure 43 shows the pattern of $\sigma^{*2}_{(x^{dp})}(t_i^{dp})$ depending on the time to maturity for 20 tick price changes. Note that the intraday seasonality coefficients are fixed to a value corresponding to 14:00 GMT while the dynamic variables are set to their unconditional expectations.[12] The highest volatility values are found within a time horizon of about 90 days to maturity. Hence, this finding seems to be caused by the roll-over from the previous contract to the front month contract which proceeds every three months. The strong increase of the volatility around the roll-over might be related to some price finding process associated with the front month contract. After 80 days to maturity, this process seems to be finished leading to a stabilization of the volatility pattern at a relatively constant level.

---

[10] Note that the BurrII(1) distribution is identical to the logistic distribution.

[11] The interpretation of single coefficients is misleading, since the seasonality pattern is based on the sum of all included terms. Retaining only significant coefficients would lead to some type of filtering, which is difficult to interpret as the expansion terms are not orthogonal.

[12] The volatilities are computed based on the estimates of the basic ACPI model (Tables 33 through 35). We also plotted the implied volatility graphs based on the estimates of the gamma compounded ACPI model (Table 37), however, did not find any significant differences. For this reason, we do not show these figures here.

Table 37: ML estimates of gamma compounded ACPI(1,2) models for Bund future price durations using 10, 15 and 20 tick price changes. Standard errors computed based on OPG estimates. Diagnostics: Log Likelihood (LL), Bayes Information Criterion (BIC), mean ($\bar{\bar{e}}_i$) and standard deviation (S.D.) of ACPI residuals.

	$dp = 10$		$dp = 15$		$dp = 20$	
	est.	S.E.	est.	S.E.	est.	S.E.
Thresholds						
μ_1^* $(\bar{x}_1 = 5')$	-6.034	0.301				
ν_2^* $(\bar{x}_2 = 10')$	-5.194	0.303	-4.590	0.439		
ν_3^* $(\bar{x}_3 = 20')$	-4.300	0.307	-3.756	0.441	-2.344	0.565
ν_4^* $(\bar{x}_4 = 40')$	-3.429	0.312	-2.900	0.446	-1.481	0.565
ν_5^* $(\bar{x}_5 = 1h)$	-2.933	0.317	-2.339	0.452	-0.910	0.567
ν_6^* $(\bar{x}_6 = 2h)$	-2.164	0.328	-1.546	0.465	-0.146	0.574
ν_7^* $(\bar{x}_7 = 3h)$	-1.772	0.335	-1.114	0.475	0.241	0.579
ν_8^* $(\bar{x}_8 = 4h)$	-1.506	0.340	-0.847	0.481	0.486	0.583
ν_9^* $(\bar{x}_9 = 6h)$	-1.119	0.350	-0.357	0.495	0.949	0.592
ν_{10}^* $(\bar{x}_{10} = 8h)$	-0.940	0.356	-0.012	0.506	1.371	0.602
ν_{11}^* $(\bar{x}_{11} = 24h)$	0.133	0.387	1.446	0.560	2.832	0.641
ν_{12}^* $(\bar{x}_{12} = 36h)$	0.451	0.397	1.923	0.581	3.523	0.666
ν_{13}^* $(\bar{x}_{13} = 48h)$	0.698	0.404	2.209	0.595	3.916	0.680
Intraday seasonalities						
δ_1^s	1.407	0.136	1.855	0.247	1.374	0.324
$\delta_{s,1}^s$	0.896	0.056	1.021	0.095	0.624	0.125
$\delta_{s,2}^s$	0.080	0.037	0.050	0.062	-0.138	0.085
$\delta_{s,3}^s$	0.054	0.033	-0.003	0.055	-0.149	0.077
$\delta_{s,4}^s$	0.114	0.030	0.092	0.053	-0.001	0.075
$\delta_{s,5}^s$	-0.077	0.030	-0.145	0.051	-0.162	0.072
$\delta_{s,6}^s$	-0.001	0.028	-0.049	0.049	-0.079	0.068
$\delta_{c,1}^s$	-0.050	0.035	0.264	0.055	0.356	0.074
$\delta_{c,2}^s$	0.170	0.030	0.365	0.052	0.280	0.069
$\delta_{c,3}^s$	0.043	0.029	0.086	0.050	-0.098	0.068
$\delta_{c,4}^s$	0.150	0.030	0.178	0.051	0.113	0.068
$\delta_{c,5}^s$	0.157	0.029	0.143	0.049	0.000	0.067
$\delta_{c,6}^s$	-0.022	0.026	-0.050	0.046	0.040	0.064
Seasonalities over the future's maturity						
$\mathbb{1}_{>150}$	0.353	0.328	0.902	0.374	1.398	0.467
δ_1^{*s}	-3.264	0.384	-2.742	0.465	-2.250	0.569
$\delta_{s,1}^{*s}$	-0.181	0.141	-0.084	0.181	0.180	0.202
$\delta_{s,2}^{*s}$	0.291	0.094	0.396	0.107	0.490	0.119
$\delta_{s,3}^{*s}$	-0.070	0.076	-0.043	0.083	0.031	0.094
$\delta_{s,4}^{*s}$	0.113	0.067	0.171	0.071	0.166	0.084
$\delta_{s,5}^{*s}$	0.041	0.062	0.066	0.064	0.107	0.078
$\delta_{s,6}^{*s}$	-0.015	0.056	-0.014	0.057	0.042	0.071
$\delta_{c,1}^{*s}$	0.981	0.087	1.010	0.122	1.005	0.116
$\delta_{c,2}^{*s}$	-0.056	0.073	-0.038	0.080	-0.084	0.080
$\delta_{c,3}^{*s}$	-0.055	0.067	-0.112	0.067	-0.198	0.076
$\delta_{c,4}^{*s}$	0.040	0.061	0.021	0.062	0.002	0.072
$\delta_{c,5}^{*s}$	-0.071	0.056	-0.133	0.058	-0.133	0.069
$\delta_{c,6}^{*s}$	0.076	0.053	0.002	0.054	0.134	0.067

Table 37 cont'd:

	$dp = 10$		$dp = 15$		$dp = 20$	
	est.	S.E.	est.	S.E.	est.	S.E.
	Heterogeneity variance					
$\text{Var}[v_i]$	0.576	0.047	0.897	0.074	1.156	0.095
	Dynamic parameters					
α_1	0.913	0.009	0.982	0.004	0.993	0.002
β_1	−0.680	0.015	−0.828	0.018	−0.882	0.022
β_2	−0.055	0.012	−0.097	0.017	−0.067	0.022
	Diagnostics					
Obs	7490		3559		2165	
LL	-14597		-7141		-4304	
BIC	-14793		-7317		-4465	
$\bar{\bar{e}}_i$	-0.812		-0.176		0.245	
S.D.	1.476		1.662		1.809	

Table 38: Test statistics for a test on serial correlation. Based on the regressions in Table 37. $\Upsilon^{(j)}$ denotes the $\chi^2(1)$ test statistic as derived in (6.37).

	$dp = 10$	$dp = 15$	$dp = 20$
$\Upsilon^{(1)}$	0.35	0.88	0.21
$\Upsilon^{(2)}$	6.22**	17.97***	6.39**
$\Upsilon^{(3)}$	0.25	0.77	0.15
$\Upsilon^{(4)}$	3.45	0.30	0.80
$\Upsilon^{(5)}$	1.94	9.52***	3.98**
$\Upsilon^{(6)}$	0.51	0.01	0.20
$\Upsilon^{(7)}$	0.64	0.04	0.95
$\Upsilon^{(8)}$	0.01	0.01	1.17
$\Upsilon^{(9)}$	0.22	1.74	0.60
$\Upsilon^{(10)}$	1.38	0.01	1.88
$\Upsilon^{(11)}$	1.18	0.00	0.68
$\Upsilon^{(12)}$	0.02	1.08	3.76*
$\Upsilon^{(13)}$	0.23	0.10	0.21
$\Upsilon^{(14)}$	1.56	0.14	2.32
$\Upsilon^{(15)}$	0.46	0.95	0.61

***, **, *: significance at the 1%, 5% or 10% level, respectively.

Furthermore, the fact that we cannot identify the beginning of the next roll-over period from this contract to the following front month contract might be explained by the theoretical finding that the stochastic properties of the future should converge to the underlying asset with decreasing time to maturity. Accordingly, the volatility pattern of the future should be mainly driven by the volatility of the underlying asset.

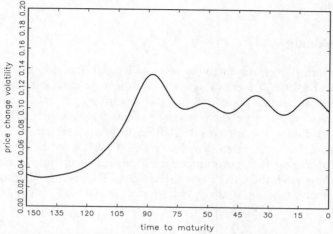

Figure 43: Price change volatility $\sigma^{*2}_{(x^{dp})}(t^{dp}_i)$ vs. time to maturity for $dp = 20$. 14:00 GMT, Bund future trading, LIFFE.

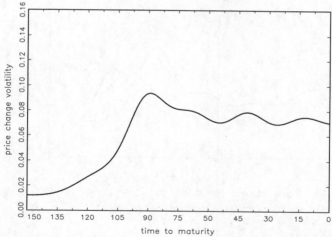

Figure 44: Price change volatility $\sigma^{*2}_{(x^{dp})}(t^{dp}_i)$ vs. time to maturity for $dp = 10$. 14:00 GMT, Bund future trading, LIFFE.

By analyzing the volatility patterns based on the price change durations for 10 tick and 15 tick price changes, we can confirm our results found for the 20 tick aggregation level. Figure 44 shows the corresponding volatility pattern based on 10 tick price changes. Note, that the similarity of volatility patterns based on price changes of different sizes means that a higher frequency of small price changes is also combined with a higher frequency of larger price changes, i.e., fluctuations around a certain price level come along with movements of the price level itself.

Intraday Seasonalities

Figures 45 and 46 depict the intraday seasonality patterns for 10 tick and 20 tick price change durations evaluated based on a fixed time to maturity of 30 days. Several studies point out that the opening of related exchanges has an important influence on the intraday seasonality pattern[13]. The main corresponding exchange apart from the EUREX in Frankfurt, where a similar contract is traded, is the Chicago Board of Trade (CBOT), which is the main exchange for futures on U.S. government fixed income securities. During the analyzed sample period, the CBOT opened at 7:20 CT corresponding to 13:20 GMT. Furthermore, trading at the NYSE started at 9:30 ET, corresponding to 14:30 GMT.

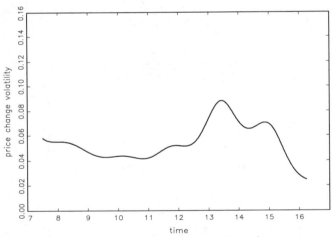

Figure 45: Price change volatility $\sigma^{*2}_{(x^{dp})}(t_i^{dp})$ vs. intraday time $dp = 10$. 30 days to maturity, Bund future trading, LIFFE.

Figure 45 shows a slight volatility spike in the beginning of the trading day which is well explained by the price finding process at the opening, where the price information conveyed by U.S. traders needs to be scrutinized. Then, the volatility reveals a decreasing pattern leading to the well known lunchtime

---

[13] See e.g. the analysis of Harvey and Huang (1991) on currency futures or the work of Gouriéroux, Jasiak, and LeFol (1999) on trading at the Paris Bourse.

effect around 11:00 GMT. However, the volatility increases sharply around 13:20 GMT which fits in well with the opening of the CBOT. After that period, the volatility does not remain on this high level, but drops continually, interrupted by an additional spike before 15:00 GMT which is obviously associated with the opening of the NYSE. The fact that the volatility level remains for the entire afternoon on a higher level than in the morning is in line with the results of Harvey and Huang (1991) who find a significantly increased volatility level, when corresponding exchanges are opened.

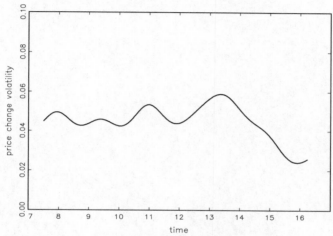

Figure 46: Price change volatility $\sigma^{*2}_{(x^{dp})}(t_i^{dp})$ vs. intradaytime for $dp = 20$. 30 days to maturity, Bund future trading, LIFFE.

The 20 tick volatility (Figure 46) reveals a less pronounced intraday pattern which confirms the findings based on the time-to-maturity volatility shapes. Hence, the intraday volatility seems to be driven mainly by more frequent but smaller price movements, while the time-to-maturity volatility patterns are dominated by lower frequent but larger price changes. Based on 20 tick price changes, the CBOT opening influence can still be identified, however, the volatility level drops significantly faster than for the five tick price changes. Hence, additional U.S. traders obviously do not generate price events above 20 ticks, after the initial information dissemination.

7

Univariate and Multivariate Dynamic Intensity Models

In this chapter, we discuss dynamic parameterizations of the intensity function. The important difference between this and the previous chapter is that here we focus on fully parametric models for the intensity function. Baseline intensity components are completely specified, and thus, a dynamic model is built directly on the *intensity* function instead of on the *integrated intensity* function. A direct parameterization of the intensity in a continuous-time framework is a valuable approach in order to account for time-varying covariates and multivariate structures.

We focus on two general ways of introducing dynamic structures in intensity processes. The first possibility is to parameterize the intensity function in terms of an autoregressive structure, which is updated at each occurrence of a new point. Following this idea, Russell (1999) proposes a dynamic extension of a parametric PI model that he calls *autoregressive conditional intensity* (ACI) model. A valuable alternative is to specify a so-called *self-exciting intensity process* where the intensity is driven by a function of the backward recurrence time to all previous points. A particular type of linear self-exciting processes is introduced by Hawkes (1971). Both types of models are presented in Section 7.1. We discuss the theoretical properties of the particular approaches and illustrate how to estimate the models. Multivariate extensions of both frameworks are presented in Section 7.2. In Section 7.3, we discuss dynamic latent factor models for intensity processes as proposed by Bauwens and Hautsch (2003). Such models are based on the assumption that the conditional intensity given the (observable) history of the process is not anymore deterministic but stochastic and follows a dynamic process. The major idea is to introduce a latent dynamic factor in both types of models, leading to a latent factor autoregressive conditional intensity (LF-ACI) and a latent factor Hawkes (LF-Hawkes) model. Section 7.4 deals with applications of the different types of dynamic intensity models to financial duration data. We illustrate the use of multivariate Hawkes models to estimate multivariate instantaneous price change volatilities. A further objective is devoted to the analysis of limit order book data. In this framework, we apply a bivariate ACI model to estimate the simultaneous buy/sell arrival intensity based on data from the ASX. Finally, we illustrate applications of univariate and multivariate LFI models for the investigation of XETRA trade durations.

7.1 Univariate Dynamic Intensity Models

In the following sections, we assume the existence of a time-varying covariate process[1] which occurs at discrete points $t_1^0, t_2^0, \ldots, t_{n^0}^0$. Then, $N^0(t)$ and $\check{N}^0(t)$ denote the corresponding right-continuous and, respectively, left-continuous counting processes associated with the arrival times of the covariate process z_i^0. Moreover, let $\{\tilde{t}_i\}$ be the pooled process of *all* points t_i and t_i^0.

7.1.1 The ACI Model

A straightforward way of specifying an autoregressive intensity process is to parameterize the intensity function in terms of a time series model. Then, $\lambda(t; \mathfrak{F}_t)$ follows a dynamic process that is updated whenever a new point occurs. As proposed by Russell (1999), the intensity is driven by three components: one component $\Psi(t)$ capturing the dynamic structure, a baseline intensity component $\lambda_0(t)$ (Russell calls it backward recurrence time function), as well as a seasonality component $s(t)$. By imposing a multiplicative structure, the ACI model is obtained by a dynamic extension of a (parametric) PI model and is given by

$$\lambda(t; \mathfrak{F}_t) = \Psi(t)\lambda_0(t)s(t). \tag{7.1}$$

By including static covariates and specifying $\Psi(t)$ in logarithmic form, $\Psi(t)$ is given by

$$\Psi(t) = \exp\left(\tilde{\Psi}_{\check{N}(t)+1} + z'_{\check{N}(t)}\gamma + z^{0\prime}_{\check{N}^0(t)}\tilde{\gamma}\right) \tag{7.2}$$

$$\tilde{\Psi}_i = \omega + \sum_{j=1}^{P} \alpha_j \check{\epsilon}_{i-j} + \sum_{j=1}^{Q} \beta_j \tilde{\Psi}_{i-j}, \tag{7.3}$$

where the innovation term can specified either as

$$\check{\epsilon}_i := 1 - \epsilon_i^* = 1 - \Lambda(t_{i-1}, t_i) \tag{7.4}$$

or, alternatively, as

$$\check{\epsilon}_i := -0.5772 - \ln \epsilon_i^* = -0.5772 - \ln \Lambda(t_{i-1}, t_i). \tag{7.5}$$

---

[1] For example, such a process might be associated with the arrival of new orders in the market.

Note that $\Psi(t)$ is a left-continuous function that only changes during a spell due to changes of the time-varying covariate process. Hence, in absence of time-varying covariates, $\Psi(t)$ remains constant between t_{i-1} and t_i, i.e.,

$$\Psi(t) = \Psi(t_i) \qquad \text{for } t_{i-1} < t \leq t_i.$$

In that case, $\Psi(t_i)$ is known instantaneously after the occurrence of t_{i-1} and does not change until t_i. Then, $\lambda(t)$ changes between t_{i-1} and t_i only as a deterministic function of time (according to $\lambda_0(t)$ and $s(t)$). Under the conditions given in Theorem 2.2 (see Chapter 2), the integrated intensity follows an i.i.d. standard exponential process, hence, the model innovations $\breve{e}_i$ are i.i.d. exponential or, respectively, extreme value variates that are centered by their unconditional mean and enter the model negatively. This leads to a positive value of α when the intensity is positively autocorrelated.

Note that in difference to the ACPI model presented in Chapter 6, the baseline intensity function $\lambda_0(t)$ is fully parameterized. Russell (1999) suggests to specify $\lambda_0(t)$ according to a standard Weibull parameterization

$$\lambda_0(t) = \exp(\omega)x(t)^{a-1}, \qquad a > 0, \tag{7.6}$$

where a value of a larger (smaller) than one is associated with an upward (downward) sloping intensity function, i.e. a "positive" or respectively "negative" duration dependence. Alternatively, a standard Burr type baseline intensity function is obtained by

$$\lambda_0(t) = \exp(\omega)\frac{x(t)^{a-1}}{1+\eta x(t)^a}, \qquad a > 0, \eta \geq 0, \tag{7.7}$$

which allows for non-monotonous hazard shapes.

Note that $\tilde{\Psi}_i$ follows an ARMA type dynamic that is updated by i.i.d. zero mean innovations. Thus, $E[\tilde{\Psi}_i] = 0$ and weak stationarity of $\tilde{\Psi}_i$ is ensured if the roots of the lag polynomial based on the persistence parameters $\beta_1, \ldots, \beta_Q$ lie inside the unit circle.

The explicit derivation of theoretical moments of the intensity function is rendered difficult due to the fact that conditional expectations of $\lambda(t; \mathfrak{F}_t)$ typically cannot be computed analytically. This is because the relationship between the intensity function at some point t_i and the expected time until the next point t_{i+1} mostly cannot be expressed in closed form. In general, the computation of the conditional expected arrival time of the next point, $E\left[t_i | \mathfrak{F}_{t_{i-1}}\right]$, is performed by exploiting the relationship $\epsilon_i := \Lambda(t_{i-1}, t_i) \sim Exp(1)$. However, in the case of a Weibull parameterization of $\lambda_0(t)$, (7.6), $\Lambda(t_{i-1}, t_i)$ in fact can be expressed in closed form. Then, under the assumption of no (time-varying) covariate arrival during the current spell, $E\left[t_i | \mathfrak{F}_{t_{i-1}}\right]$ is computed as

$$E\left[t_i | \mathfrak{F}_{t_{i-1}}\right] = t_{i-1} + E\left[\left[\frac{\epsilon_i a}{\Psi(t_i)\exp(\omega)}\right]^{1/a} \Bigg| \mathfrak{F}_{t_{i-1}}\right]. \tag{7.8}$$

Correspondingly, the conditionally expected intensity at the next point t_i is derived·as

$$\mathrm{E}\left[\lambda(t_i)\middle|\, \mathfrak{F}_{t_{i-1}}\right] = \Psi(t_i)\exp(\omega)\mathrm{E}\left[(t_i - t_{i-1})^{a-1}\middle|\, \mathfrak{F}_{t_{i-1}}\right]$$

$$= \Psi(t_i)\exp(\omega)\mathrm{E}\left[\left[\frac{\epsilon_i a}{\Psi(t_i)\exp(\omega)}\right]^{\frac{a-1}{a}}\middle|\, \mathfrak{F}_{t_{i-1}}\right]. \qquad (7.9)$$

Hence, these expressions are computed based on the conditional mean of transformations of an exponential variate. Nevertheless, the computation of autocovariances of the intensity function and of the corresponding durations x_i has to be performed numerically on the basis of simulation procedures. In order to provide insights into the dynamic properties of ACI processes, we simulate ACI(1,1) dynamics based on a constant baseline intensity function, i.e., $\lambda_0(t) = \exp(\omega)$. Figures 47 through 49 show simulated ACI(1,1) processes based on different parameter settings. The figures depict the ACF of $\lambda(t_i; \mathfrak{F}_{t_i})$ (measured at the particular points $t_1, t_2, \ldots$), as well as the ACF of the resulting duration process. We choose parameterizations associated with differently persistent intensity processes. The processes shown in Figure 47 and 48 are based on a value of $\beta = 0.97$ and thus are close to the non-stationary region. Both processes imply significantly autocorrelated durations coming along with an exponentially decay of the ACF. In particular, the ACF of the durations in Figure 47 shows a shape that is quite typical for real financial duration series. Obviously, the persistence of the duration processes strongly depends on the value of β, while the strength of the serial dependence is driven by the innovation parameter α. For example, for $\alpha = 0.05$ and $\beta = 0.7$ (Figure 49), the ACF of the duration series declines sharply and tends towards zero for higher lags. Thus, autocorrelation patterns that resemble the autocorrelation structures of typical financial duration series require values of β quite close to one.

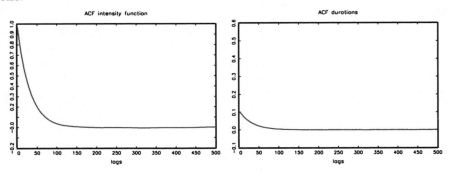

Figure 47: ACF of univariate ACI(1,1) processes. Left: ACF of $\lambda(t_i; \mathfrak{F}_{t_i})$. Right: ACF of x_i. Based on $5,000,000$ drawings. Parameterization: $\omega = 0$, $\alpha = 0.05$, $\beta = 0.97$.

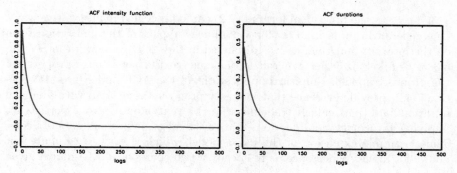

Figure 48: ACF of univariate ACI(1,1) processes. Left: ACF of $\lambda(t_i; \mathfrak{F}_{t_i})$. Right: ACF of x_i. Based on $5,000,000$ drawings. Parameterization: $\omega = 0$, $\alpha = 0.2$, $\beta = 0.97$.

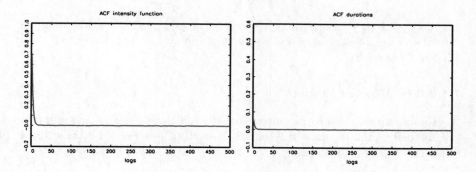

Figure 49: ACF of univariate ACI(1,1) processes. Left: ACF of $\lambda(t_i; \mathfrak{F}_{t_i})$. Right: ACF of x_i. Based on $5,000,000$ drawings. Parameterization: $\omega = 0$, $\alpha = 0.05$, $\beta = 0.7$.

The log likelihood function of the ACI model is obtained by adapting (2.22) in an obvious way. For example, by assuming a Weibull type baseline intensity function, (7.6), the integrated intensity function can be computed analytically, and the log likelihood function is obtained based on piecewise analytical integrations of the intensity function over all points $\tilde{t}_j$ with $t_{i-1} < \tilde{t}_j \leq t_i$, hence

$$\ln \mathcal{L}(W; \theta) = \sum_{i=1}^{n} \left\{ -\sum_j \left[\Psi(\tilde{t}_j) s(\tilde{t}_j) \exp(\omega) \frac{1}{a} (\tilde{t}_{j+1} - \tilde{t}_j)^a \right] \right.$$
$$\left. + \ln \left(\Psi(t_i) s(t_i) \exp(\omega) x_i^{a-1} \right) \right\}. \tag{7.10}$$

However, in the case of more sophisticated parameterizations of $\lambda_0(t)$, $\Lambda(t_{i-1}, t_i)$ has to be computed numerically.

Of course, the functional form in (7.2) can be specified in alternative ways. One generalization is to allow for a nonlinear impact of the innovation term on the intensity function, as it is discussed in the ACD framework in Section 5.3 of Chapter 5. Other generalizations concern the persistence property of the proposed model. Empirical applications of the ACI model (see also Section 7.4.2) provide evidence that intensity processes associated with financial duration data (particularly trade durations) reveals a quite strong persistence implying autocorrelation functions that reveal a low, hyperbolic rate of decay. Thus, a simple extension of the basic model, which is easy to estimate but allows for higher persistence, is to specify a two-component ACI model given by

$$\tilde{\Psi}_i = w\tilde{\Psi}_{i,1} + (1-w)\tilde{\Psi}_{i,2} \tag{7.11}$$

$$\tilde{\Psi}_{i,1} = \alpha_1 \breve{\epsilon}_{i-1} + \tilde{\Psi}_{i-1,1}$$

$$\tilde{\Psi}_{2,i} = \alpha_2 \breve{\epsilon}_{i-1} + \beta\tilde{\Psi}_{i-1,2}, \tag{7.12}$$

where $\tilde{\Psi}_{i,1}$ and $\tilde{\Psi}_{i,2}$ denote the two intensity components leading to a weighted sum of an integrated process and a weakly stationary process (see also Section 5.3.4 in Chapter 5).

7.1.2 The Hawkes Model

A valuable alternative to an autoregressive specification is to parameterize the intensity function in terms of a self-exciting process. The basic linear self-exciting process is given by

$$\lambda(t; \mathfrak{F}_t) = \exp(\omega) + \int_{-\infty}^{t} w(t-s)dN(s)$$

$$= \exp(\omega) + \sum_{i \geq 1} \mathbb{1}_{\{t_i \leq t\}} w(t-t_i), \tag{7.13}$$

where ω is a constant, and $w(\cdot)$ is some non-negative weighting function[2]. Hence, in self-exciting processes, the intensity is driven by a weighted non-increasing function of the backward recurrence time to all previous points. Hawkes (1971) introduces a general class of self-exciting processes given by

$$\lambda(t; \mathfrak{F}_t) = \mu(t) + \int_{(0,t)} \sum_{j=1}^{P} \alpha_j \exp(-\beta_j(t-u))dN(u)$$

$$= \mu(t) + \sum_{j=1}^{P} \sum_{i=1}^{\breve{N}(t)} \alpha_j \exp(-\beta_j(t-t_i)), \tag{7.14}$$

---

[2] See, for example, Hawkes (1971), Hawkes and Oakes (1974) or Cox and Isham (1980).

where $\alpha_j \geq 0$, $\beta_j \geq 0$ for $j = 1, \ldots, P$ and $\mu(t) > 0$ is a deterministic function of time. The parameter α_j determines the scale, whereas β_j determines the time decay of the influence of past points of the process. Hence, the response of a previous event on the intensity function in t decays exponentially and is driven by the parameter β_j. For $P > 1$, the Hawkes(P) model is based on the superposition of differently parameterized exponentially decaying weighted sums of the backward recurrence time to all previous points. The function $\mu(t)$ can be specified in terms of covariates $z_{\breve{N}(t)}$ and $z^0_{\breve{N}^0(t)}$ and a seasonality function $s(t)$, thus

$$\mu(t) = \exp(\omega) + s(t) + z'_{\breve{N}(t)}\gamma + z^{0\prime}_{\breve{N}^0(t)}\tilde{\gamma}. \tag{7.15}$$

Note that the non-negativity of $\mu(t)$ has to be ensured. Thus, alternatively, $\mu(t)$ might be specified in terms of a logarithmic form. Alternatively, one might assume a multiplicative impact of the seasonality function. In this context, the model is given by

$$\lambda(t; \mathfrak{F}_t) = s(t) \left\{ \mu(t) + \sum_{j=1}^{P} \sum_{i=1}^{\breve{N}(t)} \alpha_j \exp(-\beta_j(t - t_i)) \right\}. \tag{7.16}$$

Obviously, alternative functional forms for the decay function are possible. However, the choice of an exponential decay simplifies the derivation of the theoretical properties of the model. More details can be found in Hawkes (1971), Hawkes and Oakes (1974) or Ogata and Akaike (1982). Hawkes (1971) shows if $\sum_{j=1}^{P} \alpha_j/\beta_j < 1$, $\lambda(t)$ is a weakly stationary process, where for the special case $P = 1$ and $\mu(t) = \mu$, the unconditional mean of $\lambda(t; \mathfrak{F}_t)$ is given by

$$\mathrm{E}[\lambda(t_i)] = \frac{\mu}{1 - \int_0^{\infty} \alpha \exp(-\beta u) du} = \frac{\mu \beta}{\beta - \alpha}. \tag{7.17}$$

Hawkes-type intensity models can serve as epidemic models since the occurrence of a number of events increases the probability for further events. In natural sciences, they play an important role in modelling the emission of particles from a radiating body or in forecasting seismic events (see, for example, Vere-Jones, 1970). The study by Bowsher (2002) was among the first applications of Hawkes type processes to financial data. He proposes a generalization (the so-called generalized Hawkes model) that nests the basic model and allows for spill-over effects between particular trading days.

Note that the Hawkes model allows us to estimate dependencies in the intensity process without imposing parametric time series structures. The dependence is specified in terms of the elapsed time since past events. Hence, the marginal contribution of some previous event on the current intensity is independent from the number of intervening events.

As in the case of ACI processes, the computation of the conditional expectation of the next point of the process, $\mathrm{E}\left[t_i | \mathfrak{F}_{t_{i-1}}\right]$, cannot be expressed in closed form. Therefore, t_i is computed as the solution of

$$
\begin{aligned}
\epsilon_i &= \Lambda(t_{i-1}, t_i) \\
&= \int_{t_{i-1}}^{t_i} \mu(s)ds - \sum_{j=1}^{P}\sum_{k=1}^{i-1} \frac{\alpha_j}{\beta_j}\left[1 - \exp(-\beta_j(t_i - t_k))\right],
\end{aligned}
\tag{7.18}
$$

where ϵ_i denotes an i.i.d. standard exponential variate. The solution of (7.18) leads to a nonlinear function of an exponentially distributed random variable, $t_i = g_1(\epsilon_i; \mathfrak{F}_{t_{i-1}})$ and is calculated by using the law of iterated expectations (under the assumption of no time-varying covariates) as

$$
\mathrm{E}\left[t_i | \mathfrak{F}_{t_{i-1}}\right] = \mathrm{E}\left[\mathrm{E}\left[g_1(\epsilon_i; \mathfrak{F}_{t_{i-1}}) | \epsilon_i\right] \big| \mathfrak{F}_{t_{i-1}}\right].
\tag{7.19}
$$

The conditional expectation of $\lambda(t_i)$ given the information set at t_{i-1} is computed as

$$
\mathrm{E}\left[\lambda(t_i) | \mathfrak{F}_{t_{i-1}}\right] = \mu(t_{i-1}) + \sum_{j=1}^{P}\sum_{k=1}^{i-1} \alpha_j \mathrm{E}\left[\exp(-\beta_j t_i) | \mathfrak{F}_{t_{i-1}}\right] \exp(\beta_j t_k).
\tag{7.20}
$$

This expressions requires to compute the conditional expectation of a function of the time until the next point. However, this conditional expectation typically cannot be expressed in closed form and requires to use simulation methods.

In order to provide more insights into the dynamic properties of Hawkes processes, we simulate several processes under different parameterizations.

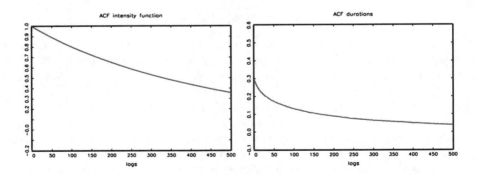

Figure 50: ACF of univariate Hawkes(1) processes. Left: ACF of $\lambda(t_i; \mathfrak{F}_{t_i})$. Right: ACF of x_i. Based on $5,000,000$ drawings. Parameterization: $\omega = \ln(0.2)$, $\alpha = 0.2$, $\beta = 0.21$.

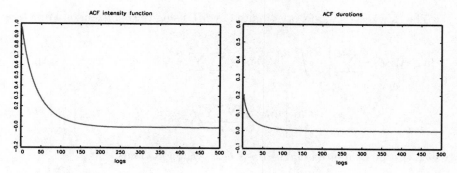

Figure 51: ACF of univariate Hawkes(1) processes. Left: ACF of $\lambda(t_i; \mathfrak{F}_{t_i})$. Right: ACF of x_i. Based on $5,000,000$ drawings. Parameterization: $\omega = \ln(0.2)$, $\alpha = 0.2$, $\beta = 0.25$.

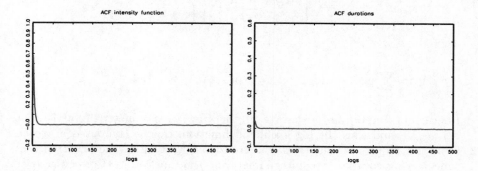

Figure 52: ACF of univariate Hawkes(1) processes. Left: ACF of $\lambda(t_i; \mathfrak{F}_{t_i})$. Right: ACF of x_i. Based on $5,000,000$ drawings. Parameterization: $\omega = \ln(0.2)$, $\alpha = 0.2$, $\beta = 0.5$.

The illustrations in Figures 50 through 52 are described as in Section 7.1.1. It is nicely shown that the persistence of the intensity process and the resulting duration process strongly depends on the ratio α/β. Figure 50 is associated with a parameterization leading to a quite persistent intensity process. The ACF of the implied duration process shows a pattern that is very realistic for financial duration series (compare to the empirical autocorrelation functions shown in Chapter 4). Hence, the Hawkes process allows for quite persistent duration processes implying a relatively slowly decaying ACF. It turns out that the persistence in the implied duration series is higher than for ACI processes even with values of β very close to one. Figure 51 shows a less persistent process coming along with autocorrelation functions that clearly decay faster than in Figure 50. In contrast, Figure 52 reveals only very weak serial dependencies.

The Hawkes(P) model is estimated by ML using the log likelihood function

$$
\ln \mathcal{L}\left(W;\theta\right) = \sum_{i=1}^{n} \left\{ - \int_{t_{i-1}}^{t_i} \mu(s)ds - \sum_{j=1}^{P}\sum_{k=1}^{i-1} \frac{\alpha_j}{\beta_j}\left[1 - \exp(-\beta_j(t_i - t_k))\right] \right.
$$

$$
\left. + \ln\left[\mu(t_i) + \sum_{j=1}^{P}\sum_{k=1}^{i-1}\alpha_j \exp(-\beta_j(t_i - t_k))\right]\right\}, \quad (7.21)
$$

which can be computed on the basis of a recursion. Thus,

$$
\ln \mathcal{L}\left(W;\theta\right) = \sum_{i=1}^{n} \left\{ - \int_{t_{i-1}}^{t_i} \mu(s)ds - \sum_{j=1}^{P}\sum_{k=1}^{i-1} \frac{\alpha_j}{\beta_j}\left[1 - \exp(-\beta_j(t_i - t_k))\right] \right.
$$

$$
\left. + \ln\left[\mu(t_i) + \sum_{j=1}^{P}\alpha_j A_i^j\right]\right\}, \quad (7.22)
$$

where

$$
A_i^j := \sum_{k=1}^{i-1} \exp(-\beta_j(t_i - t_k)) = \exp(-\beta_j(t_i - t_{i-1}))(1 + A_{i-1}^j). \quad (7.23)
$$

Hence, one important advantage of the Hawkes parameterization is that $\Lambda(t_i, t_{i-1})$, and thus the log likelihood function, can be computed in closed form and require no numerical integration. Nevertheless, the occurrence of time-varying covariates requires integrating the function $\mu(s)$ piecewise over all intervening points $\{\tilde{t}_j\}$ with $t_{i-1} < \tilde{t}_j \leq t_i$.

7.2 Multivariate Dynamic Intensity Models

7.2.1 Definitions

In order to consider multivariate point processes, the fundamental concepts presented in Chapter 2 have to be extended to the multivariate case. In the following, we define a S-dimensional point process with the sequence of arrival times $\{t_i^s\}$ of the single processes $s = 1, \ldots, S$. Accordingly, let $N^s(t)$ and $\breve{N}^s(t)$ represent the right-continuous and, respectively, left-continuous counting functions associated with s-type events. Correspondingly, $N(t)$ denotes the counting function of the *pooled* process, which pools and orders the arrival times of *all* particular processes. Let n denote the number of points in the pooled process and n^s the number of s-type elements in the sample. Furthermore, define y_i^s as an indicator variable that takes the value 1 if the i-th point of the pooled process is of type s. Moreover, x_i^s, $i = 1, \ldots n^s$, and $x^s(t)$

denote the durations and backward recurrence times associated with the s-th process.

We assume that the pooled process is orderly which implies that all single point processes are orderly, too[3]. Under this assumption, the intensity function for the s-th process is given by

$$\lambda^s(t; \mathfrak{F}_t) = \lim_{\Delta \to 0} \Delta^{-1} \text{Prob} \left[N^s(t+\Delta) - N^s(t) > 0, N^{s'}(t+\Delta) - N^{s'}(t) = 0 \,|\, \mathfrak{F}_t \right]$$

$$(7.24)$$

$\forall s \neq s'$. Note that the properties of the intensity function also hold in the multivariate case (see, for example, Brémaud, 1981 or Bowsher, 2002). Hence, for the s-type integrated intensity function, $\Lambda^s(t^s_{i-1}, t^s_i) = \int_{t^s_{i-1}}^{t^s_i} \lambda^s(u; \mathfrak{F}_u) du$, it follows that

$$\Lambda^s(t^s_{i-1}, t^s_i) \sim \text{i.i.d.} \quad Exp(1). \tag{7.25}$$

This property also holds for the integrated intensity associated with the pooled process. Since the $\mathfrak{F}_t$-intensity of the pooled process is $\lambda(t; \mathfrak{F}_t) = \sum_{s=1}^{S} \lambda^s(t; \mathfrak{F}_t)$, it is shown that[4]

$$\Lambda(t_{i-1}, t_i) := \int_{t_{i-1}}^{t_i} \sum_{s=1}^{S} \lambda^s(u; \mathfrak{F}_u) du \sim \text{i.i.d.} \quad Exp(1). \tag{7.26}$$

Theorem 5.2 in Karr (1991) (see Chapter 2) also holds in the multivariate case. Thus, the log likelihood function is given by

$$\ln \mathcal{L}(W; \theta) = \sum_{s=1}^{S} \left\{ \int_0^{t_n} (1 - \lambda^s(u; \mathfrak{F}_u)) du + \sum_{i \geq 1} \mathbb{1}_{\{t^s_i \leq t_n\}} y^s_i \ln \lambda^s(t_i; \mathfrak{F}_{t_i}) \right\},$$

$$(7.27)$$

where t_n denotes the last observable point of the pooled process.

7.2.2 The Multivariate ACI Model

A multivariate ACI model is based on the specification of the vector of S intensity functions

$$\lambda(t; \mathfrak{F}_t) = \begin{bmatrix} \lambda^1(t; \mathfrak{F}_t) \\ \lambda^2(t; \mathfrak{F}_t) \\ \vdots \\ \lambda^S(t; \mathfrak{F}_t) \end{bmatrix}, \tag{7.28}$$

---

[3] In other words, the pooled process is a *simple* point process.
[4] See also Bowsher (2002).

where each component is parameterized as

$$\lambda^s(t; \mathfrak{F}_t) = \Psi^s(t)\lambda_0^s(t)s^s(t), \qquad s = 1, \ldots S, \tag{7.29}$$

where $\Psi^s(t)$, $\lambda_0^s(t)$ and $s^s(t)$ are the corresponding s-type dynamic component, baseline intensity function and seasonality component. Russell (1999) proposes specifying $\Psi^s(t)$ as[5]

$$\Psi^s(t) = \exp\left(\tilde{\Psi}_{\tilde{N}(t)+1}^s + z_{\tilde{N}(t)}'\gamma^s + z_{\tilde{N}^0(t)}^{0\prime}\tilde{\gamma}^s\right), \tag{7.30}$$

where the vector $\tilde{\Psi}_i = \left(\tilde{\Psi}_i^1, \tilde{\Psi}_i^2, \ldots, \tilde{\Psi}_i^S\right)$ is parameterized in terms of a VARMA type specification, given by

$$\tilde{\Psi}_i = \left(A^s \check{\epsilon}_{i-1} + B\tilde{\Psi}_{i-1}\right) y_{i-1}^s, \tag{7.31}$$

with $A^s = \{\alpha_j^s\}$ denoting a $S \times 1$ vector associated with the innovation term and $B = \{\beta_{ij}\}$ a $S \times S$ matrix of persistence parameters. The innovation term $\check{\epsilon}_i$ is given by

$$\check{\epsilon}_i = \epsilon_i^{s*} y_i^s, \tag{7.32}$$

where

$$\epsilon_i^{s*} := 1 - \Lambda^s(t_{i-1}^s, t_i^s) \tag{7.33}$$

or

$$\epsilon_i^{s*} := -0.5772 - \ln \Lambda^s(t_{i-1}^s, t_i^s) \tag{7.34}$$

denotes the s-type innovation term with

$$\Lambda^s(t_{i-1}^s, t_i^s) = \sum_j \int_{\tilde{t}_j^s}^{\tilde{t}_{j+1}^s} \lambda^s(u; \mathfrak{F}_u) du, \tag{7.35}$$

and j indexes all points with $t_{i-1}^s < \tilde{t}_j \leq t_i^s$.

Since we assume that the process is orderly, and thus only one type of point can occur at each instant, the innovation $\check{\epsilon}_i$ is a scalar random variable and is associated with the most recently observed process. Note that (7.31) implies a regime-switching structure since A^s is a vector of coefficients reflecting the impact of the innovation term on the intensity of the S processes when the previous point was of type s.

Generally, it is not straightforward to establish the distributional properties of $\check{\epsilon}_i$ since $\check{\epsilon}_i$ is a mixture of the particular processes $\check{\epsilon}_i^s$, $s = 1, \ldots, S$,

---

[5] For ease of illustration, we restrict our consideration to a lag order of one. The extension to higher order specifications is straightforward.

which are exponentially or extreme value distributed, respectively. Neverthe-
less, since $\check{\epsilon}_i$ is based on a mixture of i.i.d. random variables, it follows itself an
i.i.d. process. For this reason, weak stationarity of the model depends on the
eigenvalues of the matrix B. Thus, if the eigenvalues of B lie inside the unit
circle, the process $\tilde{\Psi}_i$ is weakly stationary. Furthermore, own simulation stud-
ies strongly support the hypothesis that $\mathrm{E}[\check{\epsilon}_i] = \mathrm{E}[\epsilon_i^{s*}]$. Thus, it is reasonable
to center the innovation terms as presented in (7.33) and (7.34).

However, Bowsher (2002) proposes to construct an alternative innovation
term by exploiting the relationship given in (7.26). Hence $\check{\epsilon}_i$ can be alternatively specified as

$$\check{\epsilon}_i = \int_{t_{i-1}}^{t_i} \sum_{s=1}^{S} \lambda^s(t; \mathfrak{F}_t) ds, \qquad (7.36)$$

which is an i.i.d. exponential variate under correct model specification.

The baseline intensity functions are extensions of the univariate case. The
most simple way is to assume that $\lambda_0^s(t)$ is constant, but depends on the type
of the event that occurred most recently, i.e.

$$\lambda_0^s(t) = \exp(\omega_r^s) \, y_{\check{N}(t)}^r, \quad r = 1, \ldots, S, \quad s = 1, \ldots, S. \qquad (7.37)$$

Alternatively, the baseline intensity function may be specified as a multivariate
extension of (7.6), i.e.,

$$\lambda_0^s(t) = \exp(\omega^s) \prod_{r=1}^{S} x^r(t)^{a_r^s - 1}, \; (a_r^s > 0), \qquad (7.38)$$

or of (7.7)

$$\lambda_0^s(t) = \exp(\omega^s) \prod_{r=1}^{S} \frac{x^r(t)^{a_r^s - 1}}{1 + \eta_r^s x^r(t)^{a_r^s}}, \; (a_r^s > 0, \; \eta_r^s \geq 0). \qquad (7.39)$$

A special case occurs when the s-th process depends only on its own backward
recurrence time, which corresponds to $a_r^s = 1$ and $\eta_r^s = 0, \; \forall \, r \neq s$.

The computations of $\mathrm{E}\left[t_i^s \mid \mathfrak{F}_{t_{i-1}}\right]$ and $\mathrm{E}\left[\lambda^s(t_i) \mid \mathfrak{F}_{t_{i-1}}\right]$ are complicated
by the fact that in the multivariate case, $\lambda_0^s(t)$ depends on the backward
recurrence times of *all* S processes. In that case, there exists no closed form
solution for the integrated intensity[6], and thus, $\Lambda^s(t_{i-1}^s, t_i^s)$ as well as the
conditional expectations of t_i^s and $\lambda^s(t_i)$ have to be computed numerically.

Figures 53 through 55 show the autocorrelation functions of bivariate simulated ACI(1,1) processes. In order to illustrate the dynamic interactions between the two processes, we show the cross-autocorrelation function (CAFC)

---

[6] With the exception of the most simple case of a constant baseline intensity function.

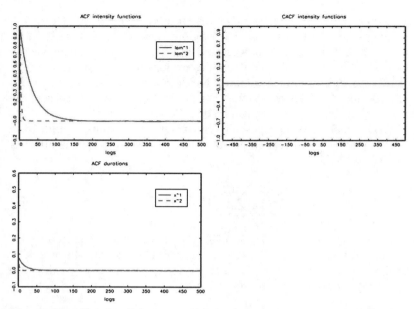

Figure 53: Simulated bivariate ACI(1,1) processes. Upper: ACF and CACF of $\lambda^{1,2}(t_i; \mathfrak{F}_{t_i})$. Lower: ACF of $x_i^{1,2}$. $\omega^1 = \omega^2 = 0$, $\alpha_1^1 = \alpha_2^2 = 0.05$, $\alpha_2^1 = \alpha_1^2 = 0$, $\beta_{11} = 0.97$, $\beta_{12} = \beta_{21} = 0$, $\beta_{22} = 0.7$. Based on $5,000,000$ drawings.

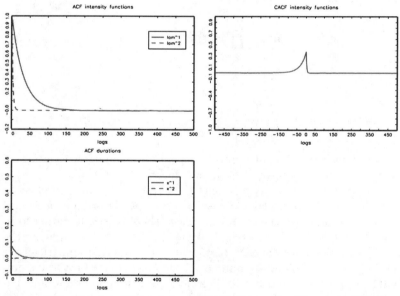

Figure 54: Simulated bivariate ACI(1,1) processes. Upper: ACF and CACF of $\lambda^{1,2}(t_i; \mathfrak{F}_{t_i})$. Lower: ACF of $x_i^{1,2}$. $\omega^1 = \omega^2 = 0$, $\alpha_1^1 = \alpha_2^2 = \alpha_2^2 = 0.05$, $\alpha_1^2 = 0$, $\beta_{11} = 0.97$, $\beta_{12} = \beta_{21} = 0$, $\beta_{22} = 0.7$. Based on $5,000,000$ drawings.

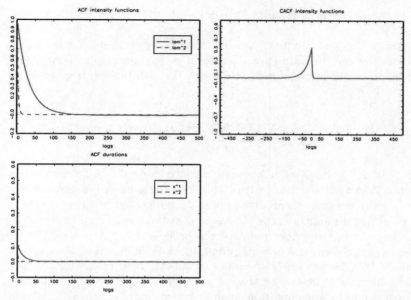

Figure 55: Simulated bivariate ACI(1,1) processes. Upper: ACF and CACF of $\lambda^{1,2}(t_i; \mathfrak{F}_{t_i})$. Lower: ACF of $x_i^{1,2}$. $\omega^1 = \omega^2 = 0$, $\alpha_1^1 = \alpha_2^1 = \alpha_1^2 = \alpha_2^2 = 0.05$, $\beta_{11} = 0.97$, $\beta_{12} = \beta_{21} = 0$, $\beta_{22} = 0.7$. Based on $5,000,000$ drawings.

as well[7]. All simulations are based on constant baseline intensity functions implying no interdependence between the processes. Figure 53 is based on a completely diagonal ACI specification. Thus, the processes do not interact, which is also revealed by the CACF. In contrast, the process in Figure 54 implies an asymmetric interdependence ($\alpha_1^2 = 0$) between the processes, while Figure 55 is associated with (symmetric) interactions in both directions ($\alpha_2^1 = \alpha_1^2 = 0.05$). The CACF shows an increase around zero, which is the more pronounced the stronger the interactions between the processes. The asymmetry of the dip results from the fact that the process-specific serial dependencies are of different strengths. Thus, the higher persistence of $\lambda^1(t_i; \mathfrak{F}_{t_i})$ implies higher values of Corr $\left[\lambda^1(t_i; \mathfrak{F}_{t_i}), \lambda^2(t_{i-j}; \mathfrak{F}_{t_{i-j}})\right]$ for $j < 0$ than for $j > 0$.

The log likelihood function of the multivariate ACI model is computed as

$$\ln \mathcal{L}(W; \theta) = \sum_{s=1}^{S} \sum_{i=1}^{n} \left\{ -\Lambda^s(t_{i-1}, t_i) + y_i^s \ln \lambda^s(t_i; \mathfrak{F}_{t_i}) \right\}, \qquad (7.40)$$

where $\Lambda^s(t_{i-1}, t_i)$ is calculated according to (7.35).

---

[7] The graphs of the CACF depict the plot of Corr $\left[\lambda^1(t_i; \mathfrak{F}_{t_i}), \lambda^2(t_{i-j}; \mathfrak{F}_{t_{i-j}})\right]$ versus j.

7.2.3 The Multivariate Hawkes Model

The univariate Hawkes(P) model, as illustrated in Subsection 7.1.2, is readily extended to the multivariate case. Hence, in a S−dimensional Hawkes(P) process, the intensity function associated with the s-th process is given by

$$\lambda^s(t;\mathfrak{F}_t) = \mu^s(t) + \sum_{r=1}^{S}\sum_{j=1}^{P}\sum_{k=1}^{\check{N}^r(t)} \alpha_j^{sr}\exp(-\beta_j^{sr}(t-t_k^r)) \tag{7.41}$$

with $\alpha_j^{sr} \geq 0$, $\beta_j^{sr} \geq 0$. Thus, in the multivariate case, $\lambda^s(t;\mathfrak{F}_t)$ depends not only on the backward recurrence time to all s-type points, but also on the backward recurrence time to all other points of the pooled process.

The conditional moments of $\lambda^s(t_i)$, as well as the conditional expectation of the arrival time of the next point of the process, $\mathrm{E}\left[t_i^s | \mathfrak{F}_{t_{i-1}}\right]$ cannot be expressed in closed form and are computed in a way similar to that shown in Subsection 7.1.2. Hawkes (1971) provides parameter restrictions for α^{sr} and β^{sr} under which the multivariate process given by (7.41) is weakly stationary. However, in general these conditions cannot be explicitly expressed in closed form and have to be verified by numerical methods.[8]

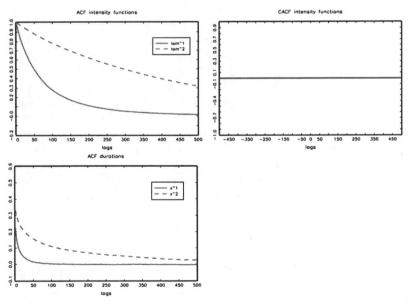

Figure 56: Simulated bivariate Hawkes(1) processes. Upper: ACF and CACF of $\lambda^{1,2}(t_i;\mathfrak{F}_{t_i})$. Lower: ACF of $x_i^{1,2}$. $\omega^1 = \omega^2 = \ln(0.1)$, $\alpha^{11} = \alpha^{22} = 0.2$, $\beta^{11} = 0.25$, $\beta^{22} = 0.21$, $\alpha^{12} = \alpha^{21} = 0$. Based on $5,000,000$ drawings.

---

[8] For more details, see Hawkes (1971).

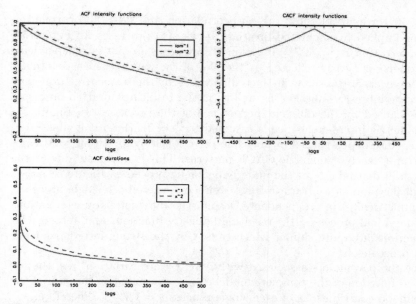

Figure 57: Simulated bivariate Hawkes(1) processes. Upper: ACF and CACF of $\lambda^{1,2}(t_i; \mathfrak{F}_{t_i})$. Lower: ACF of $x_i^{1,2}$. $\omega^1 = \omega^2 = \ln(0.1)$, $\alpha^{11} = \alpha^{22} = 0.2$, $\beta^{11} = 0.4$, $\beta^{22} = 0.21$, $\alpha^{12} = 0.1$, $\alpha^{21} = 0$, $\beta^{12} = 0.3$. Based on 5,000,000 drawings.

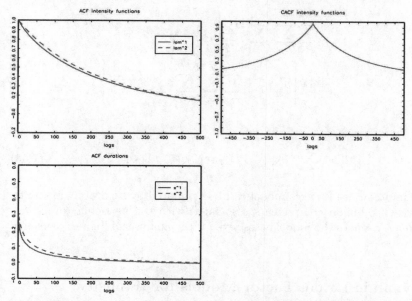

Figure 58: Simulated bivariate Hawkes(1) processes. Upper: ACF and CACF of $\lambda^{1,2}(t_i; \mathfrak{F}_{t_i})$. Lower: ACF of $x_i^{1,2}$. $\omega^1 = \omega^2 = \ln(0.1)$, $\alpha^{11} = \alpha^{22} = 0.2$, $\beta^{11} = 0.4$, $\beta^{22} = 0.25$, $\alpha^{12} = \alpha^{21} = 0.1$, $\beta^{12} = 0.3$, $\beta^{12} = 0.6$. Based on 5,000,000 drawings.

Figures 56 through 58 show the autocorrelation and cross-autocorrelation functions implied by simulated bivariate Hawkes(1) processes. In Figure 56, it is assumed that $\alpha^{12} = \alpha^{21} = 0$, i.e., both processes are not interdependent. This property is clearly reflected by the plot of the CACF, which is actually zero. Moreover, we observe distinct differences in the autocorrelation patterns of the intensity function, as well as in the resulting duration processes which is caused by the different persistence of the processes. In Figure 57, an asymmetric interdependence is assumed ($\alpha^{21} = 0$). Here, it is illustrated that even interdependencies in only one direction lead to strong contemporaneous and nearly symmetric CACF patterns. This is obviously due to the strong serial dependence of the underlying single processes. Figure 58 shows the resulting plots of a bivariate Hawkes process where both single processes are (symmetrically) interdependent. Despite the different parameterizations of the individual processes, the resulting intensity functions and autocorrelation functions are quite similar which caused by the strong interdependence of both processes.

Since the parameters associated with $\lambda^s(t; \mathfrak{F}_t)$ are variation free, the log likelihood function of the complete model can be computed as the sum of the log likelihood contributions of each single process $s = 1, \dots S$. Therefore,

$$
\ln \mathcal{L}(W; \theta) = \sum_{s=1}^{S} \sum_{i=1}^{n} \left\{ - \int_{t_{i-1}}^{t_i} \mu^s(u) du - \right. \tag{7.42}
$$

$$
\sum_{r=1}^{S} \sum_{j=1}^{P} \sum_{k=1}^{\check{N}^r(t_i)} \frac{\alpha_j^{sr}}{\beta_j^{sr}} \left[1 - \exp(-\beta_j^{sr}(t_i - t_k^r)) \right]
$$

$$
\left. + \ln \left[\mu^s(t_i) + \sum_{r=1}^{S} \sum_{j=1}^{P} \sum_{k=1}^{\check{N}^r(t_i)} \alpha_j^{sr} A_i^{j,sr} \right] \right\},
$$

where

$$
A_i^{j,sr} = \sum_{k=1}^{\check{N}^r(t_i)} \exp(-\beta_j^{sr}(t_i - t_k^r))(y_{i-1}^r + A_{i-1}^{j,sr}). \tag{7.43}
$$

Thus, a multivariate Hawkes model can be estimated by separately maximizing the log likelihood components associated with the particular processes. This property makes the model attractive for the analysis of high-dimensional point processes.

7.3 Dynamic Latent Factor Models for Intensity Processes

An important property of Hawkes processes and ACI processes as discussed in the previous subsections is that the conditional intensity function given the history of the process is completely deterministic. This is due to the fact that in both model frameworks, the intensity process is parameterized in terms of the history of *observable* factors.

However, the assumption that the intensity function is completely explained by the observable process history is rather unrealistic. As already discussed in Section 6.4.4 of Chapter 6, in classical duration literature the consideration of unobserved heterogeneity effects plays an important role to obtain well specified econometric models. Moreover, the recent financial econometrics literature witnesses a growing interest in the specification and application of latent factor models. According to the the mixture-of-distribution hypothesis (see Clark, 1973), a latent factor is economically motivated as a variable capturing the information and activity flow that cannot be observed directly, but influences the activity on a financial market. This idea serves as the economic foundation of the stochastic volatility (SV) model (Taylor, 1982) and the well-known volume-volatility relationship (Tauchen and Pitts, 1983). Following this strand of the literature, it is reasonable to assume that not only the volatility and the trading volume are driven by a latent factor, but the trading intensity itself as well.

In the subsequent sections, we discuss a framework that combines the idea of latent factor models with the concept of dynamic intensity processes. The so-called latent factor intensity (LFI) model introduced by Bauwens and Hautsch (2003) can be seen as the counterpart to the SV model (Taylor, 1982) and the SCD model proposed by Bauwens and Veredas (1999)[9]. The LFI model is based on the assumption that the conditional intensity function given the process history consists of two components: a univariate latent one and an observation driven one. In this sense, the LFI model combines features of parameter driven models and observation driven models. The observable component can be specified univariately or multivariately. In the latter case, the latent factor corresponds to a common component that captures the impact of a general factor that influences all individual point processes and interacts with their observable component. Two limit cases emerge naturally: one when the latent factor is irrelevant and the intensity is driven by the dynamics of the specific components and the other when the specific components are not relevant and the latent factor completely dominates.

Hence, the model dynamics are driven by the interaction between the latent dynamic factor which is updated by latent innovations and the observable dynamic component, which is updated by process-specific innovations. In this sense, the model can be interpreted as a dynamic extension of a doubly stochastic Poisson process (see, e.g., Grandell, 1976 or Cox and Isham, 1980).

In Section 7.3.1, we present the general structure of the LFI model and discuss the specification of the latent dynamics. Combinations of the latent factor model with univariate and multivariate ACI and Hawkes specifications of the observable component are illustrated in the Sections 7.3.2 and 7.3.3. Section

---

[9] See Section 5.3.3 in Chapter 5.

7.3.4 illustrates the dynamic properties of LFI models based on simulation results. Estimation issues are considered in Section 7.3.5. The estimation of the model is not straightforward since the latent dynamic factor is not observable and has to be integrated out. We propose a simulated maximum likelihood (SML) procedure based upon the efficient importance sampling (EIS) technique introduced by Richard (1998) and illustrate the application of this approach to the proposed model.

7.3.1 The LFI Model

The basic LFI model is based on a decomposition of the intensity function into a latent and an observable component. In this context, we define the information set $\mathfrak{F}_t$ more explicitly as $\mathfrak{F}_t := \sigma(\mathfrak{F}_t^o \cup \mathfrak{F}_t^*)$, consisting of an observable conditioning set $\mathfrak{F}_t^o$ that includes the complete *observable* history of the process up to t (inclusive possible explanatory variables) and an *unobservable* history $\mathfrak{F}_t^*$ of some latent factor $\lambda_{\check{N}(t)+1}^*$. Then, in a multivariate framework, the basic LFI model is given by

$$\lambda^s(t; \mathfrak{F}_t) = \lambda^{o,s}(t; \mathfrak{F}_t^o) \left(\lambda_{\check{N}(t)+1}^* \right)^{\delta^{s*}}, \tag{7.44}$$

where $\lambda^{o,s}(t; \mathfrak{F}_t^o)$ denotes a s-type conditionally deterministic intensity component based on the observable elements included in $\mathfrak{F}_t^o$ and δ^{s*} drives the process-specific influence of λ_i^*. The latent factor is conditionally log normally distributed and follows an AR(1) process

$$\ln \lambda_i^* = a^* \ln \lambda_{i-1}^* + u_i^*, \qquad u_i^* \sim \text{i.i.d. } N(0,1). \tag{7.45}$$

Note that the latent factor is indexed by the left-continuous counting function, i.e., it does not change during a spell. In particular, it is assumed that λ_i^* has left-continuous sample paths with right-hand limits which means that it is updated instantaneously after the occurrence of t_{i-1} and remains constant until (and inclusive) t_i. In order to obtain a valid intensity process, it is assumed that the latent innovations u_i^* are independent from the series of integrated intensities $\epsilon_i := \Lambda(t_{i-1}, t_i)$ which are i.i.d. standard exponentially distributed (under correct model specification). An important prerequisite for weak stationarity of the LFI model is weak stationarity of the latent component which is fulfilled for $|a^*| < 1$. [10] Clearly, the latent factor dynamics can be extended to ARMA(P,Q) parameterizations, however most often, a simple AR(1) dynamic has shown to be sufficient in capturing latent dynamics. Notice that we omit a constant since we assume that the observation driven component $\lambda^o(t; \mathfrak{F}_t^o)$ encloses a constant that would be not identified otherwise.

By defining $\lambda_i^{s*} := \delta^{s*} \ln \lambda_i^*$ as the latent factor that influences the s-type component, it is easy to see that

---

[10] Of course, under the normality assumption, the latent factor is even strictly stationary.

$$\lambda_i^{s*} = a^* \lambda_{i-1}^{s*} + \delta^{s*} u_i^*.$$

Hence, δ^{s*} corresponds to a scaling factor that scales the latent component influencing the s−type intensity process. This flexibility ensures that the impact of a latent shock u_i^* on the individual processes can differ and is driven by the parameter δ^{s*}.[11]

This approach can be extended by specifying δ^{s*} time-varying. Then, the process-specific scaling factor can change over time in order to allow for conditional heteroscedasticity. An important source of heteroscedasticity could be intraday seasonality associated with deterministic fluctuations of the overall information and activity flow. For example, it could be due to institutional settings, like the opening of other related markets. Hence, a reasonable specification could be to index δ^{s*} by the counting function and parameterize it in terms of a linear spline function

$$\delta_i^{s*} = s_0^* \left(1 + \sum_{k=1}^{K} \nu_{0,k}^{s*} \mathbb{1}_{\{\tau(t_i) \geq \bar{\tau}_k\}} (\tau(t_i) - \bar{\tau}_k)\right), \qquad (7.46)$$

where $\tau(t)$ denotes the calendar time at t, $\bar{\tau}_k$, $k = 1, \dots, K - 1$ denote the exogenously given (calendar) time points and s_0^* and $\nu_{0,k}^{s*}$ are the corresponding coefficients of the spline function.

A further valuable generalization of the LFI model is to allow for regime-switching latent dynamics. Thus, a more flexible LFI model is obtained by specifying the autoregressive parameter in dependence of the length of the previous spell. Such a specification is in line with a threshold model (see e.g. Tong, 1990, or Zhang, Russell, and Tsay, 2001) and is obtained by

$$\ln \lambda_i^* = a_r^* \mathbb{1}_{\{\bar{x}_{r-1} < x_{\tilde{N}(t)} \leq \bar{x}_r\}} \ln \lambda_{i-1}^* + u_i^*, \qquad r = 1, \dots, R - 1, \qquad (7.47)$$

where $\bar{x}_r$ denotes the exogenously given thresholds (with $\bar{x}_0 := 0$), and a_r^* are the regime-dependent latent autoregressive parameters.

The computation of conditional moments of $\lambda(t; \mathfrak{F}_t)$ given the *observable* information set is difficult since the latent variable has to be integrated out. In the following, we illustrate the computation of conditional moments given $\mathfrak{F}_{t_{i-1}}^o$ in a general way for arbitrary parameterizations of $\lambda^o(t; \mathfrak{F}_t^o)$. Let w_i a row of the matrix of *observable* variables W and correspondingly, we define $W_i = \{w_j\}_{j=1}^i$. Accordingly, $L_i^* = \{\lambda_j^*\}_{j=1}^i$ defines the sequence of *latent* variables until t_i. Furthermore, let $f(W_i, L_i^* | \theta)$ denote the joint density function of W_i and L_i^*, and $p(\lambda_i^* | W_i, L_i^*, \theta)$ the conditional density of λ_i^* given W_i and L_i^*. Then, the conditional expectation of an arbitrary function of λ_i^*, $\vartheta(\lambda_i^*)$ given the *observable* information set up to t_{i-1} can be computed as

---

[11] Note that δ^{s*} can be even negative. Hence, theoretically it is possible that the latent component simultaneously increases one component while decreasing the other component.

$$\mathrm{E}\left[\vartheta(\lambda_i^*)\,\middle|\,\mathfrak{F}_{t_{i-1}}^o\right] = \frac{\int \vartheta(\lambda_i^*)p(\lambda_i^*|W_{i-1},L_{i-1}^*,\theta)f(W_{i-1},L_{i-1}^*|\theta)dL_i^*}{\int f(W_{i-1},L_{i-1}^*|\theta)dL_{i-1}^*}. \quad (7.48)$$

The integrals in this ratio cannot be computed analytically, but they can be computed numerically, e.g., by efficient importance sampling (see Subsection 7.3.6). Hence, the computations of the conditional expectations $\mathrm{E}\left[t_i|\mathfrak{F}_{t_{i-1}}^o\right]$ and $\mathrm{E}\left[\lambda(t_i)|\mathfrak{F}_{t_{i-1}}^o\right]$ are conducted by exploiting the exponential distribution of the integrated intensity $\epsilon_i = \Lambda(t_{i-1},t_i)$. Then, the calculations are performed by conditioning on predetermined values of ϵ_i and λ_{i-1}^* and applying the law of iterated expectations. Hence, $\mathrm{E}\left[t_i|\mathfrak{F}_{t_{i-1}}^o\right]$ is computed as[12]

$$\mathrm{E}\left[t_i|\mathfrak{F}_{t_{i-1}}^o\right] = \mathrm{E}\left[g_1(\cdot)|\mathfrak{F}_{t_{i-1}}^o\right] \quad (7.49)$$

$$g_1(\cdot) = \mathrm{E}\left[t_i|\lambda_{i-1}^*;\mathfrak{F}_{t_{i-1}}^o\right] = \mathrm{E}\left[g_2(\cdot)|\lambda_{i-1}^*;\mathfrak{F}_{t_{i-1}}^o\right], \quad (7.50)$$

where $t_i = g_2(\epsilon_i;\mathfrak{F}_{t_{i-1}}^o,\lambda_{i-1}^*)$ is determined by solving the equation $\Lambda(t_i,t_{i-1}) = \epsilon_i$ for given values of ϵ_i, λ_{i-1}^* and $\mathfrak{F}_{t_{i-1}}^o$. Clearly, the complexity of the expression for t_i depends on the parameterization of $\lambda^o(t;\mathfrak{F}_t^o)$. However, as discussed in Section 7.1, at least for the multivariate case, closed form solutions for $g_2(\cdot)$ do not exist. After computing the conditional expectation of t_i given λ_{i-1}^* (see (7.50)), the next step is to integrate over the latent variable ((7.49)) based on the integrals in (7.48). Since the computations of $\mathrm{E}\left[\lambda(t_i)|\mathfrak{F}_{t_{i-1}}\right]$ are done similarly, we refrain from showing them here.

7.3.2 The Univariate LFI Model

In the univariate LFI model, both components $\lambda^o(t;\mathfrak{F}_t^o)$ and λ_i^* are specified univariately. Parameterizing $\lambda^o(t;\mathfrak{F}_t^o)$ as an ACI process defined by (7.1) through (7.5) results in a so-called univariate LF-ACI(P,Q) model. In this framework, the innovation term is specified either in plain or in logarithmic form, i.e., by

$$\check{\epsilon}_i = 1 - \Lambda^o(t_{i-1},t_i) \qquad \text{or} \qquad (7.51)$$

$$\check{\epsilon}_i = -0.5772 - \ln\Lambda^o(t_{i-1},t_i), \qquad (7.52)$$

where

$$\Lambda^o(t_{i-1},t_i) := \sum_j \int_{\tilde{t}_j}^{\tilde{t}_{j+1}} \lambda^o(s;\mathfrak{F}_s^o)ds = \sum_j \int_{\tilde{t}_j}^{\tilde{t}_{j+1}} \frac{\lambda(s;\mathfrak{F}_s)ds}{\lambda_i^{*\delta^*}} \quad (7.53)$$

with $j : t_{i-1} \le \tilde{t}_j \le t_i$ and $\sum_j \int_{\tilde{t}_j}^{\tilde{t}_{j+1}} \lambda(u;\mathfrak{F}_u)du = \varepsilon_i \sim$ i.i.d. $Exp(1)$. Hence, the innovation of an univariate LFI process is based on the integral over

---

[12] For ease of notation, we neglect the existence of time-varying covariates.

the *observable* intensity component $\lambda^o(t; \mathfrak{F}^o_t)$ which equals an i.i.d. standard exponential variate that is standardized by a stationary log-normal random variable. Of course, $\Lambda^o(t_{i-1}, t_i)$ is not anymore i.i.d. exponentially distributed then. Nevertheless, we center it by 1 or -0.5772, respectively, in order to make it comparable to the innovation of the pure ACI model as benchmark case. Because $\Lambda^o(t_{i-1}, t_i)$, and thus the LFI-ACI innovation depends on lags of λ^*_i, the component $\lambda^o(t; \mathfrak{F}^o_t)$ is directly affected by lags of the latent factor. Hence, λ^*_i influences the intensity $\lambda(t; \mathfrak{F}_t)$ not only contemporaneously (according to (7.44)), but also through its lags.

Due to the log-linear structure of both components $\lambda^o(t; \mathfrak{F}^o_t)$ and λ^*_i, the model dynamics are characterized by a direct interaction between the latent dynamic factor, which is updated by latent innovations u^*_i and the observable dynamic component, which is updated by the innovations $\breve{\epsilon}_i$. By excluding covariates and assuming for simplicity $\lambda_0(t) = 1$ and $s(t) = 1$, the intensity function for the LF-ACI(1,1) model is rewritten in logarithmic form as

$$
\begin{aligned}
\ln \lambda(t; \mathfrak{F}_t) &= \Psi(t) + \delta^* \ln \lambda^*_{\breve{N}(t)+1} \\
&= \alpha \breve{\epsilon}_{\breve{N}(t)} + \beta \Psi(t_{\breve{N}(t)}) + a^* \delta^* \ln \lambda^*_{\breve{N}(t)} + \delta^* u^*_{\breve{N}(t)} \\
&= \alpha \breve{\epsilon}_{\breve{N}(t)} + \delta^* (a^* - \beta) \ln \lambda^*_{\breve{N}(t)} + \delta^* u^*_{\breve{N}(t)} + \beta (\ln \lambda(t_{\breve{N}(t)}; \mathfrak{F}_{t_{\breve{N}(t)}})).
\end{aligned}
\tag{7.54}
$$

Hence, the LF-ACI model can be represented as an ARMA model for the log intensity function that is augmented by a further dynamic component.

By specifying the observation driven intensity component $\lambda^o(t; \mathfrak{F}^o_t)$ in terms of a Hawkes specification as given in (7.14), we obtain the so-called univariate LF-Hawkes(P) model. As in the LF-ACI model, the latent factor influences the intensity $\lambda(t; \mathfrak{F}_t)$ not only contemporaneously, but also by its lags that affect the component $\lambda^o(t; \mathfrak{F}^o_t)$. This is due to the fact that in the Hawkes process, $\lambda^o(t; \mathfrak{F}^o_t)$ is updated by the backward recurrence time to all previous points whose occurrences themselves depend on the history of λ^*_i.

The major advantage of the parameterizations of the LFI-ACI and LFI-Hawkes specifications is that in both approaches, the model innovations can be computed directly based on the *observable* history of the process, which allows for separate calculations of the components $\lambda^o(t; \mathfrak{F}^o_t)$ and λ^*_i.

7.3.3 The Multivariate LFI Model

By specifying $\lambda^o(t; \mathfrak{F}^o_t)$ multivariately, we obtain the multivariate LFI model. In this case, λ^*_i is updated at every point of the *pooled* process. The multivariate LF-ACI(P,Q) model is obtained by parameterizing the observable component $\lambda^o(t; \mathfrak{F}^o_t)$ as described in Section 7.2.2. Then, the innovation term is specified asinnovations of multivariate LF-ACI models

$$
\breve{\epsilon}_{\breve{N}(t)} = 1 - \Lambda^{o,s}(t^s_{\breve{N}^s(t)-1}, t^s_{\breve{N}^s(t)}) y^s_{\breve{N}(t)} \qquad \text{or} \tag{7.55}
$$

$$
\breve{\epsilon}_{\breve{N}(t)} = -0.5772 - \ln \Lambda^{o,s}(t^s_{\breve{N}^s(t)-1}, t^s_{\breve{N}^s(t)}) y^s_{\breve{N}(t)}, \tag{7.56}
$$

where

$$\Lambda^{o,s}(t_{i-1}^s, t_i^s) := \sum_j \int_{\tilde{t}_j}^{\tilde{t}_{j+1}} \lambda^{o,s}(u; \mathfrak{F}_u^o) du = \sum_j \int_{\tilde{t}_j}^{\tilde{t}_{j+1}} \frac{\lambda^s(u; \mathfrak{F}_u) du}{\lambda_j^{*\delta^{s*}}} \qquad (7.57)$$

with $j : t_{i-1}^s \leq \tilde{t}_j \leq t_i^s$ and $\sum_j \int_{\tilde{t}_j}^{\tilde{t}_{j+1}} \lambda^s(u; \mathfrak{F}_u) du = \varepsilon_i \sim$ i.i.d. $Exp(1)$. Hence, in the multivariate setting, $\Lambda^{o,s}(t_{i-1}^s, t_i^s)$ corresponds to an i.i.d. exponential variate that is piecewise standardized by a stationary log normal random variable.

Correspondingly, a multivariate LF-Hawkes(P) model is obtained by specifying $\lambda^o(t; \mathfrak{F}_t^o)$ according to (7.41). As in the univariate case, in both types of models, the latent factor affects the components $\lambda^s(t; \mathfrak{F}_t)$ not only contemporaneously (according to (7.44)), but influences also the lagged innovations of the model. For this reason, the latent factor causes cross-autocorrelations not only between $\lambda^s(t; \mathfrak{F}_t)$, but also between the intensity components $\lambda^{o,s}(t; \mathfrak{F}_t^o)$, $s = 1, \ldots, S$. This will be illustrated in more detail in the following subsection.

7.3.4 Dynamic Properties of the LFI Model

Figures 59 through 63 show the (cross-)autocorrelation functions of the individual intensity components and the corresponding duration processes based on simulated bivariate LF-ACI(1,1) processes using logarithmic innovations ((7.56)). As in Section 7.2.2, all simulations are based on constant baseline intensity functions. Figures 59 and 60 are based on LF-ACI specifications implying no interactions between the observation driven components $\lambda^{o,s}(t, \mathfrak{F}_t^o)$ and strong serial dependencies in the latent factor. Not surprisingly, it turns out that the impact of the latent factor strongly depends on the magnitude of the latent variances. In particular, the simulation in Figure 59 is based on $\delta^{1*} = \delta^{2*} = 0.01$. Here, only a very weak cross-autocorrelation between $\lambda^1(t; \mathfrak{F}_t)$ and $\lambda^2(t; \mathfrak{F}_t)$ can be identified. In contrast, in Figure 60, the impact of the latent factor is clearly stronger. Here, it causes also slight contemporaneous correlations between the *observable* components $\lambda^{o,1}(t_i; \mathfrak{F}_{t_i}^o)$ and $\lambda^{o,2}(t_i; \mathfrak{F}_{t_i}^o)$. As already discussed above, this is due to the fact that λ_i^* influences the intensity components not only contemporaneously but also through the lagged innovations $\check{\varepsilon}_i$. It is shown that the latent dynamics dominate the dynamics of the particular processes $\lambda^1(t_i; \mathfrak{F}_{t_i})$ and $\lambda^2(t_i; \mathfrak{F}_{t_i})$, as well as of x_i^1 and x_i^2, leading to quite similar autocorrelation functions. Moreover, a clear increase of the autocorrelations in the duration processes is observed.

Figure 61 shows the corresponding plots of symmetrically interdependent ACI components ($\alpha_2^1 = \alpha_1^2 = 0.05$). Here, the latent variable enforces the contemporaneous correlation between the two processes and drives the autocorrelation functions of the individual intensity components towards higher similarity. Moreover, while the CACF of $\lambda^{o,1}(t_i; \mathfrak{F}_{t_i}^o)$ vs. $\lambda^{o,2}(t_i; \mathfrak{F}_{t_i}^o)$ shows the

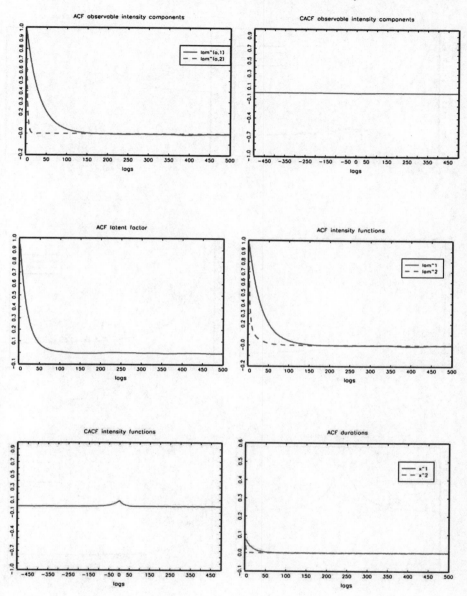

Figure 59: Simulated bivariate LF-ACI(1,1) processes. Upper: ACF and CACF of $\lambda^{o,1}(t_i; \mathfrak{F}^o_{t_i})$ and $\lambda^{o,2}(t_i; \mathfrak{F}^o_{t_i})$. Middle: ACF of λ^*_i, as well as of $\lambda^1(t_i; \mathfrak{F}_{t_i})$ and $\lambda^2(t_i; \mathfrak{F}_{t_i})$. Lower: CACF of $\lambda^1(t_i; \mathfrak{F}_{t_i})$ vs. $\lambda^2(t_i; \mathfrak{F}_{t_i})$, as well as ACF of x^1_i and x^2_i. $\omega^1 = \omega^2 = 0$, $\alpha^1_1 = \alpha^2_2 = 0.05$, $\alpha^1_2 = \alpha^2_1 = 0$, $\beta_{11} = 0.97$, $\beta_{12} = \beta_{21} = 0$, $\beta_{22} = 0.7$, $a^* = 0.95$, $\delta^{1*} = \delta^{2*} = 0.01$. Based on $5,000,000$ drawings.

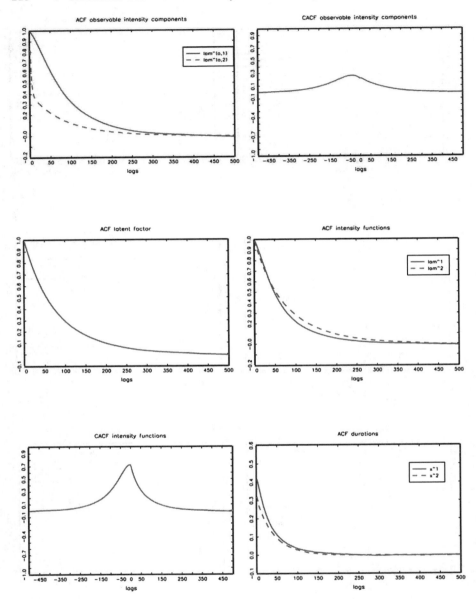

Figure 60: Simulated bivariate LF-ACI(1,1) processes. Upper: ACF and CACF of $\lambda^{o,1}(t_i; \mathfrak{F}^o_{t_i})$ and $\lambda^{o,2}(t_i; \mathfrak{F}^o_{t_i})$. Middle: ACF of λ^*_i, as well as of $\lambda^1(t_i; \mathfrak{F}_{t_i})$ and $\lambda^2(t_i; \mathfrak{F}_{t_i})$. Lower: CACF of $\lambda^1(t_i; \mathfrak{F}_{t_i})$ vs. $\lambda^2(t_i; \mathfrak{F}_{t_i})$, as well as ACF of x^1_i and x^2_i. $\omega^1 = \omega^2 = 0$, $\alpha^1_1 = \alpha^2_2 = 0.05$, $\alpha^1_2 = \alpha^2_1 = 0$, $\beta_{11} = 0.97$, $\beta_{12} = \beta_{21} = 0$, $\beta_{22} = 0.7$, $a^* = 0.99$, $\delta^{1*} = \delta^{2*} = 0.1$. Based on $5,000,000$ drawings.

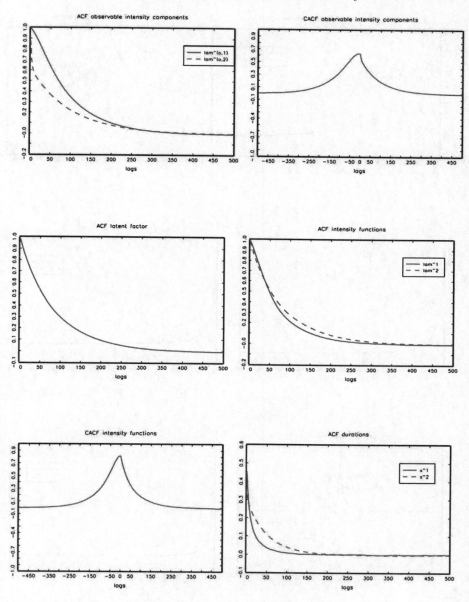

Figure 61: Simulated bivariate LF-ACI(1,1) processes. Upper: ACF and CACF of $\lambda^{o,1}(t_i; \mathfrak{F}^o_{t_i})$ and $\lambda^{o,2}(t_i; \mathfrak{F}^o_{t_i})$. Middle: ACF of λ^*_i, as well as of $\lambda^1(t_i; \mathfrak{F}_{t_i})$ and $\lambda^2(t_i; \mathfrak{F}_{t_i})$. Lower: CACF of $\lambda^1(t_i; \mathfrak{F}_{t_i})$ vs. $\lambda^2(t_i; \mathfrak{F}_{t_i})$, as well as ACF of x^1_i and x^2_i. $\omega^1 = \omega^2 = 0$, $\alpha^1_1 = \alpha^1_2 = \alpha^2_1 = \alpha^2_2 = 0.05$, $\beta_{11} = 0.97$, $\beta_{12} = \beta_{21} = 0$, $\beta_{22} = 0.7$, $a^* = 0.99$, $\delta^{1*} = \delta^{2*} = 0.1$. Based on $5,000,000$ drawings.

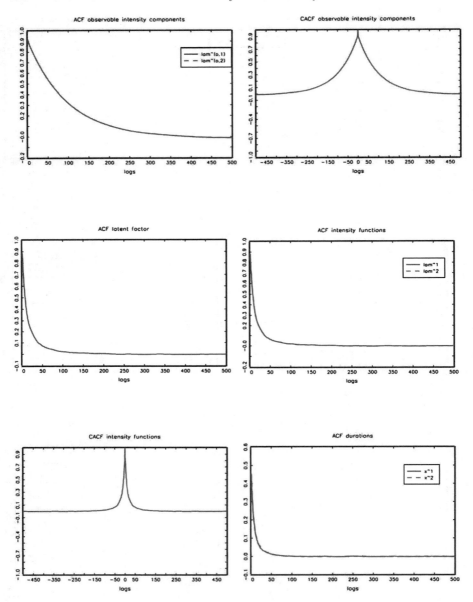

Figure 62: Simulated bivariate LF-ACI(1,1) processes. Upper: ACF and CACF of $\lambda^{o,1}(t_i; \mathfrak{F}_{t_i}^o)$ and $\lambda^{o,2}(t_i; \mathfrak{F}_{t_i}^o)$. Middle: ACF of λ_i^*, as well as of $\lambda^1(t_i; \mathfrak{F}_{t_i})$ and $\lambda^2(t_i; \mathfrak{F}_{t_i})$. Lower: CACF of $\lambda^1(t_i; \mathfrak{F}_{t_i})$ vs. $\lambda^2(t_i; \mathfrak{F}_{t_i})$, as well as ACF of x_i^1 and x_i^2. $\omega^1 = \omega^2 = 0$, $\alpha_1^1 = \alpha_2^1 = \alpha_1^2 = \alpha_2^2 = 0.05$, $\beta_{11} = \beta_{22} = 0.2$, $\beta_{12} = \beta_{21} = 0$, $a^* = 0.95$, $\delta^{1*} = \delta^{2*} = 0.5$. Based on $5,000,000$ drawings.

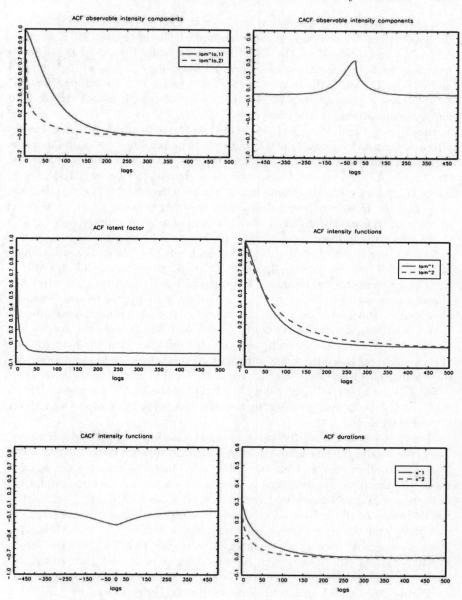

Figure 63: Simulated bivariate LF-ACI(1,1) processes. Upper: ACF and CACF of $\lambda^{o,1}(t_i;\mathfrak{F}_{t_i}^o)$ and $\lambda^{o,2}(t_i;\mathfrak{F}_{t_i}^o)$. Middle: ACF of λ_i^*, as well as of $\lambda^1(t_i;\mathfrak{F}_{t_i})$ and $\lambda^2(t_i;\mathfrak{F}_{t_i})$. Lower: CACF of $\lambda^1(t_i;\mathfrak{F}_{t_i})$ vs. $\lambda^2(t_i;\mathfrak{F}_{t_i})$, as well as ACF of x_i^1 and x_i^2. $\omega^1 = \omega^2 = 0$, $\alpha_1^1 = \alpha_1^2 = \alpha_2^1 = \alpha_2^2 = 0.05$, $\beta_{11} = 0.97$, $\beta_{12} = \beta_{21} = 0$, $\beta_{22} = 0.7$, $a^* = 0.99$, $\delta^{1*} = 0.1$, $\delta^{2*} = -0.1$. Based on $5,000,000$ drawings.

well known asymmetric shape that has been already discussed in Section 7.2.2, the joint latent factor causes a more symmetric shape of the CACF between $\lambda^1(t_i; \mathfrak{F}_{t_i})$ and $\lambda^2(t_i; \mathfrak{F}_{t_i})$. The DGP associated with Figure 62 is based on relatively weak dynamics in the individual intensity components $\lambda^{o,1}(t_i; \mathfrak{F}_{t_i}^o)$ and $\lambda^{o,2}(t_i; \mathfrak{F}_{t_i}^o)$ while the impact of the latent factor is quite strong. Here, λ_i^* completely dominates the dynamics of the joint system. It causes strong, similar autocorrelations in the components $\lambda^{o,1}(t_i; \mathfrak{F}_{t_i}^o)$ and $\lambda^{o,2}(t_i; \mathfrak{F}_{t_i}^o)$, as well as in $\lambda^1(t_i; \mathfrak{F}_{t_i})$ and $\lambda^2(t_i; \mathfrak{F}_{t_i})$. Moreover, its impact on the CACF is clearly stronger than in the cases outlined above. In particular, the contemporaneous correlation between $\lambda^1(t_i; \mathfrak{F}_{t_i})$ and $\lambda^2(t_i; \mathfrak{F}_{t_i})$ is nearly one. Nevertheless, the CACF dies out quite quickly which is due to the latent AR(1) structure. The parameterization underlying Figure 63 resembles the specification in Figure 61. However, here the scaling factors are chosen as $\delta^{1*} = 0.1$ and $\delta^{2*} = -0.1$. It is nicely shown that the latent component influences $\lambda^1(t_i; \mathfrak{F}_{t_i})$ positively while influencing $\lambda^2(t_i; \mathfrak{F}_{t_i})$ negatively which causes clear negative cross-autocorrelations between $\lambda^1(t_i; \mathfrak{F}_{t_i})$ and $\lambda^2(t_i; \mathfrak{F}_{t_i})$ as well as a flattening of the CACF between $\lambda^{o,1}(t_i; \mathfrak{F}_{t_i}^o)$ and $\lambda^{o,2}(t_i; \mathfrak{F}_{t_i}^o)$ compared to Figure 61.

The experiments in Figures 64 through 68 are based on simulated LF-Hawkes processes of the order $P = 1$. The plots in Figure 64 are based on *independent* Hawkes(1) processes for the observable component that interact with a relatively weak latent factor ($\delta^{1*} = \delta^{2*} = 0.05$). It is shown that the dynamics of the individual intensity components remain nearly unaffected, thus, the latent factor has only weak influences. Nevertheless, slight contemporaneous correlations between $\lambda^{o,1}(t_i; \mathfrak{F}_{t_i}^o)$ and $\lambda^{o,2}(t_i; \mathfrak{F}_{t_i}^o)$ as well as $\lambda^1(t_i; \mathfrak{F}_{t_i})$ and $\lambda^2(t_i; \mathfrak{F}_{t_i})$ are observed. The asymmetric dip in the CACF between $\lambda^1(t; \mathfrak{F}_t)$ and $\lambda^2(t; \mathfrak{F}_t)$ is obviously caused by the different persistence of the underlying Hawkes processes.

Figure 65 is based on the same parameterization as Figure 64, however, here, the standard deviation of the latent factor is increased to $\delta^{1*} = \delta^{2*} = 0.5$. The graphs show that in that case, the latent factor dominates the dynamics of the complete system causing significant contemporaneous correlations and cross-autocorrelations between $\lambda^{o,1}(t_i; \mathfrak{F}_{t_i}^o)$ and $\lambda^{o,2}(t_i; \mathfrak{F}_{t_i}^o)$ and drives the dynamics of the individual processes to a higher degree of similarity. The fact that even the dynamics of the components $\lambda^{o,1}(t_i; \mathfrak{F}_{t_i}^o)$ and $\lambda^{o,2}(t_i; \mathfrak{F}_{t_i}^o)$ strongly resemble each other (compared to Figure 64) illustrate the strong impact of λ_i^* on the innovations (in this context past backward recurrence times) of the model.

Figure 66 is based on interdependent Hawkes(1) processes ($\alpha^{12} = \alpha^{21} = 0.1$). Here, the latent factor clearly amplifies the contemporaneous correlation structure, but weakens the persistence in the cross-autocorrelations. Figure 67 is the counterpart to Figure 62 revealing only weak dynamics in the particular components $\lambda^{o,1}(t_i; \mathfrak{F}_{t_i}^o)$ and $\lambda^{o,2}(t_i; \mathfrak{F}_{t_i}^o)$ combined with a strong latent dynamic. Again, the resulting autocorrelation functions of the intensities, as well as of the duration processes are clearly dominated by

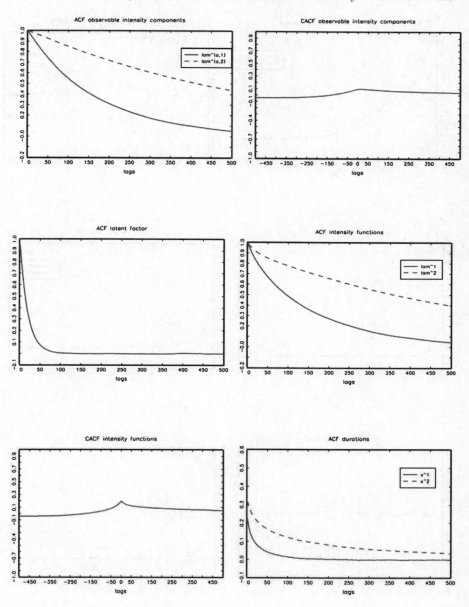

Figure 64: Simulated bivariate LF-Hawkes(1) processes. Upper: ACF and CACF of $\lambda^{o,1}(t_i; \mathfrak{F}^o_{t_i})$ and $\lambda^{o,2}(t_i; \mathfrak{F}^o_{t_i})$. Middle: ACF of λ^*_i, as well as of $\lambda^1(t_i; \mathfrak{F}_{t_i})$ and $\lambda^2(t_i; \mathfrak{F}_{t_i})$. Lower: CACF of $\lambda^1(t_i; \mathfrak{F}_{t_i})$ vs. $\lambda^2(t_i; \mathfrak{F}_{t_i})$, as well as ACF of x^1_i and x^2_i. $\omega^1 = \ln(0.4)$, $\omega^2 = \ln(0.3)$, $\alpha^{11} = \alpha^{22} = 0.2$, $\beta^{11} = 0.25$, $\beta^{22} = 0.21$, $\alpha^{12} = \alpha^{21} = 0$, $a^* = 0.95$, $\delta^{1*} = \delta^{2*} = 0.05$. Based on $5,000,000$ drawings.

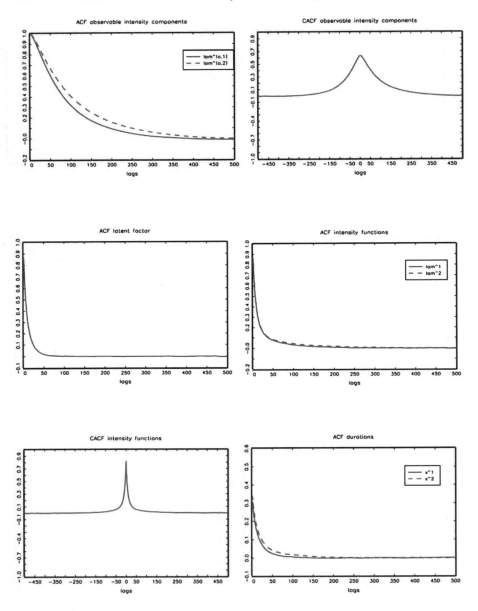

Figure 65: Simulated bivariate LF-Hawkes(1) processes. Upper: ACF and CACF of $\lambda^{o,1}(t_i; \mathfrak{F}^o_{t_i})$ and $\lambda^{o,2}(t_i; \mathfrak{F}^o_{t_i})$. Middle: ACF of λ^*_i, as well as of $\lambda^1(t_i; \mathfrak{F}_{t_i})$ and $\lambda^2(t_i; \mathfrak{F}_{t_i})$. Lower: CACF of $\lambda^1(t_i; \mathfrak{F}_{t_i})$ vs. $\lambda^2(t_i; \mathfrak{F}_{t_i})$, as well as ACF of x^1_i and x^2_i. $\omega^1 = \ln(0.4)$, $\omega^2 = \ln(0.3)$, $\alpha^{11} = \alpha^{22} = 0.2$, $\beta^{11} = 0.25$, $\beta^{22} = 0.21$, $\alpha^{12} = \alpha^{21} = 0$, $a^* = 0.95$, $\delta^{1*} = \delta^{2*} = 0.5$. Based on $5,000,000$ drawings.

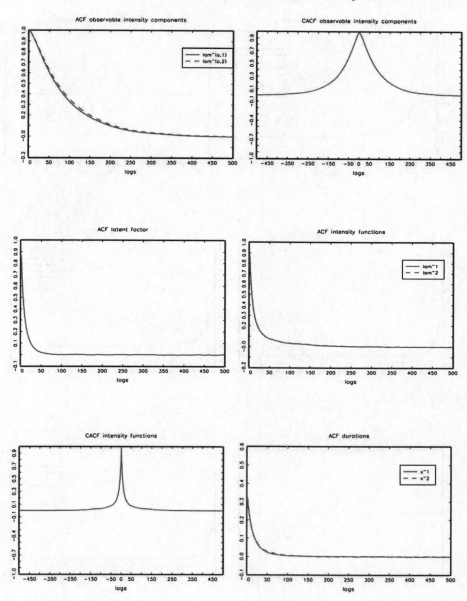

Figure 66: Simulated bivariate LF-Hawkes(1) processes. Upper: ACF and CACF of $\lambda^{o,1}(t_i; \mathfrak{F}_{t_i}^o)$ and $\lambda^{o,2}(t_i; \mathfrak{F}_{t_i}^o)$. Middle: ACF of λ_i^*, as well as of $\lambda^1(t_i; \mathfrak{F}_{t_i})$ and $\lambda^2(t_i; \mathfrak{F}_{t_i})$. Lower: CACF of $\lambda^1(t_i; \mathfrak{F}_{t_i})$ vs. $\lambda^2(t_i; \mathfrak{F}_{t_i})$, as well as ACF of x_i^1 and x_i^2. $\omega^1 = \ln(0.2)$, $\omega^2 = \ln(0.1)$, $\alpha^{11} = \alpha^{22} = 0.2$, $\beta^{11} = 0.4$, $\beta^{22} = 0.25$, $\alpha^{12} = \alpha^{21} = 0.1$, $\beta^{12} = 0.3$, $\beta^{21} = 0.6$, $a^* = 0.95$, $\delta^{1*} = \delta^{2*} = 0.5$. Based on $5,000,000$ drawings.

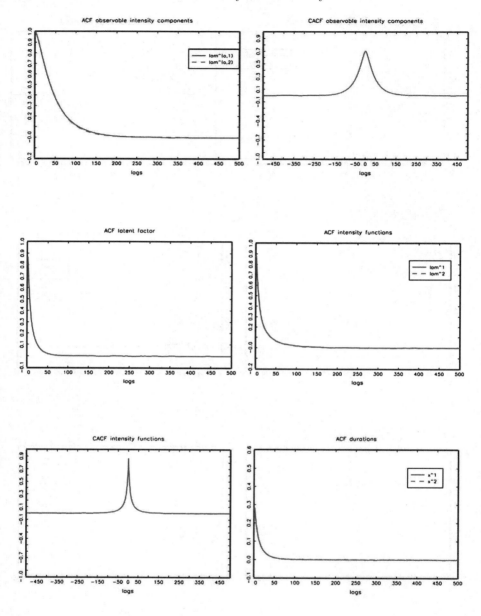

Figure 67: Simulated bivariate LF-Hawkes(1) processes. Upper: ACF and CACF of $\lambda^{o,1}(t_i; \mathfrak{F}_{t_i}^o)$ and $\lambda^{o,2}(t_i; \mathfrak{F}_{t_i}^o)$. Middle: ACF of λ_i^*, as well as of $\lambda^1(t_i; \mathfrak{F}_{t_i})$ and $\lambda^2(t_i; \mathfrak{F}_{t_i})$. Lower: CACF of $\lambda^1(t_i; \mathfrak{F}_{t_i})$ vs. $\lambda^2(t_i; \mathfrak{F}_{t_i})$, as well as ACF of x_i^1 and x_i^2. $\omega^1 = \omega^2 = \ln(0.6)$, $\alpha^{11} = \alpha^{22} = 0.2$, $\beta^{11} = \beta^{22} = 0.5$, $\alpha^{12} = \alpha^{21} = 0$, $a^* = 0.95$, $\delta^{1*} = \delta^{2*} = 0.5$. Based on $5,000,000$ drawings.

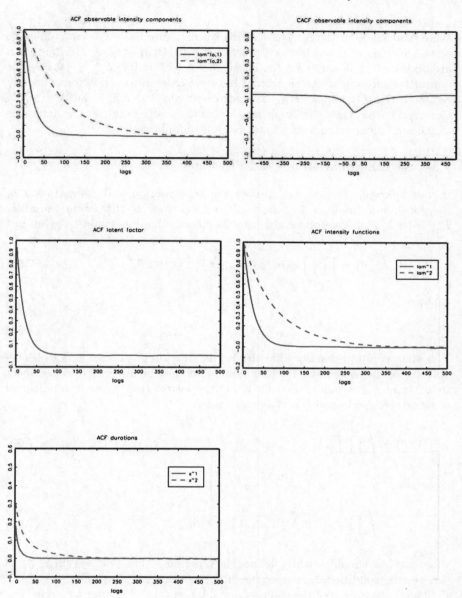

Figure 68: Simulated bivariate LF-Hawkes(1) processes. Upper: ACF and CACF of $\lambda^{o,1}(t_i; \mathfrak{F}_{t_i}^o)$ and $\lambda^{o,2}(t_i; \mathfrak{F}_{t_i}^o)$. Middle: ACF of λ_i^*, as well as of $\lambda^1(t_i; \mathfrak{F}_{t_i})$ and $\lambda^2(t_i; \mathfrak{F}_{t_i})$. Lower: CACF of $\lambda^1(t_i; \mathfrak{F}_{t_i})$ vs. $\lambda^2(t_i; \mathfrak{F}_{t_i})$, as well as ACF of x_i^1 and x_i^2. $\omega^1 = \ln(0.2)$, $\omega^2 = \ln(0.1)$, $\alpha^{11} = \alpha^{22} = 0.2$, $\beta^{11} = 0.4$, $\beta^{22} = 0.25$, $\alpha^{12} = \alpha^{21} = 0$, $a^* = 0.95$, $\delta^{1*} = 0.1$, $\delta^{2*} = -0.1$. Based on $5,000,000$ drawings.

the dynamics of the latent factor. Finally, in Figure 68, we assume different persistent independent Hawkes processes for $\lambda^{o,1}(t; \mathfrak{F}_t^o)$ and $\lambda^{o,2}(t; \mathfrak{F}_t^o)$ which are oppositely influenced by the latent factor ($\delta^{1*} = 0.1$, $\delta^{2*} = -0.1$). This parameterization leads to significant negative cross-autocorrelations not only between $\lambda^1(t; \mathfrak{F}_t)$ and $\lambda^2(t; \mathfrak{F}_t)$ but also between $\lambda^{o,1}(t; \mathfrak{F}_t^o)$ and $\lambda^{o,2}(t; \mathfrak{F}_t^o)$. However, in that case, the latent factor obviously does not drive the dynamics of the individual components towards higher similarity.

7.3.5 SML Estimation of the LFI Model

In the following section, we discuss the construction and computation of the likelihood function for the multivariate case. If the latent variables $\tilde{L}_n^* := \{\ln \lambda_i^*\}_{i=1}^n$ were observable, the likelihood function would be given by

$$\mathcal{L}(W, \tilde{L}_n^*; \theta) = \prod_{i=1}^{n} \prod_{s=1}^{S} \exp(-\Lambda^s(t_{i-1}, t_i)) \left[\lambda^{o,s}(t_i; \mathfrak{F}_{t_i}^o) \lambda_i^{*\delta^{s*}} \right]^{y_i^s}, \qquad (7.58)$$

where

$$\Lambda^s(t_{i-1}, t_i) = \lambda_i^{*\delta^{s*}} \Lambda^{o,s}(t_{i-1}, t_i). \qquad (7.59)$$

The main computationally difficulty is that the latent process is not observable, hence the conditional likelihood function must be integrated with respect to $\tilde{L}_n^*$ using the assumed distribution of the latter. Hence, the unconditional or integrated likelihood function is given by

$$\mathcal{L}(W; \theta) = \int \prod_{i=1}^{n} \prod_{s=1}^{S} \lambda_i^{*\delta^{s*} y_i^s} \exp(-\lambda_i^{*\delta^{s*}} \Lambda^{o,s}(t_{i-1}, t_i)) \left[\lambda^{o,s}(t_i; \mathfrak{F}_{t_i}^o) \right]^{y_i^s} \qquad (7.60)$$

$$\times \frac{1}{\sqrt{2\pi}} \exp \left[-\frac{(\ln \lambda_i^* - m_i^*)^2}{2} \right] d\tilde{L}_n^*.$$

$$= \int \prod_{i=1}^{n} f(w_i, \ln \lambda_i^* | W_{i-1}, \tilde{L}_{i-1}^*, \theta) d\tilde{L}_n^*, \qquad (7.61)$$

where the last equality clearly defines the function $f(\cdot)$ and $m_i^* = \mathrm{E}[\ln \lambda_i^* | \mathfrak{F}_{i-1}^*]$ denotes the conditional mean of the latent component.

The computation of the n-dimensional integral in (7.60) must be done numerically and represents a challenge. We use the efficient importance sampling (EIS) method of Richard (1998) because this method has proven to be quite efficient for the computation of the likelihood function of stochastic volatility models (see also Liesenfeld and Richard, 2003), to which the LFI model resembles.

In order to implement the EIS algorithm, the integral (7.61) is rewritten as

$$\mathcal{L}(W;\theta) = \int \prod_{i=1}^{n} \frac{f(w_i, \ln \lambda_i^* | W_{i-1}, \tilde{L}_{i-1}^*, \theta)}{m(\ln \lambda_i^* | \tilde{L}_{i-1}^*, \phi_i)} \prod_{i=1}^{n} m(\ln \lambda_i^* | \tilde{L}_{i-1}^*, \phi_i) d\tilde{L}_n^*, \quad (7.62)$$

where $\{m(\ln \lambda_i^* | \tilde{L}_{i-1}^*, \phi_i)\}_{i=1}^n$ denotes a sequence of auxiliary importance samplers for $\{\ln \lambda_i^*\}_{i=1}^n$ indexed by the parameters $\{\phi_i\}_{i=1}^n$. EIS relies on a choice of the auxiliary parameters $\{\phi_i\}_{i=1}^n$ that provide a good match between $\{m(\ln \lambda_i^* | \tilde{L}_{i-1}^*, \phi_i)\}_{i=1}^n$ and $\{f(w_i, \ln \lambda_i^* | W_{i-1}, \tilde{L}_{i-1}^*, \theta)\}_{i=1}^n$, so that the integral (7.61) can be computed reliably by importance sampling, i.e.

$$\mathcal{L}(W;\theta) \approx \hat{\mathcal{L}}_{n^{MC}}(W;\theta) = \frac{1}{n^{MC}} \sum_{r=1}^{n^{MC}} \prod_{i=1}^{n} \frac{f(w_i, \ln \lambda_i^{*(r)} | W_{i-1}, \tilde{L}_{i-1}^{*(r)}, \theta)}{m(\ln \lambda_i^{*(r)} | \tilde{L}_{i-1}^{*(r)}, \phi_i)}, \quad (7.63)$$

where $\{\ln \lambda_i^{*(r)}\}_{i=1}^n$ denotes a trajectory of random draws from the sequence of auxiliary importance samplers $\{m(\ln \lambda_i^* | \tilde{L}_{i-1}^*, \phi_i)\}_{i=1}^n$, and n^{MC} such trajectories are generated. More precisely, the "good match" criterion is the minimization of the sampling variance of the Monte Carlo importance sampling estimator $\hat{\mathcal{L}}_{n^{MC}}(W;\theta)$ by choice of the auxiliary parameters $\{\phi_i\}_{i=1}^n$. Before describing precisely how this is implemented, we explain how the importance samplers themselves are defined.

Let $k(\tilde{L}_i^*, \phi_i)$ denote a density kernel for $m(\ln \lambda_i^* | \tilde{L}_{i-1}^*, \phi_i)$, given by

$$k(\tilde{L}_i^*, \phi_i) = m(\ln \lambda_i^* | \tilde{L}_{i-1}^*, \phi_i) \chi(\tilde{L}_{i-1}^*, \phi_i), \quad (7.64)$$

where

$$\chi(\tilde{L}_{i-1}^*, \phi_i) = \int k(\tilde{L}_i^*, \phi_i) d\ln \lambda_i^*. \quad (7.65)$$

The implementation of EIS requires the selection of a class of density kernels $k(\cdot)$ for the auxiliary sampler $m(\cdot)$ that provide a good approximation to the product $f(w_i, \ln \lambda_i^* | W_{i-1}, \tilde{L}_{i-1}^*, \theta) \chi(\tilde{L}_{i-1}^*, \phi_i)$. A seemingly obvious choice of auxiliary samplers is the sequence of log-normal densities $N(m_i^*, 1)$ that is a component of the sequence of functions $f(w_i, \ln \lambda_i^* | \tilde{W}_{i-1}, \tilde{L}_{i-1}^*, \theta)$. However, this choice is not good enough because these direct samplers do not take the observed data into account.

As argued convincingly by Richard (1998), a convenient and efficient possibility is to use a parametric extension of the direct samplers, normal distributions in our context. We can approximate the function

$$\prod_{s=1}^{S} \lambda_i^{*\delta^{s*}y_i^s} \exp(-\lambda_i^{*\delta^{s*}} \Lambda^{o,s}(t_{i-1}, t_i)), \quad (7.66)$$

that appears in (7.60) by the normal density kernel

$$\zeta(\ln \lambda_i^*) = \exp(\phi_{1,i} \ln \lambda_i^* + \phi_{2,i}(\ln \lambda_i^*)^2). \quad (7.67)$$

We can also include the $N(m_i^*, 1)$ density function in the importance sampler $m(\ln \lambda_i^* | \tilde{L}_{i-1}^*, \phi_i)$. Using the property that the product of normal densities is itself a normal density, we can write a density kernel of $m(\cdot)$ as

$$k(\tilde{L}_i^*, \phi_i) \propto \exp\left((\phi_{1,i} + m_i^*) \ln \lambda_i^* + \left(\phi_{2,i} - \frac{1}{2} \right) (\ln \lambda_i^*)^2 \right) \qquad (7.68)$$

$$= \exp\left(-\frac{1}{2\pi_i^2} (\ln \lambda_i^* - \mu_i)^2 \right) \exp\left(\frac{\mu_i^2}{2\pi_i^2} \right), \qquad (7.69)$$

where

$$\pi_i^2 = (1 - 2\phi_{2,i})^{-1} \qquad (7.70)$$

$$\mu_i = (\phi_{1,i} + m_i^*)\, \pi_i^2. \qquad (7.71)$$

Hence, the integrating constant (7.65) is given by

$$\chi(\tilde{L}_{i-1}^*, \phi_i) = \exp\left(\frac{\mu_i^2}{2\pi_i^2} - \frac{m_i^{*2}}{2} \right) \qquad (7.72)$$

(neglecting the factor $\pi_i \sqrt{2\pi}$ since it depends neither on $\tilde{L}_{i-1}^*$ nor on ϕ_i).

As mentioned above, the choice of the auxiliary parameters has to be optimized in order to minimize the MC variance of $\hat{\mathcal{L}}_{nMC}(W; \theta)$. This can be split into n minimization problems of the form (see Richard, 1998, for a detailed explanation):

$$\min_{\phi_{i,0}, \phi_i} \sum_{r=1}^{n^{MC}} \left\{ \ln\left[f(w_i, \ln \lambda_i^{*(r)} | W_{i-1}, \tilde{L}_{i-1}^{*(r)}, \theta) \chi(\tilde{L}_i^{*(r)}, \phi_{i+1}) \right] \qquad (7.73) \right.$$

$$\left. - \phi_{0,i} - \ln k(\tilde{L}_i^{*(r)}, \phi_i) \right\}^2,$$

where $\phi_{0,i}$ are auxiliary proportionality constants. These problems must be solved sequentially starting at $i = n$, under the initial condition $\chi(\tilde{L}_n^*, \phi_{n+1}) = 1$ and ending at $i = 1$. In (7.73), $\{\ln \lambda_i^{*(r)}\}_{i=1}^n$ denotes a trajectory of random draws from the sequence of direct samplers $\{N(m_i^*, 1)\}_{i=1}^n$. Given that $k(\tilde{L}_i^*, \phi_i)$ is a normal density kernel (see (7.69)), the minimization problem (7.73) is solved by computing an ordinary least squares estimate.

To summarize, the EIS algorithm requires the following steps:

Step 1: Generate n^{MC} trajectories $\{\ln \lambda_i^{*(r)}\}_{i=1}^n$ using the direct samplers $\{N(m_i^*, 1)\}_{i=1}^n$.

Step 2: For each i (from n to 1), estimate by OLS the regression (with n^{MC} observations) implicit in (7.73), which takes precisely the following form:

$$\delta^{s*} y_i^s \ln \lambda_i^{*(r)} - \left(\lambda_i^{*(r)} \right)^{\delta^{s*}} \sum_{s=1}^{S} \Lambda^{o,s}(t_{i-1}, t_i) + \sum_{s=1}^{S} y_i^s \ln \lambda^{o,s}(t_i; \mathfrak{F}_{t_i}^o) \qquad (7.74)$$

$$+ \ln\left(\chi(\tilde{L}_i^{*(r)}, \phi_{i+1}) \right)$$

$$= \phi_{0,i} + \phi_{1,i} \ln \lambda_i^{*(r)} + \phi_{2,i}(\ln \lambda_i^{*(r)})^2 + u^{(r)}, \ r = 1, \ldots, n^{MC},$$

(where $u^{(r)}$ is an error term), using $\chi(\tilde{L}_n^{*(r)}, \phi_{n+1}) = 1$ as initial condition, and then (7.72).

Step 3: Generate n^{MC} trajectories $\{\ln \lambda_i^{*(r)}\}_{i=1}^n$ using the EIS samplers $\{N(\mu_i, \pi_i^2)\}_{i=1}^n$ (see (7.70) and (7.71)) to compute $\hat{\mathcal{L}}_{n^{MC}}(W; \theta)$ as defined in (7.63). Per construction of the LFI model, the computation of the terms $\Lambda^{o,s}(t_{i-1}, t_i)$ and $\lambda^{o,s}(t_i; \mathfrak{F}_{t_i}^o)$ can be done separately and is done in a step before the EIS algorithm.

As recommended by Liesenfeld and Richard (2003), steps 1 to 3 should be iterated about five times to improve the efficiency of the approximations. This is done by replacing the direct sampler in step 1 by the importance samplers built in the previous iteration. It is also possible to start step 1 (of the first iteration) with a sampler other than the direct one. This is achieved by immediately multiplying the direct sampler by a normal approximation to $\delta^{s*} \ln \lambda_i^* - \lambda_i^{*\delta^{s*}} \sum_{s=1}^{S} \Lambda^{o,s}(t_{i-1}, t_i)$, using a second order Taylor expansion (TSE) of the argument of the exponential function around $\ln \lambda_i^* = 0$. This yields

$$\delta^{s*} y_i^s \ln \lambda_i^* - \lambda_i^{*\delta^{s*}} \sum_{s=1}^{S} \Lambda^{o,s}(t_{i-1}, t_i) \tag{7.75}$$

$$\approx \text{constant} + \ln \lambda_i^* - (\ln \lambda_i^*)^2 \sum_{s=1}^{S} \Lambda^{o,s}(t_{i-1}, t_i),$$

which implies that $\phi_{1,i} = 1$ and $\phi_{2,i} = -\sum_{s=1}^{S} \Lambda^{o,s}(t_{i-1}, t_i)$ must be inserted into (7.70) and (7.71) to obtain the moments of the TSE normal importance sampler. In this way, the initial importance sampler takes into account the data and enables one to reduce to three (instead of five) iterations over the three steps.

7.3.6 Testing the LFI Model

In order to provide diagnostics with respect to the latent process, it is necessary to produce sequences of filtered estimates of functions of the latent variable λ_i^*, which take the form of the ratio of integrals in (7.48). Liesenfeld and Richard (2003) propose evaluating the integrals in the denominator and numerator by MC integration using the EIS algorithm where the parameter values θ are set equal to their ML estimates. The integral in the denominator corresponds to $\mathcal{L}(W_{i-1}; \theta)$, the marginal likelihood function of the first $i-1$ observations, and is evaluated on the basis of the sequence of auxiliary samplers $\{m(\ln \lambda_j^* | \tilde{L}_{j-1}^*, \hat{\phi}_j^{i-1})\}_{j=1}^{i-1}$ where $\{\hat{\phi}_j^{i-1}\}$ denotes the value of the EIS auxiliary parameters associated with the computation of $\mathcal{L}(W_{i-1}; \theta)$. Furthermore, the numerator of (7.48) is approximated by

$$\frac{1}{n^{MC}} \sum_{r=1}^{n^{MC}} \left\{ \vartheta \left(\ln \lambda_i^{*(r)}(\theta) \right) \prod_{j=1}^{i-1} \left[\frac{f \left(w_j, \ln \lambda_j^{*(r)}(\hat{\phi}_j^{i-1}) | W_{j-1}, \tilde{L}_{j-1}^{*(r)}(\hat{\phi}_{j-1}^{i-1}), \theta \right)}{m \left(\ln \lambda_j^{*(r)}(\hat{\phi}_j^{i-1}) | \tilde{L}_{j-1}^{*(r)}(\hat{\phi}_{j-1}^{i-1}), \hat{\phi}_j^{i-1} \right)} \right] \right\},$$

$$(7.76)$$

where $\{\ln \lambda_j^{*(r)}(\hat{\phi}_j^{i-1})\}_{j=1}^{i-1}$ denotes a trajectory drawn from the sequence of importance samplers associated with $\mathcal{L}(W_{i-1}; \theta)$, and $\ln \lambda_i^{*(r)}(\theta)$ is a random draw from the conditional density $p(\ln \lambda_i^* | W_{i-1}, \tilde{L}_{i-1}^{*(r)}(\hat{\phi}_{i-1}^{i-1}), \theta)$. The computation of the sequence of filtered estimates requires rerunning the complete EIS algorithm for every function $\vartheta(\ln \lambda_i^*)$ and for every i (=1 to n).

The sequence of filtered estimates can be computed to evaluate the series of the latent variable and to compute conditional moments (see Section 7.3.1) and forecasts.

The LFI residuals of the s-th process are computed on the basis of the trajectories drawn from the sequence of auxiliary samplers characterizing $\mathcal{L}(W; \theta)$, i.e.,

$$\hat{\epsilon}_i^{s,(r)} := \int_{t_{i-1}^s}^{t_i^s} \left[\lambda_i^{*(r)}(\hat{\phi}_i^n) \right]^{\delta^{s*}} \hat{\lambda}^{o,s}(u; \mathfrak{F}_u^o) du \qquad (7.77)$$

$$= \sum_{j=N(t_{i-1}^s)}^{N(t_i^s)} \left[\lambda_i^{*(r)}(\hat{\phi}_i^n) \right]^{\delta^{s*}} \int_{t_j}^{t_{j+1}} \hat{\lambda}^{o,s}(u; \mathfrak{F}_u^o) du.$$

The residual diagnostics are computed for each of the n^{MC} sequences separately.[13] Under correct specification, the residuals $\hat{\epsilon}_i^{s,(r)}$ should be i.i.d. $Exp(1)$. Hence, model evaluations can be done by testing the dynamical and distributional properties of the residual series using the techniques described in Sections 5.4.1 and 5.4.2 of Chapter 5.

7.4 Applications of Dynamic Intensity Models

7.4.1 Estimating Multivariate Price Intensities

The goal of the following subsection is to illustrate the application of Hawkes processes for multivariate volatility estimation and to test for the existence of contemporaneous and cross-autocorrelation structures in price intensities. We use price durations generated from a sample consisting of five NYSE stocks during the period 01/02/01 to 02/28/01: Boeing, Coca-Cola, General Electric, Home Depot and Philip Morris. The data preparation follows along the lines described in Chapter 4, while the generation of the price duration is performed

---

[13] Note that it is not reasonable to evaluate the model based on the average trajectory since dispersion and dynamic effects would eliminate each other then and would lead to non-interpretable results.

Table 39: Descriptive statistics and Ljung-Box statistics of $0.20 midquote change durations for the Boeing, Coca-Cola, General Electric, Home Depot and Philip Morris stocks traded at the NYSE. Data extracted from the TAQ database, sample period from 01/02/01 to 02/28/01.

	Boeing	Coca-Cola	General Electric	Home Depot	Philip Morris
Obs	16236	12678	17637	18789	11152
Mean	57.252	72.757	52.872	49.351	83.158
S.D.	69.222	99.296	66.507	65.094	129.665
LB(20)	2974	2248	7695	3824	3442

Overnight spells are removed. Descriptive statistics of durations in seconds.

according to the procedure described in Section 3.1.2 of Chapter 3. The price durations are computed based on midquote changes of the size $0.20. Table 39 shows the descriptive statistics of the particular price duration series. We observe average price durations between 45 and 80 seconds, which are associated with an intraday volatility measured at a very high frequency.[14]

The multivariate $0.20 price change intensity is modelled using a five-dimensional Hawkes process. According to the BIC value, we choose a lag order of $P = 1$. We jointly estimate a linear spline function for each series based on the nodes 9:30, 10:00, 11:00, 12:00, 13:00, 14:00 and 15:00. The seasonality function enters the model multiplicatively according to (7.16). The model is estimated by maximizing the log likelihood function given in (7.43) and by exploiting the property that the log likelihood components associated with the individual processes can be maximized separately. For numerical reasons, we standardize the particular series by the average duration mean of the pooled process. Note that this is only a matter of scaling and does not change the order of the processes.

Table 40 gives the estimation results of the five-dimensional Hawkes model. The entries in the table give the coefficients α^{sr} and β^{sr} associated with the impact of the r-type backward recurrence times on the s-th series. We observe highly significant estimates in most cases. In particular, the significance of the interaction coefficients $(r \neq s)$ illustrates the existence of strong contemporaneous correlations and cross-autocorrelations between the individual series. The estimated seasonalities indicate the well known inverse U-shape intraday pattern and are very similar for all series. Since this issue has already been discussed in Chapter 4 in more detail, we do not show the seasonality plots here. The Hawkes residuals are computed as $\hat{\epsilon}_i^s = \hat{\Lambda}^s(t_{i-1}, t_i)$. We compute the mean, standard deviation and Ljung-Box statistics of the residuals, as well as the test against excess dispersion. The test against excess dispersion provides satisfying results since the null hypothesis of no excess dispersion is not rejected in most cases. In contrast, the relatively high Ljung-Box

---

[14] Since the aggregation level is quite low, we neglect overnight effects.

Table 40: ML estimates of a 5-dimensional Hawkes(1) model for \$0.20-price durations of the Boeing (B), Coca-Cola (C), General Motors (G), Home Depot (H) and Philip Morris (P) stock traded at the NYSE. Data from the TAQ database, sample period 01/02/01 to 02/28/01. Standard errors based on OPG estimates.

	Boeing		Coca-Cola		GM		Home Depot		PM	
	est.	S.E.	est.	S.E.	est.	S.E.	est.	S.E.	est.	S.E.
ω	0.106	0.010	0.060	0.006	0.066	0.010	0.108	0.080	0.026	0.005
$\alpha^{\cdot B}$	0.021	0.001	0.011	0.000	0.023	0.001	0.006	0.056	0.013	0.000
$\alpha^{\cdot C}$	0.015	0.001	0.031	0.002	0.028	0.002	0.006	0.062	0.009	0.000
$\alpha^{\cdot G}$	0.017	0.006	0.015	0.004	0.021	0.006	0.012	0.796	0.003	0.001
$\alpha^{\cdot H}$	0.012	0.005	0.005	0.001	0.020	0.008	0.038	1.005	0.009	0.005
$\alpha^{\cdot P}$	0.004	0.002	0.004	0.001	0.011	0.003	0.009	0.314	0.038	0.011
$\beta^{\cdot B}$	0.016	0.009	0.210	0.110	0.176	0.070	0.554	0.386	0.419	0.152
$\beta^{\cdot C}$	0.136	0.069	0.034	0.044	0.211	0.068	0.142	1.492	0.182	0.092
$\beta^{\cdot G}$	0.303	0.190	0.188	0.326	0.013	0.005	0.108	1.998	0.019	0.013
$\beta^{\cdot H}$	0.209	0.266	0.646	0.437	0.182	0.106	0.043	0.455	0.229	0.110
$\beta^{\cdot P}$	0.121	0.190	0.729	0.579	0.123	0.089	0.101	0.550	0.034	0.020
s_1	-2.780	0.194	-1.893	0.265	-3.210	0.145	-1.898	0.210	-2.396	0.243
s_2	2.684	0.232	1.957	0.325	3.146	0.174	1.888	0.265	2.483	0.289
s_3	0.093	0.110	-0.197	0.162	0.076	0.091	-0.135	0.134	-0.292	0.149
s_4	-0.025	0.106	0.103	0.161	-0.003	0.088	0.093	0.131	0.341	0.153
s_5	0.185	0.110	0.019	0.164	0.031	0.092	0.231	0.134	-0.154	0.157
s_6	-0.272	0.110	0.251	0.168	0.000	0.093	-0.199	0.133	-0.047	0.157
s_7	0.407	0.123	-0.655	0.187	0.061	0.102	0.075	0.147	0.209	0.171
Obs	76492									
LL	-192325									

Diagnostics of Hawkes residuals										
Mean	0.99		0.99		0.99		0.99		0.99	
S.D.	0.98		1.02		0.99		0.98		1.01	
LB	42.95	0.00	26.06	0.16	104.15	0.00	39.25	0.00	48.32	0.00
Disp.	0.94	0.34	2.12	0.03	0.09	0.92	1.90	0.05	1.48	0.13

Diagnostics: Log Likelihood (LL) and diagnostics (mean, standard deviation, Ljung-Box statistic wrt 20 lags, as well as excess dispersion test (the latter two statistics inclusive p-values)) of Hawkes residuals $\hat{\epsilon}_i^s$, $s = 1, \ldots, 5$.

statistics associated with the individual residual series $\hat{\epsilon}_i^s$ indicate that the dynamics are not completely captured by the model. Thus, even though the BIC value identifies a lag order of $P = 1$ as the best specification, there is obviously some need for higher parameterized models. These results indicate that particularly in the multivariate case, there exists a clear trade-off between a satisfying goodness-of-fit and model parsimony.

To illustrate the estimated dynamics graphically, we plot the resulting autocorrelation and cross-autocorrelation functions of the estimated intensity

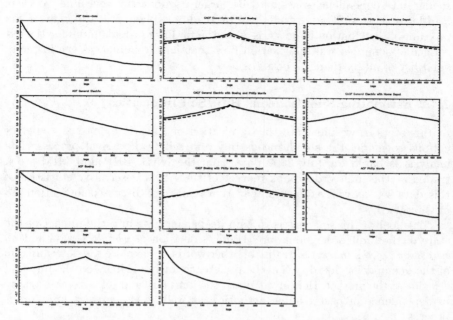

Figure 69: Autocorrelation and cross-autocorrelation functions of the estimated intensity functions based on a 5-dimensional Hawkes(1) model for $0.20-price durations of the Boeing (B), Coca-Cola (C), General Electric (G), Home Depot (H) and Philip Morris (P) stock traded at the NYSE. Data based on TAQ database, sample period 01/02/01 to 02/28/01.
First row: left: ACF of Coca-Cola, middle: CACF C-G (solid line) and C-B (broken line), right: CACF C-P (solid line) and C-H (broken line).
Second row: left: ACF of General Electric, middle: CACF G-B (solid line) and G-P (broken line), right: CACF G-H.
Third row: left: ACF of Boeing, middle: CACF of B-P (solid line) and B-H (broken line), right: ACF of Philip Morris.
Fourth row: left: CACF P-H, middle: ACF of Home Depot.

functions (Figure 69). In fact, we find strong interdependencies between the single price intensity series and thus, can provide empirical evidence for significant co-movements in price change volatility.[15] The contemporaneous correlation is around 0.6 for all series coming along with strongly persistent cross-autocorrelations in both directions. Hence, volatility shocks in one series lead to quite persistent spill-over effects in the other series causing strong multidimensional clustering structures. However, the shape of the individual (cross-)autocorrelation functions is quite similar for all series. Hence, the par-

---

[15] Recall the close relationship between the price intensity $\lambda^{dp}(t;\mathfrak{F}_t)$ and the corresponding instantaneous volatility $\tilde{\sigma}^2_{(x^{dp})}(t)$ as given in (3.8).

ticular interdependencies seem to be largely symmetric, revealing no clear lead-lag structure. These results can be readily explained in the light of the mixture-of-distribution hypothesis since the findings provide hints for the existence of some general information flow that jointly influences the intraday volatility of all particular stocks.

7.4.2 Estimating Simultaneous Buy/Sell Intensities

In this section, we illustrate the application of (bivariate) intensity specifications to model the simultaneous buy and sell trade arrival process. The analysis is based on July 2002 data from the NAB, BHP and MIM stock traded at the limit order book market of the ASX. Descriptive statistics of this data set, as well as data preparation issues are discussed in Section 4.3 of Chapter 4.

The goal of this analysis is to gain deeper insights into the impact of the state of the limit order book on a trader's decision of *when* to trade and *on which side of the market* to trade. In other words, we explore the determinants of the demand for liquidity, i.e. the preference for immediacy on the buy and sell side of the market. Hence, we model the joint buy and sell arrival intensity in dependence of past trading activities, as well as the current limit order process.

We use several variables that reflect the state of the order book with respect to market depth, tightness and the standing volume in the ask and bid queue. In order to capture the aggregated supply and demand on the market, we quantify the aggregated queued volume, as well as the volume-price quantiles on the bid and ask side. The corresponding limit prices associated with a given aggregated volume allow us to construct the (piecewise) slope of the market reaction curve and thus, enables us to measure the depth on both sides of the market. Based on this variable, it is possible to identify the potential price impact caused by a buy/sell market order of a given size which is an important determinant of market liquidity (compare the discussion in Section 3.4 of Chapter 3).

Since we are mainly interested in the question of whether traders exploit situations in which trading on one particular side of the market is favorable, we construct two types of variables. Firstly, as a measure for imbalances between the market sides, we use the difference between the standing volume in the ask and bid queue. Secondly, as a measure for differences between the depth on the buy and the sell side, we compare the piecewise slopes of the ask and bid reaction curve with respect to given volume quantiles. In particular, we construct slope differentials by computing the difference of absolute price differences between corresponding volume quantiles on each side of the market. Such as measure provides us an "average" market reaction curve indicating the relative bid depth compared to the ask depth. As a measure for the tightness of the market, we use the (inside) spread between the current best bid and best ask price. Moreover, in order to account for the influence of recent transactions and limit orders, we include the trading volume at the previous trade, the type and the quoted volume of the current limit order, as well as the previous midquote change. Table 41 gives the definition of the particular explanatory variables used in the empirical study. Note that with exception of the previous trading volume all variables change whenever a new limit order enters the market. Thus, they have to be modelled as time-varying covariates.[16]

Table 41: Definition of explanatory variables used in the empirical study.

	Variables observed at each transaction
$TRVB$	traded buy volume
$TRVS$	traded sell volume
Time-varying covariates associated with the limit order process	
$AVOL$	total volume on the ask queue
$BVOL$	total volume on the bid queue
D_VOL	AVOL-BVOL
A_x	price associated with the $x\%$ quantile of the cumulated volume on the ask queue
B_x	price associated with the $x\%$ quantile of the cumulated volume on the bid queue
MQ	midquote $((A_0 + B_0)/2)$
$DFF1$	$DFF1 = |B_{10} - mq| - |A_{10} - mq|$
$DFF2$	$DFF2 = |B_{20} - B_{10}| - |A_{20} - A_{10}|$
$DFF4$	$DFF4 = |B_{40} - B_{20}| - |A_{40} - A_{20}|$
$DFF6$	$DFF6 = |B_{60} - B_{40}| - |A_{60} - A_{40}|$
$DFF9$	$DFF9 = |B_{90} - B_{60}| - |A_{90} - A_{60}|$
$SPRD$	bid-ask spread
DMQ	midquote-change $DMQ = MQ_i - MQ_{i-1}$
$QASK$	1: if limit order is an ask, 0: if limit order is a bid
$QVOLA$	quoted volume if limit order is an ask, (0 for bids)
$QVOLB$	quoted volume if limit order is a bid, (0 for asks)

Tables 42 through 44 present the results of bivariate ACI models. The lag order is chosen according to the BIC leading to an ACI(1,1) specification for all three series. As in the previous subsection, we standardize the time scale by the average duration of the pooled process. The baseline intensity function is specified in terms of a Burr parameterization as given in (7.39). The individual seasonality functions for both series are modelled as linear spline functions based on 1 hour nodes. By estimating the ACI model based on different innovation specifications (7.33), (7.34) and (7.36), we find a clear

---

[16] The trading volume is treated as a mark.

Table 42: ML estimates of bivariate ACI(1,1) models with static and time-varying covariates for the MIM stock. Data extracted from ASX trading, sample period July 2002. (1): Model without covariates. (2): Model with time-varying covariates. (3): Model with marks. Standard errors based on OPG estimates. Coefficients of covariates multiplied by 10.

	(1) est.	(1) S.E.	(2) est.	(2) S.E.	(3) est.	(3) S.E.
	ACI parameters buy trades					
ω^B	-0.160	0.122	-0.868	0.192	-1.427	0.224
a^B_B	0.786	0.025	0.821	0.024	0.817	0.024
a^B_S	0.983	0.023	0.900	0.028	0.882	0.028
η^B_B	0.136	0.040	0.019	0.019	0.018	0.020
η^B_S	-0.012	0.009	-0.019	0.011	-0.016	0.012
α^B_B	0.260	0.031	0.202	0.023	0.190	0.028
α^B_S	-0.052	0.028	-0.103	0.023	-0.062	0.026
β_{BB}	0.686	0.058	0.844	0.033	0.656	0.066
β_{BS}	-0.038	0.038	0.014	0.019	-0.367	0.080
	ACI parameters sell trades					
ω^S	-0.164	0.103	0.153	0.120	-1.024	0.263
a^S_B	1.006	0.019	0.994	0.024	0.970	0.025
a^S_S	0.819	0.022	0.823	0.022	0.822	0.022
η^S_B	-0.010	0.003	-0.015	0.003	-0.013	0.003
η^S_S	0.026	0.024	0.012	0.022	0.001	0.019
α^S_B	0.022	0.029	-0.043	0.019	0.074	0.038
α^S_S	0.072	0.013	0.119	0.011	0.164	0.025
β_{SB}	0.005	0.048	-0.038	0.024	-0.224	0.060
β_{SS}	0.975	0.011	0.986	0.009	0.628	0.077
	Seasonality parameters buy trades					
s_1	-0.581	0.371	-0.745	0.422	-0.853	0.350
s_2	0.021	0.657	0.452	0.698	0.472	0.584
s_2	-0.081	0.503	-0.412	0.534	-0.167	0.447
s_4	0.874	0.471	1.019	0.505	0.858	0.449
s_5	0.458	0.502	0.244	0.549	0.276	0.477
s_6	0.512	0.662	0.596	0.708	0.276	0.608
	Seasonality parameters sell trades					
s_1	-0.923	0.264	-0.898	0.344	-0.984	0.332
s_2	0.628	0.456	0.674	0.533	0.660	0.551
s_2	-0.453	0.384	-0.718	0.436	-0.430	0.404
s_4	1.449	0.371	1.763	0.455	1.312	0.367
s_5	-0.663	0.381	-0.718	0.415	-0.447	0.373
s_6	2.018	0.577	2.136	0.736	1.641	0.580
	Static covariates buy trades					
$TRVB$			0.579	0.150	0.777	0.144
$TRVS$			0.097	0.155	0.267	0.151
	Static covariates sell trades					
$TRVB$			0.475	0.145	0.427	0.147
$TRVS$			0.524	0.140	0.553	0.136
	Time-varying covariates buy trades					
D_VOL			-0.337	1.281	-0.354	1.136
$DFF1$			-4.451	0.325	-3.201	0.306
$DFF2$			-0.627	0.271	-0.416	0.225
$DFF4$			-0.072	0.187	-0.027	0.155
$DFF6$			-0.214	0.126	-0.199	0.113

Table 42 cont'd:

$DFF9$	-0.175	0.137	-0.005	0.125
$SPRD$	-1.405	1.886	0.436	1.212
DMQ	-3.699	0.841	-3.891	0.526
$QASK$	3.300	2.183	2.396	2.455
$QVOLA$	-0.181	0.156	0.129	0.193
$QVOLB$	0.077	0.166	0.435	0.168
Time-varying covariates sell trades				
D_VOL	-1.888	1.696	0.583	1.148
$DFF1$	3.236	0.281	2.417	0.287
$DFF2$	1.019	0.295	0.391	0.261
$DFF4$	0.487	0.244	-0.041	0.161
$DFF6$	-0.075	0.175	-0.108	0.114
$DFF9$	-0.144	0.146	-0.131	0.126
$SPRD$	-3.782	1.420	2.013	1.174
DMQ	7.759	1.074	5.016	0.585
$QASK$	-2.626	2.249	4.218	2.362
$QVOLA$	0.232	0.183	0.074	0.167
$QVOLB$	-0.086	0.148	0.393	0.188

Obs (n)	3617	3617	3617
Obs (n^0)	9508	9508	9508
LL	-5292	-5010	-5080
BIC	-5415	-5240	-5309

Diagnostics of ACI residuals for the buy series						
	stat.	p-value	stat.	p-value	stat.	p-value
Mean	0.97		0.98		0.99	
S.D.	0.97		1.03		1.03	
LB(20)	32.29	0.04	26.03	0.16	34.77	0.02
Disp.	0.80	0.42	0.88	0.37	0.91	0.35
Diagnostics of ACI residuals for the sell series						
	stat.	p-value	stat.	p-value	stat.	p-value
Mean	1.00		1.01		1.00	
S.D.	1.02		1.06		1.04	
LB(20)	16.65	0.67	18.00	0.58	35.98	0.01
Disp.	0.81	0.41	2.23	0.02	1.32	0.18

Diagnostics: Log Likelihood (LL), Bayes Information Criterion (BIC) and diagnostics (mean, standard deviation, Ljung-Box statistic, as well as excess dispersion test) of ACI residuals $\hat{\varepsilon}_i^s$.

under-performance of specification (7.36) in terms of the BIC and the distributional properties of the resulting residuals. Thus, specifying $\breve{\varepsilon}_i$ in terms of *process-specific* innovations (as in (7.33) or (7.34)) seems to be essential to capture the dynamic properties of multivariate financial intensity processes. In particular, the best performance is found for specification (7.33)

Table 43: ML estimates of bivariate ACI(1,1) models with static and time-varying covariates for the NAB stock. Data extracted from ASX trading, sample period July 2002. (1): Model without covariates. (2): Model with time-varying covariates. (3): Model with marks. Standard errors based on OPG estimates. Coefficients of covariates multiplied by 10.

	(1)		(2)		(3)	
	est.	S.E.	est.	S.E.	est.	S.E.
ACI parameters buy trades						
ω^B	0.022	0.057	-0.135	0.071	-0.207	0.073
a^B_B	0.901	0.009	0.888	0.009	0.900	0.009
a^B_S	0.876	0.009	0.904	0.010	0.897	0.010
η^B_B	0.047	0.009	0.044	0.009	0.035	0.009
η^B_S	-0.007	0.005	-0.003	0.005	-0.002	0.005
α^B_B	0.032	0.002	0.058	0.004	0.052	0.004
α^B_S	0.020	0.004	0.040	0.005	0.035	0.005
β_{BB}	0.997	0.001	0.993	0.003	0.992	0.003
β_{BS}	-0.001	0.002	-0.003	0.004	-0.004	0.004
ACI parameters sell trades						
ω^S	-0.164	0.064	-0.380	0.075	-0.485	0.079
a^S_B	0.855	0.009	0.878	0.010	0.875	0.010
a^S_S	0.949	0.012	0.934	0.013	0.945	0.013
η^S_B	-0.014	0.004	-0.008	0.005	-0.011	0.004
η^S_S	0.067	0.010	0.059	0.010	0.056	0.010
α^S_B	0.020	0.004	0.040	0.005	0.035	0.005
α^S_S	0.053	0.005	0.072	0.006	0.071	0.006
β_{SB}	-0.009	0.002	-0.041	0.006	-0.031	0.006
β_{SS}	0.978	0.003	0.937	0.008	0.944	0.008
Seasonality parameters buy trades						
s_1	-1.009	0.152	-0.975	0.189	-1.086	0.098
s_2	0.803	0.254	0.763	0.305	0.856	0.178
s_2	-0.695	0.197	-0.658	0.233	-0.587	0.163
s_4	1.498	0.181	1.469	0.220	1.344	0.141
s_5	0.037	0.186	0.179	0.228	0.150	0.149
s_6	0.144	0.277	0.130	0.348	0.052	0.204
Seasonality parameters sell trades						
s_1	-1.010	0.174	-1.290	0.125	-1.309	0.131
s_2	0.784	0.302	1.093	0.219	1.130	0.227
s_2	-0.688	0.228	-0.547	0.173	-0.577	0.175
s_4	1.630	0.214	1.372	0.173	1.373	0.175
s_5	-0.006	0.238	-0.115	0.181	-0.076	0.183
s_6	-0.052	0.340	-0.106	0.232	-0.124	0.240
Static covariates buy trades						
$TRVB$			0.172	0.055	0.256	0.056
$TRVS$			0.189	0.056	0.166	0.056
Static covariates sell trades						
$TRVB$			0.610	0.064	0.637	0.062
$TRVS$			0.594	0.065	0.693	0.064
Time-varying covariates buy trades						
D_VOL			2.613	0.524	1.950	0.498
$DFF1$			-0.028	0.010	-0.026	0.009
$DFF2$			-0.010	0.007	-0.009	0.006
$DFF4$			-0.002	0.004	-0.005	0.004
$DFF6$			-0.011	0.004	-0.014	0.004

Table 43 cont'd:

$DFF9$			-0.005	0.003	-0.009	0.003
$SPRD$			-1.981	0.075	-1.164	0.065
DMQ			-0.730	0.018	-0.680	0.017
$QASK$			-1.249	0.675	-0.937	0.837
$QVOLA$			0.269	0.068	0.302	0.086
$QVOLB$			0.097	0.060	0.247	0.065
Time-varying covariates sell trades							
D_VOL			-0.416	0.338	-0.378	0.335
$DFF1$			0.008	0.010	0.005	0.010
$DFF2$			0.003	0.006	0.003	0.006
$DFF4$			0.004	0.003	0.004	0.003
$DFF6$			0.002	0.003	0.002	0.003
$DFF9$			-0.003	0.002	-0.003	0.002
$SPRD$			-1.655	0.081	-1.076	0.073
DMQ			0.762	0.017	0.747	0.016
$QASK$			3.410	0.760	3.138	0.893
$QVOLA$			0.034	0.076	0.043	0.088
$QVOLB$			0.508	0.067	0.470	0.075
Obs (n)	21490		21490		21490	
Obs (n^0)	42299		42299		42299	
LL	-32156		-31369		-31532	
BIC	-32306		-31648		-31811	

Diagnostics of ACI residuals for the buy series						
	stat.	p-value	stat.	p-value	stat.	p-value
Mean	1.00		1.00		1.00	
S.D.	1.00		1.01		1.01	
LB(20)	31.67	0.04	19.42	0.49	22.12	0.33
Disp.	0.21	0.82	1.08	0.27	0.91	0.35

Diagnostic ACI residuals for the sell series						
	stat.	p-value	stat.	p-value	stat.	p-value
Mean	1.00		1.00		1.00	
S.D.	1.00		1.01		1.01	
LB(20)	24.85	0.20	30.36	0.06	26.51	0.14
Disp.	0.07	0.93	1.13	0.25	1.38	0.16

Diagnostics: Log Likelihood (LL), Bayes Information Criterion (BIC) and diagnostics (mean, standard deviation, Ljung-Box statistic, as well as excess dispersion test) of ACI residuals $\hat{\varepsilon}_i^s$.

which is used in the regressions shown in the Tables 42 through 44.[17] For each bivariate series, we estimate three specifications. Column (1) gives the estimation results of an ACI model without any covariates, while column (2) gives the results of the complete model that captures the limit order arrivals as

---

[17] For ease of illustration we refrain from showing the other specifications here.

Table 44:   ML estimates of bivariate ACI(1,1) models with static and time-varying covariates for the BHP stock. Data extracted from the ASX trading, sample period July 2002. (1): Model without covariates. (2): Model with time-varying covariates. (3): Model with marks. Standard errors based on OPG estimates. Coefficients of covariates multiplied by 10.

	(1)		(2)		(3)	
	est.	S.E.	est.	S.E.	est.	S.E.
ACI parameters buy trades						
ω^B	0.341	0.045	0.012	0.059	0.039	0.059
a_B^B	0.909	0.007	0.972	0.007	0.908	0.007
a_S^B	0.933	0.007	0.880	0.008	0.947	0.008
η_B^B	0.045	0.006	0.026	0.007	0.046	0.007
η_S^B	-0.001	0.002	0.052	0.002	0.000	0.002
α_B^B	0.040	0.002	0.032	0.003	0.045	0.003
α_S^B	0.038	0.003	0.074	0.004	0.039	0.004
β_{BB}	0.994	0.001	-0.010	0.002	0.992	0.002
β_{BS}	-0.007	0.002	0.977	0.002	-0.006	0.004
ACI parameters sell trades						
ω^S	-0.002	0.053	0.908	0.075	-0.405	0.074
a_B^S	0.952	0.010	0.951	0.011	0.953	0.011
a_S^S	0.897	0.010	0.054	0.011	0.909	0.010
η_B^S	0.023	0.008	-0.000	0.008	0.036	0.009
η_S^S	0.059	0.008	0.045	0.008	0.045	0.007
α_B^S	0.037	0.004	0.032	0.004	0.038	0.004
α_S^S	0.076	0.005	0.989	0.005	0.086	0.006
β_{SB}	-0.014	0.002	-0.010	0.002	-0.020	0.003
β_{SS}	0.976	0.003	-1.310	0.003	0.956	0.005
Seasonality parameters buy trades						
s_1	-1.441	0.098	1.295	0.134	-1.232	0.143
s_2	1.452	0.178	-0.770	0.227	1.249	0.228
s_2	-0.780	0.163	1.269	0.196	-0.877	0.205
s_4	1.205	0.141	-0.160	0.176	1.396	0.190
s_5	-0.061	0.149	0.493	0.181	-0.184	0.185
s_6	0.267	0.204	-1.223	0.249	0.526	0.259
Seasonality parameters sell trades						
s_1	-1.394	0.123	0.942	0.149	-1.219	0.130
s_2	1.185	0.231	-0.438	0.267	0.954	0.236
s_2	-0.465	0.195	1.319	0.209	-0.485	0.195
s_4	1.231	0.175	-0.211	0.192	1.388	0.185
s_5	-0.175	0.200	0.401	0.214	-0.315	0.201
s_6	0.366	0.273	0.031	0.288	0.426	0.256
Static covariates buy trades						
$TRVB$			0.315	0.049	0.345	0.049
$TRVS$			0.345	0.049	0.347	0.049
Static covariates sell trades						
$TRVB$			0.617	0.062	0.634	0.059
$TRVS$			0.568	0.061	0.782	0.060
Time-varying covariates buy trades						
D_VOL			1.919	0.523	1.464	0.525
$DFF1$			-0.236	0.024	-0.201	0.023
$DFF2$			-0.011	0.018	-0.009	0.018
$DFF4$			-0.028	0.010	-0.029	0.010
$DFF6$			0.003	0.009	0.007	0.010

Table 44 cont'd:

$DFF9$	0.005	0.007	0.011	0.007
$SPRD$	-4.681	0.276	-2.707	0.218
DMQ	-0.821	0.028	-0.861	0.027
$QASK$	-0.639	0.626	-3.138	0.805
$QVOLA$	0.175	0.057	0.441	0.078
$QVOLB$	0.051	0.050	0.088	0.053
Time-varying covariates sell trades				
D_VOL	0.705	0.493	0.829	0.414
$DFF1$	0.199	0.028	0.156	0.025
$DFF2$	0.025	0.020	-0.002	0.018
$DFF4$	0.004	0.010	-0.005	0.009
$DFF6$	0.065	0.009	0.054	0.008
$DFF9$	0.027	0.008	0.023	0.007
$SPRD$	-3.549	0.328	-1.586	0.258
DMQ	2.050	0.035	2.424	0.044
$QASK$	1.093	0.773	1.466	0.903
$QVOLA$	0.092	0.071	0.148	0.084
$QVOLB$	0.271	0.061	0.268	0.067

Obs (n)	27108	27108	27108
Obs (n^0)	54555	54555	54555
LL	-40639	-39988	-39945
BIC	-40793	-40275	-40232

Diagnostics of ACI residuals for the buy series					
	stat.	p-value	stat.	p-value	stat.	p-value
Mean	1.00		1.00		1.00	
S.D.	1.00		1.01		1.02	
LB(20)	30.39	0.06	27.71	0.11	23.04	0.28
Disp.	0.79	0.42	1.57	0.11	2.17	0.02

Diagnostic of ACI residuals for the sell series					
	stat.	p-value	stat.	p-value	stat.	p-value
Mean	1.00		1.00		0.99	
S.D.	1.00		1.01		1.01	
LB(20)	25.32	0.18	21.29	0.38	25.51	0.18
Disp.	0.47	0.63	1.18	0.23	1.17	0.23

Diagnostics: Log Likelihood (LL), Bayes Information Criterion (BIC) and diagnostics (mean, standard deviation, Ljung-Box statistic, as well as excess dispersion test) of ACI residuals $\hat{\varepsilon}_i^s$.

time-varying covariates. Regression (3) is a specification that ignores the limit order arrival *during* a spell. Here, the explanatory variables enter the model not as time-varying covariates, but as marks that are only updated at each single transaction. In this specification, possible changes of the limit order book variables during a spell are ignored.

In all cases, the inclusion of covariates leads to a better BIC and to improved dynamic properties of the residual series since both regressions (2) and (3) clearly outperform specification (1). It turns out that the inclusion of time-varying covariates does not necessarily improve the econometric specification. However, with exception of the NAB stock, we find the highest BIC value for specification (2). Nevertheless, the coefficients of the covariates are relatively stable in the different specifications. For all series (with exception of the MIM buy transactions), we observe a strong persistence that is indicated by relatively small innovation coefficients (α) and persistence parameters (β) that are close to one. The Ljung-Box statistics show that the models are able to capture the dynamics in the data. In nearly all cases, the uncorrelatedness of the ACI residuals cannot be rejected. The parameters of the baseline intensity function indicate a negative duration dependence, i.e. a decreasing intensity function over time which is typical for trade duration processes. The residual diagnostics show that the unconditional mean, as well as the unconditional standard deviation of the ACI residuals are quite close to one. In particular, the hypothesis of no excess dispersion is not rejected in all cases indicating that the model fits the data quite well. Moreover, by comparing the goodness-of-fit of the estimated models with the fit of corresponding Weibull ACI parameterizations[18], we observe that the higher flexibility of the Burr parameterization is indeed needed to capture the distributional properties of the data.

We find significantly positive coefficients for both the traded buy and sell volume at the previous transaction ($TRVB$ and $TRVS$). Thus, a high transaction volume increases not only the intensity on the same side of the market, but also the intensity on the other market side. This result illustrates that this variable obviously does not serve as a proxy for buy or sell specific information, but for more general information increasing the overall trading intensity on the market. Concerning the difference between the total ask and bid volume (D_VOL), we find evidence for a significantly positive impact on the buy intensity.[19] This result provides evidence for the fact that traders exploit periods of higher liquidity on the buy side. However, the influence on the sell intensity is insignificant for most of the regressions.

Furthermore, liquidity differences between both sides of the market measured by differences between the corresponding inter-quantile ranges (DFF_x) are found to be significant in most cases. Interestingly, the buy (sell) intensity decreases with the (relative) depth on the ask (bid) side. This finding illustrates that traders seem to infer from the state of the limit order book with respect to future price movements. The higher the dispersion of limit prices in the ask (bid) queue the higher the portion of investors who expect an increase (decrease) of the price in the future. Hence, while a high price dispersion in the queues decreases market liquidity, it simultaneously reveals

---

[18] They are not shown here.
[19] At least for two of the three series.

important signals with respect to investors' expectations. Our results show that the informational value of the order book seems to dominate pure liquidity arguments.

For the inside bid-ask spread ($SPRD$), we find a significantly negative coefficient for both the buy and sell intensity. Therefore, the tightness of the market is negatively correlated with the trading intensity. This result stands in contrast to traditional information-based market microstructure models (see Chapter 3), where a high spread generally serves as a proxy for information on the market. However, here dominates a liquidity effect since a high spread makes the crossing from one market side to the other market side more expensive.

Past (signed) midquote changes (DMQ) significantly decrease (increase) the buy (sell) intensity. Note that a negative price change can only be caused by the arrival of a sufficiently large sell market order that absorbs the highest price level on the bid side or, alternatively, an ask order with a limit price that shifts the current best ask price downwards. However, both events increase (decrease) the liquidity on the buy (sell) side. On the other hand, a counterargument arises by the fact that a negative midquote change indicates "bad" expectations with respect to future price movements, and thus, should lead to a lower (higher) buy (sell) intensity. In fact, our regression results lend support to the latter hypothesis since the coefficient associated with DMQ is significantly negative (positive) for buy (sell) trades. These results are confirmed by the estimates concerning the dummy variable $QASK$, which indicates the type of the most current limit order. Hence, the arrival of an ask (bid) quote decreases (increases) the buy (sell) intensity. Thus, the type of the most current quote also reveals traders' expectations with respect to future price movements. Finally, with respect to the quoted volume of the previous limit order ($QVOLA$ and $QVOLB$), we find results similar to those for the impact of the previous transaction volume. This variable reveals no specific buy-sell specific information, but increases the intensity on both sides of the market.

Summarizing the results, we can conclude that the state of the market and the limit order book has a significant impact on the buy/sell intensity. Traders seem to observe the order book and obviously tend to exploit situations of liquidity imbalances between the two market sides, and thus, phases of lower (relative) liquidity costs. However, market participants seem to trade not only due to liquidity reasons but also due to informational aspects. Our results show that especially the influence of recent trade and quote arrivals on the buy/sell intensity is related to information rather than to liquidity. Overall, it turns out that the findings are relatively stable over all regressions, in particular, for the more actively traded NAB and BHP stock.

7.4.3 Estimating Trading Intensities Using LFI Models

The goal of this subsection is to illustrate the application of the LFI model and to empirically test for the existence of a latent factor. The major objective is whether we can identify latent dynamics in financial duration data and whether the inclusion of such a factor leads to an improved econometric specification. The following study is conducted based on financial duration series from German XETRA trading of the Allianz and BASF stock. The sample period covers two trading weeks from 11/15/99 to 11/26/99, corresponding to $6,035$ observations for the Allianz stock and $5,710$ observations for the BASF stock. The data preparation follows according to Section 4.2.2 of Chapter 4. The descriptive statistics strongly resemble the descriptive statistics shown in Section 4.3 of Chapter 4 and are not presented here. Deterministic time-of-the-day effects are modelled based on a linear spline function with nodes for every hour. Furthermore, as in the previous subsections, the single series are standardized by their average trade duration mean.[20] We estimate univariate and bivariate (LF-)ACI and (LF-)Hawkes models. The ACI and Hawkes model are estimated by standard ML, while the estimation of the LFI models requires using the EIS technique discussed in Section 7.3.5. In the ML-EIS procedure, we use $n^{MC} = 50$ Monte Carlo replications, while the efficiency steps in the algorithm are repeated 5 times. In the first iteration, we start with the TSE normal importance sampler. It turns out that the ML-EIS procedure performs very well and leads to a proper convergence of the estimation algorithm. The standard errors are based on the inverse of the estimated Hessian. The lag order P and Q is chosen on the basis of the BIC.

Table 45 shows the estimation results of univariate ACI(1,1) and LF-ACI(1,1) models for both stocks. The baseline intensity function is specified according to (7.6). Moreover, we restrict our consideration to the basic LFI specification without a time-varying latent parameter. This case will be considered below. The estimates of $\hat{\alpha}$ and $\hat{\beta}$ are typical for trade duration series. For $\hat{\alpha}$, we find a relatively low value, while $\hat{\beta}$ is very close to one, which indicates a strong persistence in the data. For the seasonality variables, we find insignificant estimates in nearly all cases. However, the resulting spline functions show plots that are well known from Chapter 4. For this reason, they are omitted here. We observe highly significant coefficients (a^* and δ^*) associated with the latent process, thus, evidence for the existence of a latent autoregressive process is provided. Furthermore, we notice a clear increase of the likelihood and the BIC values. Hence, the inclusion of a latent factor improves the goodness-of-fit of the model. It turns out that the autoregressive parameter of the observable component $\lambda^o(t; \mathfrak{F}_t^o)$ moves towards more persistence while the impact of the (observable) innovations decreases when the latent factor is taken into account. Thus, the latent

---

[20] In the bivariate models, we use the average duration with respect to the pooled process.

Table 45: ML(-EIS) estimates of univariate Weibull ACI(1,1) and LF-ACI(1,1) models for Allianz and BASF trade durations. Data based on German XETRA trading, sample period from 11/15/99 to 11/26/99. Standard errors based on the inverse of the estimated Hessian.

	Allianz				BASF			
	ACI(1,1)		LF-ACI(1,1)		ACI(1,1)		LF-ACI(1,1)	
	est.	S.E.	est.	S.E.	est.	S.E.	est.	S.E.
ω	-0.541	0.129	-0.209	0.183	-0.360	0.120	-0.132	0.150
a	0.808	0.008	0.993	0.026	0.850	0.008	0.992	0.024
α	0.047	0.006	0.031	0.006	0.047	0.005	0.036	0.006
β	0.964	0.010	0.979	0.007	0.975	0.006	0.982	0.005
s_1	1.311	0.755	1.165	1.033	0.965	0.638	0.691	0.710
s_2	-0.719	1.186	-0.680	1.596	-0.999	1.033	-0.686	1.062
s_3	-0.831	1.129	-0.335	1.391	0.167	0.969	0.231	0.923
s_4	-0.388	1.072	-1.060	1.342	-0.441	0.958	-0.609	1.071
s_5	0.530	1.018	0.275	1.168	-0.553	0.915	-0.256	1.056
s_6	1.398	1.086	2.428	1.367	2.882	0.978	2.639	1.125
s_7	-0.994	1.209	-1.339	1.531	-2.323	1.112	-2.441	1.350
s_8	0.390	1.094	0.391	1.461	1.256	1.022	1.602	1.256
a^*			0.385	0.058			0.392	0.081
δ^*			0.630	0.060			0.526	0.061
Obs	6035		6035		5710		5710	
LL	-5539		-5484		-5291		-5254	
BIC	-5591		-5545		-5343		-5315	

Diagnostics of ACI and LF-ACI residuals								
	stat.	p-value	stat.	p-value	stat.	p-value	stat.	p-value
Mean $\hat{\Lambda}^o$	1.00		1.24		1.00		1.17	
S.D. $\hat{\Lambda}^o$	1.05		1.61		1.04		1.42	
LB(20) $\hat{\Lambda}^o$	34.39	0.02	41.01	0.00	35.05	0.02	42.76	0.00
Mean $\hat{\epsilon}^{s,(r)}$			1.01				1.01	
S.D. $\hat{\epsilon}^{s,(r)}$			1.03				1.02	
LB(20) $\hat{\epsilon}^{s,(r)}$			18.63	0.55			20.99	0.40
Disp.	2.55	0.01	1.88	0.06	2.12	0.03	1.18	0.24

Diagnostics: Log Likelihood (LL), Bayes Information Criterion (BIC) and diagnostics (mean, standard deviation and Ljung-Box statistic (inclusive p-values)) of ACI residuals $\hat{\Lambda}(t_{i-1}, t_i)$, average diagnostics (mean, standard deviation, Ljung-Box statistic, as well as excess dispersion test) over all trajectories of LFI residuals $\hat{\epsilon}_i^{s,(r)}$.

innovations displace the observable ones and drive the observable component towards a more constant process. The residuals are computed based on (7.77) for the LF-ACI model and based on $\hat{\epsilon}_i = \hat{\Lambda}(t_{i-1}, t_i)$ for the ACI model. Note that in the LF-ACI framework, the latter is obviously not anymore i.i.d. exponentially distributed.

Table 46: ML(-EIS) estimates of univariate Hawkes(2) and LF-Hawkes(2) models for Allianz and BASF trade durations. Data based on German XETRA trading, sample period from 11/15/99 to 11/26/99. Standard errors based on the inverse of the estimated Hessian.

	Allianz				BASF			
	Hawkes		LF-Hawkes		Hawkes		LF-Hawkes	
	est.	S.E.	est.	S.E.	est.	S.E.	est.	S.E.
ω	0.379	0.040	0.480	0.064	0.400	0.041	0.398	0.041
α_1	0.018	0.004	0.005	0.003	0.628	0.079	0.034	0.007
β_1	0.033	0.011	0.008	0.004	2.328	0.311	0.063	0.015
α_2	0.929	0.124	0.939	0.149	0.034	0.007	0.624	0.079
β_2	2.663	0.245	2.756	0.251	0.062	0.015	2.334	0.312
s_1	-0.676	0.278	-0.365	0.476	-0.583	0.265	-0.566	0.283
s_2	0.798	0.390	0.215	0.802	0.625	0.402	0.607	0.469
s_3	-0.189	0.328	0.031	0.714	-0.035	0.342	-0.033	0.431
s_4	-0.086	0.312	-0.204	0.500	-0.142	0.333	-0.146	0.359
s_5	0.073	0.313	0.137	0.383	0.088	0.341	0.094	0.349
s_6	0.498	0.329	0.708	0.411	0.533	0.360	0.530	0.373
s_7	-0.471	0.317	-0.532	0.411	-0.694	0.342	-0.692	0.346
s_8	0.153	0.256	0.129	0.341	0.415	0.278	0.418	0.280
a^*			0.960	0.018			0.127	6.173
δ^*			0.051	0.015			0.000	0.033
Obs	6035		6035		5710		5710	
LL	-5398		-5393		-5196		-5196	
BIC	-5454		-5458		-5252		-5261	

	Diagnostics of Hawkes and LF-Hawkes residuals							
	stat.	p-value	stat.	p-value	stat.	p-value	stat.	p-value
Mean $\hat{\Lambda}^o$	1.00		1.02		1.00		1.00	
S.D. $\hat{\Lambda}^o$	1.01		1.05		0.99		0.99	
LB(20) $\hat{\Lambda}^o$	17.87	0.60	53.65	0.00	21.92	0.34	21.89	0.35
Mean $\hat{\epsilon}^{s,(r)}$			1.00				1.00	
S.D. $\hat{\epsilon}^{s,(r)}$			1.01				0.99	
LB(20) $\hat{\epsilon}^{s(r)}$			18.53	0.55			21.89	0.35
Disp.	0.32	0.75	0.71	0.48	0.36	0.72	0.36	0.72

Diagnostics: Log Likelihood (LL), Bayes Information Criterion (BIC) and diagnostics (mean, standard deviation and Ljung-Box statistic (inclusive p-values)) of ACI residuals $\hat{\Lambda}(t_{i-1}, t_i)$, average diagnostics (mean, standard deviation, Ljung-Box statistic, as well as excess dispersion test) over all trajectories of LFI residuals $\hat{\epsilon}_i^{s,(r)}$.

Nevertheless, we also state these residuals in order to illustrate the properties of the particular components of the model. The LF-ACI residuals are computed based on each single trajectory of random draws of $\lambda_i^{*(r)}$, $r = 1, \ldots, n^{MC}$. In the tables, we state the *average* values of the particular statistics over all trajectories and the corresponding p-values.[21] We find that the inclusion of the latent factor leads to a clear reduction of the Ljung-Box-statistics. In particular, the null hypothesis of no serial correlation in the residual series cannot be rejected for the LF-ACI model. Thus, it turns out that the latent factor model captures the duration dynamics in a better way. Based on the test against excess dispersion, the ACI specifications are clearly rejected. In contrast, a reduction of the test statistic is observed for the LF-ACI model, indicating an improvement of the model's goodness-of-fit.

Table 46 presents the estimation results of the Hawkes(2) and LF-Hawkes(2) processes. The use of a lag order of $P = 2$ is absolutely necessary to obtain a satisfying goodness-of-fit. In particular, Hawkes(1) specifications (which are not shown here) are not able to capture the dynamics in the data, leading to distinctly higher Ljung-Box statistics and BIC values. With respect to the latent factor, we only find mixed evidence. For the Allianz series, highly significant estimates of a^* and δ^* are observed coming along with a value of $\hat{a}^*$ close to one. For this series, the latent factor seems to capture remaining persistence in the data. However, based on the BIC, the inclusion of latent dynamics does not increase the goodness-of-fit of the model. This result is confirmed by the individual residual diagnostics that indicate no improvements of the LF-Hawkes specifications compared to the pure Hawkes model. For the BASF series, the result is even more clear-cut. Here, the latent variance σ^* tends towards zero, thus, the existence of a latent factor is clearly rejected. These results provide evidence against the need for a latent factor in a Hawkes model.

Table 47 contains the regression results of bivariate Weibull (LF-)ACI models. We estimate one joint seasonality function for both series. Columns (1) and (2) in Table 47 show the results based on an ACI(1,1) and a basic LF-ACI(1,1) specification. As a starting point, we assume identical latent standard deviations for both processes, i.e. $\delta^{1*} = \delta^{2*}$. We find clear interdependencies between the two processes, thus, the backward recurrence time of one process influences not only its own intensity, but also the intensity of the other process. Accordingly, the innovation components reveal significant interdependencies as well. Moreover, as indicated by the estimates of B, we find again a high persistence in the processes. Interestingly, the latent autoregressive process is more dominant than in the univariate case. The latent parameters are highly significant and lead to a strong decrease of the BIC. This result indicates that the latent factor captures unobserved information that *jointly* influences both transaction processes. As in the univariate models, the introduction of a latent factor increases the persistence of the observable components, while the magnitude of the ACI innovations is reduced. The diagnostic checks show that the LF-ACI model captures the dynamics in a better way than the pure ACI model leading to a

---

[21] For the excess dispersion test, we take the average value of the test statistic.

Table 47: ML(-EIS) estimates of bivariate Weibull (LF-)ACI(1,1) models for Allianz and BASF trade durations. Data based on German XETRA trading, sample period from 11/15/99 to 11/26/99. Standard errors based on the inverse of the estimated Hessian.

	(1) ACI(1,1) est.	S.E.	(2) LF-ACI(1,1) est.	S.E.	(3) LF-ACI(1,1) est.	S.E.	(4) LF-ACI(1,1) est.	S.E.
				Allianz				
ω^1	-0.900	0.100	-0.819	0.177	-1.011	0.129	-0.818	0.129
a_1^1	0.828	0.008	0.979	0.014	1.001	0.017	0.992	0.017
a_2^1	0.880	0.009	1.024	0.014	1.043	0.015	1.062	0.015
α_1^1	0.083	0.019	0.022	0.005	0.016	0.004		
α_2^1	0.061	0.014	0.014	0.005	0.010	0.004		
β_{11}	0.911	0.041	0.986	0.004	0.988	0.003		
β_{12}	-0.059	0.025	-0.010	0.004	-0.070	0.003		
				BASF				
ω^2	-0.925	0.102	-0.800	0.177	-0.962	0.131	-0.874	0.129
a_1^2	0.919	0.009	1.074	0.015	1.086	0.016	1.106	0.016
a_2^2	0.862	0.009	1.013	0.015	1.027	0.017	1.028	0.018
α_1^2	0.042	0.011	0.004	0.005	0.001	0.004		
α_2^2	0.071	0.010	0.034	0.005	0.029	0.004		
β_{21}	-0.030	0.014	-0.001	0.002	-0.010	0.002		
β_{22}	0.972	0.010	0.994	0.003	0.992	0.002		
				Seasonalities				
s_1	0.833	0.488	1.129	0.988	0.427	0.529	0.361	0.567
s_2	-0.889	0.747	-1.095	1.445	-0.277	0.632	-0.330	0.857
s_3	0.131	0.601	0.212	1.091	0.019	0.100	0.274	0.578
s_4	-0.307	0.586	-0.692	1.362	-0.679	0.489	-0.856	0.549
s_5	-0.296	0.588	-0.338	1.383	-0.110	0.548	0.047	0.905
s_6	1.764	0.622	2.676	1.373	2.362	0.729	2.072	1.062
s_7	-1.360	0.626	-2.277	1.367	-2.079	0.932	-1.653	0.847
s_8	0.761	0.535	1.531	1.146	1.394	0.871	1.123	0.700
				Latent Factor				
a_1^*			0.582	0.028	0.742	0.020	0.794	0.016
a_2^*					0.252	0.055	0.415	0.046
δ^{1*}			0.569	0.033	0.587	0.036	0.560	0.038
δ^{2*}					0.565	0.037	0.576	0.037
Obs	11745		11745		11745		11745	
LL	-18838		-18709		-18666		-18800	
BIC	-18941		-18821		-18788		-18884	

Diagnostics: Log Likelihood (LL), Bayes Information Criterion (BIC) and diagnostics (mean, standard deviation and Ljung-Box statistic (inclusive p-values)) of ACI residuals $\hat{\Lambda}(t_{i-1}, t_i)$, average diagnostics (mean, standard deviation, Ljung-Box statistic, as well as excess dispersion test) over all trajectories of LFI residuals $\hat{\epsilon}_i^{s,(r)}$.

Table 47 cont'd:

	(1)		(2)		(3)		(4)	
	\multicolumn{8}{c}{Diagnostics of ACI and LF-ACI residuals for the Allianz series}							
	stat.	p-value	stat.	p-value	stat.	p-value	stat.	p-value
Mean $\hat{\Lambda}^o$	1.00		1.23		1.06		1.10	
S.D. $\hat{\Lambda}^o$	1.05		1.57		1.38		1.45	
LB(20) $\hat{\Lambda}^o$	31.23	0.05	51.65	0.00	60.52	0.00	292.88	0.00
Mean $\hat{\epsilon}^{s,(r)}$			1.01		1.01		1.02	
S.D. $\hat{\epsilon}^{s,(r)}$			1.05		1.05		1.05	
LB(20) $\hat{\epsilon}^{s,(r)}$			26.32	0.16	24.35	0.23	87.75	0.00
Disp.	2.90	0.00	2.67	0.01	2.80	0.01	2.94	0.00
	\multicolumn{8}{c}{Diagnostics of ACI and LF-ACI residuals for the BASF series}							
	stat.	p-value	stat.	p-value	stat.	p-value	stat.	p-value
Mean $\hat{\Lambda}^o$	1.00		1.27		1.10		1.13	
S.D. $\hat{\Lambda}^o$	1.04		1.61		1.41		1.50	
LB(20) $\hat{\Lambda}^o$	38.59	0.01	51.60	0.00	57.96	0.00	481.19	0.00
Mean $\hat{\epsilon}^{s,(r)}$			1.01		1.01		1.02	
S.D. $\hat{\epsilon}^{s,(r)}$			1.05		1.05		1.06	
LB(20) $\hat{\epsilon}^{s,(r)}$			26.64	0.15	24.67	0.21	134.19	0.00
Disp.	2.35	0.02	2.68	0.01	2.61	0.01	3.12	0.00

Diagnostics: Log Likelihood (LL), Bayes Information Criterion (BIC) and diagnostics (mean, standard deviation and Ljung-Box statistic (inclusive p-values)) of ACI residuals $\hat{\Lambda}(t_{i-1}, t_i)$, average diagnostics (mean, standard deviation, Ljung-Box statistic, as well as excess dispersion test (the latter two statistics inclusive p-values)) over all trajectories of LFI residuals $\hat{\epsilon}_i^{s,(r)}$.

significant reduction of the Ljung-Box statistics. However, in contrast to the univariate case, the introduction of the latent factor does not improve the distributional properties of the model. In particular, it leads to a slightly higher excess-dispersion.

In specification (3), we estimate an extended LF-ACI model based on process-specific latent variances and a regime-switching latent autoregressive parameter. The parameter a^* is specified according to (7.47) based on $R = 2$ regimes with an exogenously given threshold value $\bar{x} = 1$. In fact, we find a significantly higher estimate for the autoregressive parameter associated with small (previous) durations (a_1^*). Hence, the serial dependence of the latent factor obviously decreases over the length of the previous spell, and thus, the importance of latent information declines the longer it dates back. The estimates of δ^{1*} and δ^{2*} are not significantly different, hence, it turns out that the latent factor influences both components in a similar way. We find a clear increase of the BIC and slight improvements in the residual diagnostics indicating a better overall goodness-of-fit

252 7 Univariate and Multivariate Dynamic Intensity Models

Table 48: ML(-EIS) estimates of bivariate (LF-)Hawkes(2) models for Allianz and BASF trade durations. Data based on German XETRA trading, sample period from 11/15/99 to 11/26/99. Standard errors based on the inverse of the estimated Hessian.

	(1) LF-Hawkes		(2) LF-Hawkes		(3) LF-Hawkes		(4) LF-Hawkes	
	est.	S.E.	est.	S.E.	est.	S.E.	est.	S.E.
Allianz								
ω^1	0.143	0.190	0.162	0.023	0.155	0.020	0.152	0.020
α_1^{11}	0.442	0.442	0.459	0.049	0.443	0.047	0.449	0.048
β_1^{11}	1.441	1.381	1.501	0.147	1.547	0.151	1.515	0.150
α_1^{12}	0.289	0.390	0.298	0.042	0.287	0.041	0.289	0.041
β_1^{12}	1.771	2.572	1.920	0.296	1.995	0.311	1.987	0.310
α_2^{11}	0.009	0.019	0.008	0.001	0.008	0.001	0.009	0.001
β_2^{11}	0.017	0.039	0.014	0.003	0.015	0.003	0.015	0.003
BASF								
ω^2	0.162	0.186	0.182	0.022	0.176	0.020	0.180	0.021
α_1^{21}	0.016	0.034	0.014	0.003	0.014	0.003	0.014	0.003
β_1^{21}	0.030	0.066	0.023	0.006	0.024	0.005	0.024	0.005
α_1^{22}	0.181	0.299	0.180	0.032	0.172	0.031	0.171	0.031
β_1^{22}	1.702	2.906	1.858	0.333	1.898	0.342	1.907	0.345
α_2^{22}	0.300	0.352	0.309	0.038	0.299	0.037	0.294	0.038
β_2^{22}	1.168	1.672	1.193	0.173	1.261	0.181	1.318	0.200
Seasonalities								
s_1	-0.650	1.84	-0.726	0.205	-0.695	0.192	-0.709	0.191
s_2	0.735	2.75	0.768	0.318	0.758	0.280	0.776	0.279
s_3	-0.120	2.32	-0.058	0.277	-0.119	0.236	-0.137	0.231
s_4	-0.0827	2.25	-0.109	0.256	-0.117	0.238	-0.090	0.228
s_5	0.0662	2.31	0.062	0.258	0.183	0.239	0.162	0.239
s_6	0.421	2.42	0.454	0.272	0.395	0.246	0.400	0.246
s_7	-0.474	2.28	-0.506	0.261	-0.533	0.241	-0.532	0.238
s_8	0.210	1.83	0.224	0.210	0.238	0.197	0.231	0.194
Latent Factor								
a_1^*			0.928	0.053	0.985	0.015	0.981	0.016
a_2^*					0.511	0.211	0.446	0.244
δ^{1*}			0.047	0.028	0.058	0.018	0.052	0.018
δ^{2*}							0.074	0.026
Obs	11745		11745		11745		11745	
LL	-18578		-18576		-18564		-18564	
BIC	-18682		-18688		-18681		-18685	

Diagnostics: Log Likelihood (LL), Bayes Information Criterion (BIC) and diagnostics (mean, standard deviation and Ljung-Box statistic (inclusive p-values)) of ACI residuals $\hat{\Lambda}(t_{i-1}, t_i)$, average diagnostics (mean, standard deviation, Ljung-Box statistic, as well as excess dispersion test) over all trajectories of LFI residuals $\hat{\epsilon}_i^{s,(r)}$.

Table 48 cont'd:

	(1)		(2)		(3)		(4)	
	Diagnostics of Hawkes and LF-Hawkes residuals (Allianz)							
	stat.	p-value	stat.	p-value	stat.	p-value	stat.	p-value
Mean $\hat{\Lambda}^o$	1.00		1.01		0.98		0.99	
S.D. $\hat{\Lambda}^o$	1.01		1.03		1.01		1.01	
LB(20) $\hat{\Lambda}^o$	20.58	0.42	23.59	0.26	23.09	0.28	22.25	0.33
Mean $\hat{\epsilon}^{s,(r)}$			1.00		1.00		1.00	
S.D. $\hat{\epsilon}^{s,(r)}$			1.01		1.01		1.01	
LB(20) $\hat{\epsilon}^{s,(r)}$			20.11	0.45	20.47	0.43	20.59	0.42
Disp.	0.49	0.63	0.47	0.64	0.52	0.60	0.49	0.63
	Diagnostics of Hawkes and LF-Hawkes residuals (BASF)							
	stat.	p-value	stat.	p-value	stat.	p-value	stat.	p-value
Mean $\hat{\Lambda}^o$	1.00		1.01		0.98		0.98	
S.D. $\hat{\Lambda}^o$	0.99		1.01		0.99		1.00	
LB(20) $\hat{\Lambda}^o$	24.64	0.22	28.13	0.11	26.93	0.14	28.00	0.11
Mean $\hat{\epsilon}^{s,(r)}$			1.00		1.00		1.00	
S.D. $\hat{\epsilon}^{s,(r)}$			0.99		0.99		0.99	
LB(20) $\hat{\epsilon}^{s,(r)}$			25.18	0.19	21.77	0.35	21.11	0.39
Disp.	0.44	0.66	0.44	0.66	0.39	0.70	0.35	0.72

Diagnostics: Log Likelihood (LL), Bayes Information Criterion (BIC) and diagnostics (mean, standard deviation and Ljung-Box statistic (inclusive p-values)) of ACI residuals $\hat{\Lambda}(t_{i-1}, t_i)$, average diagnostics (mean, standard deviation, Ljung-Box statistic, as well as excess dispersion test (the latter two statistics inclusive p-values)) over all trajectories of LFI residuals $\hat{\epsilon}_i^{s,(r)}$.

of this specification. Finally, column (4) gives the estimation result for a LF-ACI specification without any dynamics in the observable component, thus $\lambda^o(t; \mathfrak{F}_t^o) = 1$. In that case, the joint latent component captures the dynamics of both processes, which is associated with an increase of a_1^* and a_2^*, while the latent variances are nearly unaffected. However, it turns out that according to the BIC, specification (4) outperforms specification (1) which illustrates the usefulness of modelling the system dynamics by a *joint* latent component. However, we observe that the dynamic specification of the latent factor is not sufficient to completely capture the dynamics of the model. In terms of the residual diagnostics, specification (4) underperforms a fully parameterized LF-ACI specification. This result is not surprising since a simple AR(1) dynamic in the latent factor is probably not appropriate for modelling the persistence in the process.[22]

---

[22] Note that we also estimated a LF-ACI model based on heteroscedastic latent variances that are specified according to (7.46) based on one hour nodes. It turns out that the seasonality effects are not individually or collectively significant,

Table 48 displays the estimation results of bivariate (LF-)Hawkes(2) processes. As in the univariate case, the choice of a lag order $P = 2$ is absolutely necessary to obtain a satisfactory specification of the model. However, according to the BIC, we find a restricted version[23] of a (LF-)Hawkes(2) process as the best and most parsimonious specification. Column (1) and (2) show the results based on the basic Hawkes(2) model, as well as the basic LF-Hawkes(2) specification. As for the LF-ACI estimates, specification (2) is based on the restriction $\delta^{1*} = \delta^{2*}$. The dynamic parameters α^{sr} and β^{sr} are highly significant in most cases, yielding evidence for clear interdependencies between both processes. For the latent factor, we obtain highly significant estimates with a value of $\hat{a}^*$ close to one and a quite low standard deviation. Hence, as in the univariate case, the latent factor seems to capture remaining persistence in the data. Nonetheless, the latent factor's influence is clearly weaker than in the LF-ACI model. This is confirmed by the reported BIC value and the residual diagnostics that do not indicate improvements of the goodness-of-fit of the LF-Hawkes model compared to the pure Hawkes model. In the columns (3) and (4), we present generalized LF-Hawkes(2) specifications based on process-specific latent variances and a regime-switching latent dynamic parameter. Again, we observe clear empirical evidence for a significantly lower serial dependence in the latent component when the previous spell was long. The fact that the highest BIC value and thus, the best goodness-of-fit is found for specification (3), indicates that this higher flexibility in the LF-Hawkes specification is supported by the data. This finding is confirmed by the residual diagnostics that report slight improvements of the dynamical and distributional properties at least for the BASF series. Nevertheless, as in the LF-ACI model, the use of of process-specific latent variances leads to no improvement of the model specification. In particular, the influences of the latent factor on the individual intensity components do not differ significantly.

Summarizing, our results provide evidence for the existence of an unobservable autoregressive component that is common to both duration series. In the ACI model, the introduction of a latent factor leads to a better model performance, as witnessed by a clear increase of the maximized likelihood and the BIC value, as well as improved residuals diagnostics. In contrast, in the Hawkes model, the inclusion of a latent factor appears to be less important. In this framework, only slight improvements of the model's goodness-of-fit can be observed. Obviously, Hawkes processes provide an overall better goodness-of-fit compared to the ACI model since no strong evidence for omitted (dynamic) factors can be found based on our data.

---

thus, we find no evidence for seasonality effects driving the variance of the latent processes. For ease of exposition we refrain from showing these results here.

[23] The second order cross-terms are set to zero, i.e. $\alpha_2^{12} = \alpha_2^{21} = \beta_2^{12} = \beta_2^{21} = 0$.

8

Summary and Conclusions

The analysis of financial data on the lowest aggregation level is an ongoing topic in the recent financial econometrics literature. The availability of financial transaction databases has created a new exciting field in econometrics and empirical finance which allows us to look at old puzzles in finance from a new perspective and to address a variety of new issues. Transaction data - sometimes refereed to as ultra-high-frequency data - can be seen as the informational limiting case where all transactions are recorded and information with respect to the complete transaction and order flow on a financial market is collected. The main property of transaction data is the irregular spacing in time. For this reason, they have to be considered as marked point processes. Since the seminal work of Engle and Russell (1998), much research effort has been devoted to the specification and application of financial point process models.

The goal of this monograph was to provide a detailed overview of dynamic duration and intensity models for the analysis of financial point processes. We systematized the recent developments in the literature and discussed strengths and weaknesses of alternative frameworks. Moreover, we further developed existing models and proposed new approaches. A further objective was to discuss different types of financial point processes and to examine their economic implications and statistical properties. By using data from different exchanges and trading systems, we illustrated the application of the proposed methodological framework and explored issues of volatility, liquidity and market microstructure analysis.

Three major econometric frameworks for the analysis of dynamic point processes have been considered in much detail: As a starting point, we focussed on the modelling of point processes in a discrete-time duration framework. The most common type of duration model is the autoregressive conditional duration (ACD) model proposed by Engle and Russell (1998). The fundamental building block of the ACD model is the specification of the conditional duration mean. The validity of this conditional mean restriction is the basic assumption behind the ACD model and is the prerequisite to ensure the QML properties of the Exponential ACD (EACD) specification. By addressing this issue, we discussed several generalizations with respect to the functional form of the basic specification. In order to test for the validity of the conditional mean restriction, we suggested the framework of (integrated) conditional moment tests. Applications of this testing framework to NYSE trade and price durations indicated that more simple models, as linear or logarithmic parameterizations, lead to clear violations of the conditional mean restriction. It turned out that augmented ACD specifications are required that imply nonlinear news impact functions. Nevertheless, the major strength of the ACD model is its easy implementation and nice practicability. However, one drawback of this approach is that the treatment of censoring mechanisms is rather difficult. In the case of censoring, the exact timing of particular points of the process cannot be observed, but can only be approximated by a corresponding interval. Such effects lead to problems in an autoregressive framework because the information about the exact length of a spell is needed for the sequel of the time series.

For this reason, we presented dynamic extensions of the semiparametric proportional intensity (PI) model as an alternative framework. The PI model assumes a multiplicative relationship between a baseline intensity component and a function of covariates. In this sense, it can be interpreted as the counterpart to an accelerated failure time model to which the ACD model belongs to. We suggested a reformulation of the PI approach in terms of a log-linear model of the integrated intensity function as a valuable starting point for a dynamic extension. Since the baseline intensity remains unspecified, the integrated intensity function is unknown and the model corresponds to a latent variable approach. The so-called autoregressive conditional proportional intensity (ACPI) model is built on an observation driven ARMA-type dynamic for the (unknown) integrated intensity function. The model is estimated by adopting a categorization approach which allows for a simultaneous estimation of the dynamic model parameters and discrete points of the baseline survivor function. Since the dynamics of the ACPI model are driven not by the durations themselves, but by a function of the conditional survivor function, it allows for a straightforward consideration of censoring structures in a dynamic framework.

Thirdly, we discussed dynamic parameterizations of the intensity function. The major strength of dynamic intensity models compared to autoregressive duration approaches is that they allow to model point processes in a continuous-time framework. This property is of great importance whenever it is required to account for information events that occur during two points of the process, as in the case of time-varying covariates or in a multivariate framework. We illustrated the theoretical properties of autoregressive conditional intensity models (Russell, 1999), as well as self-exciting intensity processes (Hawkes, 1971). In this context, we focused on the dynamic properties of the particular models and the relationship between the intensity process and the corresponding duration process. Furthermore, we discussed dynamic extensions of doubly stochastic Poisson processes. The major idea is to combine the framework of latent factor models with the framework of intensity processes leading to a so-called latent factor intensity (LFI) model. We illustrated the introduction of a dynamic latent factor in both types of the aforementioned models leading to a latent factor ACI (LF-ACI) specification, as well as a latent factor Hawkes (LF-Hawkes) model. Empirical applications of univariate and multivariate LFI models to financial duration data indicated that the inclusion of latent dynamics in intensity specifications leads to an improvement of the econometric specification.

Besides the provision of an appropriate methodological framework, a further objective of this monograph was to gain deeper insights into the statistical properties of financial duration processes based on various assets and exchanges. In this context, we focused on data from the New York Stock Exchange (NYSE), the German Stock Exchange, the London International Financial Futures Exchange (LIFFE), the EUREX trading platform in Frankfurt, as well as from the Australian Stock Exchange (ASX). By investigating the distributional and dynamical properties of various types of financial durations, we found clear differences between the particular duration series with respect to their statistical features. Furthermore, we provided evidence that the form of the particular trading system has an influence on the dynamic properties of the particular series.

The econometric framework was applied to several economic issues. One objective was devoted to the analysis of trade durations. By applying the ACD model to quantify the determinants of the trading intensity based on various assets, we found a weak confirmation of theoretical market microstructure hypotheses. Furthermore, by modelling the simultaneous buy and sell arrival intensity in the limit order book of the ASX on the basis of bivariate ACI models, we found the state of the limit order book to be an important influential factor. Evidence was provided that traders' preference for immediacy depends on the market depth and the costs of liquidity on the particular sides of the market, as well as on the informational value of the limit order book.

Furthermore, we illustrated how to build volatility estimators based on price durations. This concept was applied to the analysis of volatility in LIFFE Bund future trading. Here, we explicitly focused on price durations at high aggregation levels which requires to account for censoring structures. By applying the ACPI model, we estimated volatility shapes based on price changes of different sizes. This analysis allowed us to analyze the magnitude and the frequency of fluctuations of the price process. We observed that the shape of intraday price change volatility is dominated largely by higher frequent but smaller price changes while time to maturity seasonalities are affected mainly by price changes of larger size. Moreover, an analysis of multivariate volatility patterns was conducted based on a multivariate Hawkes model. By estimating simultaneous price intensities based on five NYSE stocks, we found evidence for strong contemporaneous correlations, as well as spill-over effects between the individual volatility components.

A further objective was devoted to the analysis of illiquidity risks. Using NYSE trade and quote data, we explored the dynamic properties of the time-varying excess demand for liquidity and its relationship to market depth. We found that excess volume durations and thus, periods of asymmetric information and high inventory risk are predictable based on market microstructure variables reflecting the past trading process. By measuring the absolute price change over an excess volume spell, we obtained a proxy for realized market depth. Strong evidence for a positive correlation between the price impact associated with a given excess volume and the length of the underlying spell was found.

Summing up, our investigations illustrated that point process models are not only useful approaches to account for the sampling randomness in econometric models for irregularly spaced data but provide also a valuable framework to analyze financial market activity from different viewpoints. As discussed in much detail, measuring the intensity of the trading event arrival allows us not only to explore the timing structures of the intraday trading process, but enables us also to construct alternative volatility and liquidity measures under explicit consideration of the passing trading time. In this sense, applications of financial duration and intensity models could play an important role in the framework of time deformation models, like, for example in the intrinsic time approach as discussed e.g. by Müller et al (1995), Dacorogna et al (1996) or Carr and Wu (2004) in the context of time-deformed Lévy processes. In these models, time is not measured physically but in terms of passing trading opportunities, like, for instance, in terms of transactions, price changes or cumulative trading volumes. The relationship between intrinsic time and physical time is determined by the frequency or, respectively, by the intensity of the passing events. In this sense, the framework outlined in this book provides an important basis to explore the relationship between calendar time and alternative time concepts in more detail. Ultimately, such studies should lead to better understanding of traders' perception of time and its implications for asset pricing and financial risk management.

A

Important Distributions for Duration Data

In the following, the gamma function $\Gamma(m)$ is defined as

$$\Gamma(m) = \int_0^\infty x^{m-1} \exp(-x) dx$$

for $m > 0$.

A.1 The Poisson Distribution

Acronym: $x \sim Po(\lambda)$.
   Probability function:

$$P(X = x) = \frac{\exp(-\lambda)\lambda^x}{x!}, \qquad x = 0, 1, 2, \ldots, \lambda > 0.$$

   Mean:

$$E[x] = \lambda.$$

Variance:

$$\mathrm{Var}[x] = \lambda.$$

A.2 The Lognormal Distribution

Acronym: $x \sim LN(\mu, \sigma^2)$.
   Probability density function:

$$f(x) = \frac{1}{x\sigma\sqrt{2\pi}} \exp\left(-\frac{1}{2\sigma^2}(\ln(x) - \mu)^2\right), \qquad x > 0.$$

Cumulative density function:

$$F(x) = \Phi\left(\frac{\ln(x) - \mu}{\sigma}\right),$$

where $\Phi(\cdot)$ denotes the c.d.f. of the standard normal distribution.
Uncentered moments:

$$\mathrm{E}[x^s] = \exp\left(s\mu + \frac{\sigma^2 s^2}{2}\right).$$

Variance:

$$\mathrm{Var}[x] = \exp\left(2\mu + \sigma^2\right)\left(\exp(\sigma^2) - 1\right).$$

Remark: If $x \sim LN(\mu, \sigma^2)$, then $\ln(x) \sim N(\mu, \sigma^2)$.

A.3 The Exponential Distribution

Acronym: $x \sim Exp(\lambda)$.
   Probability density function:

$$f(x) = \frac{1}{\lambda}\exp\left(-\frac{x}{\lambda}\right), \qquad \lambda > 0.$$

Cumulative density function:

$$F(x) = 1 - \exp\left(-\frac{x}{\lambda}\right).$$

Uncentered moments:

$$\mathrm{E}[x^s] = \lambda^s \Gamma(1 + s).$$

Variance:

$$\mathrm{Var}[x] = \lambda^2.$$

Hazard function:

$$\tilde{\lambda}(x) = 1/\lambda.$$

A.4 The Gamma Distribution

Acronym: $x \sim \mathcal{G}(\lambda, m)$.
  Probability density function:

$$f(x) = \frac{x^{m-1} \exp(-x/\lambda)}{\lambda^m \Gamma(m)}, \qquad x > 0, \lambda > 0, m > 0.$$

Uncentered moments:

$$\mathrm{E}[x^s] = \frac{\lambda^s \Gamma(m+s)}{\Gamma(m)}, \qquad m + s > 0.$$

Variance:

$$\mathrm{Var}[x] = m\lambda^2.$$

Remark: If $x \sim \mathcal{G}(\lambda, m)$, then $x/\lambda \sim \mathcal{G}(1, m)$. A $\mathcal{G}(\lambda, 1)$ distribution is equivalent to an $Exp(\lambda)$ distribution.

A.5 The Weibull Distribution

Acronym: $x \sim W(\lambda, a)$.
  Probability density function:

$$f(x) = a\lambda^{-a} x^{a-1} \exp\left[-\left(\frac{x}{\lambda}\right)^a\right], \qquad x > 0, \lambda > 0, a > 0.$$

Cumulative density function:

$$F(x) = 1 - \exp\left[-\left(\frac{x}{\lambda}\right)^a\right].$$

Uncentered moments:

$$\mathrm{E}[x^s] = \lambda^s \Gamma(1 + s/a).$$

Variance:

$$\mathrm{Var}[x] = \lambda^2 \left[\Gamma(1 + 2/a) - \Gamma(1 + 1/a)^2\right].$$

Hazard function:

$$\tilde{\lambda}(x) = \frac{a}{\lambda} \left(\frac{x}{\lambda}\right)^{a-1}.$$

Remark: A $W(\lambda, 1)$ distribution is equivalent to an $Exp(\lambda)$ distribution. A Weibull distribution with $\lambda = 1$ is called standard Weibull distribution.

A.6 The Generalized Gamma Distribution

Acronym: $x \sim \mathcal{GG}(\lambda, a, m)$.
  Probability density function:

$$f(x) = \frac{a}{\lambda^{am}\Gamma(m)} x^{ma-1} \exp\left[-\left(\frac{x}{\lambda}\right)^a\right], \qquad x > 0, \lambda > 0, a > 0, m > 0.$$

Uncentered moments:

$$\mathrm{E}[x^s] = \lambda^s \frac{\Gamma(m + s/a)}{\Gamma(m)}.$$

Variance:

$$\mathrm{Var}[x] = \lambda^2 \left[\frac{\Gamma(m + 2/a)}{\Gamma(m)} - \left(\frac{\Gamma(m + 1/a)}{\Gamma(m)}\right)^2\right].$$

Remark: A $\mathcal{GG}(\lambda, 1, m)$ distribution is equivalent to a $\mathcal{G}(\lambda, m)$ distribution.
A $\mathcal{GG}(\lambda, a, 1)$ distribution is equivalent to a $W(\lambda, a)$ distribution.

A.7 The Generalized F Distribution

Acronym: $x \sim GF(\lambda, a, m, \eta)$.
  Probability density function:

$$f(x) = \frac{a x^{am-1}[\eta + (x/\lambda)^a]^{(-\eta-m)}\eta^\eta}{\lambda^{am}\mathcal{B}(m, \eta)}, \qquad x > 0, \lambda > 0, a > 0, m > 0, \eta > 0.$$

where $\mathcal{B}(\cdot)$ describes the complete beta function with $\mathcal{B}(m, \eta) = \frac{\Gamma(m)\Gamma(\eta)}{\Gamma(m+\eta)}$.
  Uncentered moments:

$$\mathrm{E}[x^s] = \lambda^s \eta^{s/a} \frac{\Gamma(m + s/a)\Gamma(\eta - s/a)}{\Gamma(m)\Gamma(\eta)}, \qquad s < a\eta.$$

Remark: For $\eta \to \infty$, the $GF(\lambda, a, m, \eta)$ distribution converges to a $\mathcal{GG}(\lambda, a, m)$ distribution. A $GF(\lambda, a, 1, 1)$ distribution is equivalent to a log-logistic distribution.

A.8 The Burr Distribution
(according to Lancaster, 1997)

Acronym: $x \sim Burr(\lambda, a, \eta)$.
Probability density function:

$$f(x) = \frac{a}{\lambda}\left(\frac{x}{\lambda}\right)^{a-1}\left[1 + \eta\left(\frac{x}{\lambda}\right)^a\right]^{-(1+\eta^{-1})}, \qquad x > 0, \lambda > 0, a > 0, \eta > 0.$$

Cumulative density function:

$$F(x) = 1 - (1 + \eta\lambda^{-a}x^a)^{-1/\eta}.$$

Uncentered moments:

$$\mathrm{E}[x^s] = \lambda^s\frac{\Gamma(1 + s/a)\Gamma(\eta^{-1} - s/a)}{\eta^{1+s/a}\Gamma(1 + \eta^{-1})}, \qquad s < a/\eta.$$

Hazard function:

$$\tilde{\lambda}(x) = \frac{a}{\lambda}\left(\frac{x}{\lambda}\right)^{a-1}(1 + \eta\lambda^{-a}x^a)^{-1}.$$

Remark: A $Burr(\lambda, a, 1)$ distribution is equivalent to a log-logistic distribution. For $\eta \to 0$, the $Burr(\lambda, a, \eta)$ distribution converges to a $W(\lambda, a)$ distribution.

A.9 The Extreme Value Type I Distribution
(Gumbel (minimum) distribution)

Acronym: $x \sim EV(\lambda, m)$.
Probability density function:

$$f(x) = \frac{1}{\lambda}\exp\left(\frac{x - m}{\lambda} - \exp\left(\frac{x - m}{\lambda}\right)\right).$$

Cumulative density function:

$$F(x) = \exp\left(-\exp\left(\frac{x - m}{\lambda}\right)\right).$$

Mean:

$$\mathrm{E}[x] = m + \lambda \cdot 0.5772.$$

Variance:

$$\mathrm{Var}[x] = \frac{\lambda^2\pi^2}{6}.$$

Hazard function:

$$\tilde{\lambda}(x) = \exp\left(\frac{x - m}{\lambda}\right).$$

Remark: An $EV(1, 0)$ distribution is called standard extreme value type I distribution (standard Gumbel (mimimum) distribution).

A.10 The Burr Type II Distribution

(according to Johnston, Kotz and Balakrishnan, 1994)

Acronym: $x \sim BurrII(\eta)$.
Probability density function:

$$f(x) = \frac{\eta \exp(-x)}{[1 + \exp(-x)]^{\eta+1}}, \qquad \eta > 0.$$

Cumulative density function:

$$F(x) = [1 + \exp(-x)]^{-\eta}.$$

Remark: For $\eta \to \infty$, the $BurrII(\eta)$ distribution converges to an (type I) $EV(1,0)$ distribution. A $BurrII(1)$ distribution is equivalent to a logistic distribution.

A.11 The Pareto Distribution

Acronym: $x \sim P(\nu, m)$.
Probability density function:

$$f(x) = \frac{\nu m^\nu}{x^{\nu+1}}, \qquad x > m > 0.$$

Cumulative density function:

$$F(x) = 1 - \left(\frac{m}{x}\right)^\nu.$$

Uncentered moments:

$$\mathrm{E}[x^s] = \frac{\nu m^s}{\nu - s}, \qquad \nu > s.$$

Variance:

$$\mathrm{Var}[x] = \nu m^2 \left(\frac{1}{\nu - 2} - \frac{\nu}{(\nu - 1)^2}\right).$$

Hazard function:

$$\tilde{\lambda}(x) = \frac{\nu}{x}.$$

B

List of Symbols (in Alphabetical Order)

a, a^+, a^-	distribution parameters
a^*, a_j^*	autoregressive parameters of a dynamic latent factor
a_i	ask price at time t_i
a_r^s	distribution parameter
$\mathfrak{a}_j$	boundary for a space for infinite sequences
A°	$A^\circ := n^{-1} \sum_{i=1}^n \mathrm{E}[\tilde{h}_i(\theta_0)]$
$\hat{A}^\circ$	estimator of A°
A^s	innovation parameter vector of a multivariate ACI process
A_i^k, $A_i^{j,sr}$	recursion terms in univariate and multivariate Hawkes processes
$\mathcal{A}(\cdot)$	arbitrary polynomial function
$\mathfrak{A}$	constant
α_0	"true" autoregressive innovation parameter
α_j, $\tilde{\alpha}_j$	autoregressive parameters
α_j^v	autoregressive parameters of a GARCH specification for log durations
α_j^s, $\alpha_j^{s,r}$	parameters of univariate or multivariate Hawkes processes
α_j^+, α_j^-	regime dependent autoregressive parameters of a NPACD model
$\alpha_j^{(r)}$	regime dependent autoregressive parameters
α_{1j}, α_{2j}	autoregressive parameters of a smooth transition ACD model
$\alpha(L)$	lag polynomial
b	autoregressive parameter
b_i	bid price at time t_i
$\mathfrak{b}_j$	boundary for a space for infinite sequences
B	persistence parameter matrix in a multivariate ACI process
B°	$B^\circ := n^{-1} \sum_{i=1}^n \mathrm{E}[s_i(\theta_0)s_i(\theta_0)']$
$\hat{B}^\circ$	estimator of B°
$\mathfrak{B}$	constant
$\mathcal{B}(m, \eta)$	Beta function with parameters m and η
β_0	"true" autoregressive persistence parameter
β_j, β_{ij}, $\tilde{\beta}_j$	autoregressive parameters
β_j^v	autoregressive parameters of a GARCH specification for log durations

β_j^{sr}	parameters of Hawkes processes
$\beta_j^{(r)}$	regime dependent autoregressive parameters
$\tilde{\beta}_a, \tilde{\beta}_0$	regression coefficients
$\beta(L)$	lag polynomial
$c, \tilde{c}$	autoregressive parameters
c_i	censoring indicator
C	boundary of a critical region of a statistical test
$\mathcal{C}(\cdot)$	arbitrary polynomial function
χ^2	χ^2 statistic
$\chi(\cdot)$	integrating constant in the EIS procedure
d	constant or fractional integration parameter
d_i	price change between consecutive trades
dp	predetermined size for price changes
dp_i	price or midquote change measured over a given duration x_i
$dspr_i$	bid-ask spread change
dv	predetermined cumulated volume
dv_i	effectively measured cumulated volume over a given duration x_i
δ_j	power exponents in a Box-Cox specification
δ^*	parameter of a latent factor
δ^{s*}	s-type specific latent standard deviation in a LFI model
$\delta^s, \delta_{s,j}^s, \delta_{c,j}^s$	parameters of a Fourier series approximation
Δ	arbitrary time interval
e_i	generalized error
$e_{i,r}$	regime dependent generalized error
$\tilde{e}_i$	conditional expectation of some transformation of ϵ_i^*
$EV(\lambda, m)$	extreme value distribution with parameters λ and m
$Exp(\lambda)$	exponential distribution with parameter λ
ϵ_i	$\epsilon_i := \Lambda(t_{i-1}, t_i)$
ϵ_i^*	$\epsilon_i^* := \ln \Lambda(t_{i-1}, t_i)$
ϵ_i^{s*}	$\epsilon_i^{s*} := \ln \Lambda^s(t_{i-1}^s, t_i^s)$
$\breve{\epsilon}_i$	$\breve{\epsilon}_i := -0.5772 - \epsilon_i^*$
$\varepsilon_i, \tilde{\varepsilon}_i$	error terms of a regression model
$\varepsilon_i^{(k)}$	error terms of a regime switching regression model
$\bar{\varepsilon}_k$	breakpoints of ε_i associated with different regimes
η_i	martingale difference
η, η_r^s	distribution parameters
$f(x)$	probability density function (p.d.f.)
$F(x)$	cumulative density function (c.d.f.)
$F^{(r)}(x)$	state-dependent c.d.f. associated with some state r
$\mathfrak{F}$	Borel field of subsets of Ω
$\mathfrak{F}_t$	history up to and inclusive t
$\mathfrak{F}_t^N$	$\mathfrak{F}_t^N := \sigma(t_{N(t)}, z_{N(t)}, \ldots, t_1, z_1)$
$\mathfrak{F}_{t_i}^d$	$\mathfrak{F}_{t_i}^d := \sigma(x_i^d, z_i, \ldots, x_1^d, z_1)$

$\mathfrak{F}_t^o$	observable history of some process
$\mathfrak{F}_t^*$	history of some unobserved process
$g(\cdot)$	arbitrary continuous function
$g(x;\theta)$	$g(x;\theta) = (m(x;\theta)\,s(x;\theta))$
$\mathcal{G}(\lambda, m)$	gamma distribution with parameters λ and m
$\mathcal{G}^{-1}(\lambda, \alpha)$	α-quantile function of the $\mathcal{G}(\lambda, \lambda)$ distribution
$\mathcal{GG}(\lambda, a, m)$	generalized gamma distribution with parameters λ, a and m
$GF(\lambda, a, m, \eta)$	generalized F distribution with parameters λ, a, m and η
γ	coefficient vector associated with marks
γ^v	coefficient vector associated with volatility covariates
$\tilde{\gamma}$	coefficient vector associated with time-varying covariates
$\Gamma(m)$	Gamma function at m
$\hat{\Gamma}_j$	$\hat{\Gamma}_j = n^{-1} \sum_{i=j+1}^{n} s_i(\hat{\theta}) s_i(\hat{\theta})'$
h_i^v	conditional variance of log durations
$h_i(\theta)$	i-th contribution to the Hessian matrix
$\tilde{h}_i(\theta)$	$\tilde{h}_i(\theta) := \mathrm{E}\left[h_i(\theta)\middle|\,\mathfrak{F}_{t_{i-1}}\right]$
H_0	null hypothesis
$H(m, x)$	$H(m, x) := \mathcal{G}^{-1}(m, \Phi(x))$
i, i', i''	indices
I	identity matrix
$\mathcal{I}(\theta)$	information matrix
ι	vector of ones
j	index
J	number of unconditional moment restrictions or truncation lag order
k	index
$k(L_i^*, \phi_i)$	density kernel
K, K^-, K^+	number of categories in a categorization approach
$\kappa(s_1, s_2)$	$\kappa(s_1, s_2) := \int_{s_1}^{s_2} u f_{\epsilon^*}(u)du$
$l_i(\theta)$	log likelihood contribution of the i-th observation
L	predetermined number of lags or lag operator
L^*	matrix or vector of latent variables
L_i^*	$L_i^* = \{\lambda_j^*\}_{j=1}^{i}$
$\tilde{L}_i^*$	$L_i^* = \{\ln \lambda_j^*\}_{j=1}^{i}$
$LN(\mu, \sigma^2)$	lognormal distribution with parameters μ and σ^2
$\mathcal{L}(W; \theta)$	likelihood function depending on W and θ
$\mathcal{L}^*(\cdot)$	likelihood function of a latent model
λ	intensity parameter of a homogeneous Poisson process
λ_i^*	latent intensity component
λ_i^{s*}	s-type latent intensity component
$\lambda_i^{*(r)}$	random draw of r-th trajectory of λ_i^*
$\lambda^*(t)$	continuous-time latent intensity component
$\lambda(t; \mathfrak{F}_t)$	$\mathfrak{F}_t$-intensity function

$\bar{\lambda}(t)$	discrete approximation of an intensity function
$\tilde{\lambda}(s)$	hazard function function
$\lambda_0(t)$	baseline intensity function
$\lambda_{0,r}(t)$	regime dependent baseline intensity function
$\lambda^s(t;\mathfrak{F}_t)$	s-th component of a multivariate intensity $\lambda(t;\mathfrak{F}_t)$
$\lambda^o(t)$	observable component of a LFI model
$\lambda^{o,s}(t;\mathfrak{F}_t)$	s-th observable component of a LFI model
$\lambda^{dp}(t;\mathfrak{F}_t)$	intensity function associated with cumulated dp-price changes
$\lambda_0^s(t;\mathfrak{F}_t)$	s-th component of a multivariate baseline intensity function at t
$\bar{\lambda}_0(t)$	discrete approximation of a baseline intensity function
Λ_i	parameter matrix in a VAR approach
$\Lambda_0(t_{i-1},t_i)$	$\Lambda_0(t_{i-1},t_i) := \int_{t_{i-1}}^{t_i} \lambda_0(u;\mathfrak{F}_u)du$
$\Lambda_{0,r}(t_{i-1},t_i)$	$\Lambda_0(t_{i-1},t_i) := \int_{t_{i-1}}^{t_i} \lambda_{0,r}(u;\mathfrak{F}_u)du$
$\Lambda(t_{i-1},t_i)$	$\Lambda(t_{i-1},t_i) := \int_{t_{i-1}}^{t_i} \lambda(u;\mathfrak{F}_u)du$
$\Lambda^s(t_{i-1}^s,t_i^s)$	$\Lambda^s(t_{i-1}^s,t_i^s) := \int_{t_{i-1}^s}^{t_i^s} \lambda^s(u;\mathfrak{F}_u)du)$
$\Lambda^o(t_{i-1},t_i)$	$\Lambda^o(t_{i-1},t_i) := \int_{t_{i-1}}^{t_i} \lambda^o(u;\mathfrak{F}_u^o)du$
$\Lambda^{o,s}(t_{i-1}^s,t_i^s)$	$\Lambda^{o,s}(t_{i-1}^s,t_i^s) := \int_{t_{i-1}^s}^{t_i^s} \lambda^{o,s}(u;\mathfrak{F}_u^o)du$
m	distribution parameter
m_i	$m_i = (m_{1,i}\, m_{2,i})'$
$m_{1,i},\, m_{2,i}$	standard normal random variables
m_i^*	conditional mean of latent factor λ_i^*
$m_j(x_i;\theta)$	unconditional j-th moment restriction for i
$m_j(x;\theta)$	vector of the j-th unconditional moment restriction
$m(x;\theta)$	$m(x;\theta) = (m_1(x;\theta)\ \ldots\ m_J(x;\theta))$
$\{m(\cdot)\}_{i=1}^n$	sequence of auxiliary importance samplers
M	dimension of the parameter vector θ
M^0	dimension of the parameter vector θ_0
M^a	dimension of the parameter vector θ_a
M^{ICM}	Monte Carlo replications for the computation of Υ_{ICM}
$M_n(t)$	unconditional moment restriction of an ICM test
$M_n^*(t)$	approximation function for $M_n(t)$
mq_i	midquote at t_i
μ_i	$\mu_i := (\phi_{1,i} + m_i^*)\,\pi_i^2$
μ^{*l}	$\mu^{*l} := \mu_k^*$ if $x_i^{d,l} = k+1$
μ^{*u}	$\mu^{*u} := \mu_k^*$ if $x_i^{d,u} = k$
μ_k^*	$\mu_k^* := \ln \Lambda_0(t_{i-1}, t_{i-1} + \bar{x}_k)$
$\mu_{k,r}^*$	$\mu_{k,r}^* := \ln \Lambda_{0,r}(t_{i-1}, t_{i-1} + \bar{x}_k)$
$\mu(t), \mu^s(t)$	deterministic functions of time
n	sample size
n_k	number of observations in some category k
n^s	number of observations of the s-th process

n^0	number of observations of a (time-varying) covariate process
n^{dp}	sample size of a thinned point process of dp-price durations
n^Δ	sample size of Δ-min intervals
n'	sample size of some prediction period
n^{MC}	Monte Carlo replications
n_j^Δ	number of events in the interval $[j\Delta, (j+1)\Delta)$
$N(t)$	$N(t) := \sum_{i \geq 1} \mathbb{1}_{\{t_i \leq t\}}$
$N^s(t)$	$N^s(t) := \sum_{i \geq 1} \mathbb{1}_{\{t_i^s \leq t\}}$
$N^0(t)$	$N^0(t) := \sum_{i \geq 1} \mathbb{1}_{\{t_i^0 \leq t\}}$
$N^{dp}(t)$	counting function associated with cumulative dp price changes
N_j^Δ	$N_j^\Delta := N((j+1)\Delta) - N(j\Delta), \qquad j = 1, 2, \ldots$
$\breve{N}(t)$	$\breve{N}(t) := \sum_{i \geq 1} \mathbb{1}_{\{t_i < t\}}$
$\breve{N}^s(t)$	$\breve{N}^s(t) := \sum_{i \geq 1} \mathbb{1}_{\{t_i^s < t\}}$
$\breve{N}^0(t)$	$\breve{N}^0(t) := \sum_{i \geq 1} \mathbb{1}_{\{t_i^0 < t\}}$
$\tilde{N}(t)$	$\tilde{N}(t) := N(\tau_t)$
ν, ν_i	autoregressive or distribution parameters
$\nu_{i,k}$	$\nu_{i,k} := \mu_k^* - \phi_i$
$\nu_0, \nu_{0,k}$	coefficients of a linear spline function for $\lambda_0(t)$
$\nu_{0,k}^{s*}$	coefficients of a linear spline function for δ_i^{s*}
ν_i^l	$\nu_i^l := \mu^{*l} - \phi_i$
ν_i^u	$\nu_i^u := \mu^{*u} - \phi_i$
$o(\Delta)$	remainder term with $o(\Delta)/\Delta \to 0$ as $\Delta \to 0$
$\omega, \tilde{\omega}, \omega_r^s, \omega^r$	constant terms
ω_0	"true" constant term
ω^v	constant of GARCH specification for log durations
$\omega^{(k)}$	regime dependent constant
Ω	sample space
p	power exponent
p_i	transaction price at observation t_i
p_k^*	probability to observe an outcome in the category k
p_{ij}^*	elements of a transition matrix P^*
$p(t)$	transaction price at some point in time t
$p(\lambda_i^*|W_i, L_i^*; \theta)$	conditional density of λ_i^* given W_i and L_i^*
pv_{ICM}	critical value of an ICM test
P	lag order
P^v	lag order of GARCH specification of log durations
P^*	transition matrix of a Markov chain
$Po(\lambda)$	Poisson distribution with parameter λ
$\mathcal{P}$	probability measure on $\mathfrak{F}$
π_i^2	$\pi_i^2 = (\sigma_i^{*-2} - 2\phi_{2,i})^{-1}$
ϕ_i	dynamic function or auxiliary parameter
$\phi_{0,i}, \phi_{1,i}, \phi_{2,i}$	auxiliary parameters

$\phi(L)$	$\phi(L) := 1 - \alpha(L) - \beta(L)$
$\phi(\mathfrak{F}_{t_i}; \theta)$	(time-varying) function depending on $\mathfrak{F}_{t_i}$ and θ
$\Phi(x)$	c.d.f. of the standard normal distribution
$\psi_i, \psi_{1,i}, \psi_{2,i}$	autoregressive parameters
Ψ_i	conditional duration mean
$\Psi_{i,0}$	specification of Ψ_i under some null hypothesis
$\Psi_{1,i}, \Psi_{2,i}$	conditional mean components of a component model
$\Psi_i^{(r)}$	regime specific conditional mean function
$\Psi(t), \Psi^s(t)$	functions capturing dynamics in intensity models
$\breve{\Psi}_i$	seasonal adjusted function capturing dynamics in ACD models
$\tilde{\Psi}_i$	recursive term in some intensity specification
$\tilde{\Psi}_{1i}, \tilde{\Psi}_{2i}$	intensity components of a two-component intensity model
q_i	probability integral transform
Q	lag order or degree of some polynomial
Q^v	lag order of GARCH specification of log durations
$\mathcal{Q}$	parameter range in a Monte Carlo study
r	regime index
r_i	simple net return between t_{i-1} and t_i
r_k^*	threshold value of some regime
$\bar{r}_i$	conditional mean of r_i
$r(x_i; \theta)$	conditional moment function for i
$r^\Delta(t)$	Δ-min return at t
R	number of (observable or unobservable) regimes
R_i	regime indicator for observable regimes
R^2	coefficient of determination
R_i^*	regime indicator for unobservable regimes
$\bar{R}^2$	adjusted coefficient of determination
ρ_i	i-th order autocorrelation function (ACF)
$\bar{\rho}_i$	i-th order autocorrelation function (ACF) of categorized data
s	index
s_0^*	spline parameter for latent standard deviations
$s_i(\theta)$	i-th contribution to the score vector
$s(\theta)$	score matrix; $s(\theta) = (s_1(\theta)', \ldots, s_n(\theta)')$
$s(t), s^s(t)$	deterministic seasonality functions
$s^2(t)$	asymptotic variance of $M_n(t)$
S	dimensions of a multivariate point process
$S(x)$	survivor function
$S_0(x)$	baseline survivor function
$S_{0,r}(x)$	regime dependent baseline survivor function
$\mathfrak{S}$	$\mathfrak{S} = \{t_1 \in \mathbb{R} : \mathrm{E}[\varrho \exp(t_1 u)] = 0\}$
σ_ε^2	variance of $(\varepsilon_i - 1)^2$
$\hat{\sigma}_{\hat{\varepsilon}}^2$	sample variance of residuals $\hat{\varepsilon}_i$, $i = 1, \ldots, n$
$\sigma^2(t_i)$	conditional volatility per time at t_i, based on trade-to-trade returns

$\sigma^2_{(r^\Delta)}(t)$	conditional volatility per time at t, based on Δ-min returns
$\sigma^2_{(d,x)}(t_i)$	conditional volatility per time at t_i, based on d_i and x_i
$\sigma^2_{(x^{dp})}(t_i)$	conditional volatility at t_i, based on dp-price durations
$\sigma^{*2}_{(x^{dp})}(t_i)$	conditional dp-price change volatility at t_i
$\tilde{\sigma}^2(t)$	instantaneous volatility at t
$\tilde{\sigma}^2_{(x^{dp})}(t)$	instantaneous dp-price change volatility at t
$\Sigma(\theta)$	variance covariance matrix depending on θ
spd_i	bid-ask spread at t_i
t	calendar time
t_i	arrival time of the i-th event
t_n	last observed point of the point process
t_i^{dp}	arrival times of cumulative price changes of size dp
t_i^l, t_i^u	lower and upper boundary of a censoring interval around t_i
t_i^s	arrival of the i-th s-type event in a multivariate point process
t_i^0	arrival of the i-th event of a (time-varying) covariate process
$\check{t}_i$	pooled process of all points t_i and t_i^0
$\check{t}_i^s$	pooled process of all points t_i^s and t_i^0
$\bar{t}$	normalized calendar time, $\bar{t} \in [0;1]$
$\tilde{t}_i$	points associated with $\tilde{N}(t)$
$\mathbf{t}$	vector of nuisance parameters $\mathbf{t} = (\mathbf{t}_1, \dots, \mathbf{t}_d)$
τ_t	stopping time at t; solution to $\int_0^{\tau_t} \lambda(s; \mathfrak{F}_s)ds = t$
$\tau(t)$	clock time at t
$\bar{\tau}_k$	exogenously given clock time points
θ	parameter vector of a regression model
θ_0	parameter vector under some null hypothesis
$\hat{\theta}^0$	estimate under some null hypothesis
θ_ε	vector of distribution parameters
θ_a	vector of additional (omitted) variables
$\theta_\varepsilon^{(r)}$	vector of regime switching distribution parameters
Θ	parameter space
u, u_i	random variables
$U(t)$	arbitrary measurable point process
Υ	test statistic
$\tilde{\Upsilon}_{m,ICM}$	simulated ICM test statistic for the m-th n-tuple of $\varpi_{\cdot,m}$
v, v_i	random variables
vol_i	transaction volume at t_i
$\overline{vol}_i$	average transaction volume measured over a certain duration i
ϱ	arbitrary random variable with $\mathrm{E}[|\varrho|] < \infty$
$\varrho(\mathcal{A}(\cdot))$	largest eigenvalue in absolute terms of $\mathcal{A}(\cdot)$
ϖ_i	bounded i.i.d. random variables
$\varsigma(\cdot)$	transition function

$\vartheta(\cdot)$	arbitrary continuous function
w	weights of a component model
w_i	i-th element of W
$w_j(\cdot)$	jth weighting function of $m_j(x_i; \theta_0)$
$VNET_i$	VNET measure evaluated at t_i
W	matrix of observable data
W_i	$W_i := \{w_j\}_{j=1}^i$
$W_n(\mathbf{t})$	$W_n(\mathbf{t}) = M_n(\mathbf{t})/(s^2(\mathbf{t}))^{1/2}$
$W(\lambda, m)$	Weibull distribution with parameters λ and m
x_i	$x_i := t_i - t_{i-1}$
$\hat{x}_i$	predicted duration x_i
x_i^*	$x_i^* := \ln \Lambda_0(t_{i-1}, t_i)$
x_i^s	$x_i^s := t_i^s - t_{i-1}^s$
x_i^l, x_i^u	lower and upper boundary of a censored duration
x_i^{dp}	$x_i^{dp} := t_i^{dp} - t_{i-1}^{dp}$
x_i^d	$x_i^d := k \cdot \mathbb{1}_{\{\bar{x}_{k-1} < x_i \le \bar{x}_k\}}$
$x_i^{d,dp}$	$x_i^{d,dp} := k \cdot \mathbb{1}_{\{\bar{x}_{k-1} < x_i^{dp} \le \bar{x}_k\}}$
$x_i^{d,l}, x_i^{d,u}$	lower and upper bounds of censored categorized durations
$\bar{x}_k$	category bounds of categorized durations, $k = 1, \dots, K-1$
$\bar{x}_{(dp)}$	mean price duration with respect to the aggregation level dp
$\breve{x}_i$	seasonal adjusted duration
$x(t)$	backward recurrence time at t
$x^s(t)$	s-type backward recurrence time at t
$\xi(\cdot)$	one-to-one mapping from $\mathbb{R}$ to a compact bounded subset of $\mathbb{R}$
Ξ	compact subset of $\mathbb{R}^d$
y_i^b	buy/sell indicator ($y_i^b = 1$ for buys, $y_i^b = -1$ for sells)
y_i^s	indicator variable for multivariate point processes
$y_{i,1}^v, y_{i,2}^v, y_{i,3}^v$	volume indicator variables
z_i	vector of marks at t_i
$z_{a,i}$	vector of (omitted) marks at t_i
z_j^Δ	covariates associated with N_j^Δ
$z_{\breve{N}^0(t)}^0$	discretized covariate process at t
$z(t)$	vector of time-varying covariates at t
$\mathcal{Z}_i(x)$	path of $Z(t)$ from $Z(t_{i-1})$ to $Z(t_{i-1} + x)$
$\mathcal{Z}_i(x_1, x_2)$	path of $Z(t)$ from $Z(t_{i-1} + x_1)$ to $Z(t_{i-1} + x_2)$
$\zeta(a, m, \eta)$	mean of a $GF(1, a, m, \eta)$ distribution
$\zeta(\lambda_i^*)$	normal density kernel in dependence of λ_i^*

References

AALEN, O. (1978): "Nonparametric Inference for a Family of Counting Processes," *The Annals of Statistics*, 6(4), 701–726.

ADMATI, A., AND P. PFLEIDERER (1988): "A Theory of Intraday Patterns: Volume and Price Variability," *Review of Financial Studies*, 1, 3–40.

AÏT-SAHALIA, Y., AND P. A. MYKLAND (2003): "The Effects of Random and Discrete Sampling when Estimating Continuous-Time Diffusions," *Econometrica*, 71(2), 483–549.

AÏT-SAHALIA, Y. (1996): "Testing Continuous-Time Models of the Spot Interest Rate," *Review of Financial Studies*, 9, 385–426.

ALBERT, J., AND S. CHIB (1995): "Bayesian Residual Analysis for Binary Response Regression Models," *Biometrika*, 82, 747–759.

AMIHUD, Y., AND H. MENDELSON (1980): "Dealership Markets: Market Making with Inventory," *Journal of Financial Economics*, 8, 31–53.

ANDERSEN, T. G., AND T. BOLLERSLEV (1998a): "Answering the Skeptics: Yes, Standard Volatility Models Do Provide Accurate Forecasts," *International Economic Review*, 39 (4), 885–905.

ANDERSEN, T. G., AND T. BOLLERSLEV (1998b): "Deutsche Mark-Dollar Volatility: Intraday Activity Patterns, Macroeconomic Announcements, and Longer Run Dependencies," *Journal of Finance*, 53, 219–265.

ANDERSEN, T. G. (1996): "Return Volatility and Trading Volume: An Information Flow Interpretation of Stochastic Volatility," *Journal of Finance*, 51, 169–204.

BAGEHOT, W. (1971): "The Only Game in Town," *Financial Analysts Journal*, 27, 12–14, 22.

BAILLIE, R. T., T. BOLLERSLEV, AND H.-O. MIKKELSEN (1996): "Fractionally Integrated Generalized Autoregressive Conditional Heteroskedasticity," *Journal of Econometrics*, 52, 91–113.

BAILLIE, R. T. (1996): "Long Memory Processes and Fractional Integration in Econometrics," *Journal of Econometrics*, 73, 5–59.

BALL, C. (1988): "Estimation Bias Induced by Discrete Security Prices," *Journal of Finance*, 43, 841–865.

BANGIA, A., F. DIEBOLD, T. SCHUERMANN, AND J. STROUGHAIR (2002): "Modeling Liquidity Risk, with Implications for Traditional Market Risk Measurement and Management," in *Risk Management: The State of the Art*, ed. by S. Figlewski, and R. Levich. Kluwer Academic Publishers, Amsterdam.

BARCLAY, M. J., AND J. B. WARNER (1993): "Stealth Trading and Volatility," *Journal of Financial Economics*, 34, 281–305.

BAUWENS, L., F. GALLI, AND P. GIOT (2003): "The Moments of Log-ACD Models," Discussion Paper 2003/11, CORE, Université Catholique de Louvain.

BAUWENS, L., P. GIOT, J. GRAMMIG, AND D. VEREDAS (2000): "A Comparison of Financial Duration Models Via Density Forecasts," *International Journal of Forecasting, forthcoming*.

BAUWENS, L., AND P. GIOT (2000): "The Logarithmic ACD Model: An Application to the Bid/Ask Quote Process of two NYSE Stocks," *Annales d'Economie et de Statistique*, 60, 117–149.

BAUWENS, L., AND P. GIOT (2001): *Econometric Modelling of Stock Market Intraday Activity*. Kluwer.

BAUWENS, L., AND P. GIOT (2003): "Asymetric ACD Models: Introducing Price Information in ACD Models with a Two State Transition Model," *Empirical Economics*, 28 (4), 1–23.

BAUWENS, L., AND N. HAUTSCH (2003): "Dynamic Latent Factor Models for Intensity Processes," Discussion Paper 2003/103, CORE, Université Catholique de Louvain.

BAUWENS, L., AND D. VEREDAS (1999): "The Stochastic Conditional Duration Model: A Latent Factor Model for the Analysis of Financial Durations," Discussion Paper 9958, CORE, Université Catholique de Louvain, forthcoming *Journal of Econometrics*, 2003.

BERAN, J. (1994): *Statistics for Long-Memory Processes*. Chapman & Hall, New York.

BESSEMBINDER, H. (2000): "Tick Size, Spreads, and Liquidity: An Analysis of Nasdaq Securities Trading Near Ten Dollars," *Journal of Financial Intermediation*, 9(3), 213–239.

BIAIS, B., P. HILLION, AND C. SPATT (1995): "An Empirical Analysis of the Limit Order Book and the Order Flow in the Paris Bourse," *The Journal of Finance*, 50, 1655–1689.

BIERENS, H. J., AND W. PLOBERGER (1997): "Asymptotic Theory of Integrated Conditional Moment Tests," *Econometrica*, 65, 1129–1151.

BIERENS, H. J. (1982): "Consistent Model Specification Tests," *Journal of Econometrics*, 20, 105–134.

BIERENS, H. J. (1990): "A Consistent Conditional Moment Test of Functional Form," *Econometrica*, 58, 1443–1458.

BLACK, F. (1971): "Toward a Fully Automated Exchange, Part I, II," *Financial Analysts Journal*, 27, 29–34.

BLACK, F. (1976): "Studies in Stock Price Volatility," in *Proceedings of the 1976 Business Meeting of the Business and Economic Statistics Section, American Statistical Association*, pp. 177–181.

BLUME, L., D. EASLEY, AND M. O'HARA (1994): "Market Statistics and Technical Analysis," *Journal of Finance*, 49 (1), 153–181.

BOLLERSLEV, T., AND J. WOOLDRIDGE (1992): "Quasi-Maximum Likelihood Estimation and Inference in Dynamic Models with Time Varying Covariances," *Econometric Reviews*, 11, 143–172.

BOLLERSLEV, T. (1986): "Generalized Autoregressive Conditional Heteroskedasticity," *Journal of Econometrics*, 31, 307–327.

BOUGEROL, P., AND N. PICARD (1992): "Stationarity of GARCH Processes," *Journal of Econometrics*, 52, 115–127.

BOWSHER, C. G. (2002): "Modelling Security Markets in Continuous Time: Intensity based, Multivariate Point Process Models," Discussion Paper 2002-W22, Nuffield College, Oxford.

BOX, G. E., AND G. M. JENKINS (1976): *Time Series Analysis: Forecasting and Control.* Holden Day, San Francicso.

BRÉMAUD, P. (1981): *Point Processes and Queues, Martingale Dynamics.* Springer, New York.

BRESLOW, N. E. (1972): "Contribution to the Discussion of the Paper by D. R. Cox," *Journal of the Royal Statistical Society, Series B*, 34, 216–17.

BRESLOW, N. E. (1974): "Covariance Analysis of Censored Survival Data," *Biometrics*, 30, 89–100.

BREUSCH, T. S., AND A. R. PAGAN (1979): "A Simple Test for Heteroskedasticity and Random Coefficient Variation," *Econometrica*, 47, 203–207.

BREUSCH, T. S. (1978): "Testing for Autocorrelation in Dynamic Linear Models," *Australian Economic Papers*, 17, 334–355.

CAMERON, A. C., AND P. K. TRIVEDI (1998): *Regression Analysis of Count Data.* Cambridge University Press, Cambridge.

CARRASCO, M., AND X. CHEN (2002): "Mixing and Moment Properties of various GARCH and Stochastic Volatility Models," *Econometric Theory*, 18(1), 17–39.

CARR, P., AND L. WU (2004): "Time-Changed Lévy Processes and Option Pricing," *Journal of Financial Economics, forthcoming.*

CHAN, L. K. C., AND J. LAKONISHOK (1995): "The Behaviour of Stock Prices around Institutional Trades," *Journal of Finance*, 4, 1147–1173.

CHRISTIE, A. (1982): "The Stochastic Behavior of Common Stock Variances," *Journal of Financial Economics*, 10, 407–432.

CLARK, P. (1973): "A Subordinated Stochastic Process Model with Finite Variance for Speculative Prices," *Econometrica*, 41, 135–155.

CONROY, R. M., R. S. HARRIS, AND B. A. BENET (1990): "The Effects of Stock Splits on Bid-Ask Spreads," *Journal of Finance*, 45, 1285–1295.

COPELAND, T. E., AND D. GALAI (1983): "Information Effects and the Bid-Ask Spread," *Journal of Finance*, 38, 1457–1469.

COX, D. R., AND V. ISHAM (1980): *Point Processes.* Chapman and Hall, London.

COX, D. R., AND D. OAKES (1984): *Analysis of Survival Data.* Chapman and Hall, London.

COX, D. R., AND E. J. SNELL (1968): "A General Definition of Residuals," *Journal of the Royal Statistical Society*, B30, 248–265.

COX, D. R. (1972): "Regression Models and Life Tables," *Journal of the Royal Statistical Society, Series B*, 34, 187–220.

COX, D. R. (1975): "Partial Likelihood," *Biometrika*, 62, 269.

COX, D. R. (1981): "Statistical Analysis of Time Series: Some Recent Developments," *Scandinavian Journal of Statistics*, 8, 93–115.

DACOROGNA, M. M., C. L. GAUVREAU, U. A. MÜLLER, R. B. OLSEN, AND O. V. PICTET (1996): "Changing Time Scale for Short-Term Forecasting in Financial Markets," *Journal of Forecasting*, 15, 203–227.

DACOROGNA, M. M., R. GENCAY, U. A. MÜLLER, R. B. OLSEN, AND O. V. PICTET (2001): *An Introduction to High-Frequency Finance.* Academic Press.

DAVIS, R. A., T. H. RYDBERG, N. SHEPHARD, AND S. B. STREETT (2001): "The CBin Model for Counts: Testing for Common Features in the Speed of Trading, Quote Changes, Limit and Market Order Arrivals," Discussion paper.

DEMPSTER, A. P., N. M. LAIRD, AND D. B. RUBIN (1977): "Maximum Likelihood from Incomplete Data via the EM Algorithm," *Journal of the Royal Statistical Society Series B*, 39, 1–38.

DEMSETZ, H. (1968): "The Cost of Transacting," *Quarterly Journal of Eonomics*, 82, 33–53.

DE JONG, R. M. (1996): "The Bierens Test under Data Dependence," *Journal of Econometrics*, 72, 1–32.

DIAMOND, D. W., AND R. E. VERRECCHIA (1981): "Information Aggregation in a Noisy Rational Expectations Economy," *Journal of Financial Economics*, 9, 221–235.

DIAMOND, D. W., AND R. E. VERRECCHIA (1987): "Constraints on Short-Selling and Asset Price Adjustment to Private Information," *Journal of Financial Economics*, 18, 277–311.

DIEBOLD, F. X., T. A. GUNTHER, AND A. S. TAY (1998): "Evaluating Density Forecasts, with Applications to Financial Risk Management," *International Economic Review*, 39, 863–883.

DING, Z., AND C. W. J. GRANGER (1996): "Modeling Volatility Persistence of Speculative Returns: A New Approach," *Journal of Econometrics*, 73, 185–215.

DROST, F. C., C. A. J. KLAASSEN, AND B. J. M. WERKER (1997): "Adaptive Estimation in Time Series Models," *The Annals of Statistics*, 25, 786–818.

DROST, F. C., AND T. E. NIJMAN (1993): "Temporal Aggregation of GARCH Processes," *Econometrica*, 61(2), 909–927.

DROST, F. C., AND B. J. M. WERKER (1996): "Closing the GARCH Gap: Continuous Time GARCH Modeling," *Journal of Econometrics*, 74, 31–57.

DROST, F. C., AND B. J. M. WERKER (2001): "Semiparametric Duration Models," Discussion Paper 2001-11, CentER, Tilburg University.

DUFOUR, A., AND R. F. ENGLE (2000a): "The ACD Model: Predictability of the Time between Consecutive Trades," Discussion paper, ISMA Centre, University of Reading.

DUFOUR, A., AND R. F. ENGLE (2000b): "Time and the Impact of a Trade," *Journal of Finance*, 55, 2467–2498.

EASLEY, D., AND M. O'HARA (1987): "Price, Trade Size, and Information in Securities Markets," *Journal of Financial Economics*, 19, 69–90.

EASLEY, D., AND M. O'HARA (1992): "Time and Process of Security Price Adjustment," *The Journal of Finance*, 47 (2), 577–605.

EFRON, B. (1986): "Double Exponential Families and their Use in Generalized Linear Regression," *Journal of the American Statistical Association*, 81, 709–721.

ELYASIANI, E., S. HAUSER, AND B. LAUTERBACH (2000): "Market Response to Liquidity Improvements: Evidence from Exchange Listings," *Financial Review*, 35, 1–14.

ENGLE, R. F., AND J. LANGE (2001): "Predicting VNET: A Model of the Dynamics of Market Depth," *Journal of Financial Markets*, 2(4), 113–142.

ENGLE, R. F., AND A. LUNDE (2003): "Trades and Quotes: A Bivariate Point Process," *Journal of Financial Econometrics*, 1(2), 159–188.

ENGLE, R. F., AND V. K. NG (1993): "Measuring and Testing the Impact of News on Volatility," *Journal of Finance*, 48, 1749–1778.

ENGLE, R. F., AND J. R. RUSSELL (1994): "Forecasting Transaction Rates: The Autoregressive Conditional Duration Model," Discussion Paper 4966, University of California San Diego.

ENGLE, R. F., AND J. R. RUSSELL (1997): "Forecasting The Frequency of Changes in Quoted Foreign Exchange Prices with Autoregressive Conditional Duration Model," *Journal of Empirical Finance*, 4, 187–212.

ENGLE, R. F., AND J. R. RUSSELL (1998): "Autoregressive Conditional Duration: A New Model for Irregularly Spaced Transaction Data," *Econometrica*, 66, 1127–1162.

ENGLE, R. F. (1982): "Autoregressive Conditional Heteroscedasticity with Estimates of the Variance of United Kingdom Inflation.," *Econometrica*, 50, 987–1006.

ENGLE, R. F. (1984): "Wald, Likelihood Ratio and Lagrange Multiplier Tests in Econometrics," in *Handbook of Econometrics, Vol. II*, ed. by Z. Griliches, and M. D. Intriligator, chap. 13, pp. 775–826. Elsevier Science.

ENGLE, R. F. (1996): "The Econometrics of Ultra-High Frequency Data," Discussion Paper 96-15, University of California San Diego.

ENGLE, R. F. (2000): "The Econometrics of Ultra-High-Frequency Data," *Econometrica*, 68, 1, 1–22.

FERNANDES, M., AND J. GRAMMIG (2000): "Non-parametric Specification Tests for Conditional Duration Models," ECO Working Paper 2000/4, European University Institute.

FERNANDES, M., AND J. GRAMMIG (2001): "A Family of Autoregressive Conditional Duration Models," Discussion Paper 2001/31, CORE, Université Catholique de Louvain.

FINUCANE, T. J. (2000): "A Direct Test of Methods for Inferring Trade Direction from Intra-Day Data," *Journal of Financial and Quantitative Analysis*, 35, 553–676.

FLEMING, M., AND E.-M. REMOLONA (1999): "Price Formation and Liquidity in the U.S. Treasury Market: The Response to Public Information," *Journal of Finance*, 54, 1901–1915.

FRANKE, G., AND D. HESS (2000): "Electronic Trading versus Floor Trading," *Journal of Empirical Finance*, 7, 455–478.

FRENCH, K. R., W. SCHWERT, AND R. F. STAMBAUGH (1987): "Expected Stock Returns and Volatility," *Journal of Financial Economics*, 19, 3–30.

GALLANT, R. A. (1981): "On the Bias in Flexible Functional Forms and an Essential Unbiased Form: The Fourier Flexible Form," *Journal of Econometrics*, 15, 211–245.

GARMAN, M. (1976): "Market Microstructure," *Journal of Financial Economics*, 3, 257–275.

GERHARD, F., AND N. HAUTSCH (2002a): "Semiparametric Autoregressive Proportional Hazard Models," Discussion Paper W2, Nuffield College, Oxford.

GERHARD, F., AND N. HAUTSCH (2002b): "Volatility Estimation on the Basis of Price Intensities," *Journal of Empirical Finance*, 9, 57–89.

GERHARD, F., AND W. POHLMEIER (2002): "On the Simultaneity of Absolute Price Changes and Transaction Durations," Discussion paper, University of Konstanz.

GERHARD, F. (2001): "A Simple Dynamic for Limited Dependent Variables," Discussion Paper W11, Nuffield College, Oxford.

GHYSELS, E., C. GOURIÉROUX, AND J. JASIAK (1998): "Stochastic Volatility Duration Models," Discussion paper, York University, forthcoming *Journal of Econometrics*, 2003.

GHYSELS, E., AND J. JASIAK (1998): "GARCH for Irregularly Spaced Financial Data. The ACD-GARCH Model," *Studies in Nonliear Dynamics and Econometrics*, 2(4), 133–149.

278   References

GIOT, P., AND J. GRAMMIG (2002): "How Large is Liquidity Risk in an Automated Auction Market?," Discussion Paper 2002-23, University of St. Gallen.

GLOSTEN, L. R., AND L. E. HARRIS (1988): "Estimating the Components of the Bid/Ask Spread," *Journal of Financial Economics*, 21, 123–142.

GLOSTEN, L. R., AND P. R. MILGROM (1985): "Bid, Ask and Transaction Prices in a Specialist Market with Heterogeneously Informed Traders," *Journal of Financial Economics*, 14, 71–100.

GLOSTEN, L. R. (1994): "Is the Electronic Open Limit Order Book Inevitable," *The Journal of Finance*, 49, 1127–1161.

GODFREY, L. G. (1978): "Testing against General Autoregressive and Moving Average Error Models when the Regressors include Lagged Dependent Variables," *Econometrica*, 46, 1293–1302.

GONZALEZ-RIVERA, G. (1998): "Smooth transition GARCH models," *Studies in Nonlinear Dynamics and Econometrics*, 3, 61–78.

GØRGENS, T., AND J. L. HOROWITZ (1999): "Semiparametric Estimation of a Censored Regression Model with an Unknown Transformation of the Dependent Variable," *Journal of Econometrics*, 90, 155–191.

GOTTLIEB, G., AND A. KALAY (1985): "Implications of the Discreteness of Observed Stock Prices," *Journal of Finance*, 40, 135–154.

GOURIÉROUX, C., J. JASIAK, AND G. LEFOL (1999): "Intra-Day Market Activity," *Journal of Financial Markets*, 2, 193–226.

GOURIÉROUX, C., A. MONFORT, E. RENAULT, AND A. TROGNON (1987): "Generalised Residuals," *Journal of Econometrics*, 34, 5–32.

GOURIÉROUX, C., A. MONFORT, AND A. TROGNON (1985): "A General Approach to Serial Correlation," *Econometric Theory*, 1, 315–340.

GRAMMIG, J., AND K.-O. MAURER (2000): "Non-Monotonic Hazard Functions and the Autoregressive Conditional Duration Model," *Econometrics Journal*, 3, 16–38.

GRAMMIG, J., AND M. WELLNER (2002): "Modeling the Interdependence of Volatility and Inter-Transaction Duration Process," *Journal of Econometrics*, 106, 369–400.

GRANDELL, J. (1976): *Doubly Stochastic Poisson Processes*. Springer, Berlin.

GRANGER, C. W. J., AND T. TERÄSVIRTA (1993): *Modelling Nonlinear Economic Relationships*. Oxford University Press, Oxford, UK.

GREENE, J., AND S. SMART (1999): "Liquidity Provision and Noise Trading: Evidence from the "Investment Dartboard" Column," *Journal of Finance*, 54, 1885–1899.

GRITZ, R. M. (1993): "The Impact of Training on the Frequency and Duration of Employment," *Journal of Econometrics*, 57, 21–51.

HAGERUD, G. E. (1997): "A New Non-Linear GARCH Model," Discussion paper, EFI, Stockholm School of Economics.

HAJIVASSILIOU, V. A., AND P. A. RUUD (1994): "Classical Estimation Methods for LDV Models Using Simulation," in *Handbook of Econometrics*, ed. by R. F. Engle, and D. L. McFadden, chap. 40, pp. 2383–2441. Elsevier Science B.V.

HAMILTON, J. D., AND O. JORDA (2002): "A Model for the Federal Funds Rate Target," *Journal of Political Economy*, 110, 1135–1167.

HAN, A., AND J. A. HAUSMAN (1990): "Flexible Parametric Estimation of Duration and Competing Risk Models," *Journal of Applied Econometrics*, 5, 1–28.

HARRIS, L. (1990): "Estimation of Stock Variances and Serial Covariances from Discrete Observations," *Journal of Financial and Quantitative Analysis*, 25, 291–306.

HARVEY, A. C. (1990): *The Econometric Analysis of Time Series*. P. Allan, Hemel Hempstead, 2nd. edn.

HARVEY, C. R., AND R. D. HUANG (1991): "Volatility in the Foreign Currency Futures Market," *Review of Financial Studies*, 4, 543–569.

HASBROUCK, J., G. SOFIANOS, AND D. SOSEBEES (1993): "New York Stock Exchange Systems and Trading Procedures," Discussion Paper 93-01, New York Stock Exchange.

HASBROUCK, J. (1991): "Measuring the Information Content of Stock Trades," *The Journal of Finance*, 46, 179–207.

HAUTSCH, N., AND D. HESS (2002): "The Processing of Non-Anticipated Information in Financial Markets: Analyzing the Impact of Surprises in the Employment Report," *European Finance Review*, 6, 133–161.

HAUTSCH, N., AND W. POHLMEIER (2002): "Econometric Analysis of Financial Transaction Data: Pitfalls and Opportunities," *Allgemeines Statistisches Archiv*, 86, 5–30.

HAUTSCH, N. (1999): "Analyzing the Time Between Trades with a Gamma Compounded Hazard Model. An Application to LIFFE Bund Future Transctions," Discussion Paper 99/03, Center of Finance and Econometrics, University of Konstanz, Konstanz.

HAUTSCH, N. (2001): "Modelling Intraday Trading Activity using Box-Cox ACD Models," Discussion Paper 02/05, CoFE, University of Konstanz.

HAUTSCH, N. (2003): "Assessing the Risk of Liquidity Suppliers on the Basis of Excess Demand Intensities," *Journal of Financial Econometrics*, 1 (2), 189–215.

HAWKES, A. G., AND D. OAKES (1974): "A Cluster Process Representation of a Self-Exciting Process," *Journal of Applied Probability*, 11, 493–503.

HAWKES, A. G. (1971): "Spectra of Some Self-Exciting and Mutually Exciting Point Processes," *Biometrika*, 58, 83–90.

HECKMANN, J. J., AND B. SINGER (1984): "Econometrics Duration Analysis," *Journal of Econometrics*, 24, 63–132.

HEINEN, A., AND E. RENGIFO (2003): "Multivariate Autoregressive Modelling of Time Series Count Data Using Copulas," Discussion Paper 2203/25, CORE, Université Catholique de Louvain.

HELLWIG, M. (1982): "Rational Expectations Equilibrium with Conditioning on Past Prices: A Mean-Variance Example," *Journal of Economic Theory*, 26, 279–312.

HENDRY, D. F. (1995): *Dynamic Econometrics*. Oxford University Press, Oxford.

HENTSCHEL, L. (1995): "All in the Family: Nesting Symmetric and Asymmetric GARCH Models," *Journal of Financial Economics*, 39, 71–104.

HONORÉ, B. E. (1990): "Simple Estimation of Duration Model with Unobserved Heterogeneity," *Econometrica*, 58, 453–473.

HOROWITZ, J. L., AND G. R. NEUMANN (1987): "Semiparametric Estimation of Employment Duration Models," *Econometric Review*, 6(1), 5–40.

HOROWITZ, J. L. (1996): "Semiparametric Estimation of a Regression Model with an Unknown Transformation of the Dependent Variable," *Econometrica*, 64, 103–137.

HO, T., AND H. STOLL (1981): "Optimal Dealer Pricing Under Transactions and Return Uncertainty," *Journal of Financial Economics*, 9, 47–73.

HUJER, R., S. VULETIC, AND S. KOKOT (2002): "The Markov Switching ACD Model," Discussion Paper 90, Johann Wofgang Goethe-University, Frankfurt.

JASIAK, J. (1998): "Persistence in Intertrade Durations," *Finance*, 19(2), 166–195.

KALBFLEISCH, J. D., AND R. L. PRENTICE (1980): *The Statistical Analysis of Failure Time Data*. Wiley.

KAPLAN, E. L., AND P. MEIER (1958): "Nonparametric Estimation from Incomplete Observations," *Journal of the American Statistical Association*, 53, 457–481.

KARR, A. F. (1991): *Point Processes and their Statistical Inference*. Dekker, New York.

KEIM, D. B., AND A. MADHAVAN (1996): "The Upstairs Market for Large-Block Transactions: Analysis and Measurement of Price Effects," *The Review of Financial Studies*, 9, 1–36.

KEMPF, A., AND O. KORN (1999): "Market Depth and Order Size," *Journal of Financial Markets*, 2, 29–48.

KEYNES, J. (1930): *Treatise on Money*. MacMillan.

KIEFER, N. M. (1988): "Economic Duration Data and Hazard Functions," *Journal of Economic Literature*, 26, 646–679.

KYLE, A. S. (1985): "Continuous Auctions and Insider Trading," *Econometrica*, 53 (6), 1315–1335.

LANCASTER, T. (1979): "Econometric Methods for the Duration of Unemployment," *Econometrica*, 47 (4), 939–956.

LANCASTER, T. (1997): *The Econometric Analysis of Transition Data*. Cambridge University Press.

LAWRENCE, A. L., AND P. A. LEWIS (1980): "The Exponential Autoregressive-Moving Average EARMA(P,Q) Model," *The Journal of the Royal Statistical Society, B*, 42, 150–161.

LEE, C. M. C., AND M. J. READY (1991): "Inferring Trade Direction from Intraday Data," *Journal of Finance*, 46, 733–746.

LEE, S., AND B. HANSEN (1994): "Asymptotic Theory for the GARCH(1,1) Quasi-Maximum Likelihood Estimator," *Econometric Theory*, 10, 29–52.

LIESENFELD, R., AND W. POHLMEIER (2003): "Ein dynamisches Hürdenmodell für Transaktionspreisveränderungen auf Finanzmärkten," in *Empirische Wirtschaftsforschung: Methoden und Anwendungen*, ed. by W. Franz, M. Stadler, and H. J. Ramser. Mohr Siebeck.

LIESENFELD, R., AND J.-F. RICHARD (2003): "Univariate and Multivariate Stochastic Volatility Models: Estimation and Diagnostics," *Journal of Empirical Finance*, 10, 505–531.

LIESENFELD, R. (1998): "Dynamic Bivariate Mixture Models: Modeling the Behavior of Prices and Trading Volume," *Journal of Business & Economic Statistics*, 16, 101–109.

LUNDBERGH, S., T. TERÄSVIRTA, AND D. VAN DIJK (2003): "Time-Varying Smooth Transition Autoregressive Models," *Journal of Business and Economic Statistics*, pp. 104–121.

LUNDBERGH, S., AND T. TERÄSVIRTA (1999): "Modelling Economic High-Frequency Time Series with STAR-STGARCH Models," Discussion Paper 99-009/4, Tinbergen Institute.

LUNDE, A. (2000): "A Generalized Gamma Autoregressive Conditional Duration Model," Discussion paper, Aarlborg University.

MADHAVAN, A. (2000): "Market Microstructure: A Survey," *Journal of Financial Markets*, 3 (3), 205–258.

MANGANELLI, S. (2002): "Duration, Volume and Volatility Impact of Trades," Discussion Paper 125, European Central Bank, Frankfurt.

MANRIQUE, A., AND N. SHEPHARD (1998): "Simulation Based Likelihood Inference for Limited Dependent Processes," *Econometrics Journal*, 1, 147–202.

MCCALL, B. P. (1996): "Unemployment Insurance Rules, Joblessness and Part-Time Work," *Econometrica*, 64 (3), 647–682.

MEDDAHI, N., E. RENAULT, AND B. J. M. WERKER (1998): "Modelling High-Frequency Data in Continuous Time," Discussion paper, Uninversité de Montréal.

MEYER, B. D. (1990): "Unemployment Insurance and Unemployment Spells," *Econometrica*, 58 (4), 757–782.

MOFFITT, R. (1985): "Unemployment Insurance and the Distribution of Unemployment Spells," *Journal of Econometrics*, 28, 85–101.

MOKKADEM, A. (1990): "Propriétés de mélange des modèles autoregressifs polynomiaux," *Annales de l'Institut Henri Poincaré*, 26, 219–260.

MÜLLER, U. A., M. M. DACOROGNA, R. D. DAVÉ, O. V. PICTET, R. OLSEN, AND J. R. WARD (1995): "Fractals and Intrinsic Time - A Challenge to Econometricians," Discussion paper, Olsen & Associates.

NELSON, D. (1991): "Conditional Heteroskedasticity in Asset Returns: A New Approach," *Journal of Econometrics*, 43, 227–251.

NEUMANN, G. (1997): *Search Models and Duration Data*, chap. 7, pp. 300–351, Handbook of Applied Econometrics. M.H. Pesaran, ed. (Basil Blackwell), Oxford.

NEWEY, W. K., AND K. D. WEST (1987): "A Simple, Positive Semidefinite, Heteroskedasticity and Autocorrelation Consistent Covariance Matrix," *Econometrica*, 55, 703–708.

NEWEY, W. (1985): "Maximum Likelihood Specification Testing and Conditional Moment Tests," *Econometrica*, 5, 1047–1070.

NICKELL, S. (1979): "Estimating the Probability of Leaving Unemployment," *Econometrica*, 47(5), 1249–1266.

OAKES, D., AND L. CUI (1994): "On Semiparametric Inference for Modulated Renewal Processes," *Biometrika*, 81, 83–90.

OAKES, D. (2001): "Biometrika Centenary: Survival Analysis," *Biometrika*, 88, 99–142.

OGATA, Y., AND H. AKAIKE (1982): "On Linear Intensity Models for Mixed Doubly Stochastic Poisson and Self-exciting Point Processes," *Journal of the Royal Statistical Society, Series B*, 44, 102–107.

O'HARA, M. (1995): *Market Microstructure*. Basil Blackwell, Oxford.

ORBE, J., E. FERREIRA, AND V. NUNEZ-ANTON (2002): "Length of Time Spent in Chapter 11 Bankruptcy: A Censored Partial Regression Model," *Applied Economics*, 34, 1949–1957.

PRENTICE, R. L., AND L. A. GLOECKLER (1978): "Regression Analysis of Grouped Survival Data with Application to Breast Cancer Data," *Biometrics*, 34, 57–67.

RABEMANANJARA, J., AND J. M. ZAKOIAN (1993): "Threshold ARCH Models and Asymmetries in Volatility," *Journal of Applied Econometrics*, 8, 31–49.

RENAULT, E., AND B. J. M. WERKER (2003): "Stochastic Volatility Models with Transaction Risk," Discussion paper, Tilburg University.

RICHARD, J.-F. (1998): "Efficient High-dimensional Monte Carlo Importance Sampling," Discussion paper, University of Pittsburgh.

RIDDER, G. (1990): "The Non-Parametric Identification of Generalized Accelerated Failure-Time Models," *Review of Economic Studies*, 57, 167–182.

ROLL, R. (1984): "A Simple Implicit Measure of the Effective Bid-Ask Spread in an Efficient Market," *Journal of Finance*, 39, 1127–1139.

ROMEO, C. J. (1999): "Conducting Inference in Semiparametric Duration Models and Inequality Restrictions on the Shape of the Hazard Implied by Job Search Theory," *Journal of Applied Econometrics*, 14, 587–605.

ROSENBLATT, M. (1952): "Remarks on a Multivariate Transformation," *Annals of Mathematical Statistics*, 23, 470–472.

RUSSELL, J. R., AND R. F. ENGLE (1998): "Econometric Analysis of Discrete-Valued Irregularly-Spaced Financial Transactions Data Using a New Autoregressive Conditional Multinominal Model," Discussion Paper 98-10, University of California San Diego.

RUSSELL, J. R. (1999): "Econometric Modeling of Multivariate Irregularly-Spaced High-Frequency Data," Discussion paper, University of Chicago.

RYDBERG, T. H., AND N. SHEPHARD (2003): "Dynamics of Trade-by-Trade Price Movements: Decomposition and Models," *Journal of Financial Econometrics*, 1(1), 2–25.

SNYDER, D. L., AND M. I. MILLER (1991): *Random Point Processes and Time and Space*. Springer, New York, 2 edn.

SPIERDIJK, L., T. E. NIJMAN, AND A. H. O. VAN SOEST (2002): "The Price Impact of Trades in Illiquid Stocks in Periods of High and Low Market Activity," Discussion Paper 2002-29, CentER, Tilburg University.

SPIERDIJK, L. (2002): "An Empirical Analysis of the Role of the Trading Intensity in Information Dissemination on the NYSE," Discussion Paper 2002-30, CentER, Tilburg University.

STINCHOMBE, M. B., AND H. WHITE (1992): "Consistent Specification Tests Using Duality and Banach Space Limit Theory," Discussion Paper 93-14, University of California, San Diego.

STOLL, H. (1978): "The Supply of Dealer Services in Securities Markets," *The Journal of Finance*, 33, 1133–1151.

STRICKLAND, C. M., C. S. FORBES, AND G. M. MARTIN (2003): "Bayesian Analysis of the Stochastic Conditional Duration Model," Discussion Paper 14/2003, Monash University, Australia.

SUEYOSHI, G. T. (1995): "A Class of Binary Response Models for Grouped Duration Data," *Journal of Applied Econometrics*, 10, 411–431.

TAUCHEN, G. E., AND M. PITTS (1983): "The Price Variability-Volume Relationship on Speculative Markets," *Econometrica*, 51, 485–505.

TAYLOR, S. J. (1982): "Financial Returns Modelled by the Product of Two Stochastic Processes - a Study of Daily Sugar Prices," in *Time Series Analysis: Theory and Practice*, ed. by O. D. Anderson. North-Holland, Amsterdam.

TERÄSVIRTA, T. (1998): "Modelling Economic Relationships with Smooth Transition Regressions," in *Handbook of Applied Economic Statistics*, ed. by A. Ullah, and D. E. A. Giles, pp. 507–552. Marcel Dekker, New York.

TERÄSVIRTA, T. (1994): "Specification, Estimation, and Evaluation of Smooth Transition Autoregressive Models," *Journal of the American Statistical Association*, 89, 208–218.

THOMPSON, W. A. (1977): "On the Treatment of Grouped Durations in Life Studies," *Biometrics*, 33, 463–470.

TONG, H. (1990): *Non-linear Time Series: A Dynamical System Approach*. Oxford University Press, Oxford.

VAN DEN BERG, G. J., AND B. VAN DER KLAAUW (2001): "Combining Micro and Macro Unemployment Duration Data," *Journal of Econometrics*, 102, 271–309.

VEREDAS, D., J. RODRIGUEZ-POO, AND A. ESPASA (2002): "On the (Intradaily) Seasonality, Dynamics and Durations Zero of a Financial Point Process," Discussion Paper 2002/23, CORE, Université Catholique de Louvain.

VERE-JONES, D. (1970): "Stochastic Models for Earthquake Occurrence," *Journal of the Royal Statistical Society, Series B*, 32, 1–62.

WHITE, H. (1982): "Maximum Likelihood Estimation of Misspecified Models," *Econometrica*, 50(1), 1–25.

WINKELMANN, R. (1997): *Econometric Analysis of Count Data*. Springer, Berlin.

YASHIN, A., AND E. ARJAS (1988): "A Note on Random Intensities and Conditional Survival Functions," *Journal of Applied Probability*, 25, 630–635.

ZAKOIAN, J. M. (1994): "Threshold Heteroskedastic Models," *Journal of Economic Dynamics and Control*, 18, 931–955.

ZHANG, M. Y., J. RUSSELL, AND R. S. TSAY (2001): "A Nonlinear Autoregressive Conditional Duration Model with Applications to Financial Transaction Data," *Journal of Econometrics*, 104, 179–207.

Index

Printing and Binding: Strauss GmbH, Mörlenbach

Cognitive Science
A Developmental Approach to
the Simulation of the Mind

Bruno G. Bara
Centre for Cognitive Science,
University of Turin, Italy

LAWRENCE ERLBAUM ASSOCIATES, PUBLISHERS
Hove (UK)   Hillsdale (USA)

First published in Italian in 1990 by Bollati Boringhieri Editore.
Translated from the Italian by John Douthwaite.

Lawrence Erlbaum Associates Ltd., Publishers
27 Church Road
Hove
East Sussex, BN3 2FA
UK

British Library Cataloguing in Publication Data

A catalogue record for this book is available from the British Library

    ISBN 0-86377-362-1

Printed and bound by BPC Wheatons Ltd., Exeter

To Giulio
Simona
Elena

salt of the earth

Contents

Acknowledgements

The real reason why a person writes a book is because he knows that at the end he will be able to thank those who have helped him—a refined and long-lasting pleasure.

The following people will be quoted many times for their scientific merits, but I would like to begin by remembering them as companions-in-arms in founding the Artificial Intelligence Research Unit in 1980: Gabriella Airenti and Marco Colombetti, invaluable friends and colleagues. Later but no less important additions to the Unit were Antonella Carassa and Giuliano Geminiani.

Philip Johnson-Laird enabled me to understand the international world of science, working with me continuously on reasoning, from the first time I met him in England as the fortunate recipient of a research award for postgraduates. I virtually wrote the entire book thinking over and reutilising the work carried out together with these five people, in the incredibly long period of our collaboration, in which we shared not only our work, but also our joys, conflicts, holidays, loves, and dreams of glory.

A small group of PhD students created a fertile soil and an impassioned working milieu. The most significant contributions from this group were furnished by Enrico Blanzieri, Monica Bucciarelli, Astrid Ricci, and Maurizio Tirassa.

Research is done well only if the environment allows it, and I have had the opportunity of exploiting two excellent places: first, the Institute of Psychology at the Medical Faculty at the University of Milan. The open-mindedness of the Head of the Institute, Marcello Cesa-Bianchi, permitted the development of artificial intelligence and cognitive science in years when the fashion was anything but these fields. Second, the Applied Psychology Unit of the Medical Research Council in Cambridge, directed by Alan Baddeley, represented a reference point, a place for reflection, a lay monastery with no commitments save those devoted to pure research, where I exploited computers and discussions in the tea-breaks.

In 1991 the Artificial Research Unit of the University of Milan closed, to give life to the more ambitious project of the Center for Cognitive Science, inaugurated in 1993 at the University of Turin. Practically all the researchers working at the Unit transferred to the Center to start this new adventure.

I now realise that my fortune has been that of having always found myself in working environments which were extremely rich from a human point of view, and which enabled me to withstand the anxiety connected with an activity so uncertain as is pure research, with years of one's life invested on the basis of intuitions as rarefied as they were potentially crucial.

The manuscript was read by experts, collaborators who have saved me from countless errors and suggested a corresponding number of improvements. Critical advice has been generously given by the philosopher of science, Roberto Cordeschi; the psychologist Giuliana Mazzoni; the psychiatrist, Giorgio Rezzonico; and the neuroscientists, Anthony Harris and Steven Small.

Agnese Incisa, Caroline Osborne, and Maria Antonietta Schepisi gave me all the editorial assistance I needed with understanding, charm and competence.

The present English version would never have existed if I had not met John Douthwaite, who teaches English at the University of Udine. He began by translating the book competently, and ended up affectionately rewriting the book with me.

My debts of affection: both of my daughters, aged 8 and 6, have a file on the computer, in order to help me when I am tired and continue the work in my stead. Their presence filled me with enthusiasm at all times.

Marcella kept them away from me, and happily accepted the many sacrifices involved, of spending her holidays alone, with the little terrors on the rampage, and of the numerous weekends spent in Milan instead of in places which were less grey. Even more, with loving trust, she never asked me if it was worthwhile.

B.G.B
Milan, February 1995

PART ONE
Methodology

CHAPTER ONE

Introduction

This book has essentially three objectives:

1. An historical description of the birth of cognitive science: which disciplines created the conditions for its development, and which at present foster its growth.
2. An outline of the method of inquiry which characterises cognitive science, and through which it is defined. The method employed is constructivist, and it is carried out by means of the techniques of artificial intelligence. This amounts to assuming that mental activity is in principle reproducible by a computer program.
3. A description of the state of the art in the relevant subdisciplines.

The first objective, a history of the subject, does not present insurmountable difficulties if one has access to a good library and if a large number of the founders of the subject are still active and invitable to dinner.

The second objective, an account of cognitive methodology, is my major goal. As will become clear later, cognitive science is composed of a set of at least six different subjects, some of which may be subdivided even further. Besides the constructivist method, each discipline which goes to make up cognitive science adopts particular techniques which are suitable to the specific area treated, and which are different in linguistics, anthropology or psychology. The simulation approach has to specialise in specific applications which must be explained and justified. This special form of understanding is crucial if the spirit of cognitive science is to be penetrated in order to comprehend its themes and its intra- and interdisciplinary debates.

A further reason for dwelling on the methodological aspects lies in their stability in the face of rapid change typical of research. And in fact the third objective, to furnish an up-to-date state of the art, has been pursued only in part; first, science continually progresses, books and articles accumulate, and it is an illusion that one can keep up with everything in an exhaustive fashion. Second, the state of the art in certain areas, such as the neurosciences, would of itself take up all the space available. Third, excellent texts on specific fields—such as perception and memory—already exist.

With regard to the third objective, the state of the art, the choice has therefore fallen on devoting greater space to exploring dark areas, of which little has been written, and to let the illuminated zones speak for themselves by referring readers to books which are easy to consult. I have not left aside, at least not voluntarily, any of those areas which are at present considered significant; but it seemed wise to concentrate attention, not on the officially entrenched topics, but on the most promising developments. I have also devoted space to areas which are not in great favour today, such as time and emotion, but which are nevertheless recognised as being of extreme importance for the understanding of the human mind, and which are increasingly becoming the objects of attention.

The refrain of my project is, despite the titles of the various chapters, not *what* cognitive science is, but rather *how* cognitive science is done. The constant internal cross-references, together with the concentration on the framework of cognitive science as a whole, at the cost of sacrificing the details, may be attributed to this global stance. My particular viewpoint as a research worker has clearly determined the perspective adopted in this volume: A developmental approach is privileged right from the subtitle of the book, and each chapter emphasises the fact that the developmental aspects are essential to comprehend the ultimate mode of functioning that is achieved once the system has reached total maturity.

Cognitive science bears the same relationship to traditional psychology, namely clinical and experimental psychology, that the combustion engine bore to animal-drawn carts. It certainly does not solve all of mankind's problems; furthermore it creates new, unexpected ones. But it does change our life style, and even if we enjoy pretending we are nostalgic over bygone times, at bottom no one has ever regretted the disappearance of the stagecoach. What is more, we all hope that someone will invent a less polluting and more efficient engine, or better still an alternative, revolutionary method of transport: a mode of achieving motion which does not require energy, for example. Similarly, cognitive scientists hope that someone will invent an instrument which requires less effort in its use than the computer, and which will furnish results corresponding more closely to the complexity of the mind than those that a computer with its crude, hypersimplified procedures can yield.

Until a new methodology which is more effective than the constructivist approach is invented, computer simulation, despite all its limits, will remain the cardinal method for conducting scientific investigation into the human being. How such research is carried out is the subject of the first part of the book, while the second part will examine the results of this research: fairly meagre, it might be said, in relation to the vast area of mental life that remains unexplored; fairly rich if we compare the progress made in the last 20 years with the sluggish, uncertain advances of the preceding period.

Human beings have always had two major interests: understanding and controlling the external world, and understanding and controlling themselves. The study of the mind goes back to ancient times, and it has produced fascinating results both in the oriental and in the western tradition. The oriental tradition has favoured the global approach to comprehending mind and nature, rejecting the sectorial and obviously limited approach of the scientific and rationalist standpoint. The western tradition, from the Greeks on, in whose culture a great deal of our present knowledge is rooted, has, on the other hand, chosen an analytical approach, accepting the scientific method with its theoretical limitations and its applicational advantages. Publicly recognising its limits is an extremely recent luxury, one which science has been able to allow itself only because it no longer needs to struggle to affirm itself socially. In the field of human sciences this is tantamount to recognising that the complexity of the mind does not admit the functional decomposition employed to investigate it up to the present time.

Stated differently, a qualitative change in the comprehension of mental processes will take place when we are able to consider the mind as a single entity, and not as decomposed into independent functional units such as perception, thought, memory and so on. Buddha had already said as much 2500 years ago, warning us against the hope of understanding the world, and ourselves, by means of reason and the paucity of instruments that this generates; we reach the same conclusion from exactly the opposite direction, thanks indeed to increasingly sophisticated products, among which the computer occupies pride of place.

That a person should hope to capture and convey the flavour of such a wide and complex area as cognitive science is in itself sound evidence of hypomania. I will therefore not add the presumptuousness of objectivity. My personal opinions have openly conditioned the choice of what is important and thus requires extensive treatment, and what is less important and may for this reason be condensed without harm to the reader. For example, the space conceded to mental models bears witness to the role they can play in all the psychic functions; but this is more a bet than proven reality. In the end, the book does not so much mirror cognitive science as it is, as how I would like it to be or hope it will become: an interpretation, therefore, more than a photograph.

A second difficulty, and one which is more insidious because it is not always evident even to myself, is generated by the varying degrees of my ignorance in the various specific sectors. My competence is greater in the fields of thought and language compared to memory and perception, and hence I can define what I do not know with greater accuracy in the former than in the latter. When one is not an expert in a field, discerning one's limits in that area as well as the limits of the knowledge acquired is an arduous task. Popular writers always tend to believe everything is clear and already known. If we add to this the extreme complexity of the object of our study, the human brain, the picture to be depicted can hardly be said to be banal. I have tried to do so without tricks and embellishments, declaring both what is uncertain and what is unknown, in a language which is deliberately non-esoteric. Finally, the technical parts are introduced without presuming that the reader has a larger culture than that of a high school student.

A final word about references. Whenever possible, they are to books and not to single papers. I was careful to mention books written by the authors themselves as the longer version of their original works. The reason is that books are more carefully elaborated than papers, and they can be easily found in any decent library.

CHAPTER TWO

What is cognitive science?

A satisfactory answer to this question can only be given at the end of the entire book; for the time being we shall have to make do with a first approximate definition, which we shall try to perfect as we move forward. The principal difficulty in explaining what cognitive science really is consists in the recursiveness which ties it to artificial intelligence, in as much as the latter is both part of cognitive science (the *definiendum*), and part of the criteria which define cognitive science itself (the *definiens*).

Nevertheless, we may initially define cognitive science as that set of subjects which study the human mind, accepting simulation as their unifying method, a method which is typical of artificial intelligence. The so-called cognitive hexagon, illustrated in Fig. 2.1, is composed of: psychology, philosophy, linguistics, anthropology, neuroscience, artificial intelligence.

We shall now analyse how the constitution of this subject was rendered possible, reconstructing remote and recent causes, and then tracing its development. The point of view adopted is that of the cognitive scientist, and no pretence to objectivity or comprehensiveness is made with regard to the six areas quoted. This will grant us a certain degree of freedom in interpreting a series of significant events in other subjects, underlining those aspects which are important to cognitive science, without our having to respect criteria determining the specific importance of individual phenomena within their own discipline. The

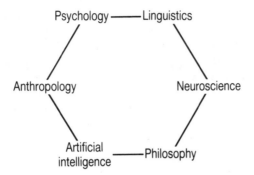

FIG. 2.1. The cognitive hexagon.

objective in reinterpreting history in this way, namely of failing to respect the absolute conventions and boundaries of each science, is that it will help to achieve an understanding of the evolution of the reasons which first enabled and then determined the birth of cognitive science.

ANCIENT HISTORY

Interest in the human being cannot be limited exclusively to the disciplines contained in the cognitive hexagon, since it has been a constant concern throughout human history in a myriad of fields of investigation: some still flourish today, others have ceased to exist in a more or less recent past. The art of magic kindled the energies of numberless scholars and practitioners for many centuries, only to see its influence slowly weaken to the point that it has virtually disappeared today. Other arts and other sciences centred on the human being have instead retained and increased their social esteem, and we shall be referring to them in the following part of the chapter; in particular, we shall concentrate our attention on philosophy and psychology, those disciplines that have always been best qualified to speak about the mind.

Philosophy
Cognitive science is engaged in the study of subjects which humanity has been inquiring into for thousands of years: the nature of knowledge and how it is acquired, the relationship between mind and body, how

emotion and cognition interact, and so forth. Each area we shall tackle has already been extensively analysed by philosophers, even though the instruments employed differ from those adopted here. This makes it impossible to be fully explicit about the debts that psychology and cognitive science owe philosophy. I will therefore limit myself to outlining the contribution of a single author, a direct forerunner of our subject. However, when the results achieved by a philosopher are still fundamental today, as is the case with Aristotle with regard to formal thought, these will be summarised in the pertinent chapter.

René Descartes

The two crucial moments in the history of the human sciences are on the one hand the affirmation that the human body may be studied by empirical means, and is thus not to be treated as sacred, and on the other hand the intuition that the same assumption may be applied to the mind. In his treatise, *De homine*, published in Leiden only in 1662, 12 years after his death, Descartes was the first to carry out an analysis of the human body by means of the hypothetical reconstruction of an animate statue, based on the model of a machine. On the basis of results obtained from the dissection and vivisection of animals, the French philosopher affirmed that the vital functions of an animal were the result of heat and of the movement of liquids within the body. At a later stage, the Cartesian hypothesis on fluids was extended to man himself: "And indeed the nerves may well be compared to the sides of machines ... his muscles and his tendons to the various other devices and springs ... furthermore breathing and other similar natural actions of this machine ... can be compared to the movements of a watch and of a windmill".

Dissatisfied with the uncertainty which characterised philosophy, from which it ensued that no result could be held to be definitively certain, Descartes worked out a method based on systematic doubt, summarised in the aphorism *Cogito, ergo sum* (I think, therefore I am). Thus, he decided to eliminate everything he could not say he was absolutely certain of, limiting himself to the awareness of doubt. The capacity to doubt was based on the capacity to think, and this was Descartes' new point of departure from which he set out to construct his philosophical system. The central point in the system was the mind, because the first knowledge man could be certain of only existed there: that is, the self-science of thinking, and therefore of being. Within the mind, a privileged position was assigned to those ideas referring to something immutable and eternal, as in the case of arithmetic and of geometry, whose existence was independent of that of similar entities in the external world, and so self evident as to be considered as generated by the mind itself. The body was altogether different. It was under the constant control of the mind, even if it was far from clear how such control came to be exercised, nor even how any form of interaction between the two systems was possible. The body was in fact equatable with a car: It could be broken down into pieces, and these could be disassembled, modified and replaced without bringing about any fundamental change. In contrast to this, the mind could not be broken down into separate parts, but had to be dealt with as a single complex entity, equipped with special characteristics that could not be reduced to anything else. Among these stood out the capacity to transmit thoughts to other minds through language, a means which, furthermore, could not be reproduced mechanically.

We may not agree today with many of the solutions proposed by Descartes, but it cannot be denied that the problems he tackled are to a large extent those we face today, a fact which testifies to his genius in identifying the aspects on which it is interesting and important to focus attention. From the methodological point of view, his modernity is astounding—hypothesising the construction of an automaton, with the aim of carrying out investigations into human beings, even if only with regard to their bodies, corresponds exactly to the present dictates of artificial intelligence. A further aspect that must be underlined is his considering the mind as a single, unfragmentable entity, an assertion that is extremely close to the approaches based on complexity and development.

In conclusion, if ever a philosopher is to be considered as the precursor of cognitive science, this philosopher is definitely Descartes. The list of problems he drew up to reform philosophy corresponds largely to what modern science believes its present agenda to be—Descartes anticipated the objects of investigation of the four centuries to come. The relationship between mind and body, the confrontation between rationalism and empiricism, the relationship between philosophical reflection and scientific research—all these questions are still pertinent today and are approached in practically the same way Descartes formulated them, though, obviously, they have been enriched by the contributions of thinkers who came after him: from Hume to Kant to Wittgenstein.

La Mettrie and the French Revolution
The first official martyr of artificial intelligence was Julien Offray de La Mettrie, who took up the Cartesian theme of the study of the human being as a machine, and who was persecuted for his materialism much more so than his more illustrious colleague had been. Exiled by the French monarchy, he emigrated to Leiden, where he managed to publish

L'homme machine in 1748, a book which earned him the esteem of the revolutionaries, but which cost him a second exile, from Holland. The inquisitor burnt his books in public, as a warning to those who spread atheism. La Mettrie attempted to furnish a model of the human being which comprised the rules governing the way humans work; but his ingenuous reductionism led him to postulate one single property, namely movement, as the essential component which explained how the human machine thinks, feels and acts: "We therefore conclude courageously that man is a machine, and that in the entire universe there is only one substance which is modified in different fashions".

According to the French physician, the fact that a constant correlation may be demonstrated between psychic states and bodily states justifies our considering psychic events not as the product of an immaterial soul, but as the result of cerebral functions, even if we have so far been unable to discover how psychic phenomena depend on physical phenomena. It was La Mettrie's persistence in applying the Cartesian method not only to the body, but also to the human psyche, that caused the church and the ruling class to loathe him so strongly that he suffered continual persecution. A few years later, his theories were openly taken up by the *méchaniciens*, who were thus defined by Denis Diderot and Jean le Rond d'Alembert in the *Encyclopédie*: "This is the name given to those modern physicians who, after the discovery of the circulation of the blood and the spread of the philosophy of Descartes, have thrown off the yoke of authority and have adopted the method of the Geometricians in the research they have carried out with regard to animal husbandry ... The body of an animal, and consequently the human body, is here considered as a machine to all effects and purposes".

At the same time another important figure—Jacques Vaucanson—was working all over Europe with great success. Vaucanson produced physical replicas of the discoveries others were making about animals and man. He thus built a series of extraordinary automata, including a duck and a flautist, each part of which corresponded exactly to what was known in anatomy and physiology. Vaucanson, who wished to be considered a modern scientist, half-way between medicine and engineering, explained every detail of his constructions scrupulously so that they would not appear mysterious. The ingeniousness of the human being supplanted magic in becoming the object of wonder. The most ambitious project of this second Prometheus, as he was enthusiastically defined by La Mettrie, namely to build an artificial human being, complete with blood flowing and capable of speaking just as the flautist played, was never to be realised, but it was already a significant fact that it was spoken of as a project which was possible, and no longer as a dream.

By encouraging this type of study and exporting its methods and results throughout Europe, the French Revolution once and for all sanctioned the principle that the human body and the human mind could be investigated; they were no longer sacred and inviolable. The goals of science were the discovery of the truth and the improvement of the human condition, not the glorification of God and his official representatives on earth—state and church. These were the new foundations which gave birth to the new science of the human being, of which cognitive science is only the latest expression.

Cybernetics

Cybernetics has a special relationship with cognitive science and artificial intelligence because it foreshadows both, the former since its wide-ranging ambitions are not restricted within the confines of any subject, the latter because of the similarity in the methods employed. The best definition of cybernetics refers to the one method it employs to study both artificial and living systems. The same theory explains the mechanisms accounting for human and robot behaviour, treating them as basically identical. What all machines—living or artificial—have in common is the concept of feedback, which allows them to receive information from the external world, information which is employed to achieve an internal goal. Cybernetics extends the use of computers to spheres which are far removed both from computer science and from its traditional applications, tackling such heterodox topics as neuronal networks, the rationalisation of the traffic, ballistic missile control systems, the mystery of the disappearance of the dinosaurs.

Cybernetics had its forerunners too, and once again these may be traced to that extraordinary period of social and intellectual upheavals at the end of the 18th century. Within the framework of scientific progress that was deemed fundamental for the achievement of social welfare following the revolution, the French government commissioned the engineer Gaspard de Prony to produce the monumental work, and one which was indispensable for the rise of industrialisation which was then emerging, of compiling the logarithmic and trigonometric tables from 1 to 200,000. In the first chapter of his work, *An inquiry into the nature and causes of the wealth of nations* (1776), the economist Adam Smith expounded his theory of the division of labour, describing the production of pins as an illustration of his thesis. Prony, inspired by this text which he chanced to be reading, had the idea of "putting logarithms into production like pins".

Transferring the model of the division of labour to mathematical calculations did not require too great an effort. The essential step had already been achieved, and consisted in Smith's realisation of the importance of the organisation of work with a view to promoting efficiency, and in his having outlined the principles on which this could be accomplished. Though it did not comprise physical parts, the system conceived by Prony to compute the logarithm tables may really and truly be considered to be a calculating machine. Analogously to what was happening in manufacturing, this machine employed human beings as constituent parts too: the intellectual division of labour parallelled that occurring at the industrial level.

In manufacturing, the collective labour of several workers had the advantage of greater potential productivity compared to the labour of a single craftsman, even if it had the opposite effect of progressively limiting the global capacities of each individual, as Karl Marx (1867) was to point out later. In the same way, the human machine devised by Prony made it possible to carry out the work which a single individual brain would never have been able to do. It therefore had to be divided into various levels of difficulty and executed by many people organised hierarchically. It must further be noted that the simplest and most repetitive operations, those typical of the earliest calculating machines—such as addition and subtraction—were entrusted to people with a low level of education, the real mathematicians being assigned only to tasks of coordination and formulation of the solutions to problems.

It should be remembered that Prony succeeded brilliantly in his undertaking, bringing ever-increasing credit to the theories of the division of labour, and well deserving the trust of the revolutionary government. And to tell the truth, the tables he produced were extremely effective in carrying out their task for at least a century, before acting as a stimulus for a new creative leap which led to the conception of the first calculating machine in a proper sense. The ingenious aspect of Prony's work consists on the one hand in having decomposed a complex calculation into its elementary constituent operations, and on the other in having sought the laws determining their recomposition at a higher level—starting from operations on numbers, since the rules governing their mechanical reproduction were already known, the goal of formalising operations on operations was achieved.

The automation of simple calculations, however, was not an invention of the 18th century. Blaise Pascal had already built an adding machine as early as 1642, and in 1671 Wilhelm Leibniz, whom Wiener held to be the patron saint of cybernetics, had constructed a machine for multiplication sums. Leibniz had also introduced two concepts which were crucial for the development of computer science: a universal symbolism and a definition of a calculus of reasoning, both of which were based on logic, and which were the precursors to formal logic, which

Boole was to introduce two hundred years later. For those readers interested in the subject, Vernon Pratt furnishes an accurate history of the beginnings of automatic calculation in his book, *Thinking machines* (1987).

Charles Babbage's analytical engine

The next step consisted in conceiving a machine which would be able to handle complex information, one which would not be limited to manipulating figures. This advance was achieved in 1883 by the mathematician Charles Babbage, who proposed his analytical engine, the prototype of the general purpose computer, that is a calculating machine which is not designed for a specific function, but having multiple objectives. The machine thus became programmable. Babbage was convinced he could place each specific series of operations under the total control of a program on a punched card. By changing the program, the analytical engine would have been able to execute different tasks. Herein lay its versatility in comparison with preceding artefacts. In this way, it would be possible to solve any function that could be computed, any problem that could be expressed in mathematical form, and the machine would have even been capable, at least in theory, of playing chess (Morrison & Morrison, 1961).

With regard to programming, present-day computers correspond to what Babbage was the first to have had an insight into: from machines built to carry out a single type of operation, we move to machines which may be programmed for operations of a general type, namely, machines which are capable of executing diverse operations, and for which the sole restriction consists in the flexibility and ingenuity of the program. Even the general structure of the analytical engine is very similar to that of a computer: it includes a windmill which will grind data and a warehouse which will store the data, which correspond respectively to the modern concepts of the processing unit and memory; with regard to punched cards, these were adopted by computer science without any modification being undertaken, not even in the name.

The philosophical importance of the analytical engine was immediately clear, but Babbage never managed to find the funds necessary to move from the drawing board to the fully operational machine, one reason being that the high precision technology, indispensable for the construction of a machine whose mechanics are so highly sophisticated as to require a punched card for its management, was not completely available at the time. Faced with the lack of concrete results, the debate foundered on the sterile subject of what should be considered to be the limits of this type of machine, and as always insurmountable limitations were established only to be overcome no sooner had they been fixed, with little regard, moreover, for the arrogance with which they had been announced. Incidentally, this constitutes one of the most amusing aspects of artificial intelligence—the pleasure of doing something that some official authoritative thinker had previously established was absolutely impossible. The advantage the 19th century had over the previous centuries was that one was no longer risking one's life when trying to automate thought—the religion-based ban had been transformed into social disapproval.

As a matter of fact Babbage and his most faithful collaborator, Lady Ada Lovelace—daughter of the poet Lord Byron and an extremely precocious student of the logician Augustus De Morgan, who was in his turn paving the way for George Boole's logical revolution—did not have a particularly easy life, but this was due solely to the economic problems that the attempt to construct the various versions of the analytical engine caused them. When the British government, which had allocated the generous sum of £17,000 (approximately six to seven million dollars today) refused to grant further funds, Babbage embarked on a series of adventurous attempts to raise the money he needed to continue his project. After Ada Lovelace had pawned her family jewels twice, her mother, Lady Byron, having redeemed them twice in secret, the two scholars developed an infallible system to win at the horse races, a system which soon placed them at the mercy of unscrupulous and pitiless bookmakers, who persecuted them with continual threats of scandal until both their deaths.

Besides being the most intelligent publicist of the principles underlying the construction of the various versions of the analytical engine, Ada Lovelace earned a lasting place in our memory by being the first person to make the distinction between the calculator and the program for the calculator, between hardware and software.

Babbage was an honoured member of the Royal Society, which had given its full support to his plans even when the government had ceased to finance them, and he occupied the prestigious chair of mathematics at Cambridge which had already been held by Newton. Despite these facts, the failure of the public to view his work with favour, together with his own, mistaken, judgement that his work was unsuccessful, ruined the final years of his life, throwing him into a profound state of bitter frustration which alienated the affection of his contemporaries even further.

The British government's interest in the analytical engine lay principally in the need to have exact mathematical tables available, indispensable not only for industrial exigencies but also, and very importantly, to ensure safe navigation. A naval power such as 19th century imperial Britain took no heed of costs when it came to improving the efficiency of the merchant and the royal navies, and the astronomical tables which were available at that time were based on logarithms calculated by hand, with the result that the arithmetical and typographical errors which hand reckoning entailed were not visible on paper but had disastrous consequences on navigation. The work which Prony had so brilliantly accomplished was no longer sufficient to satisfy the needs for precision which the approaching 20th century demanded for its industrial development. The official goal of that substantial public investment was never achieved, because of the lack of an adequate technological environment favouring development, but Babbage's work transformed precision manufacturing to such a degree—by accelerating the conversion of manual labour into automation—that the government believed that the sum granted had been gainfully employed. Babbage's work also inaugurated the concept of the useful failure, still a mainstay today in artificial intelligence as it is in general in all the sciences which are not characterised by normal developmental processes, but which require continual technological and theoretical revolutions to make progress.

George Boole's laws of thought

In the meantime, another mathematician, one who was less interested than Babbage in the practical and worldly aspects of research, was preparing to lay the foundations of modern logic, which were later to be employed to formalise computer science as well as in the far less successful attempt to do the same for human sciences. In his *An investigation of the laws of thought* (1854), Boole defines his objectives thus: "The aim of this treatise is to investigate the fundamental laws governing those operations of the mind by which reasoning takes place; to express them in the symbolic language of a calculus and to institute, on these bases, the science of logic building up the method characterizing it ..., to draw from the various elements of truth brought to light in the course of this investigation probable indications as regards the nature and constitution of the human mind".

Boole marks a crucial turning point in 19th century logic. Logic was set free from individual interpretations by rendering it formal. What this means is that, by following a general method of derivation, conclusions which are definitely true may be reached independently of the specific content of the object of reasoning. The criterion of logical truth thus shifted from meanings, from the interpretations placed on symbols and connectives, to the relations and abstract rules that must be followed. The syntax of logic prevailed over semantics, opening up the way to the developments of modern formal logic and its applications of an automatic type. The algebra of logic, and hence the laws of thought themselves, determined which operations were lawful on abstract symbols. Boole's mathematical education helped him to separate logic from philosophy, and to channel it back to the mathematical canons of abstraction and independence from specific content matter.

The increasing psychologism in Boole's work is reflected in the subtitle of his book, which refers to the "laws of thought, on which are founded the mathematical theories of logic and probability"; his interests move from logic to cognitive laws on which he believes logic itself is founded. This type of psychologism is the opposite of the anti-philosophical attitude that induces us today to consider Boole to be the founder of modern logic, and has been viewed with a critical eye both by logicians and by psychologists.

The long-term objective the English mathematician set himself was the definition of the laws that govern the human mind. His ambition was to demonstrate that his logic corresponded exactly to the logic that human beings employ when they think. Boole is the most authoritative representative of the approach to human thought which in Chapter 7, where I deal with reasoning, we shall call mental logic. The fundamental thesis of this approach is that there exists a logic of the mind which corresponds to formal logic. This hypothesis was to be taken up by many psychologists, and refuted by an equal number of opponents, as will be seen later. What may be stated with certainty is that the psychological part of his work is the most obsolete today, while the validity of his contribution in the field of logic is unquestionable.

Norbert Wiener, cybernetician

The figure who gathered round him the first interdisciplinary group of scholars, the founders of cybernetics, and the direct forerunners of computer science, artificial intelligence and cognitive science, was another mathematician by training, interested in logic. Norbert Wiener was a restless soul and a many-sided genius, who arrived at the Massachusetts Institute of Technology after having been a pupil of the logicians Bertrand Russell at Cambridge and David Hilbert at Göttingen. He began to build up an informal discussion group on scientific methodology around 1940. His permanent collaborators from the very beginning were the Mexican physiologist Arturo Rosenblueth and the electronic engineer Julian Bigelow. Later he was joined by the mathematical logician Walter Pitts and by the neurophysiologist Warren McCulloch, who were working on problems concerning the properties of neurons and the synapses in the cerebral cortex. Gradually other members joined the team.

Together with John von Neumann, the father of digital computers, and with the blessing of Claude Shannon, the creator of information theory, Wiener organised the first meeting on cybernetics at Princeton in 1944. During this conference, it emerged that there existed a common ground of ideas shared by the research workers in the different fields—mathematics, logic, physiology, electronic engineering, and so forth—and that these ideas were organised around the concept of feedback. Once the difficulties of the war had been overcome, von Foerster and McCulloch reorganised the group, extending it to include specialists in psychology, anthropology and sociology. Invitations were extended among others to the psychologist Kurt Lewin, the anthropologists Gregory Bateson and Margaret Mead, and the sociologist Oskar Morgenstern, the originator, in collaboration with John von Neumann, of game theory, which aimed at explaining social and economic organisation employing principles which may be traced back to concepts associated with cybernetics.

These top-level figures represented a formidable collection of knowledge and intelligence, and it was thanks to the meeting of their competencies that cybernetics was enriched and took off. Later everyone was to follow his own way independently, cybernetics having fertilised all the sciences which had participated in its making.

In *Cybernetics* (1948), whose subtitle is *Control and communication in the animal and in the machine*, Wiener explains how the new science sprang from the widespread awareness that the set of problems concerning communication and control, in machines as in living systems, were all essentially part of a single problem. The dissatisfaction with the terminology available led to the coining of a neologism which was derived from the Greek word, κυβερνήτης, meaning helmsman.

The first results obtained by cybernetics were tied to the war effort: The progress achieved in aeronautics in the 1930s with regard to speed and manoeuvrability had made it clear that traditional methods of manual aiming were obsolete, and that the necessary calculations had to be automated. The prestige of the German air force, skilfully underlined by a series of demonstrations which the Luftwaffe put on before the war began, had strongly emphasised the need to improve anti-aircraft guns. What was new in the problem was that, given the velocity of the target, the projectile would have to be launched not at the target itself, but at some point in space in which missile and target could meet at some future moment. Bigelow and Wiener set about studying how the future path of the aircraft could be predicted. The simplest system consists in extrapolating the present trajectory along a straight line. But as soon as the first shot explodes, the pilot ceases to follow a linear path in order to execute the diversionary manoeuvres that the particular flight conditions permit. Predicting the curvilinear path of flight is a more promising method. But forecasting the future of a curve means carrying out mathematical calculations on its past to identify regularities or a pattern. Achieving this in real time requires an extremely powerful computer. Bigelow and Wiener's ideas were not followed up because the calculating speed of the computers that could be installed in the fire direction centres was insufficient.

Wiener did not seem to be overly discontent with regard to this partial lack of success. As a good patriot, he had made his contribution, and as a convinced anti-militarist he was pleased his suggestions had not been put to concrete use. Researchers in cybernetics always devoted great attention to the applications that could be developed from their research. After the consequences of Hiroshima and Nagasaki were made public, however, they refused to collaborate with the war effort any further. Instead they increasingly turned their attention to fields of medicine, psychology, and social development.

Another idea of Bigelow and Wiener's was destined to have a longer lease of life. Both the gunners and the pilots acted as integral parts of the projectiles and the targets respectively. Probing this intuition further yielded the concept of *feedback*. As a first approximation, it may be said that each time one intends to carry out a movement in accordance with a given model, the difference between the model and the movement as it is actually carried out is employed as a new signal which regulates the movement itself so as to keep it as close as possible to the model.

The simplest example of mechanical retroaction is the thermostat, the device which regulates heating in the home. Once the thermostat has been set to a given temperature, a valve increases the flow of heat when the room temperature moves below the programmed temperature (–). As soon as the thermometer indicates that the desired temperature has been exceeded (+), the valve interrupts the flow of heat. In this way, the real temperature oscillates around the desired value, as is shown in Fig. 2.2.

This type of feedback is called *negative* in as much as it tends to stabilise the situation around the desired point by opposing the action of the system—if the thermostat provides too much heat, the valve closes, and if it furnishes too little heat, the valve opens.

On the contrary, *positive* feedback tends instead continually to increase the oscillation of the system, since each piece of information is interpreted as a signal to increase the action already being undertaken at that point even further. Thus an external control must be employed which can intervene in order to interrupt the entire circuit when a pre-set limit is reached. If no external control system is adopted, positive feedback always leads either to a continuous increase in the system resulting in its explosion, or to a continuous decrease resulting in its disintegration.

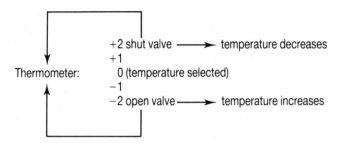

FIG. 2.2. The thermostat: the retroactive circuits direct action back from the action taken to the effects obtained.

Neurophysiology offers many examples of negative feedback in which behaviour tends to reduce progressively the discrepancy between the initial state and the final state aimed at. When we pick up a pen, or when we lift a glass to our lips, we are continuously checking that the sum of the flexions and extensions of the muscles assigned to carrying out the desired action is such as to ensure the hand does gradually approach the pen, and the cup does reach the lips little by little. Of course, control is not exerted at the level of the individual muscle, but the global action is governed by the negative feedback which prevents the hand moving beyond the pen and the glass finishing up against the teeth, as a result of an action which is excessively prolonged with respect to the goal.

From the cybernetic standpoint, the human system survives thanks to a large number of retroactive physiological and neural circuits which ensure that homeostasis is maintained. *Homeostasis* represents the optimum state of the system, from which the system tries to deviate as little as possible, reacting to variations in the environment by taking the appropriate countermeasures in order to re-establish the state of equilibrium. Internal homeostasis is achieved by means of a series of processes tending to maintain the organism in a state of well-being, by keeping all the physiological values within the narrow limits that must be respected if the organism is to remain healthy. For example, body temperature must not vary by more than half a degree centigrade, a higher degree of variation being pathological; indeed, a mere variation of 5° is incompatible with survival.

The introduction of the concept of feedback makes it possible to speak of goals not only with respect to living organisms but also with regard to machines, provided these are equipped with the means to exert control over their behaviour during performance, in order to modify it as appropriate. Machines equipped with retroactive circuits nevertheless exhibit teleological behaviour: they may possess and pursue specific goals. In particular, cybernetic automata possess sensory organs designed to detect variations in the environment, organs which can perform actions, and circuits that transmit information to each other. This led Wiener to believe that these machines are functionally equivalent to human beings, and that it is therefore possible to define laws which are valid for both. However, we have to distinguish between two types of intentionality, intrinsic and as–if, in Searle's terminology (*The rediscovery of the mind*, 1992). *Intrinsic intentionality* may be ascribed only to humans and certain other animals that possess real, intentional mental states. An example is:

[1]   I begin to be rather tired, and I need a rest.

As–if intentionality may be used only metaphorically: the system it is ascribed to presents no kind of mental phenomena. An example of this is:

[2]   The car engine begins to be rather tired, and it needs a rest.

The price paid for forgetting this distinction is that of extending mental properties to everything that exists in the world.

Information theory

The most important formal justification for the thesis that the human being and the machine are equivalent from the point of view of communication was provided by the mathematical theory of information. This theory was developed by Claude Shannon and Warren Weaver in *The mathematical theory of communication* (1949) and had an enormous impact on sciences in every human and mechanical discipline. Shannon, who had already advanced the fundamental theorems on which the theory was based, came from the Bell Telephone Laboratories, that is from the most important American telephone company attempting to discover the laws governing the transmission of information from one system to another. The main problem was to measure the quantity of information. A first approximation was achieved thanks to the concept of the maximum amount of information that may be transmitted via a given channel. It immediately became clear, however, that a series of disturbances in the communication channel, as well as unpredictable accidents that might have occurred during transmission, made the concept of maximum quantity one of dubious value.

It may be well to underline immediately the fact that we are speaking of quantity, and not of quality. The meaning of a message cannot be measured and does not therefore belong to the domain of the mathematical theory of communication. The amount of information contained in a message may be defined as a constant characteristic of that message, no matter how it is encoded, and it is independent both of the mode of transmission and of the systems of emission and reception. For there to be information, there must be a variation in the signal. The most elementary variation is constituted by the difference between presence and absence, between yes and no, on and off, zero and one.

We may now define the unitary unit of information, the *bit*, an abbreviation for *binary unit*. A bit is the amount of information required to distinguish between two equally likely alternatives. For example, to discover whether the person reading this page now is male or female, I would need one bit of information. Discriminating between four alternatives requires two bits. One bit halves the alternatives, the other bit enables you to choose between the two remaining possibilities. Similarly, selecting among eight alternatives requires three bits. Each bit reduces the alternatives by half (from eight to four to two to one).

If we call the information transmitted I, then we obtain:

[3] $I = \log_2$ (number of alternatives)

Formula [3] means that the informational content of a message corresponds to the logarithm to base 2 of the number of possible alternatives. Given that the probability of a message $P(m)$ is inversely proportional to the number of possible alternatives, we obtain

[4] $I = -\log_2 P(m)$

Formula [4] is tantamount to saying that the quantity of information contained in a message is measured by the negative logarithm to base 2 of the probability of the message itself.

Not only does the value of the information (I) vary from message to message, but it also varies within each message itself since the value of each component is not the same, tied as it is to the factor of novelty, that is to how much the receiver can foresee it. For example, in a word the first few letters are highly informative, the following letters decreasing progressively in value as they become more predictable. In "Chicago", the initial letter "C" conveys a greater amount of information than the final "o", which could be omitted without creating any real difficulty since it may be recovered quite easily. Shannon resorted to probability theory to calculate the different informative values of the parts of a message. In his basic theorem, the message is considered to be a sample extracted from a statistical body of messages which may be generated from a source, and its information content depends on how likely it is that a particular source will emit that particular signal.

This probabilistic type of approach is similar to the one adopted in statistical thermodynamics. Both make use of the concept of *entropy*, namely the amount of disorder within a given system. Entropy is taken by Shannon as being equivalent to information content. Thus, the greater the independence of the various components of a message, the greater is the amount of information contained in that message. A predictable and orderly message has a lower degree of entropy than an unpredictable, confused message in which each constituent is not clearly connected to the others. Entropy is at its height when all the messages, as well as all the constituents of those messages, are equally likely to occur.

Taking a wider view of this theory, a message may be considered to be a model distributed over time, one whose information content is proportional to the extent that the message constitutes news. News is defined as a message the receiver was not expecting. In other words, the greater the probability the information is new, the greater the amount of information a message contains. Vice versa, the more the message is expected, the less news the message contains. For example, a postcard extending Christmas greetings transmits very little information even when the message is extremely long. On the contrary, a two-line telegram which announces an unexpected inheritance contains a great deal of information, despite the brevity of the text. In this sense, the entropy of the receiving system decreases upon receipt of an informative message because this diminishes the uncertainty of the receiver with regard to the world.

Before examining the applications of information theory to the human sciences, brief reference must be made to an important area of research connected to this field. At the same time as Shannon and Weaver were proceeding with their work, developments were also taking place in cybernetics in the area of neurology, a field which has found a new lease of life today. Initially work in the two fields was carried out independently. Later, developments in the two areas took place along the same lines. In their book *A logical calculus of the ideas immanent in nervous activity* (1943), McCulloch and Pitts demonstrated that any function that can be computed may be realised by an appropriate network of idealised neurons. Idealised neurons are threshold detection elements whose properties may be reasonably attributed to real neurons.

The problem consists in discovering a general principle determining the way the neural network organises itself in its attempt to solve certain prototypical problems such as pattern recognition. This general principle could then constitute the basis of a special type of learning based on a single neuron. It must however be stressed that we are talking about an idealised neuron, one which is hypersimplified since its essential property is to transmit information of a binary type, 0 or 1, in line with the mathematical theory of communication expounded above. This type of approach—similar to connectionism—simulates the characteristics of the nervous tissue in an attempt to derive mental functioning, rather than theorising directly about the mental functions themselves.

The first neurological theory of learning was advanced by the Canadian psychologist Donald Hebb. In his book, *Organization of behavior* (1949), he advanced the hypothesis of cell assembly, thus enabling a connection to be established between neural organisation and psychological concepts. Cell assemblies are groups of idealised neurons within which information circulates. The creation of an assembly represents learning, and the flow of information constitutes memory. A particular assembly is responsible for a simple behavioural act, while a particular sequence of assemblies accounts for a complex behavioural act. This type of work inspired not only many projects in artificial intelligence but also the connectionist research paradigm (see Chapter 3) which wholeheartedly embraces the approach of the simulation of a neural network.

The difficulties

We have already spoken of the generosity with which cybernetics scholars spread their activity and their knowledge over various fields, from medicine to anthropology, from psychology to economics, from computer science to psychiatry, from engineering to linguistics. Such a wide scattering of energies over fields which at times shared little in common meant that a stable nucleus capable of acting as a coherent critical mass of research workers failed to emerge. This led to the cyberneticians—the masters of almost all the scientists of the following generation (*quorum ego*)—being paradoxically left without proselytes since the latter preferred to join sectors whose characteristics were more clearly defined.

In the physical and engineering sciences the results obtained by cybernetics remain milestones. Evaluating their impact upon the human sciences is, however, a more difficult business. Having nevertheless acknowledged the historical function of the splendid parabola of cybernetics, the crucial question today is whether the cybernetic approach is still valid in the human sciences.

Two fundamental criticisms may be levelled against cybernetics:

Quantitative vs. qualitative information. Conceptualising communication on the basis of the principles of entropy and information theory is not the correct approach in cognitive science. First of all, human communication is based on content (quality) and not on statistical probability (quantity). A message with a high degree of entropy has high informative value in quantitative terms, but this by no means signifies that it has a correspondingly high value in qualitative terms. One need only reflect on the fact that a list of 100 numbers in random order, one therefore in which each element in the message is independent of the others, has greater entropy than Shakespeare's "Devouring Time, blunt thou the lion's paw". The Shakespearian sonnet, with all its formal constraints of rhyme and structure, exhibits a high degree of predictability, a low degree of entropy, and, in contrast to the previous point, high qualitative value.

Shannon saw this quite clearly. He thus insisted that his theory could not be applied to the contents of the message. However, pseudotheories transferring laws that are valid only when applied to the formal quantitative field may be found applied to content. This is especially true of applications in the field of human communication, whether it be normal or pathological.

Second, from a psychological point of view, measuring the degree of probability of an incoming piece of information is not sufficient, even though this might represent an important datum, as in the example of the telegram announcing an inheritance. The crucial point instead is how meaningful the information is to the system. The expected probability is of no use here. What counts is the knowledge of the internal goals of the system.

Consider the following conversation:

[5]   A: Do you love me?
      B: Yes.
      A: Tell me you love me.
      B: I love you.
      A: But do you love me or do you really love me?

B's replies are not only 100% predictable, but they are also highly meaningful to A, much more so than a business conversation in which

A receives new and improbable information but which is of no affective importance. In this case, the tools furnished by cybernetics manage neither to capture the quality of the information nor to comprehend the message from the internal viewpoint of the system. The system remains a black box that cannot be explored except by analysing the actions it takes in response to change.

Homeostasis and self-perturbation.   The concept of homeostasis assumes that the system can be clearly separated from the environment from which it receives information and which it transforms through its actions, in its attempts to maintain a stable equilibrium. Equilibrium is a static state, one that must be defended against possible imbalances caused by external stimuli. It is as if the system were sitting on a chair taking great care not to move in order to avoid falling off the chair.

Instead, since human beings are complex systems, they are continuously interacting with the environment, as happens in a dance where each dancer responds to the movement of the other. Such reciprocal influence ruling out the possibility of stability, the life of the human system is characterised by constant change. The external environment does not furnish the stimuli for change directly. Rather, modifications come about as a result of the internal construction the system has created of the world. Equilibrium is a dynamic state which is being constantly adjusted to balance the forces at work. It is like riding a bicycle—if one stops one falls off the bicycle.

Wiener's homeostatic mind plays a passive role in relation to the world, in the same way that a telephone rings when someone dials a number. The mind seen as a complex system, however, actively creates its own world, just as a child creates its own world when it plays with a doll.

We have already seen which tools were employed by cybernetics. Like all tools, they may be put to certain uses and not to others. To utilise them in the analysis of human communication, going against the indications provided by the authors themselves, is like cutting a cake with a scythe—since the latter is used to cut hay, why not use it to cut cakes? Even Wiener was wrong on this point, since he was too caught up in the new and incredible perspectives opened up by his discipline to understand its structural limits. Cybernetics represents an attempt to find universal laws which apply to all fields, human as well as artificial. The connectionists, with their search after simple and general laws, were to adopt this approach. Classic cognitive science, instead, was to devote greater attention to complexity and detail, developing tools which were more deterministic than probabilistic.

Psychology

Psychology has contributed to cognitive science continuously throughout its existence. Here I shall refer to the most important historical contributions, outlining only the essential aspects of their development. Contemporary work in this field will be taken up in the second part of the book. Readers wishing to know more about the psychological origins of cognitive science, origins which are rooted basically in Gestalt psychology and cognitive psychology, may consult seminal texts like Donald Broadbent's *Perception and communication* (1958) and Ulric Neisser's *Cognitive psychology* (1967).

We have already mentioned the psychologist, Donald Hebb, who was inspired by the cyberneticians and who inspired research workers in computer science and artificial intelligence. But it was to be three other writers, who also drew their inspiration from the font of cybernetics, who were destined to write the book which would have the most enduring influence over the path which leads from psychology to cognitive science: Miller, Galanter, and Pribram.

The TOTE unit

A psycholinguist, George Miller, a mathematical psychologist, Eugene Galanter, and a neuropsychologist, Karl Pribram, worked together for only one year at the Center for Advanced Study in the Behavioral Sciences at Stanford, California. In this year they produced *Plans and the structure of behavior* (1960), without doubt the most important text in preparing the turning point in psychology, the switch to computational methodology, a change which was to culminate 17 years later in the birth of cognitive science.

Miller, Galanter, and Pribram began by recognising that behaviour is a process that is organised on several levels, all of which must be investigated before the meaning of a given action may be said to have been understood. A plan is defined as a hierarchical process which controls the order in which a sequence of operations must be carried out. Essentially, a plan for the organism is the equivalent of a computer program. It is what tells the organism second by second what it must do. In the hierarchical organisation of behaviour, molar units (strategies) are distinguished from molecular units (tactics). If a plan is to be executed successfully, the organism must exert effective control at all levels over the sequence of operations to be performed. Finally we have an image, namely all the knowledge that the organism has accumulated and organised about itself and its world. The components we have just discussed represent, in the opinion of the three authors, the constituents defining the basic unit of behaviour—a feedback circuit called TOTE (an acronym from Test-Operate-Test-Exit), the scheme of which is reproduced in Fig. 2.3.

The starting point is the system asking itself if the state of affairs is optimal. If the result of the test is negative, if, that is, the current situation does not correspond to the desired situation, then the organism will take steps to attempt to diminish the discrepancy between the present state and the target state until the latter is achieved. At this point, a positive response is received, thus permitting exit, which corresponds to a halt in the action undertaken.

Information and control flow along the arrows in the figure in the direction signalled, in line with the standpoint adopted by cybernetics. This represents the first explicit step towards the constitution of the paradigm that sees the human being as an information processing system. The best known example of a TOTE unit is the action of driving a nail into a wall, shown in Fig. 2.4.

The person continues hammering until the nail has reached the desired position, at which point the action ceases since the objective has been achieved. TOTE units may exhibit a hierarchical organisation reflecting the structure of the feedback circuit. In fact, Miller, Galanter, and Pribram believed the feedback circuit was the essential element in explaining human behaviour, at all levels. The importance of their thesis does not lie in the model, which is reductionist and clearly simplistic when invoked to explain cognitive processes, but in the methodological backwash the effects such a model created in psychology. The TOTE was readily translatable into a program, and the road to computer simulation of human behaviour could thus be taken.

Concepts and categories

The psychologist who comes closest in spirit to cognitive science is Jerome Bruner. His work ranges from perception and learning to applications of these subjects in the educational field, prompting and guiding anthropological and linguistic research, and inspiring

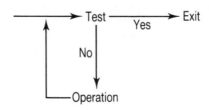

FIG. 2.3. The TOTE unit.

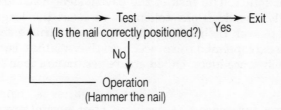

FIG. 2.4. The TOTE unit for driving in a nail.

computational programs based on his theories. Bruner was educated at Harvard and Oxford which, together with California, he claims in his fascinating autobiography (*In search of mind*, 1983) are the only places in the world where one can truly "do science". His earliest work was in the area of perception, where he attributed greater importance to individual expectations and social learning compared to the objective characteristics of the stimulus. In a series of famous experiments, Bruner demonstrated, for example, that poor children perceived images of coins as being larger than the images perceived by children from richer classes. Furthermore, the greater the value of the coin, the larger was the image perceived of the size of the coin. Analogously, Jewish children at the end of the war overestimated the dimensions of a swastika compared to non-Jewish children (see Chapter 6).

A study of thinking (1956), written together with Jacqueline Goodnow and George Austin, was his first attempt at developing an integrated theory on thought processes based on experimental research, the results of which were to be confirmed by his transcultural investigations employing a developmental approach in *Studies in cognitive growth* (1966). The main subject of this work is the study of the process of concept comprehension, and in particular of how strategies which are useful for learning new concepts are formed and applied. Bruner tackled the problem of concept acquisition by confuting the previous probabilistic approach which linked learning to a mechanical process of calculating, based on objectivising categories such as the degree of certainty of the forecast or the percentage of environmental stimuli hitting upon a subject.

On the contrary, in Bruner's view, the starting point was the intuition that a set of characteristics form a concept, following which an attempt is made to decompose it into its elementary constituents. Comprehension of a concept implies a chain of successive decisions in which the first decisions decrease the degree of freedom possible for the decisions that follow. The task of the psychologist is to bring to light, experimentally, the decisions taken, or at least the largest number possible, and to seek the regularities which can furnish the base from which may be extrapolated the procedures that enable a subject to form a concept. Such procedures, called strategies, have the following main aims:

1. To ensure the concept is acquired after a minimum number of encounters with pertinent cases.
2. To ensure that a concept will be attained with certainty, independently of the number of cases required to construct it.
3. To reduce the cognitive effort that must be borne by the process of inferencing and by the memory to a minimum, compatibly with the formation of the concept.
4. To reduce to a minimum the number of erroneous categorisations made before the concept is acquired.

Bruner constructed a series of ideal strategies which were perfectly rational according to logical criteria, and then began experimenting carefully in order to decide which of these strategies corresponded most closely to those employed by subjects. Over and above the specific results obtained as regards the strategies actually used, his merit lies in having demonstrated that the strategies utilised to form concepts may be isolated and described in a systematic fashion, both with regard to the particular objectives each strategy sets itself and with regard to the successive stages that are necessary for a concept to be attained. The attention he devoted to the subject of individual development, which was to bring him into contact with Piaget, enabled him to adopt a developmental approach in his experiments, following a path which cognitive science is only just beginning to tread—starting from the child and reaching the adult in one continuous line.

After Bruner's research, concept formation became one of the classic subjects for those working in the field of learning in artificial intelligence. In fact one of the very first cases of cognitive simulation was precisely Earl Hunt's brilliant reproduction on computer of the strategies identified by Bruner (Hunt, *Concept learning*, 1962). One important continuation of this subject is the research presently being carried out on the creation of prototypes. The concepts Bruner worked on are defined by means of a set of necessary and sufficient conditions which an example must conform to in order to be admitted to that category. For example, Swiss citizenship is an abstract legal concept which possesses a series of defining features enabling one to determine in every situation whether a person is a Swiss citizen or not. But the more a concept is concrete, the more difficult it becomes to find defining attributes which are exhaustive. The dictionary provides definitions with conjunctive categories such as

[6] bachelor = unmarried male

which is perfectly acceptable in normal conditions, but which causes a few problems when we ask if the pope may be considered a bachelor, since he is male and unmarried.

Everyday concepts cannot be dealt with in terms of necessary and sufficient conditions because none of the defining attributes are invariably present in each instance analysed. In "Natural categories" (1973), Eleanor Rosch demonstrated that categories are not logical entities definable by means of a simple set of essential characteristics which it is sufficient an instance possess, in order to be entitled to membership of the corresponding category. Instead natural categories are structured around prototypes (exemplary cases), non-prototypical members tending to be ordered in a scale which goes from high typicality to low family resemblance. For instance, not all birds may claim equal status membership of the category "birds": canaries and eagles are more typical than penguins and ostriches (Fig. 2.5).

FIG. 2.5. Which of these birds exhibits more of the features typifying the category "bird"?

Similarly, pianos and guitars are more satisfactory prototypes of the category "musical instrument" than are triangles and cymbals. But are cannon to be included in this category? Yet Tschaikovsky included them in the orchestra for his 1812 Overture!

The human being as information processor

That branch of cognitive psychology which has the greatest affinity with computer science reached its apex of computational enthusiasm in the constitution of the paradigm of the human being as an information processor. This approach, which has exercised great influence over both psychology and cognitive science, views the human being as a structure for coding information reaching it from the external world, processing this information internally, and then transmitting it again to the external world.

The metaphor of the computer has been applied quite literally. If the sole functions of the human organism are to receive, manipulate and retransmit information, then it is practically the equivalent of a calculating machine which has been inserted into a robot. That such functions are realised on a biological basis instead of an electronic one becomes a matter of small importance. Despite its extremist stance, this approach has become so widespread that it is impossible to quote even the main researchers. I will therefore cite a textbook which has come out in several editions and which thousands of psychology students have studied: *Human information processing* (1972) by Peter Lindsay and Donald Norman.

It was the promoters of the approach to the human being as an information processor that produced the change in mentality in psychology which enabled cognitive science to assert itself. Naturally the promoters then joined the ranks of the newly formed discipline. The research work of these scholars will be taken up in the second part of this book. Unquestionably, they exerted great importance in imposing cold mental processes as fundamental for the comprehension of the human being. However, certain gross oversimplifications and methodological naivety must not be accepted uncritically by cognitive science. There are basically three weak points: first, the reification of the concept of information (i.e. the computer's input and output consist of concrete, pre-packaged blocks, while a living system constructs the information it requires to interact with the environment); second, the interpretation of all mental activity as consisting exclusively of the manipulation of symbols; and third, the denial of the brain's role in sustaining any kind of mental activity. These points will be discussed in the second part of the book.

THE BIRTH OF COGNITIVE SCIENCE

The formal act of the birth of cognitive science may be considered to be the first issue of the journal *Cognitive Science*, which was first published in 1977, and which was to become the official organ of the society of the same name.

Its first birthday coincided with the completion of a report on the state of the art in cognitive science. Although this report was never published, it was distributed throughout the world in manuscript form, giving rise to considerable criticism and widespread interest. It had been commissioned by the Alfred P. Sloan Foundation which, on the basis of the report, decided to grant 15 million dollars (later increased to 20) for a period of 5 years (later extended to 7) for research projects in the field of cognitive science. With such a generous grant, the Sloan Foundation became the principal supporter of the new science, enabling it to take its first steps in relative ease, without having to worry excessively about possible applications.

The event which heralded the definitive consolidation of cognitive science as a respectable subject in the academic world was the first conference of the Cognitive Science Society, held at the University of California at San Diego in August, 1979. The conference was an international success, and remains in my memory as a succession of papers and discussions, each of which was more exciting and explosive than the preceding one. The main organiser of the event was Donald Norman, who prepared the conference with the utmost care. Each participant read and commented on the works of all the other contributors, thus enabling an interdisciplinary amalgamation to be achieved which had hitherto been deemed impossible. The final result was made possible by the collective enthusiasm of the scholars who formed the original group of the society, all of whom were fully aware that they were attempting to achieve something that was potentially revolutionary, and were consequently ready for once not to spare any effort. To name but some of those invited to speak, I will mention the neuroscientist Geschwind, the linguists Lakoff and Winograd, the philosopher Searle, the psychologists Johnson-Laird and Rumelhart, and the scholars of artificial intelligence Minsky, Newell, Schank and Simon.

Something extremely nebulous—the group of people who wished to be part of the group itself—had materialised, entering the reign of existence. From the La Jolla Conference of 1979 onwards, the problem became how to best modify something which indisputably existed already. This enabled all those scholars interested to bring about a change in their status since they could now legitimately claim they belonged to a prestigious community which convened regularly, which produced magazines and book series, and which finally had university departments and degree courses whose aim it was to produce the new, first real generation of cognitive scientists.

Very quickly, what might at an individual level have remained an unrealisable dream or the umpteenth version of a *petitio principii* towards a tired interdisciplinary stance, was transformed into an organic whole capable of growing into a new scientific paradigm to all effects and purposes. The philosopher of science, Thomas Kuhn, renowned for his concepts of normal science and revolutionary science (*The structure of scientific revolutions*, 1962), was invited to speak at one of the society's conferences. On this occasion, he underlined the revolutionary character of the movement, pointing out the crucial aspects that constituted a break with preceding tradition, the redefinition that had come about of the objectives of the discipline, and the methodological innovation that had been introduced. Kuhn himself, in his book *The essential tension* (1977), advances a well thought out definition of a scientific paradigm, in which he emphasises the importance of the scientific community in the transition phase from what is initially a transgression of the rules governing a certain area of research, to the acceptance of that transgression as the new norm in that area.

It is always difficult to record the breaking point of a tradition, since this revolves around the relationships among scientists and is thus more a subjective than an objective question. In other words, it is related more to the problem of what representation scientists create for themselves of the change that is imperceptibly taking place. Rendering the process explicit to the external world, by means of proclamations and declarations of war, is always a subsequent phase, one which takes place after things have happened, when one has the time to reflect over how, when, and why they came about. In the case of cognitive science, the resistance put up by the disciplines involved was more for the sake of form. It certainly cannot be compared to the clamorous battles that have taken place in the history of physics, as the ruptures produced by Galileo and by Einstein demonstrate. In particular, the subject that should have been threatened most by the rise of this new science, psychology, was in the 1970s in a situation of stalemate. Incompatible theories, such as psychoanalysis, Gestalt, behaviourism, cognitivism, vied for the favour of the experts in an unspoken climate of unease, the feeling being that those who had identified the important objectives irremediably lacked valid means, while those who had developed an adequate methodology were, on the other hand, incapable of setting themselves relevant goals. That the time was more than ripe is also indicated by the average age of the revolutionaries, all well over 40, therefore older than the stereotype lays down.

Cognitive science employed the same instrument that had allowed artificial intelligence to take off 20 years earlier: the computer. This time, however, the computer was also equipped with a special software, one that had been developed in the field of cognitive artificial intelligence (see Chapter 3); the methodology that was thus made available was that of simulation, and will be described in Chapter 4.

Simulation enabled each individual discipline not only to penetrate the sphere of the unobservable, typical of mental processes, but also to relate to all the other disciplines since they utilised the same methodology to validate their results. Computational methodology apart, each discipline continued to employ its specific techniques of investigation and analysis. Thus, linguists, anthropologists and neuroscientists maintained the techniques they had developed within their own particular subject, but now had one point in common—the fact that the theories they each produced singly had, at least in theory, to be capable of being reproduced in a program.

The fact that simulation was carried out on a computer improved things from the point of view of the newly-born community. Computers were complex and costly, and scholars of the human sciences were not particularly accustomed to such complex and expensive tools. In this way, it was as if cognitive scientists possessed a sort of secret weapon, one in which the spectacular results that could be obtained from its use were coupled with its considerable psychological effects. These effects were clearly more profound in those years in which the personal computer had not yet been invented, and computers were the trappings of a limited circle of users who were economically privileged and were prepared to devote a long period of study in order to be able to utilise such a promising tool.

Cognitive science takes up the traditional objectives of philosophy and psychology—studying, analysing, explaining everything that has to do with the mind. What brings about the change in the situation is the methodology of simulation. This guarantees a return to complete freedom with regard to the object under investigation while contemporaneously imposing rigid constraints on the methodology adopted. Obviously it continues to be easier to carry out research on problem solving than on the interpretation of dreams. Nevertheless, there are no theoretical prohibitions or limits to the research objectives scholars may set themselves.

If initially the definition of cognitive science was purely extensional, namely the list of disciplines that carried out investigation into the human being, subsequently a more correct definition was to emerge, based on the one hand on the objective, the study of humans, and on the other hand on the methodology, the computational method.

Figure 2.1 must therefore be seen as a dynamic model of the situation. The disciplines which make up cognitive science consciously rely increasingly on the methods employed by artificial intelligence, modifying their specific disciplinary status. The hexagon imports knowledge and exports methodology. As simulation becomes more widespread, this exchange becomes that much richer and growth continues.

CHAPTER THREE

Artificial intelligence

Artificial intelligence plays a fundamental role in cognitive science because it enables disciplines which would otherwise remain irremediably independent and detached bodies, despite their attempts at interdisciplinary integration, to achieve unity through a common computational methodology.

In this chapter we will tackle artificial intelligence first as an autonomous discipline and then as a discipline ancillary to cognitive science. We will discuss its essential characteristics and the limits that have so far been brought to light, both by scientists who champion the computational paradigm and by its detractors. Before reaching the heart of the matter, it will be useful to provide certain basic notions of computer science, without a knowledge of which many technical details would appear to be ambiguous, especially to the reader lacking a solid background in the subject. Those wishing to go beyond my brief account may refer to the excellent introduction to the subject by Ray Curnow and Susan Curran, *The Penguin computing book* (1983).

COMPUTER SCIENCE

Computer science deals with the construction of computers, and with the theories linked to their development and employment. Artificial intelligence is one of the disciplines which makes up computer science. We have already spoken about the beginnings of the mechanisation of work, both as regards manual labour—as in textiles—and as regards intellectual labour—as in the calculation of logarithms—precisely the field from which the computer evolved. But what is a computer exactly? It is a machine built to carry out mathematical operations and, in its more recent versions, logical operations as well.

The most commonly used computers today are digital computers with sequential architecture. The term *digital* (from the Latin *digitum*, figure) may be contrasted with analog to indicate that figures are used to carry out programmed operations. These figures constitute an enumerating system (for example counting up how many lines the page you are reading is made up of), or the quantification of a property through the use of a scale of values (for example measuring the length of that same page). Enumeration is discrete—the corresponding figure represents the property with absolute precision (the page contains exactly 43 lines). Quantification of a continuous property is, on the contrary, always approximate, and depends on the precision of the scale employed (the page may be 22.9cms long, or to be more precise 22.94cms long, or more accurately still 22.944cms long).

The opposite of digital is analog. The thermometer is a simple analog system. It is used to convert a change in temperature into a linear scale. The more heat increases, the more the mercury expands. Similarly, the analog computer relates variations in one type of size with variations in another, on the basis of a direct physical representation, that is without carrying out numerical calculations and without the final result being expressed in figures.

Sequential architecture refers to the fact that in a traditional computer operations are carried out one after another, in a sequence determined by the central control unit. No matter how fast the machine is, sequentiality inevitably causes a bottleneck in the system, since each operation must wait to begin until the previous operation has been concluded. Sequential architecture is also known as von Neumann architecture, after the name of the constructor of one of the earliest computers, the EDVAC, the design of which made explicit the principles later employed in digital computers.

An alternative approach which has recently come back into favour is the *parallel architecture* machine, which will be discussed later. These machines employ a high number of independent processing units, offering unquestionable advantages both in speed and in robustness.

One final important distinction is that between hardware and software. In outlining this difference, it is worth noting that this distinction is gradually becoming less clear as developments in computer science tend to include a series of logical operations in the physical structure of the computer. *Hardware* is the physical structure of the machine, the circuits it is composed of, circuits which tend to be suitable to any general purpose, on condition that it can be materially executed. *Software* is the sequence of instructions given to the machine in the form of a *program*. The execution of the program temporarily renders the machine specialised in the realisation of a specific goal.

The hardware is the physical support which is acted upon by the software to achieve a pre-established objective by means of the execution of an algorithm, which is specified by the program. But the distinction we have made vanishes into thin air as soon as we begin to consider machines specially built for a specific purpose. These machines comprise a series of instructions on how to act directly in the hardware in order to speed up computation and simplify the task of the programmer. Thus the latter no longer needs to specify the algorithms written into the physical structure of the machine—the computer already knows what to do in a broad number of cases provided for by the constructor.

Programming languages

The concrete realisation of the concept of software comes about by means of programs which specify the algorithms the machine must carry out. These programs are written using special programming languages. A programming language consists of a set of symbols and rules which act as an intermediary giving access to the *machine language*, that is the instructions written in the machine code setting off a specific succession of operations within the hardware. Originally programs were written in machine language, laying down every single step the computer had to carry out, including the specification of the memory space that was to be occupied by each given datum. Later, languages were developed that were capable of summarising a series of operations in one single instruction, thereby completely automating other instructions, such as the search for memory space, which were irrelevant to the programmer.

Among the various types of languages, those which come closest to machine language are called *assembly languages*. Each instruction corresponds to one machine operation. More efficient are the *low level languages* such as BASIC where a single instruction already stands for a number of operations. These languages are, however, repetitive, extremely detailed and, since virtually every step must be specified, those instructions which are quite complex become long and decidedly risky.

The most important programming languages in artificial intelligence are the *high level languages* which comprise symbols that call up a complex set of instructions, thus dispensing the programmer from having to specify the details of the operations. It is the machine itself which—in the usual impeccable manner peculiar to inanimate artefacts—sees to the translation of high-level instructions into operations in machine language. This allows the development of more complicated algorithms, which express in a concise fashion what the programmer wishes, thereby diminishing the possibility of error. The following are examples of high-level programming languages:

LISP: an acronym from List Processor, designed by John McCarthy, which is the most widely used language in artificial intelligence.

PROLOG: an acronym from Programming in Logic, invented in France and adopted by the Japanese for their fifth generation computer project.

The development of the computer

To carry out the calculations, the first generation of computers, developed in the 1940s, employed thermionic valves, capable of acting as amplifiers of electric power. The second generation, in the 1950s, utilised the technology of semi-conductors to introduce the transistor, which increased the reliability of the systems as well as reducing costs and size. Curnow and Curran estimate that there were about a hundred computers in operation in the whole world in the mid-1950s. In 1959 this figure had already risen to 4000 in the United States and to approximately 1000 in Europe.

The third generation, in the 1960s, replaced transistors with integrated circuits. These grouped together the functional equivalent of a large number of transistors, with their respective links, in a single *chip*, usually made of silicon. Thanks to the chip, within ten years the number of computers in use had risen from 10,000 to 100,000.

The fourth generation, which came into being in the 1970s and is still with us, substituted integrated circuits first with Large Scale Integration (LSI), which concentrated hundreds of silicon components in a single chip, and later with Very Large Scale Integration (VLSI), chips with thousands of components. Finally came Ultra Large Scale Integration (ULSI), which can place tens of thousands of components on a chip. Since this type of chip caused prices to slump and improved quality, it brought about the social and economic revolution which has raised the number of computers in use to many millions.

The fifth generation, which will be reached by the end of the century, should furnish us with computers that are a hundred to a thousand times more powerful than those in existence, with millions of components on each chip, modifying yet again the world balance in the computer field, and in all those sectors in which computers have by now become an integral part, from industry to administration, from health to education.

The other possibility for the fifth generation is parallel computers. Here, the architecture differs from a von Neumann computer which has only one central processor. The design of these machines is based on the human brain, with a high number of totally independent processing units, on a pattern inspired by neurons. The first computer of this type was the *connection machine* (Hillis, 1985). This employs 64,000 processors, though of a traditional type since they do not yet operate in complete autonomy—the real joys of massively parallel computers are still to come. Further technological advances, such as computers with a biological rather than a metallic base, are still at the design stage.

Each successive generation has reduced size and costs, and increased reliability and processing power. Progress has been so marked in the industrial field that its effects in the social sphere are great enough to justify the term "computer revolution".

The universal calculating machine

The universal calculating machine is an abstract concept, one which is useful in understanding the philosophy—and its relative simplicity—underlying the theory of computability in computer science and artificial intelligence.

In his paper "On computable numbers, with an application to the Entscheidungsproblem" (1936), the logician Alan Turing, one of the pioneers of computer science, describes a virtual machine which is to become famous as the *Turing machine*, or the universal machine. This ideal machine is an *automaton* that is operated by a tape, that has virtually no limits, and that contains symbols which at the same time constitute the input data, the memory, and the results of processing the data, or the output. The automaton can carry out three operations, and these are executed only on that portion of tape that is under the control of the finite state part of the machine (the control mechanism), that is to say the read/write head. The three operations are:

(a) reading the symbols that appear on the tape (for example, a "1" or a "0");

(b) replacing one symbol with another (for example, a 1 can be rewritten as a 0 or as a 1 again; similarly a 0 can become a 1 or can remain a 0);

(c) moving the tape one step forward or backward.

This type of machine is called a *finite state* machine in as much as it is always in a finite number of states, and these are totally determined by the initial departure state and by the symbol which it is reading. Each instruction specifies four points—the present state of the machine, the symbol being read, the action to be carried out and the state in which the machine will find itself once that action has been carried out. Turing demonstrates that if we add an unlimited tape to an automaton exhibiting extremely simple characteristics of the type listed above, then this machine can reproduce the behaviour of any other digital computer, no matter how complex or powerful the latter might be. This is how we obtain the so-called *universal machine*.

The essential procedures of the digital computer are comparable to those of a Turing machine. Even though other types of computer carry out the task in different ways, ways which are faster and more efficient, the result will nevertheless be the same. In other words, every digital computer is nothing but a universal machine with a few special features added. Therefore any operation carried out by any computer can be reproduced by a Turing machine. This conclusion establishes an important principle of equivalence among different types of machines.

A second fundamental point is to establish whether any function whatsoever may be computed. The problem consists in finding a procedure to decide whether a given problem is computable or not. Demonstrating whether a problem may be computed is the equivalent of establishing whether the problem may, in principle, be solved. The difficulty lies in the fact that no real method exists for showing that, given any function, it is computable. This does not rule out the possibility that it can be shown that certain specific functions may be computed. Indeed, this is precisely what we try to do in concrete cases. Furthermore, even if we have already established that a given function is computable, this function is generally partial, that is it is not necessarily defined with reference to all possible inputs, and no reliable method exists for determining whether computation will reach a successful conclusion independently of the input. Yet again, it can be demonstrated that a function is successfully computed in many specific cases. Stated differently, it is not always possible to determine beforehand whether the universal machine will continue the search for an impossible solution *ad infinitum*, impossible since the solution is not computable. Nor is it always possible to establish whether the machine will continue endlessly in its search for a possible solution of a function which is in principle computable.

Two conclusions may be drawn from the concept of the universal machine. The first is the theoretical equivalence of different types of machine, since all computers employ procedures which may be equated to the procedure utilised by a universal machine. The second is that it is not possible to establish *a priori* that any problem may be solved by means of computing procedures. Certain functions exist which—in theory or in practice—cannot be computed. Each problem must be faced experimentally. The universal machine can do an extraordinary number of things, despite its simplicity, but it cannot do everything.

HARD ARTIFICIAL INTELLIGENCE

Although artificial intelligence (AI) uses the computer as a system for processing information, it also enters the realm of activities which are deemed to be typically human, that is intelligent, and which traditional computer science had ignored. AI thus focuses its attention on areas such as automatic theorem proving, problem solving, chess and draughts. This is what is conventionally called *hard* AI. The objective of hard AI is, thus, to reproduce the results which the human being is capable of achieving in these areas, and to improve on them if possible by rendering them more rapid, more precise and error-proof. To avoid terminological problems, Fig. 3.1 lists the various names which have been used to refer to the two fundamental types of AI.

In order to clarify the methodological differences between the two types of AI, we may begin with what is considered to be the ideal validation test for a theory of hard artificial intelligence. The test was introduced by Alan Turing in "Computing machinery and intelligence" (1950) to answer the question of how we can decide if a machine is intelligent. Given the methodological importance and the complexity of the argumentation, I will quote his famous imitation game, nowadays known as The Turing test, verbatim.

> I propose to consider the question, "Can machines think?" This should begin with definitions of the meaning of the terms "machine" and "think". The definition might be framed so as to reflect so far as possible the normal use of the words, but this attitude is dangerous. If the meaning of the words "machine" and "think" are to be found by examining how they are commonly used it is difficult to

AI	AI
Hard	Soft
Technology	Cognitive
	Simulation of behaviour
	Cognitive simulation

FIG. 3.1. Synonyms employed in the definition of the two fundamental types of artificial intelligence.

escape the conclusion that the meaning and the answer to the question, "Can machines think?" is to be sought in a statistical survey such as a Gallup poll. But this is absurd. Instead of attempting such a definition I shall replace the question by another, which is closely related to it and is expressed in relatively unambiguous words.

The new form of the problem can be described in terms of a game which we call the "imitation game". It is played with three people, a man (A), a woman (B), and an interrogator (C) who may be of either sex. The interrogator stays in a room apart from the other two. The object of the game for the interrogator is to determine which of the other two is the man and which is the woman. He knows them by labels X and Y, and at the end of the game he says either "X is A and Y is B" or "X is B and Y is A". The interrogator is allowed to put questions to A and B thus:

C: Will X please tell me the length of his or her hair?

Now suppose X is actually A, then A must answer. It is A's object in the game to try and cause C to make the wrong identification. His answer might therefore be:

"My hair is shingled, and the longest strands are about nine inches long". ... The object of the game for the third player (B) is to help the interrogator. The best strategy for her is probably to give truthful answers. She can add such things such as "I am the woman, don't listen to him!" to her answers, but it will avail nothing as the man can make similar remarks.

We now ask the question, "What will happen when a machine takes the part of A in this game?". Will the interrogator decide wrongly as often when the game is played like this as he does when the game is played between a man and a woman? These questions replace our original, "Can machines think?".

... Some other advantages of the proposed criterion may be shown up by the specimen questions and answers. Thus:

Q: Please write me a sonnet on the subject of the Forth Bridge.

A: Count me out on this one. I never could write poetry.

Q: Add 34957 to 70764.

A: (Pause about 30 seconds and then give as answer) 105621.

Q: Do you play chess?

A: Yes.

Q: I have K at my K1, and no other pieces. You have only K at K6 and R at R1. It is your move. What do you play?
A: (After a pause of 15 seconds) R-R8 mate.

The question and answer method seems to be suitable for introducing almost any one of the fields of human endeavour that we wish to include …

The game may perhaps be criticised on the ground that the odds are weighted too heavily against the machine. If the man were to try and pretend to be the machine he would clearly make a very poor showing. He would be given away at once by slowness and inaccuracy in arithmetic. May not machines carry out something which ought to be described as thinking but which is very different from what a man does? This objection is a very strong one, but at least we can say that if, nevertheless, a machine can be constructed to play the imitation game satisfactorily, we need not be troubled by this objection.

The great innovation lies in indicating a simpler method, one that can be realised at least in principle, for the validation of a model. Just compare the replies, suggests Turing, and we will have an objective criterion by which to judge the effectiveness of a model. It must be noted that one of the replies furnished by the computer is deliberately wrong in order to lay a subtle trap to deceive the judge—the result of the sum is incorrect.

However, carrying out Turing's test presents two types of problem. One is of a practical nature, the other is methodological. The concrete problem stems from the fact that the machine cannot reply to any question whatsoever the judge may wish to set it, without any sort of limit being imposed on the question. Were this possible, it would mean that artificial intelligence had already solved practically all the problems it is dealing with, from mundane knowledge to the model of the partner implicit in every conversation. The methodological problem is that even if the machine were to furnish satisfactory answers and were therefore to obtain amazing success in terms of hard artificial intelligence, this would not constitute proof with regard to soft artificial intelligence, since it would leave unanswered the question of the procedure followed by the machine in furnishing the replies.

Among the variants the Turing test has given rise to, I shall recall the weakened version advanced by Abelson (1968), and Gunderson's (1964) sarcastic proposal, which in asking the question "Can rocks imitate humans?" requires a subject to distinguish between a pain caused by a toe stepped on by a human foot and a pain caused by a falling rock.

Early results

I shall now trace the development of hard artificial intelligence. The reader who wishes to go more deeply into the argument may consult the manual by Rich and Knight, *Artificial intelligence* (1991). The mandatory references for specialists remain, despite the enormous problems of updating besetting all compilations, the monumental three-volume work by Barr, Cohen, and Feigenbaum, *The handbook of artificial intelligence* (1982), and the two-volume *Encyclopedia of artificial intelligence* (1990), edited by Shapiro. From a historical point of view, the best work available is *Machines who think* (1979) by Pamela McCorduck, who drew up a series of individual scientific biographies of the founding fathers of this field.

The official date of birth of artificial intelligence is deemed to be 1956, when the Rockefeller Foundation financed a summer seminar at Dartmouth College, Hanover, a quiet town in New Hampshire with a brilliant scientific tradition. The main organiser was John McCarthy, who at that time taught mathematics at Dartmouth College itself, not yet having left it to colonise computer science in the far West by founding the Artificial Intelligence Laboratory at Stanford University in California, which was to become one of the major centres in the world in this field. The participants invited to the seminar included Marvin Minsky, who was later to found another important AI centre, the MIT Artificial Intelligence Laboratory, in the other Cambridge, the one near Boston; Ray Solomonoff and Oliver Selfridge from MIT itself; Nathaniel Rochester and Arthur Samuel from IBM; Claude Shannon from Bell Telephone Laboratories; and finally the pair that were to create the third historic AI centre, the CMU Laboratory of Artificial Intelligence (Carnegie Mellon University, in Pittsburgh, Pennsylvania), Allen Newell and Herbert Simon.

In the project sent to the Rockefeller Foundation, McCarthy wrote: "The seminar on artificial intelligence must work on the presupposition that each aspect of learning and any other characteristic of intelligence can, in principle, be described in such a precise manner that it may be simulated by a machine".

During the seminar the two basic tendencies that were to dominate artificial intelligence in the next 20 years emerged. The most important representatives of one approach were McCarthy (Stanford) and Minsky (MIT), while the other approach was championed by Newell and Simon (CMU). It was immediately obvious that the conjecture expressed by McCarthy in his project offered those scientists who wished to proceed along those lines only two alternatives. Either the hypothesis had to be made that if one wished to reproduce any aspect whatsoever of intelligence then it was essential to concentrate attention on the intelligent system by self-definition, namely the human being, or this type of anthropocentrism had to be totally abandoned in favour of a completely independent approach. McCarthy and Minsky, great disparagers of psychology, set themselves an objective which was the equivalent of inventing the wheel: despite the fact that the wheel does not originate from the way the human being moves, nevertheless it certainly constitutes a means of movement on the earth that is of similar, and sometimes even superior, efficiency compared to the traditional two legs.

Newell and Simon adopted the opposite approach and tried continually to find inspiration for their programs in the human being's mode of proceeding. This makes it difficult to classify those authors in terms of the dichotomy hard/soft AI, since it is often difficult to understand whether their principal interest lies in the human being or in the machine. Both of these research approaches have yielded excellent results. Often they have interweaved and fertilised each other reciprocally, thanks to researchers who passed from one tendency to another. Hence it is pointless to take up sides today in a panorama which has many more shades of grey, yet at the same time is clearer from a methodological point of view than what it was in the 1950s.

The Dartmouth seminar discussed methods, ideas, research programs—no results were available at the time. The sole exception was the presentation made by Newell and Simon (1956) of a version of Logic Theorist which had only partially been implemented on computer. Since Logic Theorist is the first example of an artificial intelligence program in the strict sense of the term, it deserves a brief description. It adopts a heuristic method of theorem proving and not an exhaustive method in its search for solutions to problems. In this, it is explicitly inspired by the work of George Polya, a mathematician who was interested both in *heuristics*, a term employed to designate general methods for the solution of formal problems, and in pedagogic techniques which are effective in teaching students heuristics themselves (*How to solve it*, 1945).

Newell and Simon called the mechanically exhaustive method of proving mathematical theorems by making a complete search for all the possible combinations of the symbols given the *British Museum algorithm*. The name derives from the analogy with a situation in which one waits patiently until a number of monkeys that have been placed in front of typewriters press the keys in random fashion and reproduce all the books contained in the British Museum by pure chance. This strategy may be contrasted with the *heuristic method*, which consists in proving a theorem by trying to guess the nature of the solution and therefore demonstrating that the "bet" one has made is correct. Newell and Simon's program generated plausible bets on the basis of extremely general criteria, and therefore attempted to prove the validity of what Polya, referring to human beings, called an intuition of the solution. The ethologist Richard Dawkins (1985) has calculated that the probability of a monkey writing out a phrase from Shakespeare's *Hamlet* without making a single mistake is one in 2728. The sample phrase consists of 28 keystrokes: "Methinks it is like a weasel" (Hamlet to Polonius). A monkey with a typewriter with an English keyboard (26 letters plus the space bar) may range over a possible combination of 27 (keys) raised to the power of 28 (the number of letters and blank spaces in the phrase). If the monkey is not particularly fortunate in its random selection, then it can take longer than the time that has passed from the creation of the universe to the present. This prediction of the time required is drastically reduced—to a few hours—if instead, as a result of a precise intuition, the monkey adopts an adequate selection criterion, such as keeping each variation which brings the phrase nearer to the original model.

The field in which Logic Theorist was applied was the demonstration of the theorems taken from *Principia mathematica* by Whitehead and Russell (1910). The truly ingenious part of the *Principia*, in which the fundamental correlations between logic and mathematics are established, does not lie in the proofs, but in having singled out the essential problems, the theorems on which attention was to be concentrated. In spite of their relative simplicity, the proofs are beyond the capacities of a university student of mathematics, and therefore *a fortiori* also of the vast majority of the population. However, Logic Theorist managed to prove 38 of the 52 theorems in the second chapter, and even found a proof of theorem 2.85 that was shorter and more elegant than that provided in *Principia*. One of Simon's famous anecdotes (in *Models of my life*, 1991) tells that while Bertrand Russell wrote him a letter of congratulation, the *Journal of Symbolic Logic*, showing less humour and foresight, refused to publish the article containing the proof in which Logic Theorist figured as co-author.

A further achievement of the program was that to realise it the authors, with the aid of a great computer expert, Cliff Shaw, invented an original programming tool, called *list processing*, which turned out to be crucial to the development of symbolic as opposed to numerical programming. Symbolic programming, which is useful when dealing not only with numbers but above all with symbols representing abstract concepts, is essential to any project in artificial intelligence. This first success led to the development of LISP, which we have seen is the language favoured by traditional artificial intelligence.

Logic Theorist has many interesting features. First, as a typical example of Newell and Simon's approach to artificial intelligence, it is characterised by a strong interest in the way human beings deal with problems. The authors have always held the view that the real significance of Logic Theorist lies in how it generates proof, and not in the actual proof that it produces. From this standpoint it may be considered a first approximation to a model simulating human behaviour. In fact, in their following works Newell and Simon increasingly adopted the viewpoint of the human being as an information processor. Their ambition was to produce something interesting both for the scholars of the human being and for those of the computer. The latter function was brilliantly realised by the list processing technique, which constitutes the second important feature of Logic Theorist, one which belongs to the technological side of the matter.

Finally the significance of the original proof of theorem 2.85 must not be underestimated, since it was the first time the experimental importance of programming had been concretely demonstrated. In fact, while it is theoretically true that the behaviour of the computer may be predicted, in as much as it is deterministically dominated by the program, as soon as the program becomes slightly complicated this predictability vanishes. The proof of theorem 2.85 had not been explicitly programmed, nor had it been predicted in any way by the authors. What emerges quite clearly here is that any simulation of complex activities must not be taken as a routine exercise, one in which the results are obvious right from the beginning, but as an experiment where the behaviour of the computer and the relevant output can only be predicted along very general lines—the details can never be foreseen.

To continue with a detailed account of the history of artificial intelligence goes beyond the scope of this book. However, in an attempt to identify at least the initial trends as precisely as possible, I shall refer to what for many years was the only book, properly speaking, in the field, the text which all the second generation scholars used to approach the subject. In 1963 Edward Feigenbaum and Julian Feldman collected the papers of those scientists who were active in the then limited area of artificial intelligence, and published them in a book entitled *Computers and thought*, destined to be a scientific best seller for 30 years.

This historic text is divided into three main parts: artificial intelligence, the simulation of cognitive processes, and a review of approaches with an annotated bibliography. The third part of the book consists of Minsky's "Steps toward artificial intelligence", which furnished the entire discipline not only with a comprehensive general framework but also with an agenda of what had to be done, a list that remained valid for two decades. Together with the illustrated bibliography, compiled by Minsky himself, the article summarises in 70 pages the roots of AI and the fundamental developmental paths researchers of the first generation were to follow courageously, albeit with varying degrees of success.

Being the excellent interpreter of the scientific trend of his time that he was, Minsky deemed five areas to be of fundamental importance: research, since even when we do not know how to solve a problem we can program a machine to look for a solution; pattern recognition techniques, which enable the identification of the essential characteristics of an object in order to be able to operate on them with the maximum efficiency; learning, thanks to which efficiency is increased further, directing research in line with previous experience; planning, which enables blind research to be replaced with more appropriate exploration, based on the analysis of the given situation; finally induction, in order to acquire more global concepts regarding the means by which to obtain more intelligent behaviour from a machine.

Everything is seen as a problem to be solved, and this becomes the principal research paradigm at this time—finding solutions to an ever-increasing number of problems, finding common solutions to various classes of problems, building techniques to improve the solutions themselves. The domain of abstraction and the tendency towards generalisation are the hallmarks of the first 20 years of artificial intelligence, and the distinctive feature of the best works of the period.

The methodological and theoretical foundations of the discipline are explored in the carefully structured collection, *The foundations of artificial intelligence* (1990), edited by Derek Partridge and Yorick Wilks.

Expert systems

Some of the results obtained by hard artificial intelligence are so important socially that nothing can justify ignoring them, not even the fact that the scholar may be interested only in the study of the human being, and not the computer. Furthermore, given that the precise objective of expert systems is to take over the place of humans in certain tasks, it is important to underline the concept that such systems belong to the field of computer science, not to the science of the mind. We shall now examine the nature of such systems and of what value they may be.

An expert system may be defined as a computer program which is capable of achieving performances in its sphere of competence that are qualitatively equal, if not superior, to those of human experts in that field. The distinguishing trait of expert systems, compared to the other domains of artificial intelligence, consists in the choice of sphere of competence—this must be easy to delimit, of considerable importance, and difficult to acquire.

First, for the system to be realised, it must be possible to separate it clearly from the rest of knowledge. If this is not so, then the solution of the problem being tackled would require the program to contain all the general knowledge about the world possessed by a human being. It is thus possible to construct an expert system that deals with a very complicated problem indeed, provided it falls within a restricted area, one which is as independent as possible from the rest of knowledge. For example, it makes sense to build an expert system for the interpretation of cardiac images obtained by angiocardiography using radionuclides. Angiocardiography is a sophisticated technique employed by nuclear medicine that consists in injecting a radioactive substance into the blood, in order to identify the morphology and functioning of cardiac chambers thanks to a special detection instrument. On the contrary, it would make no sense to build an expert system for the diagnosis of schizophrenia, since this presupposes a general knowledge of psychiatry and society that is so vast and interdependent as to render the undertaking absolutely impossible.

Second, the sphere of application must be of vital scientific, economic or social importance, otherwise building an expert system would not be worthwhile since it would then be of no interest to anyone. This criterion justifies the construction of an expert system aiming at discovering mineral deposits on the basis of geological data, while a system aiming at perfecting the method of dressing salad would be of no value.

The final criterion is that the acquisition of the competence necessary to carry out the pre-set task be both difficult and out of the ordinary. If this were not the case then producing this competence on the computer would be superfluous. This is the reason why extremely specialised areas are chosen, where human experts are rare, and therefore their knowledge is precious and not readily available to society. An expert system for the diagnosis of prenatal heart malformations would be precious, whereas an expert system for the diagnosis of exanthematic illnesses would be uneconomical. Neonatal cardiologists are rare and concentrated in a few, advanced hospitals, while our grandmothers can recognise the measles.

The knowledge that an expert system possesses must be extracted from one or more human experts in the field, and formalised by technicians of artificial intelligence called knowledge engineers, people who are capable on the one hand of getting human experts to make both the things they know and how they use them explicit, and on the other hand of putting all of this knowledge into a computer program. The goal is that later, other human users, who are experts neither in that particular field nor in computer science, can avail themselves of this specialised competence.

The two most well-known expert systems are:

1. DENDRAL and its successor, Meta-DENDRAL (Buchanan & Feigenbaum, 1978). Starting from data on a chemical compound obtained by means of mass spectrography, these systems automatically determine the formula of the molecular structure of the compound.
2. PROSPECTOR (Duda, Gaschnig, & Hart, 1979). This system acts as a geological consultancy service. It is employed to decide whether an area may be fruitful from a mineral point of view, and therefore whether it is worth continuing exploration by drilling the ground.

One field that has always seemed to be particularly favourable terrain for the development of expert systems is that of medicine. There are some ultra-specialised sectors where knowledge is evolving continuously, rendering it difficult to keep up to date. Nevertheless, these areas are useful for the non-specialist and possibly even vital for the patient. The idea of providing the general practitioner with a set of computer specialists is intellectually attractive, socially useful and makes economic sense. Even if in actual practice the idea has turned out to be more difficult to realise than had been imagined, the project deserves careful consideration. Assisted in this way, the doctor could make a relatively safe diagnosis (provided the clinical picture were sufficiently unambiguous), or exclude a particular pathology, or, in case of doubt, refer the patient to a human specialist who, while still retaining the aid of machine, would be better able than it to decide on the nature of the illness.

MYCIN (Shortliffe, 1976) is an expert system for the diagnosis of infectious diseases and the first to have been clinically tested. Since this system has risen to the rank of paradigm and its constituents are common to the vast majority of expert systems, we shall examine these constituents in detail. We have seen that the essential feature of an expert system consists in possessing such a large quantity of knowledge, over a limited domain, as to render it competitive with a human expert. The procedures for managing knowledge are not overly sophisticated, and are normally based on production rules, which will be described in Chapter 5. An example of a rule employed by MYCIN is presented in Fig. 3.2.

It is important that the rules which manipulate knowledge be as independent of each other as possible, for two reasons:

	1.	The infection that requires therapy is meningitis
	2.	Organisms were not seen in the strain of the culture
IF	3.	The type of infection is bacterial
	4.	The patient does not have a head injury defect, and
	5.	The age of the patient is between 15 and 55

THEN   The organisms that might be causing the infection are diplococcus-pneumoniae (0.75) and neisseria-meningitides (0.74).

FIG. 3.2. A typical rule from MYCIN; the pertinent probabilities of the diagnoses are in brackets.

Errors: Some rules might turn out to be wrong or might otherwise provoke consequences in the course of processing so as to render necessary their elimination or their substitution with different and more functional rules. The effects of the elimination or replacement of a rule must be as circumscribed as possible to avoid affecting contiguous rules.

Probabilistic reasoning: The inferential engine employed in expert systems is based on Bayes' theorem for dealing with probabilistic knowledge. This theorem requires that each piece of information utilised be independent of the others.

A further reason in favour of the maximum simplicity in the inferential mechanisms is that often the human user interacting with the expert system is not satisfied simply with getting the final answer but wishes to know how the program came up with that result. This need for transparency is easier to meet if it is possible to trace a line of reasoning founded on a series of linear steps rather than if the procedures are algorithmically so intricate that they are meaningless to the human user.

In order to achieve the best interface possible between the user and MYCIN, a system called TEIRESIAS was developed (Davis, 1980). The essential function of this system is to explain to the doctor who wants such information how and why MYCIN reached a given conclusion, whether it is a final or intermediate conclusion (namely, the diagnosis or the request for further information and tests). Unfortunately TEIRESIAS came into such strong conflict with the rigidity typical of inferential production rules that its efficacy was drastically limited. Furthermore, the lack of flexible interaction between user and computer frustrated some of the optimistic forecasts of the medical world, which often became disillusioned with a system that was perfectly capable of carrying out diagnoses but incapable of providing explanations and of discussing any line of reasoning failing to correspond with its own.

What makes expert systems useful also makes them difficult to build. The cooperation between the knowledge engineer and the human expert is arduous, and the programs require a long period of testing and adjustment before they become reliable. It is, however, possible to design a system having only the inferential engine and the interface with the user, which serves as the basic framework for the construction of other expert systems—the so-called *shell*, an example of which is EXPERT (Weiss & Kulikowski, 1979). But the efficiency of a shell (to which must obviously be added the specific knowledge base relevant to the applicational domain) is unsatisfactory. An optimum solution would be to build systems that can learn autonomously, but we are still so far from comprehending learning mechanisms that this option remains a dream rather than a concrete path.

Once the experimental stage has been overcome, the practical value of expert systems is potentially immense. The performance of DENDRAL equals that of a chemist; PROSPECTOR has contributed to the discovery of an important deposit of molybdenum; and other systems are being currently used in research and in industry. With regard to the medical field, the picture is more complicated: MYCIN delivers performances on a par with those of a specialist in infectious diseases, and well above those of a general practitioner. Nevertheless, it has never been employed routinely in any clinic once it passed the experimental phase at Stanford University Medical School. Of course, the prime and obvious reason is the fact that it requires a powerful dedicated computer, and not every hospital can afford such an investment, let alone the general practitioner for whom it is out of the question. Less immediate reasons include responsibility for the diagnosis and keeping the system up to date.

The relative simplicity of the structure rules out the possibility of an expert system being equipped with a valid general knowledge of the world and of the rules that govern it. The expert system is totally incapable of making even the most elementary deduction beyond its limited domain of knowledge. It is a kind of caricature of a human superspecialist, the union of a profound specialistic competence and an equally profound general ignorance. The most undesirable of the consequences of this situation is the possibility of error due to the expert system's complete lack of common sense, coupled with the impossibility of making it reason, that is of providing it with information that would enable it to modify significantly its performance in real time on a specific case. An expert system furnishes a solution and the justification for this solution, but it cannot behave like a colleague called in for a consultation, with whom one may discuss the diagnosis and the therapy—it does not cooperate: it gives orders. The responsibility for the action taken lies therefore in the hands of the doctor, it cannot be accepted by the system. Nor are doctors willing to run risks unless they are absolutely certain of what they are doing—the motto *primum non nocere* has been taught for five hundred years. How can one trust an *idiot savant*, who gives orders without answering to any possible consequences and with whom one cannot hold a debate?

Expert critiquing systems, devised by Perry Miller (1986), offer an interesting alternative. The most important example of this type of system is ATTENDING. While the classic expert system typically furnishes a diagnosis and a therapy, a critiquing system levels a set of criticisms at the therapeutic plan formulated by the doctor. In other words, the system says what it thinks about the strategy the user has decided upon, its behaviour thus being more in line with its role as a consultant without responsibility. ATTENDING furnishes for example criticism on the treatment of patients suffering from essential hypertension. After having input information on the case, the doctor proposes treatment with hypotensive drugs. The system discusses the advantages and disadvantages of the proposal, and where necessary suggests alternatives.

Updating the systems constitutes a second weak point in expert systems. Rapid progress is constantly being made in clinical knowledge, above all in superspecialistic areas, whereas expert systems do not move once they are implemented on computer. Adding new knowledge is no easy matter since this must be fully integrated with pre-existing information in the program. Furthermore, the computer is incapable of the most important form of learning, that connected to examining the consequences caused by decisions taken. That is, the computer does not know what effects the therapy it suggested has had, nor can it modify the way it works in response to the success or failure of the therapy. Until expert systems develop an efficient method of adding information and an intelligent method of learning from experience, their use in the medical field will not increase, for reasons which can in no way be criticised.

The points discussed in the previous paragraph are extremely important from the social and economic point of view, but they are of no specific interest at the cognitive level. Expert systems represent the most extreme case of focusing on the results, explicitly renouncing the adoption of human procedures. As we shall see when we deal with education, this type of choice means that expert systems are not particularly suited to didactic interaction with human beings. For example, MYCIN cannot be employed in teaching medicine for the very reason that its procedures are totally alien to the way a medical student thinks. The point still remains, however, that once the problems mentioned above have been solved, expert systems could become one of the most powerful tools that artificial intelligence can hope to make available to individual users, and thus bring about widespread social benefits, as soon as the user can exploit a valid advisory instrument, namely one of proven experience and capable of respecting its subordinate role. This product too obeys the laws of computer science, a science which is extraordinarily brilliant at creating servants, but which is a failure each time it proposes machines to take the role of the protagonist, thereby inverting the correct relationship between humans and machines. The reader wishing to investigate the subject further may consult Peter Jackson's *Introduction to expert systems* (1990).

SOFT ARTIFICIAL INTELLIGENCE

The great new advance for the human sciences is the entry on the scene of soft artificial intelligence. This second type of artificial intelligence uses the human system as its constant reference point. It may be defined as the reproduction of human mental processes, and consequently of the corresponding behaviour, by means of the computer. It must be noted that the fundamental feature of simulation is that the procedures are far more important than the final results. While in hard AI the method adopted to reach the goal makes no difference (it does not matter what colour the cat is, as long as it catches the mouse), in soft AI it is precisely the intermediate steps that constitute the most important object of investigation. The criterion of success consists in achieving the greatest similarity possible—namely equivalence—with the corresponding human method of processing the datum. These rigid imitational constraints must produce step by step a final result which is identical, and not superior, to that achieved by humans. A soft AI program must reproduce any errors the human subject engaged in a given task may make. On the contrary, in hard AI programs which aim at success, such errors are to be avoided. Figure 3.3 summarises the fundamental features that distinguish hard from soft artificial intelligence.

According to the computational approach, the essential part of mental activity may be reproduced by a computer program. This offers two interesting aspects for cognitive modelling. First, it makes explicit the mechanism through which mental states and processes determine the behaviour of the system. This point implies an important methodological constraint. The generation of a behaviour being investigated must be reproduced by an effective procedure which can be

	Hard AI	Soft AI
Results (compared to humans)	Better	Same
Procedures (compared to humans)	No constraint but efficiency required	Identical
Errors	To be avoided	To be reproduced (if present in humans)

FIG. 3.3. Criteria distinguishing hard from soft artificial intelligence.

implemented on the computer. Thus scholars speak of a constructive explanation, in as much as demonstrating that a theory does explain a given mental activity requires the building of an artefact capable of exhibiting the corresponding behaviour.

Expressing psychological theories of cognition in the form of actual procedures means utilising tools and methods which were almost exclusively developed within the area investigated by artificial intelligence. When this method achieves a high degree of integration with the experimental method, it becomes a powerful validation technique for theories of the mind. Let us examine what justifications are offered for this stance.

First, the reconstruction of the object being analysed is sufficient to guarantee that the theory which generated it is totally explicit and non-contradictory. The characteristic of being fully explicit is not the exclusive property of simulation, but holds for any type of formal approach whatever the method employed may be, such as axiomatisation in mathematics and in physics. Computer implementation is tantamount to axiomatisation of the theory. The lack of contradictions is not guaranteed in absolute terms. This criterion is valid in principle, but in practice it depends on how the implemented model has been derived. As we shall see shortly, a program does not reproduce the complete theory, but a partial model of the theory. If, therefore, a model fails to cover all the essential points of a theory, it may be that the theory has significant internal contradictions that the implemented model does not reveal. This criterion must therefore be reformulated, in the sense that the non-contradictory nature of the theory is guaranteed only if the model adopted has been correctly derived and may be considered as being complete.

Second, a program implies that the theory deals with events that may be reproduced, events that may be dealt with using a scientific method.

True, the mental activity being studied cannot be observed. However, managing to reconstruct it completely, from the input to the intermediate processing operations to the final output, brings us within the bounds of the operational sphere. A computer program constitutes for a theory of the mind such strong evidence in its favour as to enrich its heuristic power to the point that only the most significant experimental tests could equal it.

Third, a program that has been implemented may be falsified, as laid down by Popperian epistemology. Stated differently, it may eventually be proved that the behaviour produced by the program differs from the expectations; performance does not coincide with the predictions made, thus confuting the theory. This also satisfies the requirement prescribed by the epistemologist Imre Lakatos (1978) concerning the productivity of the research program. On this view, a research program should be progressive, that is, capable of generating ideas, experiments and theories which continually increase our understanding of the phenomena being studied. In Lakatos' terms, it is not a question of falsifying a theory because it is unsatisfactory in an absolute sense, but of establishing whether the theory is more or less satisfactory compared to the other approaches it is competing with. Building a program is such an intellectual challenge, such a conspicuous revealer of weaknesses that it obliges the scholar to carry out continual revisions of the theory, exploring it in progressively greater depth.

Fourth, a program is an efficient and fast generator of predictions. Testing a complex theory in an experimental situation implies executing an extremely long series of predictions, of calculations, which, if not automated, would require an enormous quantity of time, with many inevitable mistakes. The computer rapidly generates all the forecasts that the theory makes during the test, as well as all the variants that may enter the researcher's mind during experimentation. This guarantees both an accurate check is carried out on the data, and an occasion is available to modify the theory itself on the basis of the evidence gathered.

Finally, a program is capable of making time pass quickly, simulating developmental processes that cannot be observed in humans. This feature is of special interest in a developmental approach to the study of mind, in which it is important to reproduce not only a function, but also to discover how that function originated in the child and evolved from initial to steady state. Similarly, this approach opens up the possibility in psychopathology of simulating etiopathogenetic processes that lead to the various mental disturbances, tracing the illness from causes to symptoms.

We shall now analyse the essence of the validation procedure for a theory, according to simulative methodology as adopted by cognitive science, an outline of which is provided in Fig. 3.4. Starting from the top, a model is derived from a theory (*Step 1*). This step has many consequences, since by definition the importance of a theory always outweighs that of any of its empirical embodiments. A model is a particular interpretation of a theory—the construction of a model entails an act of choice, of selecting the parts deemed significant, ignoring (at least this time, in this model) other parts of the theory. The very same theory may therefore give rise to more than one model, and the models derived may not all have the same degree of success; indeed, it is probable that model M1 illustrates the theory better than does M2, while it is perhaps not as empirically applicable in a given area as is model M3. This is why the failure of one specific model never of itself entails the abandoning of the corresponding theory. Much more evidence must be accumulated against a theory before it has to be discarded.

It is always only a model that can be embodied by a program (*Step 2*). Thus a program is worth what the model is worth. Since many models can be derived from a single theory, that same theory can give rise to more than one program. The program contains further arbitrary elements compared to the model. In fact, with regard to all those aspects which are not constrained by the theory, programmers are free to act as they see fit, solving the sub-problems in the manner most congenial to them. In principle, the characteristics of the implementation should not exert a significant influence over the simulation, since their insertion

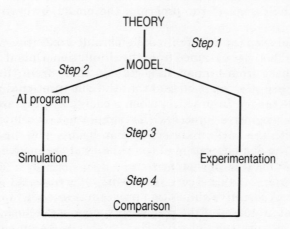

FIG. 3.4. The validation of a theory in cognitive science.

in the program comes about at a level of detail that is irrelevant from a theoretical point of view. In practice, the program is already conditioned by the choice of language, which in turn conditions the simulation. The tool is not neutral with respect to the use made of it, nor to the product obtained from it.

One of the consequences of what has been said is that no one specific implementation of a model can by itself prove a theory. It may well turn out that the program does not represent all the essential points of the theory. Where this occurs, further evidence in favour of the theory will be necessary and, as is normally the case, where to stop is a convention among scientists. It is a question of convincing the community, not of proving something in a mathematical or logical sense. Simulation involves a problem of *a priori* choice as regards what is fundamental and what is not, in a given process. Here in fact the act of defining the problem to be solved is made fully explicit by the decision concerning what has to be included since it is essential and what, instead, may be omitted since it is incidental—nothing may be considered as being obvious.

The following step (*Step 3*) consists in testing the predictions obtained from the theory by experimenting both on human subjects and with computers. In the latter case, the output of the program corresponds to the performances simulated. It is at this point that the theory may begin to be falsified, by both types of experiment. If the results obtained from the subjects in the experiment differ from those predicted, the theory must be able to explain the discrepancy. In parallel fashion, if the performance of the program does not correspond to expectations, this challenges the design of the program, the model, or even the theory itself.

In the final step (*Step 4*) the results obtained from the experiments are compared to those obtained from the simulation. Only if the two sets of results, those from human subjects and those from the computer, correspond perfectly, can it be said that the computational method has validated the theory. In practice, when a computer program simulates a particular cognitive process, the simulation obtained must be compared with the performance of human beings who carried out the same task. The demonstration of the validity of the model depends on the results of this type of comparison.

How programs may be correlated with experimental results is a crucial point. A cognitive simulation program is intended to correspond to the mental processes of the people carrying out the task, not simply to the final results they obtain. The criteria to be employed in the comparison between human and artificial performances must therefore be based both on the output and on the procedures.

Identity of the final results

The first criterion that constrains the building of a simulation program is that of identity of results. The program must reproduce, at the target level of detail, the performance obtained by humans in the same task. In the case of cognitive models, identical results is a necessary but not sufficient condition, and must be reinforced by more severe constraints. In other words, another criterion has to be introduced dealing with how the results are generated.

Equivalence of procedures

Since procedures are not directly observable, the problem arises of deciding how the equivalence criterion invoked above may be checked. Zenon Pylyshyn (*Computation and cognition*, 1984) suggests taking into consideration the intermediate processing states and response times.

Relevant knowledge states. The program must simulate knowledge states corresponding to intermediate processing stages, stages occurring between the initial and final stages and believed to be significant from the theoretical standpoint. If, for example, it has been found experimentally that human subjects solve a problem in three steps, then the program must reproduce the two intermediate steps too, and not just the solution that is the third and final step.

The proportionality of computation time. The program's computing time must be proportional to the processing time required by human subjects to carry out a given operation or a significant part of that operation. For example, if it is found experimentally that human subjects are slower at comprehending phrases in the passive form than equivalent phrases in the active form, a program for the comprehension of natural language must employ a proportionately greater amount of time to understand phrases in the passive. Proportionality in computation time must be applied not only to the final results but also to the relevant intermediate stages. The time taken to execute each of the intermediate processing stages identified has to be calculated and virtually reproduced in the program. It must be noted that computation time is not the same as the real time taken by the computer, the latter depending on factors that are extraneous to the model, such as the programming language and the type of computer actually employed. What is defined instead is a simulated virtual time, calculated and updated in accordance with criteria that will be derived from the model.

All the steps defined in Fig. 3.4 run in both directions. At each phase of the chain that links the theory to the program, there is feedback towards the top which tends to modify the theory, the model and the program on the basis of the experimental results obtained and of the comparisons made between experimentation and computation. This spiral procedure enables qualitatively different types of information to be used, from information furnished by the experiments or by the observation of individual cases, to information stemming from and highlighted by the implementation phase. All this information, however, converges while checking and perfecting the initial theory.

Thanks to its potential for reconstructing unobservable phenomena, soft artificial intelligence is to be classed as a tool for analysing procedures (which is what counts in the study of mind) and not simply observable behaviour.

CONNECTIONISM

Connectionism, the latest development in artificial intelligence, has already brought about significant changes in computer science, and it is a revolutionary turning point for cognitive simulation. No distinction will be made here between terms like connectionism, neural networks and parallel distributed processing. To know more about this subject, the reader may consult the two-volume work by David Rumelhart and James McClelland, *Parallel distributed processing* (1986), which is completed by *Explorations in parallel distributed processing* (McClelland & Rumelhart, 1988), a handbook of programs and exercises. A more introductory work is the one by William Bechtel and Adele Abrahamsen (1991), *Connectionism and the mind*. Among the many journals devoted to connectionism, at least two must be quoted: *Neural Computation* and *Neural Networks*.

Classic artificial intelligence employs computers with the so-called von Neumann architecture. This type of architecture distinguishes between a static memory containing the information, i.e. the data, and the central processing unit which operates on a limited set of active data. No matter how fast this central processing unit is made to operate, it represents an inevitable bottleneck, obliging memory and all those functions that are temporarily inactive to lie idle.

In contrast, the connectionist approach assumes that information processing comes about through the interaction of a large number of simple processing entities called *units*, each of which sends excitatory or inhibitory signals to the other units to which it is connected. Referring explicitly to the architecture of the human brain, these units simulate the neurons, assumed to be structures which are not particularly intelligent when taken singly, but which together constitute an intelligent entity. In the connectionist approach, intelligence is an *emergent property* of the global system. Even if each single processor has limited capacity, the network connecting the processors will exhibit impressive cerebral power. The high number of processing units that are active contemporaneously amply compensate for the low speed of each single operation, thereby eliminating the bottleneck of von Neumann architecture. This ensures a spectacular gain from the computer science point of view, as was well demonstrated by parallel computers once they were put on the market.

Not only would this type of computer be fast, but it would also be much more reliable. Since more than one unit can carry out the same task, the machine would continue working even if a part of it were to go wrong. While a classic architecture computer grinds to a halt if the central processing unit is not working properly, a parallel architecture computer merely produces a worse performance—it may lose in precision, but it does not break down. This gradual worsening in performance is very similar to what happens to the human brain where the death of the individual neuron is not even noticed, precisely because its function is taken over by other neurons. The possibility of *lesioning* a neural network, that is of simulating the ageing process and other neurological pathologies, is an important characteristic of neural networks. It allows the construction of a bridge which it had never before been possible to build between psychological and neuropsychological models.

The structure of an extremely simple neural network, in three layers, is illustrated in Fig. 3.5. The input units represent the entry to the system. They are activated by stimuli that come from outside the network. The hidden units receive signals from the input units and transmit signals to the output units by means of the connections available to the system. Finally, the output units constitute the exit from the system. Activating them produces the response to the stimulus received at the beginning of the process. The accuracy of the response furnished by the system, represented by the state of the output units, varies. At the outset, the response is generally casual. Once the network has been subjected to many experiences, however, its units modify the weight attributed to the activating or inhibiting signals transmitted through the connections with the other units, until optimal performance is achieved.

In other terms, networks organise themselves, redistributing activation and inhibition until a stable and effective response is realised: In a sense, they learn how to behave.

Changes in processing or knowledge structures may be realised in three ways:

Output units

Hidden units

Input units

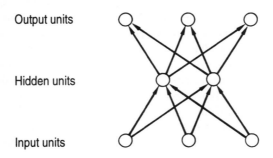

FIG. 3.5. The structure of a three-layer neural network.

1. Development of new connections.
2. Loss of existing connections.
3. Modifications in the strength of already existing connections.

Since cases 1 and 2 may be considered as extreme possibilities of case 3, connectionism has concentrated mainly on the latter. The two most important learning strategies are called *Hebb's rule* and *back propagation*. The former is a case of *unsupervised learning*, in that the final output of the network is not defined *a priori*. On the contrary, back propagation is a case of *supervised learning*, because the desired final output is defined before the training stage.

Hebb's rule is inspired by the work of Donald Hebb, already cited in Chapter 2. In its simplest formulation it may be stated as: When unit A and unit B are simultaneously excited, the strength of the connection between them increases. Hebb's proposal was meant to simulate learning at the neuronal level. In order to do this, a suitable method is to strengthen the connections between the units of the network only when there is a simultaneous activation both of presynaptic and postsynaptic elements.

The second widely adopted learning strategy employed in connectionism is called back propagation, that is error is propagated backwards. It is achieved by presenting the network both with the input stimulus and the desired scheme for the activation of the output units. The network reacts to the stimulus and compares its response with the response it has been furnished as a prototypical paradigm. Errors are then calculated. An error is the difference between the excitation or inhibition of a unit and the optimal value assigned to that unit by the prototypical paradigm. The size of the error is employed to modify the weights of the excitations reaching the unit under consideration. The attempt to reduce gradually the error is propagated backwards, influencing all the internal layers of the network. The method of back propagation of the error leaves the responsibility to the network on how to learn to furnish the desired performance. In other words, the only way the network may be influenced is through its pre-established connections and the weights given initially to the connections between the units. In contrast to the classic structures of artificial intelligence, a neural network is not programmed to carry out a task. The ability of the network to organise itself corresponds to the ability to self-program.

The fact that a neural network can find the best representation of a problem in such an autonomous manner has given rise to both great optimism and some debate. How can we tell that the constraint of reproducing identical procedures has in fact been respected if we have no idea of how the network reached a solution? From the standpoint of simulation, interest centres more on understanding the mechanisms by which humans solve certain problems, and less on the actual solutions to those given problems. In connectionism, we do not know *how* the computer has organised itself, and this leaves some scholars of the human being perplexed. It is easier for the researchers of artificial intelligence, for whom *how* is not a pertinent question, to allow themselves to be drawn into the net of the results to be obtained with parallel machines.

As we shall see in Chapter 5, connectionism excludes internal representations of concepts from the picture. This is a salutary antidote to the ingenuities of cognitivism. Inside the human brain, as inside a neural network, the only thing to be found is a more or less active set of neurons. There are certainly no representations, in any form whatsoever. The neurons realise what is mentally construed as images, words and forms, but none of these entities have physical existence. This should be common knowledge to all cognitive science, but in practice, by dint of speaking of knowledge representations, a confusion has gradually been created between the fact that knowledge is represented symbolically in programs and the misleading supposition that the symbols themselves exist physically in the brain. The fact that one can mentally construe the symbol conveying a question "?" does not mean that a question mark takes physical form in one's brain. The connectionist approach has done us a useful service in ridding us of such ingenuous and misleading representationalism.

One proposal advanced by numerous cognitive scientists may be termed *limited connectionism*, and will be gone into in detail in Chapter 4. For an example, the reader may refer to the controversial paper by Michael McCloskey: "Network and theories: The place of connectionism in cognitive science" (1991). This stance posits connectionism at the low levels of cognition (*microcognition*) and seriality at the high levels of cognition (*macrocognition*). It is the level of the global path that has to be described serially, necessarily using symbols. Once a theory of the mind has specified the entire procedure, each single step can be realised in parallel. When how to proceed is known, the computer follows the same procedure serially. At each step, if the question of how an operation is executed is of no importance to the level of detail identified, the machine carries out the transition in the parallel mode.

Limited connectionism maintains the methodological promise of parallelism, namely that of achieving total simulation, both of the organ as well as of the function. The mind, which is the function of the organ called the brain, is reproduced by means of programs running on a computer respecting cerebral architecture, these programs being composed of neural networks. The objective is to avoid the reductionist risk of identifying the mind solely with the neurological or artificial structures which realise it. The philosophical foundations of connectionism are explored by Clark in *Microcognition: Philosophy, cognitive science, and parallel distributed processing* (1989).

THE LIMITS OF ARTIFICIAL INTELLIGENCE

We shall now examine what may be considered to be the theoretical limits of simulation, a subject which has roused the passions of scientists and philosophers from the time when Lady Lovelace used to discuss what Babbage's analytical engine could and could not have done. We shall distinguish between technological limits and methodological limits.

With regard to technological limitations, these are determined by the constraints imposed by the hardware and software available. For example, an efficient artificial vision program should process the input in parallel mode, and the intermediate stages in serial mode, in order to achieve results comparable to those of humans. But at present no computer is capable of working in parallel mode in real time. Overcoming these technological constraints, by means of massively parallel architecture computers, will enable all those theories that postulate parallel data processing to develop in a more satisfactory fashion. Generally speaking, the technological limit represents a clear barrier, one which is contingent and which is continually being reduced.

The intrinsic limits of the computational method are a subject of greater interest, since they are theoretically insurmountable. The cause of all such limits may be traced back to the classic definition common to both the human being and the computer, considered as systems for the processing of symbols that convey meaning (Newell & Simon, 1976). We shall therefore discuss the similarities and differences between human brains and artificial brains in terms of their respective capacities for manipulating symbols referring to entities existing in the world, and for relating to the world itself.

Symbols and meanings

I shall introduce the problem, which may be defined as that of symbol and meaning, through the work of the scholar who has brought the problem most clearly into focus, John Searle, thanks to his mental experiment on the Chinese room (*Minds, brains and programs*, 1980). I quote the abstract of the paper, where the objectives and the basic assumptions are specified:

> This article may be considered as an attempt to explore the consequences deriving from two statements:
> (a) intentionality in human beings (and in animals) is a product of causal features inherent in the brain. I believe this to be an empirical fact regarding actual causal relations between mental processes and the brain. It means quite simply that certain processes of the brain are sufficient for intentionality;
> (b) implementing a computer program is never in itself a sufficient condition of intentionality. The main assumption in this article aims to establish this affirmation. I shall achieve this by demonstrating how a human agent could implement a program and nevertheless not have the pertinent intentionality.

Bearing in mind that for Searle intentionality is a characteristic of certain mental states by which they are directed towards the world or concern objects and states of affairs in the world (hence in his view beliefs and wishes are intentional, while anxiety and depression are not), I shall summarise his *Gedankenexperiment* and the conclusions that may be drawn from it.

The author's point of departure is the programs developed by Schank and Abelson (1977) to simulate human capacity to understand stories. Their work is based on a particular stereotyped representation of knowledge, termed "script" (see Chapter 5), which allows a computer to answer questions on a story, even if those questions are set on aspects which are not specifically mentioned in the story itself. Let us take the following story as an example: "A man went into a restaurant and ordered a hamburger. When the hamburger arrived it was all burnt and the man left the restaurant in a rage, without paying or leaving a tip". The computer is capable of giving questions such as "Did the man eat the hamburger?" the same answers that a human being would give, thanks to the stereotyped knowledge it possesses on restaurants and to the rules that permit it to tackle this type of question. Searle sets out to prove that programs of the Schank and Abelson type cannot really understand the story and that they do not explain a human being's ability to comprehend stories and provide answers to pertinent questions.

Searle develops his argument by replacing the computer with a man who speaks English, but who does not know Chinese. To this person, ideograms are meaningless doodles. If this person is shut up in a room and denied contact with the outside world, and receives a story in English, then he is obviously capable of answering questions on the story because he has understood it. Five structured sets of materials are now introduced:

A: a first batch of Chinese writing (a script);
B: a second batch of Chinese writing (a story);
C: a third batch of Chinese writing (questions);
D: a series of symbols in Chinese (answers to those questions);
P: instructions in English (a program).

Suppose now that the man in the room is given three batches of writing, A, B and C, as well as a packet, P, containing a set of rules in English on how to manipulate the ideograms, which to him, we must remember, are meaningless symbols. The rules furnished in P enable him to correlate A to B, and consequently, if he follows the appropriate instructions, to produce certain types of symbols in Chinese (D), related to the symbols in the third packet of sheets (C). If the instructions in P are sufficiently detailed, and if the man follows them diligently, the result will be that his replies, that is the series of symbols D that he has produced, will be indistinguishable from those a mother tongue Chinese speaker could have provided. From the point of view of an external observer, the answers the man provides in English and those he furnishes in Chinese are equally acceptable. But in contrast to the answers provided to the English text, in the case of the answers furnished to the Chinese text, the person produced the replies manipulating symbols without interpreting them. That is to say, he behaved like a computer program, carrying out predetermined operations on formally specified elements.

The point about Searle's demonstration is that the computer limits itself to manipulating symbols—it is unable to interpret them.

Therefore it never has access to their meaning in the world. It is in this sense that the computer can simulate intentional states but cannot possess them. It can behave "as if" it possessed mental states, but it cannot act directly on the world, as do human beings, since the latter have causal powers, claims Searle, in so far as they are materially equipped with a brain. His polemic is directed against theoretical extremists such as McCarthy and Minsky, who hold that the computer really can possess mental states, and not simply simulate them.

To tell the truth, the mental experiment is of interest to both types of artificial intelligence—hard and soft—since it throws light on the absolute limits of the computational approach as a whole. The conclusion that can be drawn is that unless the computer can also reproduce the biological substratum of the brain in toto, then it cannot generate complete imitations, but only partial simulations of mental processes, a point subscribed to by connectionists. In fact, the machine does not have access to meanings, but only to their symbols, whereas when human beings manipulate symbols they simultaneously have access to the meanings those symbols have in the world.

If we take it as proven that a program is necessarily partial, that is it can simulate only a specific mental function, and that it cannot comprehend the meaning of the symbols it manipulates, we can now ask ourselves if computational reconstruction of the process under examination can at least be complete and correct. This question brings us directly onto the terrain of complexity.

Complexity is one of the crucial points for all sciences studying the human being, and is therefore treated in the first section of the next chapter, where I deal with the epistemological problems that cognitive science has to face. For the impatient reader I may, however, anticipate that the answer is no—we cannot reach both completeness and correctness.

CHAPTER FOUR

The study of the mind

In this chapter we shall analyse the epistemological bases on which cognitive science is founded and which enable us to penetrate the realm of the mind, and hence of the unobservable *par excellence*, employing a method that is as powerful as those adopted by the physical sciences. Then we shall deal with a problem which a science of the mind must, if not solve, at least bear in mind, namely that of the complexity of the human being as a system, and of how that complexity may be handled. Finally we shall discuss the different types of theory that may be produced within the computational paradigm, comparing their defects and their merits. For a more detailed discussion, the reader may refer to Margaret Boden's *Computer models of mind* (1988).

EPISTEMOLOGICAL PROBLEMS:
FROM OBSERVATION TO REPRODUCTION

In the 1950s psychology began its attempt to become master again of its natural objective—the study of mental processes—an objective that positivists and behaviourists had worked together to outlaw. The justification that lay behind this ostracism was the "fact" that only the input and output may be kept under observation—the mental processes that connect the two escape observation; hence, behavioural scientists had to limit their work to the analysis of stimuli and responses. To overcome this objection, it was necessary to adopt a methodology which would satisfy the scientific criteria of the time and get round the obstacle of the impossibility of observing mental processes directly. It was held that once the epistemological problem had been solved, a specific tool that would have satisfied the methodological requirements would certainly have been invented sooner or later. This is in fact what happened, thanks to computer science which produced the necessary tool, and to artificial intelligence which developed the complementary procedures to make it work. Before the philosophical and technical revolution that I am about to describe took place, no way existed of investigating the mind that was both scientifically acceptable and not so crude (as was psychometry, for example) as to invalidate the entire study.

Artificial intelligence allowed the essential operational criterion, namely the reconstruction of the object under investigation, to be applied also to mental life. It is true that the mental function being examined still remained unobservable, but managing to reproduce it completely—from the input, through the intermediate processing stages, to the final output—permitted the introduction of scientific operational criteria that could replace the positivist criteria based essentially on observability. However, it must be borne in mind that a thought or a piece of knowledge cannot be subjected to direct scrutiny—they cannot be photographed, recorded or preserved in any way: they may only be validated by observables.

The behaviourist solution—to infer thought and knowledge from behaviour—is wrong, since the same act may be motivated by opposite intentions: John the Apostle kissed Jesus during the Last Supper, as did Judas a few hours later in Gethsemane. The relationships between mental states and behaviour cannot be claimed to be linear. Quite the contrary—they are so rich that any attempt to simplify such relationships is doomed to failure the minute the laboratory situation is left behind and an attempt is made to explain an action in a natural environment.

The opposite choice—based on introspection—is methodologically weak: there are too many mental states to which we do not have access to make the approach of observing what we think or feel ourselves credible. Centuries of reflection have demonstrated that seeing inside ourselves is a task beset with insurmountable obstacles: I can only find what I expect to find, nor am I able to check the accuracy of my intuitions.

What must be noted is that with regard to the topic under discussion, the fact that important insights into oneself can be achieved thanks to an introspective procedure is of no importance. The real point is that there is no way of rendering the knowledge thus gained acceptable from a scientific standpoint, without employing a different method to validate it. Epistemological correctness guarantees the road one has followed as a scientist is correct. But science is not the sole method of acquiring knowledge—poets and novelists captured fundamental aspects of life long before scientists, and with a degree of communicative efficacy that is totally alien to science.

An even more powerful means of acquiring knowledge is living life, as long as possible, and in an interesting and critical manner. The wise man of the village, the wandering Zen monk, the doyen of a corporation, the master of an art—these are all people who possess extensive knowledge of the world and of human beings, although they are not scientists. Their knowledge is not to be taken as antithetical to scientific knowledge merely because it belongs to a different domain. It has been acquired in an alternative fashion, one which is as integrated as the scientific method is analytical. An enlightening discussion of these two ways of acquiring knowledge is offered by Francisco Varela, Evan Thompson, and Eleanor Rosch in *The embodied mind: Cognitive science and human experience* (1991).

Operational epistemology abandons the observable, to concentrate on the reproducible. The Nobel laureate for physics, Percy Bridgman, states in *The logic of modern physics* (1927) that a given phenomenon has been comprehended when its essential features can be reproduced. Observation becomes one of the many techniques supporting reproduction, together with intuition, the reading of books, experimentation. Anything which is useful and practicable in the given context may be used. For example, it can be said that a person has understood what a petrol engine car is when he is able to build one, and not if he untiringly observes them for years or if he reproduces part of its behaviour by making horses draw a cart.

The ultimate aim of operational methodology is to trace every type of activity—physical, linguistic or mental—back to the mental operations on which they are based. The operational definition of anything consists in the set of instructions that enables anyone, in standard conditions, to reconstruct that particular phenomenon. A recipe is a perfect example of an operational analysis. A person really has understood what a *paté de foie gras* is when he is able to prepare it, provided he has all the ingredients available and he has the basic capacities necessary to carry out the operation.

According to the golden rule of operational methodology, defining any object or entity that is to be analysed is equal to being able to both decompose and recompose that entity. Precisely the same rule applies both to physical objects and to mental entities. Having established reconstructability as the fundamental feature of operational methodology, it is possible to define further operational criteria that are useful for shedding greater light on the procedure that the activity known as science is called upon to respect. Two of these features are essential:

Intersubjectivity.   Anybody must be able to repeat an experiment. The latter must therefore not be tied to special, personal, non-communicable characteristics of the experimenter. This is what separates science from magic. In the latter activity, it is precisely the personal activities of the agent that makes phenomena occur. The difference between the best cook ever, Gualtiero Marchesi, and the wizard Merlin is that one can learn to cook from the former (even if the great cook prefers showing off his dishes, the results he produces, rather than his recipes, the procedures he employs), whereas the latter intrinsically possesses the ability to make what he desires happen, thus rendering any attempt at imitation on the part of third parties vain.

Replication.   It must be possible to repeat an experiment at will, varying all the conditions that have not been explicitly set as essential. In science, no effect can exist which cannot be explained—a result that is obtained once in a while and that cannot be duplicated in another laboratory is discarded as an experimental artifice.

A third criterion frequently figures in the literature: research must concentrate on one single unknown variable at a time, whereas the other parameters must be assumed to be uninfluential, with the proviso that in their turn they too will become the object of future investigation. This spiral procedure enables the scientist to remain on the solid ground of entities that may be checked, and certainly represents normal practice in any well-conducted series of experiments.

Yet, this is more a strategy than an unrelinquishable principle, one reason being that some current methods of investigating nature are founded on complex explanatory schemes which are aimed at maintaining a plurality of interaction among variables. The chemist and Nobel laureate, Ilya Prigogine (*From being to becoming: Time and complexity in the physical sciences*, 1978), does away with the illusion of the simplicity of the microscopic level, and proposes an approach to nature that does not seek to reduce everything that is observable to a few fundamental laws, but rather attempts to produce laws that genuinely reflect the degree of complexity exhibited by the phenomena they interpret.

Not all of the philosophical problems concerning reproducibility of objects, above all mental objects, have been solved. In particular, how to decide when a given phenomenon has been reproduced remains an open question. An important clarification is provided by the philosophical school known as *functionalism*. In the opinion of one of its eminent representatives, Hilary Putnam (*Mind, language and reality,* 1975), the physical structure that produces a given phenomenon is irrelevant, provided no significant differences exist between the original and its reproduction at the level of analysis preselected for the comparison. As a result, the fact that the process is carried out by a system which has a biological substratum such as the brain (*wetware*), or by a system that has an electronic substratum such as the computer (*hardware*), does not automatically imply that there is some intrinsic difference between the two processes. This point, it must be stressed, refers only to the level of detail chosen beforehand as significant for the reproduction of the phenomenon. It is obvious that if the level of analysis changes, differences which cannot be eliminated will be found, the most banal being that the computer uses electric energy while the brain uses biochemical energy. But if, for example, we are studying two-dimensional vision, the important point is that the machine be able to recognise drawings and photographs, not that it be dry while the brain is moist.

As a result of having directed attention to the correct level at which a comparison must take place, the functionalist approach swept aside many of the ingenuous objections launched at artificial intelligence, in particular those playing upon the immutable difference between the nervous substratum and an artificial substratum, whether this be electric, electronic, or, as might even happen at some future date, biochemical. But the solution advanced opens up new difficulties, which we shall return to after having spoken about the modularity of the mind.

METHODOLOGICAL PROBLEMS: COMPLEXITY AND DEVELOPMENT

One problem that has historically beset the analysis of the mind is the much-debated question of whether the mind can be decomposed into different faculties. The most outspoken supporter of the hypothesis of the modularity of the mind is the philosopher, Jerry Fodor (*The modularity of mind*, 1983). The definition of a *cognitive module*—that is to say an independent cognitive component—is based on five points:

(a) it is domain specific, that is its operations do not cross over several functional domains;

(b) it is innately specified, that is its structures do not depend on any form of learning;

(c) its virtual architecture maps directly onto its neural implementation, that is it is not assembled by putting together more elementary sub-processes;

(d) it is hardwired, that is it is associated with specific, localised and elaborately structured neural systems, and not with relatively equipotential neural mechanisms, each of which can perform different tasks;

(e) it is computationally autonomous, that is it does not share any of its resources (memory, attention and so forth) with any other cognitive subsystem.

All scholars agree as to the existence of cognitive modules for the execution of specific local tasks requiring only low-level processing, such as the initial stages of perception. The debate flares up as soon as the notion of modules dedicated to non-elementary functions is hypothesised. The assumption that the mind is modular enables the researcher to investigate it by separating it into distinct parts, without this decomposition leading to the loss of any of its characteristics or essential properties. If instead we accept the opposite thesis, according to which breaking up an entity that works in a fully integrated manner involves a significant loss, we find ourselves at a dead end. In fact, when we wish to avoid eliminating any significant component, the overall working of the mind becomes far too complex to be tackled.

Two alternatives are possible:

First, unacceptable simplifications may be made by separating what should remain unified to be comprehended. An example of this would be building theories of perception that are independent of theories of emotion and vice versa. The result: laboratory models that are valid only under special circumstances when it really is possible to minimise the influences that other functions presumably exert over the function under investigation. Structurally, this type of theory lacks a solid ethological base. It is therefore unsuited to explaining the corresponding human behaviour in natural conditions, for the obvious reason that the object it studied (i.e. laboratory behaviour) was different. Such theories do not therefore constitute, to return to our previous example, theories on human perception, since real theories on this subject would be obliged to analyse how perception interacts with affect, culture, previous experience, emotional states, and other pertinent matters. Instead, such theories constitute theories on "perception in the laboratory", their aim being to render the performances of the subjects as aseptic as possible, officially in order to manage to purify observed phenomena of contingent elements, in actual fact in order to avoid a shattering comparison with the everyday reality of mental phenomena. The ultimate consequence is that theories of complex processes taking a strong modular stance are wrong from the very outset, by definition.

Second, the entire mind may be taken as the object of study. This alternative is impracticable for science. There is no way the entire integrated structure of the mind may be grasped, not even if it is engaged on a specific task. It is unrealistic to believe that any project will manage to explain the global working of something that is so complicated that it has to date eluded all attempts at achieving a satisfactory level of comprehension even of only one single part. A global theory, one that attempts to make all the relationships existing between the different functions explicit, as well as explaining the specific procedures each function adopts in executing its tasks, is doomed to remain a frustrating and foolish aspiration. The result of a more ethological approach therefore seems to be acknowledged impotence—there are too many unknown quantities, and virtually no certainties to act as departure points.

The tactical solution to this dilemma continues to be that adopted by all those researchers who do not share the strong modularity hypothesis. It consists in continuing to work on individual mental phenomena as if they were modular, though being fully aware that this is not so. This might appear to be meaningless hairsplitting. Instead, it has vital consequences. First, the scholar is alive to the fact that the theory worked out to explain a given phenomenon is certainly false, since it excludes what should be included. This implies that any attempt at extending the theory to any significant degree would lead in practice to its abandonment. The scholar must consequently be prepared to change the original structure, in as much as comprehending new phenomena means reconstructing the theory in greater depth. The more distant the new domains are from the original field, the greater the depth must be. For example, it is impossible to extend a theory on formal reasoning so that it also manages to explain everyday reasoning. Rather, a second theory must be developed in such a way as to explain both types of reasoning. The viewpoint that must be adopted is therefore that of continual reconstruction, not progressive extension.

Second, if we wish to consider together phenomena belonging to different domains such as, for example, perception and language, it is methodologically wrong to try to do so linking up the two pertinent theories, even if an adequate interface is used. Instead, in this case too, a new object must be created, one that manages to comprehend the aspects that have already been established both in the field of perception and in that of language, employing the sectorial theories that have already been worked out, but re-working them into a new and original construct.

Stated differently, it is a question of being aware of the limits of what one is doing. With the methods that are available today, it is not possible to construct a good theory about mind by first building independent mini-theories and later assembling them into a single whole, in the way one composes a puzzle by utilising already structured pieces and fitting them into one another. Instead each new theoretical conquest involves a radical remake of what has already been done, in a reconstruction which might employ the knowledge that has already been acquired, but will do so in a new logical structure. Thus, if the construction of a single building from two separate but contiguous edifices is desired, it is not sufficient (even if is undoubtedly better than nothing) to establish a passage connecting them. A good architect should dismantle them instead, in order to re-utilise the constituent parts as much as possible to create a harmonious whole.

Until a global picture of the human mind has been drawn, we shall have to resign ourselves to the fact that none of the parts that have already been analysed can be considered to be definitive, or unquestionably true. In a puzzle, each element has its own independent existence—although it can gain its full meaning when connected with the other pieces, the form and structure of each individual piece cannot be modified. On the contrary, in a construction which is continuously being restructured, no element exists in its own right. Every time the theory as a whole makes a significant step forward, its constituents are challenged, and are not necessarily utilised in the same way they were employed in the previous version.

In brief, the difference between the cognitive scientist working in a given area employing the strong modularity hypothesis and the cognitive scientist working in the same field but who rejects the strong modularity hypothesis is that the former builds theories that are partially wrong, believing they are potentially right, whereas the latter builds theories that are partially wrong, knowing they are wrong. The key difference lies in the latter possessing awareness that no limited theory can ever be true.

Is the architecture of the mind serial or parallel?

Will the introduction of the parallel architecture computer, which we spoke of in Chapter 3, lead to significant changes in the field of simulation? To be consistent with the assumption of limited connectionism, the answer to this question is affirmative when we are considering microcognition, and negative when we are dealing with macrocognition.

Parallel computers reproduce some of the characteristics of the brain at a physical level. In principle, each neuron should have its own correspondent independent processor, which is connected to other processors exhibiting exactly the same principles for the activation of connections that neuron is supposed to have with other neurons in the human brain. This should yield an extremely precise reproduction of the characteristics of the brain, including presumably those functions whose activity is determined by specific, individual neurons or by cerebral areas whose location has been accurately identified. In practice, neuropsychology could avail itself of parallel computers to investigate the pathological processes of the central nervous system, where illnesses and lesions have a precise location, as is testified by precise clinical evidence.

Enormous advantages are to be obtained from studying the normal or pathological functioning of the brain by employing a machine exhibiting an analogical correspondence with the brain itself, namely one enabling the researcher to exploit the physical correspondence between neurons and processors, damaging the latter exactly as one hypothesises the damage has occurred to the neurons, and then analysing the decaying performance of the machine caused by the simulated lesion. A machine built rigorously reflecting the structure of the brain permits refinements which cannot be achieved by a machine, such as the serial computer, which bears no structural resemblance to the brain whatsoever.

The same argument applies to microcognition in those situations, where the behaviour of a single neuron, or a small network of neurons, may be identified and is recognisable as being relatively autonomous. This is the case with perception, for example, where virtually each neuron in the cerebral cortex has its own precise task and can be simulated with a certain degree of precision. Here too, physical correspondence is an enormous help in establishing functional equivalence.

The picture changes, however, when we come to macrocognition, to those processes, such as reasoning and communication, where analysis does not take place at the level of the single processing unit. At these high level cognitive domains, analysis and simulation do not concern the individual behaviour of each neuron, but the functioning of the mind as a whole. Although it is clearly still true that it is the single neurons which allow the human to think or communicate, this time the neurons are considered as an entire network, one that cannot be divided up into pieces. At the level considered in macrocognition, whether the reproduction of the function under examination takes place on a serial machine, a parallel machine, or on a machine that has still to be invented, is irrelevant. What counts is the theory that is being experimented on, not the physical realisation of the theory itself. Any technological advance is welcome, but it is not the decisive factor.

The fact that in the final analysis it is still the neurons and their connections that do everything is true but not significant. If we wish to avoid falling into a new version of neurological reductionism, what matters is the preselected level of analysis, and at the level of mental functions the paradigm of parallelism seems equivalent to classic serialism.

Towards the development of the mind

The radical solution advanced earlier to avoid both Scylla (modularity when there is no doubt as to it being incorrect) and Charybdis (the intractability of absolute complexity) consists in employing a developmental approach to the study of the mind. This allows the principle of complexity to be maintained, at least theoretically, without remaining paralysed in the face of the insurmountable difficulty of having to deal simultaneously with all the aspects of the mind. It involves facing anew the classic themes of cognitive science from a developmental standpoint. This means studying mental processes not only as if they were steady states, an approach which generally takes into consideration exclusively the final stage, but rather concentrating on how a given function develops from the infant to the child, through adolescence, up to adulthood, and finally decays in old age.

Important results have been achieved utilising the strategy of considering the mind as a static system, but we have probably squeezed out all that can be obtained through this procedure. In his latest book, *The society theory of mind* (1986), Marvin Minsky espouses a similar perspective, attributing the greatest importance to the developmental aspect of the study of the mind.

From an epistemological standpoint, we have already seen that cognitive science may be characterised from the operational perspective, as a science in which definitions are furnished by specifying the operations that constitute a given state of affairs. The same principle, namely the reconstruction of what cannot be observed, must now be adopted not only at a general epistemological level, but also as a specific research strategy in which phenomena are explained by tracing their development.

Comprehension of a phenomenon existing in the adult cognitive system must be achieved by discovering how that phenomenon acquired its final structure over time. An analysis of the genesis of a process often leads to the elucidation of the nature of the process, whereas a historically decontextualised investigation into the same process leaves it incomprehensible.

For example, if you, the reader, look at Fig. 4.1, can you guess which horse seems likely to win the race?

On the contrary, the sequence in Fig. 4.2 leaves no doubt as to how the race will end, even if the last frame is exactly the same as the one in Fig. 4.1.

The difference between the two is that a static representation does not furnish us sufficient information as to how we reached a given state, whereas if we are aware of the stages preceding that situation, we can understand it better, and sometimes even predict how it will evolve with a reasonable degree of conviction. In our case, Fig. 4.3 represents the future. This development is inaccessible from the analysis of Fig. 4.1, while it is easy to foresee if we posses developmental knowledge of the situation, such as that supplied in the sequence depicted in Fig. 4.2.

A developmental procedural method renders standard laboratory techniques unviable, in as much as the nearer we come to the initial stages of mental development, the more difficult it becomes to raise the usual barriers between the diverse processes. If our point of departure is the newborn child, it is obvious that the beginning of any process whatsoever, from memory to motor control, is connected to the relationship that ties the child to the mother.

Inborn capacities have to develop and to create functional relations among basic structures. Proceeding along this line of reasoning, a theory that wishes to account for the logical capacities of the adult would be obliged to start from the links between the love of the mother and the first, elementary inferential abilities, and to trace them until they achieve full specialisation in the adult. The initial phase in reasoning has more to do with the affective bond between mother and child than with the laws of logic.

As is common practice in the study of the brain, it would also be useful in the study of mental functions to refer to an initial situation where potential capacity is at its greatest, and which then develops towards the construction of specific abilities. The structure and functioning of such capacities cannot be clearly established until the individual pathways which led to the building of each function identified in the adult organism have been uncovered and traced. In brief, the reconstruction of mental phenomena should simulate how the phenomena themselves came into being in the system. The developmental approach to research is consistent with the epistemological roots of cognitive science. It has already produced good results in other sciences, and it is after all the method that is adopted with some success in life outside the laboratory.

FIG. 4.1. The ahistoric present.

FIG. 4.2. The present as a function of one's development.

FIG. 4.3. The evolution of the future from the present.

THE SIMULATION OF THE MIND

We shall now return to a theme first begun in Chapter 3, and to which we already know the answer provided by Turing: when is it that we can say that the computer has satisfactorily simulated the human subjects it is carrying out the comparison with?

First, what does simulating the human mind mean? It means that a given mental phenomenon, at a given level of detail, has been reproduced by a computer program.

The more widespread and more crucial the phenomenon, the more interesting and important will its simulation be. The processes that can be reproduced may differ in their importance, as the following examples show: learning a foreign language, memorising a telephone number, feeling happy because it is spring. In any case, if the method employed is that of soft artificial intelligence, and hence if human mental processes are reproduced, we may speak of simulation.

In the simulative approach, no phenomenon is of little importance since nothing that the human being does mentally is obvious or banal, from the point of view of understanding and explaining it. A computer program capable of adding up three-figure numbers has no relevance to artificial intelligence, but an analogous simulation program able to explain how human beings manage additions, how they learn to do them as children, what difficulties are involved, and why errors take place would be a sensational success.

I spoke earlier of level of detail. Each simulation must explicitly state what level it takes place at, since the same phenomenon may be studied in terms of neurophysiology, of individual mental processes, of social structures. If, for example, we examine the phenomenon of the choice of a sexual partner, this may be investigated in terms of hormones and amines, or in terms of the correspondence between the image perceived and prototypical models stored in memory, or in terms of the social norms that direct amorous interaction in a specific culture. No one level is better than the others, once having renounced dull biological reductionism. Each level—the neurophysiological, the psychological, the anthropological—offers an equally relevant view of what is being studied. What is vital is to state which viewpoint has been adopted.

It may also be possible to build a program, or better a machine that implements a program, which faithfully reproduces the phenomenon being analysed, both at a physiological level and at a psychological level. This achievement would require the integration of the two levels into the same program, the target the connectionist paradigm has set itself, with its aspiration to grasp mental phenomena while simultaneously reproducing the behaviour of the neural substratum.

In principle, even greater heights may be reached by hypothesising total reproduction, including that of the physical level, of any phenomenon whatsoever. In this case we consider *complete imitation*. This consists of a replica of the object being studied, one in which every detail is reproduced to perfection. In practice, this corresponds to a painstaking facsimile both of the brain and of the human mind, coming closer to the construction of an artificial double of the human being than to the objectives of cognitive science. Complete imitation is a mythological theme. Indeed, in the *Iliad* Homer had already raised his voice in song to the visionary creatures created by Vulcan:

> ... gold maids
> In appearance no different from living young maidens
> Advanced bearing their lord;
> What a feeling of life they bear in their breast,
> And they possess strength and words, and in good works
> They have been instructed by the immortal Gods.

Competence

A first reply to the problem of which criteria render a computational model acceptable is furnished by the competence approach, whose best known exponent is the linguist, Noam Chomsky. It was he who introduced the term *competence* in his authoritative work, *Syntactic structures* (1957). By this term he intended the set of abstract capacities possessed by a system, independent of how these capacities are put to actual use. In linguistic terms this means dealing with the structures that make the comprehension and generation of sentences possible, and neglecting the problem of establishing whether humans really do possess precisely those hypothesised structures and whether these do explain their linguistic behaviour, deeming the question non-essential. In this view, language is seen as being completely independent of the mental functions, such as memory and thought, which enable humans to use it.

The self-sufficiency of the competence approach allows Chomsky to stress the syntactic aspect of language, governed by rules and principles, and ignore the use of language as a means of conveying mental content. For example, if a comparison of two automatic translation systems is made exclusively on the basis of competence, then the sole aspect that will have to be taken into consideration will be the output, the final result. What are evaluated therefore are the surface structures. Consequently, comprehension of the procedures human translators follow when they face a similar task is of no interest. This method is reminiscent of hard artificial intelligence, an approach which received treatment in Chapter 3.

Performance

Contrasting with competence is *performance*, which may be defined as the set of abilities actually exhibited by a system in action, that is its behaviour in real situations. The concept of performance is a familiar one in psychology, since it is a science which is concerned with behaviour occurring in real life more than with potential capacities. The methodological difficulty that performance presents us with is that it comes too close to complete imitation, which is unattainable because the prerequisite for its accomplishment is the ability to reproduce the real behaviour of a system under any condition. Returning to the previous example, a performance model of a human translator must be able to simulate the translation that person carries out even when the person in question is tired, or euphoric, or has drunk too much champagne. All these tasks imply control not only over language, but also over those areas which interact with language.

Furthermore, performance varies in accordance with individual differences. It is indissolubly linked not to the average subject, whose existence is of a purely statistical nature, but on the contrary to the irreducible individuality of any performance which really takes place. What is easy for one subject may be difficult for another. Everyone is influenced in a different way by the particular circumstances of the situation. Hence all individual differences must be included, explained and simulated. These differences include age, culture, personality, and so forth, and there are too many of them for a single performance model, which is intended after all to explain one phenomenon at a time, to hope to dominate them all. A simulation of performance implies that the entire mind has been analysed and comprehended—an undertaking, as we already pointed out when speaking about complexity, which cannot be tackled with the tools science has available at present.

To escape from the two extreme situations described above, both of which are unsatisfactory, the former for epistemological reasons, the latter because of concrete limitations, the only path that can be trod is that which leads to the more flexible alternative offered by the notion of simulation model.

Simulation model

Refining Chomsky's distinction for the domain of perception, Marr (1977) hypothesised three levels of explanation: computational theory, algorithm, and hardware implementation (see Chapter 6). Computational theory corresponds to what has to be computed and it is linked to Chomsky's notion of competence. An algorithm corresponds to how the operations specified by the theory are represented and accomplished. Finally, implementation by the hardware corresponds to the way the operations are physically realised (in the brain or in the computer). The level of the algorithm is analogous to my concept of the simulation model, and therefore to a position which tends more towards performance than to competence. I prefer to avoid the term algorithm after the criticisms that connectionism advanced against the idea that mental processes may be specified by a set of pre-determined operations. At the same time, this position accepts the fact that the reproduction of performance lies beyond the present powers of science.

The line that emerges from these considerations is that of building dynamic programs tending towards the simulation of performance. This acquires the same status as an asymptotic global simulation—it is useful as a reference point to indicate the direction research should proceed in, but it does not aspire to absolute accuracy, as it would if it were to be interpreted in its strict sense.

A simulation model may be defined as that set of abstract capacities possessed by a system, constrained by the knowledge scientists have already acquired and can utilise. What is involved here is first hypothesising the competence of a system, and then moving closer and closer to its actual performance, introducing the highest possible number of constraints that may be deduced from everything that is known about the function under investigation.

Let us take the example of constructing a simulation model of human language ability. We may take a competence model as our departure point. We gradually introduce limits generated also by other areas of cognitive science. First, neuroscience warns us that the human mind is finite, thus ruling out both the possibility of an infinite body of knowledge and the use of algorithms having exponential features, since the execution of this type of computation would call either for a greater number of neurons than it has been calculated the brain possesses, or for a length of time exceeding the human life span. Second, psychology requires that such a model should not go against what is known about the language learning process, or about the relationship between language and perception. Other disciplines may add further constraints, the respect of which transforms what was initially a competence model into a simulation model properly speaking.

Performance continues to evade our reach, but at this point it is more a question of quantity than of quality: what becomes important here is the intention of theoreticians, what they want to do, and above all the dynamic potential of the model, namely to what degree it can be developed as an increasing number of constraints are added.

The constraints introduced progressively increase the complexity of the model, since they oblige it to take stock of a larger quantity of information. This often forces the scholar to give up looking for a possible linear solution to a problem and seek a more complex solution, for the simple reason that other, independent considerations rule out the possibility that humans can behave in the first, straightforward fashion. Admittedly, to each complex problem may be found a simple solution that is also wrong. However, the advantages of the more complicated solution might not be immediately obvious, or not obvious to that particular scholar. While the traditional reference to Occam's razor may be taken as being literally true (*entia non sunt multiplicanda sine necessitate*), it must be used with care: the necessity invoked by the medieval philosopher sometimes becomes clear only when a correlation is established between one process and other psychic functions, thereby widening the field of the phenomena taken into consideration.

Finally, Fig. 4.4 shows the links between the epistemological concepts we have explored and which we shall find in the various domains of cognitive science.

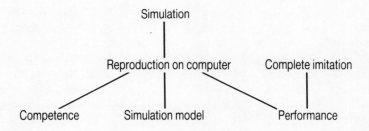

FIG. 4.4. The relationships between the key epistemological concepts in cognitive science.

We may conclude this first part by stating that cognitive science is possible. That it is also the best choice is more difficult to demonstrate. Hence this claim will only be made in the final chapter.

PART TWO
State of the art

CHAPTER FIVE

Knowledge

A description of the state of the art makes sense if it is reasonably up to date when it is read and not just when it is written. In a rapidly expanding sector, a photograph of the facts becomes obsolete almost instantaneously. Hence it does not constitute the most sensible road to take. I shall therefore attempt to follow a different route by presenting a picture of the currents of thought that have animated cognitive science in the last few decades, together with a description of the trends which presently traverse the field. My hope is that this type of strategy will render this account more resistant to time.

It would be useful to begin by introducing the outline of an integrated theory of the mind to which to refer when attempting to correlate the various subjects that will be treated in the course of this second part of the book. It would indeed be an excellent idea, if such a theory existed. Not that the need for such a theory has escaped the acumen of researchers. Unfortunately, all attempts at constructing one have so far ended in failure. We shall therefore have to make do with a less detailed scheme (Fig. 5.1), which we may assume most cognitive scientists would agree on.

The existence of a common centralised knowledge, which does not exclude the possibility of a series of local stores of specific data and procedures, guarantees that the interdependence among the various functions runs smoothly and effectively, thereby complying simultaneously with the reasonable criterion of the economic management of

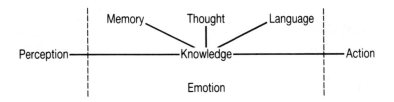

FIG. 5.1. Scheme of the human mind.

resources, in as much as data are often general while procedures tend to be more specific. For example, it is probable that the information relating to a given object is available to the processes of perception, reasoning and planning. The very same data may be employed and updated by all the processes that make use of them.

Likewise, knowledge acquired through any procedure whatsoever is immediately available to the other procedures. If I read the description of a poisonous mushroom and its properties, I shall then be able to recognise it and perhaps even know what to do in case I mistakenly ingest one. Language furnishes knowledge which perception and planning can utilise and, in their turn, modify if necessary. Nevertheless, there still remains the proven fact that some functions maintain local nuclei of knowledge which guarantee them a limited degree of autonomy. These separate stores, which duplicate a part of general knowledge for specific ends, will not receive detailed treatment here.

TYPES OF KNOWLEDGE

In tracing a first draft of a map of the structure of human knowledge, we shall divide the subject up into three interacting subsystems, each of which handles one type of knowledge: explicit, tacit and model (Fig. 5.2). The "K" that precedes many terms stands for *knowledge*.

Explicit knowledge
Explicit knowledge (*K-explicit*) is a theory of the world, conceived of as a set of all the conceptual entities describing, in the form of propositions, classes of objects (cup), relationships (the wine is in the cup), processes (the ripening of grapes), behavioural norms (one does not drink with one's mouth full) etc. Typically, it may be represented by formal logic constituting an adequate description of *knowing-that*, where the data

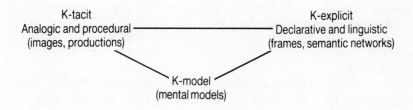

FIG. 5.2. The structure of human knowledge. The standard modes of representation are in brackets.

correspond to axioms and are handled by a limited number of general inferential rules. One concept which is virtually identical is that of *declarative* knowledge, which is typical of linguistic knowledge, and which comprises what may be expressed in words about the object or about the state of affairs at issue. Geometrical definitions offer a paradigmatic example of this point, as with the following case:

[1] Isosceles triangle = a triangle having two equal sides.

Explicit knowledge represents what people know they know about any given entity in the world, including other people. Such knowledge can be expressed verbally, and it may be reflected upon voluntarily. Such knowledge is transparent, in the sense that it can be read by the rest of the system's procedures, and if activated intentionally it may become conscious. For example, people familiar with the definition of "mammal" are also aware that they know the definition if they focus their attention on it.

This does not imply that this knowledge actually corresponds to external reality. Rather it corresponds to what people believe external reality to be. The relationship between explicit knowledge and the world is exactly the same as that existing between a theory and its model. The external world cannot be known, but human beings are capable of constructing a theory of it and employing this theory to interact with it. The degree of success achieved depends on the quality of the theory itself. All our educational establishments and a large part of professional training are centered on explicit knowledge, even if this has greater weight in the training of a mathematician than in that of a ballet dancer. The same goes for the body of social knowledge—books, laws, regulations. It must, however, be noted that this latter body of knowledge becomes individual knowledge only when it is internalised, since simply the existence of a book explaining the theory of relativity is not enough for me to acquire this knowledge cognitively—to do so I must read and understand the book.

Tacit knowledge

Tacit knowledge (*K-tacit*) refers to the knowledge that a system possesses and that enables it to interact effectively with the world, even though such information is not represented in an explicit fashion so as to be directly readable by other parts of the system. The standard mode of representation for this type of knowledge is by means of production rules, which enable the system to manage *knowing-how* to act. Among those scholars who have underlined the importance of this type of knowledge, special mention must be made of the epistemologist, Michael Polanyi (*The tacit dimension*, 1966), the first to gain an insight into its function and to understand that it could not be reduced to other types of knowledge.

The transparent part of K-tacit corresponds to the images and to the productions which translate explicit knowledge, and in particular those pertaining to change, into procedural terms. A spontaneous change in a state in the world comes about when a precise set of natural preconditions is satisfied, thereby generating a new state in the world, called effect. Examples of this include the rotation of the moon around the earth, where the preconditions correspond to the laws of universal gravitation, and the blossoming of a flower, where the preconditions are those expressed by specific botanic laws. In the case of intentional changes, however, no precondition may be considered to be a sufficient cause. What is required is the existence of an agent executing a clearly defined sequence of operations in order to generate the desired effect. An example of this type of change is going to a wedding reception, where the preconditions consist in there being a wedding with relevant reception, the agent is a guest, the operations to be executed are those that permit the guest to go to the reception, and the effect is that among the participants at the party there be the agent herself. This part of K-tacit is transparent because the system has free access to such knowledge.

A large part of K-tacit is, instead, constituted by opaque procedures—methods of acting that are triggered automatically, without requiring control or attention. Typical examples are riding a bicycle, feeling embarrassed or at ease, distinguishing the different types of cigars by their aroma—all things that an expert knows well and can do perfectly, but which can only be translated into verbal terms with gross approximations. Hence the impelling need for metaphors when talking about such subjects as love, and the tremendous number of neologisms when the taste buds are involved. For example, one book on Italian wines has this to say of the famous Piedmontese wine, Dolcetto: "Usually fairly tannic yet soft with a broad, grapey, mouth-filling flavour, Dolcetto has been hailed as Piedmont's answer to Beaujolais, though it is much too chauvinistic to follow a French lead". Unfortunately, it is impossible to capture the taste of that wine from such a description, unless one is already acquainted with it. However, if the taste is familiar, the description appears suggestively adequate. This bears witness to just how difficult it is to render opaque knowledge transparent. Opaque procedures fall by definition outside the bounds of awareness. They operate unconsciously and their reconstruction can take place only *a posteriori*, in an incomplete and arbitrary manner. Indeed, bringing a series of movements to the level of consciousness may worsen the performance quite noticeably. This explains why artists in the field of the figurative arts hate being observed while they are in the act of creation, as is borne out by many anecdotal examples, from Michelangelo to the master of calligraphy, Kosen, anxious as they always were to escape the sharp eyes of their disciples.

Summing up, tacit knowledge corresponds to knowing how to act in a given situation, living the situation from within, and not standing outside of it to speak about it or to reflect on it. It thus overlaps perfectly with the concept of *procedural knowledge*, which is typical of those cases in which a person knows how to act, but is not necessarily capable of making explicit what activated the appropriate reaction to the given situation. For example, experienced psychotherapists immediately recognise the impalpable signs of a neurosis, but may find some difficulty in referring them to a third person, and often even to themselves. Or, and this is the most frequent case, people know what they are doing but do not find it easy to put into words. Good surgeons can carry out an appendectomy practically without needing to look, but they will have a hard time explaining to an apprentice how their hands move, almost as if these parts of the body were fully autonomous and possessed an ability all their own.

Such characteristics have induced some writers, among whom is the philosopher, John Searle (*Intentionality*, 1983), to make the claim that one can no longer speak of knowledge, but rather of *background capacities*, capacities which are not—and perhaps cannot be—represented employing the methods presently available to science. It is of course possible that modern representational techniques are inadequate when it comes to accounting for the background, but once again one must not confuse present difficulty with theoretical impossibility.

The relationship between K-tacit and K-explicit is complex and hard to define, corresponding broadly to the relationship between experience and symbol. To infer the explicit knowledge equivalent of a piece of tacit knowledge corresponds to building a theory of a phenomenon, with the proviso that the phenomenon is introspective instead of pertaining to the external world. On a more general level, it may, however, be admitted that opaque procedures generate explicit results. For example, a person might have no idea why she finds one person attractive and another loathsome. Nevertheless, she is still fully aware of the final result, for example, that Ulysses is fascinating and Achilles is unbearable. As always happens when tacit and explicit knowledge interact, any combination is possible. Some people possess such extensive powers of self-analysis that they are capable of analysing procedures and results (they know the reasons why they like Ulysses more than Achilles). Finally there are people who are incapable of gaining awareness even of the results of procedures, even though these are normally explicit (they do prefer Ulysses to Achilles, but they do not know it).

Just as a propositional theory may be constructed starting from tacit knowledge, so may procedural knowledge be built from explicit knowledge. For example, I can read a book on the natural method in horse riding and then do my utmost to apply those principles to the way I ride, trying to keep the procedures thus generated under strict control until they are efficiently stored and coordinated.

Model knowledge

Model knowledge (*K-model*) corresponds to a specific model, built by integrating the two types of knowledge examined previously, which are taken as being theories. K-model may be considered as a set of partial configurations of theoretical knowledge expressed by K-explicit and K-tacit. The standard mode of representation for this form of knowledge is by means of mental models. These bring together data and procedures, knowing-that and knowing-how. K-knowledge constitutes that part of knowledge that the system is actually using at a given moment.

THE REPRESENTATION OF KNOWLEDGE

The human mind operates employing knowledge in an integrated manner. Knowledge cannot be rigidly encapsulated into one of the schemes which will be explored later. An effective structure representation must be able to account for two types of knowledge, declarative and procedural knowledge. This led to the traditional division into data and the procedures that manage the data. We shall begin by examining the systems that respect this division, gradually proceeding towards those that do not, but merge them into a single structure containing both of them. Having established the general dichotomy between declarative and procedural knowledge, we may now pass on to a detailed examination of how these types of knowledge are stored and utilised.

Only seven basic methods have been advanced for the representation of knowledge. They derive from logic, psychology or computer science. Consequently, they have different spheres of application, spheres which may be extremely wide, as is the case with formal logic, or narrow, as is the case with mental images. The types of knowledge on which the seven methods focus also differ slightly, and it may be said of each of these methods that in at least one area it is more efficient and convincing than the others. All of these methods are held to be not simply a notational device, valid only for artificial systems, but also the mode, or at least one of the modes, in which knowledge really is stored in the human mind. We shall discuss this position, which is the one which interests us most, after having presented each of these methods in depth.

Formal logic

From Boole to Piaget, there has never been a lack of illustrious supporters of the thesis that the human mind operates in accordance with the rules of classic logic. Chapter 7, which is devoted to thought, will deal with this subject extensively. The present section will only tackle the problem of employing logic to represent knowledge, avoiding the subject of reproducing the functioning of mental processes. More than in cognitive science, logic has been utilised in artificial intelligence. Since, however, those methods employing logic that have been utilised in artificial intelligence have also influenced scholars of psychology and cognitive science, it will therefore be worthwhile examining them briefly. As my reference point, I shall take the seminal work by McCarthy and Hayes ("Some philosophical problems from the standpoint of artificial intelligence", 1969).

The most elementary type of formal logic is the *propositional calculus*, whose key concepts are *truth-values* (a properly formed proposition has one of the two possible truth values: true or false), *axioms* (the content that expresses what can be formalised) and *rules of inference* (that state which can be deduced if certain axioms are taken to be true). Starting from the axioms and systematically applying rules of inference, theorems such as the following may be demonstrated:

[2] $(p \lor q) \& \sim p \rightarrow q$

One interpretation of [2], in which p = it is black and q = it is white, is:

[3]   If it is black or white, and it is not black, then it is white.

Knowledge representation by means of systems of logic gains tremendous importance when the propositional calculus is enriched by the addition of the possibility of making assertions concerning specific objects or *individuals*. As statements about individuals are called predicates, we obtain the *predicate calculus*. For example:

[4]   X is deadly.

By adding the notion of operators, or *functions*, and the predicate of equals to the predicate calculus, we obtain *first order logic*, which may be considered the standard methodology of logic for knowledge representation. The main feature of first order logic is essentially its power—it is in fact sound and complete. Completeness means that any true assertion can be proved, and soundness that any assertion that is not true cannot be proved. This power also constitutes the principal limit of first order logic, since it implies the fact that the moment the original database is extended, this technique becomes extremely difficult to handle. In particular, it is practically impossible to distinguish between those propositions that are true and significant, and those propositions that are true but of little interest.

A second characteristic of logic is that it is grounded on inference rules operating on data, from which it remains rigidly separate. It takes on a representation structure that is termed the *propositional* type. This type of representation bears no resemblance to the state of the things being represented. For example, the symbol indicating a triangle does not possess the features characterising "triangularity", nor does the symbol indicating a horse possess those of "horseness". The relationship between what is represented and its representation is established only when the representation is interpreted, when it is made explicit that a given symbol in a given context indicates a specific state of affairs. The independence of the structure from the world which it is intended to represent is the characteristic feature of propositional representations, the one which differentiates them from analogic representations. The internal structures of the latter reproduce the salient structural characteristics of the object being represented.

Classic logic possesses the property termed *monotonicity*; any further information added to the knowledge store can never invalidate a previously reached conclusion. Monotonicity offers numerous formal advantages, but it turns into a hindrance when dealing with incomplete knowledge, as is the case of the knowledge each living system has of the world it lives in. Despite the incompleteness of their knowledge, and despite the fact there are no guarantees that the knowledge one possesses has no internal contradictions, human beings manage to behave with incredible efficacy, taking decisions and making inferences which are often accurate and correct.

In the attempt to capture this type of reasoning, a branch of logic has been developed called *non-monotonic*, for the precise reason that it lacks the property of monotonicity. In non-monotonic logic, subsequent information may invalidate an established conclusion, reached when the crucial information was missing. Non-monotonic logic is a formalisation of reasoning *by default*, by means of which it is possible to reach conclusions grounded on what usually happens, in standard conditions, basing oneself precisely on the fact that one does not possess direct knowledge of the state of affairs in question. Let us consider, for instance, the statement:

[5]   Every bird flies.

This is the equivalent of saying that a normal bird in standard conditions possesses the capacity to fly. Reasoning by default may be formalised, in the opinion of Ray Reiter ("A logic for default reasoning", 1980), by means of default rules of the type:

[6] $\dfrac{\alpha : M\,\beta}{\omega}$

Rule [6] is to be read as follows: if α is true according to the knowledge of the system, and β is consistent with it, then ω may be assumed to be true by default. In contrast with classic rules of inference, rules of default may be blocked. Given a database containing α, rule [6] enables ω to be deduced. Statement [5] may be formalised thus:

[7] $\dfrac{\text{bird }(x) : M\text{ flyer }(x)}{\text{flies }(x)}$

But if the negation of β is explicitly added to the knowledge base, then the default rule can no longer be applied. Continuing with the previous example, the knowledge base may contain information such as:

[8]   Ostriches are birds.

[9]   Ostriches cannot fly.

Let s be a specific ostrich; we may now write:

[10]   Ostrich (s).

If we apply [8] and [9] together, then we derive:

[11]   ~ flyer (s)

which blocks rule [7], not consenting the derivation of ω, that is flies (s).

The general assertion [5] may be expressed informally within a framework of reasoning by default as:

[12] Every bird flies, unless special information is provided concerning its inability to fly.

Additional knowledge might in fact exist on a specific bird (say a seagull with a broken wing), or on one specific type of bird, as in our example of ostriches, birds [8] which do not fly [9]; on such occasions rule [7] is blocked in this specific case, while maintaining its general validity.

Another interesting type of non-monotonic logic is *autoepistemic logic*, introduced by Robert Moore (1985) and developed by Kurt Konolige (1988), which refers to what subjects know about their own knowledge. In other terms, explicit knowledge of what one knows and of what one does not know allows one to draw correct conclusions. Let us consider the following two premises as an example:

[13] I know all my children.

[14] I have no information as to whether Georgia is my daughter.

From these two premises we may derive the correct conclusion that:

[15] Georgia is not my daughter.

Non-monotonicity in this case consists in the fact that the meaning of the premises depends on the context. If we were to be provided with evidence in favour of the fact that Georgia is my daughter, then we would conclude that premise [13] was employed in an incorrect fashion when inference [15] was drawn. Premise [14], which was correct at the moment the inference was made, is now contradicted by the new information received concerning Georgia. Hence premise [14] will have to be duly amended into:

[16] I know that Georgia is my daughter.

At this point premise [13] again becomes correct (at least until it is superseded by events), in as much as I am again in a position to state that I know all my children. Autoepistemic logic lies at the roots of lines of reasoning such as:

[17] If *x* were true, then I would be aware of the fact.

On the grounds of this type of reasoning we may exclude the following statements as being true:

Saddam Hussein graduated from Berkeley.
Elizabeth II was once engaged to Bill Clinton.
All those whose names begin with P are exempt from taxation.

The employment of non-monotonic forms of logic is quite promising in those areas requiring the formalisation of everyday knowledge in order to draw inductions as opposed to making deductions. Classic logic cannot be applied in this type of situation, but it is useful to have some form of logic on which to ground simulation.

Advantages: It is the standard instrument for the formalisation of any theory, in particular at the competence level. Not only does it make implementation a straightforward step, but it can also substitute it when a computer program proves to be impossible to write.

Drawbacks: It is so powerful and error-proof as to become a candidate which is not credible as a representative of human knowledge, in particular at the performance or algorithmic level. Intolerably complicated when applied to large bases of data.

Semantic networks

These consist of a set of *nodes*, typically expressing concepts, connected by *oriented arcs* (or *links*) which univocally represent the relationships between nodes. They were originally developed by Quillian (1968) for the analysis of natural language, and in particular to highlight the relationships existing between the meanings of terms that were clearly interrelated. Semantic networks were methodologically redesigned by William Woods (1975) to eliminate a number of ambiguities and inaccuracies connected to the need to furnish an independent definition of the links utilised and the properties that are attributed to them. This made it possible to extend their use to logic, thanks to the demonstration provided by Lenhart Schubert (1976) of the fact that their properties can be rendered equivalent to the properties of first order logic. In this case, however, we also inherit all the disadvantages typical of logic itself when it is employed as a notational formalisation of human cognition—in particular its cumbersome nature.

A classic example of a semantic network is presented in Fig. 5.3. Each relationship expressed in Fig. 5.3 is read off starting from the node of origin and going towards the destination node. This therefore yields:

[19] Cat IS-A mammal

[20] Tail PART-OF cat

The arcs are signalled graphically in various ways, so that their different nature is explicitly made clear. The specificity of their behaviour has to be formalised each time a new arc is introduced. For example, the arc IS-A makes the node of origin inherit all the characteristics exhibited by the destination node. Since the cat is hierarchically subordinate, it will possess all the features of mammals, such as warm bloodedness, breast feeding the young, etc. On the contrary, the arc PART-OF does not possess this property. In fact the tail does not inherit the characteristics of the cat (tails do not go "miaow", nor do they love fish). Moreover, if thanks to the link PART-OF all cats have tails, we would not wish the link IS-A to enjoy the same status, in order to avoid being obliged to deduce that every mammal shares the same features as cats. In fact not all mammals possess vibrissae or retractile claws.

Despite the improvements to Quillian's original networks made by Woods, this type of formalism remains an easy prey to the so-called symbolic fallacy. This fallacy consists in assuming that as soon as a given concept has been represented by means of symbols typical of semantic networks, then the deep meaning of the concept has been comprehended. Translation into symbols is, instead, insufficient to

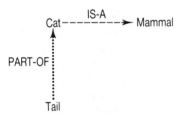

FIG. 5.3. An example of a semantic network with two different types of arcs.

grasp the semantic meaning of a concept. Connecting up two nodes with an arc does not automatically correspond to grasping the meaning of the two concepts, nor of the relationship which links them. Consider the assertion:

[21] Stella is sexier than Monica but less brilliant.

It is not sufficient to connect up the nodes Stella and Monica with two arcs pointing respectively in the direction of "sexier" and of "brilliant". Before being in a position to provide an analysis of the semantic content of the given phrase, one must first have independent definitions of "sexy" and of "brilliant", as well as of the comparative of these adjectives. Semantic networks can succumb to the temptation of avoiding some complication by finding an ingenious label for some arc, thus pretending one has made the corresponding relations between concepts explicit. The principal virtue of semantic networks, their transparency, makes them difficult to use when one is not in full possession of the relevant knowledge or when this knowledge is not totally explicit, a common situation when dealing with the reproduction of mental functions. Today semantic networks are no longer considered a realistic model of human knowledge, even though the original claim by their supporters was that they were. Semantic networks have proved their worth in representing human performance in the domains of language and explicit knowledge, but the difficulties they have encountered in handling dynamic knowledge directed at grasping and causing change in the world has limited their usefulness.

Advantages: Effective in the field of declarative (knowing-that) knowledge, power equalling that of first order logic, complete transparency; extremely useful in representing hierarchically ordered knowledge, as with taxonomies.

Drawbacks: Insufficient incisiveness in describing and handling of procedural knowledge.

Production rules

These consist of systems of rules, each of which is as simple as possible, with one part expressing a condition (IF) and another part expressing the action to be effected (THEN). The conditional part—if—describes the conditions under which the rule applies. The action part—then—specifies the appropriate action the system must undertake internally or with regard to the outside world. Production systems were introduced by Newell and Simon (1972) in the area of problem solving, and have enjoyed a recent boom thanks to their utilisation in expert systems, as was seen in Chapter 3. An example illustrating the concept is:

[22] IF   agoraphobia with panic attacks is diagnosed
      THEN   anti-depressive pharmacotherapy is advised.

There can be no doubt that a part of human knowledge may adequately be described in terms of production rules, above all in the realm of action-oriented knowledge, both with regard to its abstract aspects such as problem solving and the working of the memory, and its concrete aspects such as planning and the control of movement. Authors such as Anderson (1983) and Newell (1990) even go so far as to maintain that production systems represent the real cognitive architecture of the mind. This thesis, however, comes up against a series of difficulties, all of which may be traced to the inefficient management of all kinds of declarative knowledge. Nevertheless, within the more limited domain of memory, Anderson's theory (see pp. 125–133) remains an excellent case of integration between declarative and procedural knowledge.

Advantages: modularity, uniformity, usefulness in handling action-oriented knowledge (knowing-how).

Drawbacks: opaqueness, bad strategic management, difficult to organise hierarchically.

Frames

Frames constitute an attempt to overcome the dichotomy declarative knowledge–procedural knowledge. A frame may be defined as a data structure for the representation of a stereotyped situation. Attached to each frame are several kinds of information. The main types of knowledge contained are those that indicate which frame must be utilised, those that specify what may be expected to happen, and those that explain what must be done if the events predicted do not take place. Frames were introduced into artificial intelligence by Marvin Minsky, who took up the notion of *schema* originally introduced into psychology by Bartlett (1932), initially applying it to the problem of artificial vision ("A framework for representing knowledge", 1975). Examples of stereotypical situations include a visit to the dentist, going to a birthday party, taking a university examination.

Figure 5.4, taken from Minsky's original work, presents a classic example of a frame for the perception and representation of a three-dimensional figure from various viewpoints. In the figure there is a three-frame system allowing the viewer to perceive a cube, to make inferences concerning the hidden sides, and to carry out actions corresponding to a 45° rotation of the cube to the left or to the right.

Frames are divided into two levels: a top level and a lower level. The top level is fixed, and contains assertions that are always true about the situation. This level may be considered as an intensional definition of

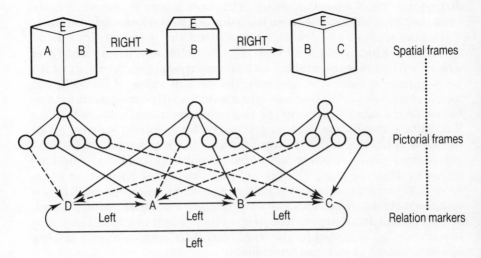

FIG. 5.4. A frame system for the representation of a cube. *Source:* Minsky (1975).

the corresponding concept. By way of illustration, at each birthday party there is at least one person whose birthday it is that day and at least one person celebrating that birthday. And at each wedding there are at least two people who are married by a third.

Instead, the lower level contains terminals, slots which are not predetermined and which must be filled with the data pertaining to the specific event. Taking up the previous examples, the host may serve champagne or orangeade, and marriages may be held in church or at the registry office. Each frame, however, has from the outset a set of prototypical values which fill the slots, assigned as default values, values that are assigned when no other specification is present and taken as valid unless evidence to the contrary emerges. Thus, even when explicit information is not provided, the system will be able to infer a series of details concerning the situation involved based upon its general knowledge. For example, to the frame "birthday party" may be added values such as "merry atmosphere" and "gifts for the person in whose honour the party is being held" by default, even though it takes no feat of the imagination to conjure up a birthday party in which the atmosphere is bleak and the guests have not come laden with gifts. Likewise, default values for a wedding might include "betrothed in love" and "happy families", but everybody is fully aware of the existence of marriages of convenience and of marriages opposed by the parents.

Every time precise information is available on how to fill a slot, this replaces the default value, thereby allowing a specific frame to distinguish itself from the others. The more a specific frame departs from standard values, the more it stands out in the knowledge store.

Thanks to their two levels, frames blend the power of intensional definitions, which include all and only those characteristics identifying a class, with the extensional breadth of prototypical definitions, in which no element considered in isolation has crucial value of its own, but together with the other features forms a sufficiently broad panorama as to enable the identification of the class a phenomenon belongs to. This flexibility immediately seemed promising for cognitive science, given the human being's ability to work both on the intensional and on the extensional plane. Humans are indeed capable of handling situations requiring them to grasp the intensional aspect of a concept, as when having to explain to a child that a square is a geometrical figure composed of four equal sides and four right angles. Other times, instead, it is the extensional aspect that is crucial, as when one tries to explain to a child what is meant by the word "furniture", which makes the use of instantiating prototypes unavoidable.

Another important point is that the lower level slots are assigned values by means of filling procedures. While the frame is a typically declarative structure, the filling procedures are equivalent to production rules, thereby allowing frames to acquire potential enabling them to represent procedural knowledge. Filling procedures operate with rules of the following type:

[23]   IF α is the default value for slot s
        AND IF $\sim \alpha \to \beta$ occurs
        THEN   replace α with the observed value β.

In fact frames are the first knowledge representation structure to manage to integrate effectively the exigencies of a declarative type, satisfied by basic formalism, with the needs of a procedural type, guaranteed by slot filling.

The concept of frame was investigated in greater depth from a psychological point of view by Roger Schank and Robert Abelson, who renamed the concept *script*, thereby tying it to a cognitive standpoint, applying it above all to simulations of episodic memory (*Scripts, plans, goals and understanding*, 1977). A script is a predetermined and stereotyped sequence of events and actions defining a well-known situation. It is, however, incapable of self-modification, nor can it tackle situations that are totally new.

Figure 5.5 illustrates the concept of script with a paradigmatic example: the restaurant. This describes what may be expected to happen and how to behave in various types of restaurant. As is obvious

Script: RESTAURANT
Track: Coffee shop

Props: Tables   Roles: Customer
            Menu   Waiter
            Food   Cook
            Check   Cashier
            Money   Owner

Entry conditions: Results:
  The customer is hungry.   The customer has less money.
  The customer has money.   The owner has more money.
                                      The customer is not hungry.
                                      The customer is pleased (optional).

Scene 1: Entering
Understanding what actions are necessary for a customer to enter and sit down at a
table.

Scene 2: Ordering
Understanding what actions are necessary for a customer to obtain the menu, choose
the food he desires to eat, and inform the waiter. Then those actions necessary for the
waiter to transmit the order to the cook, and those necessary for the cook to prepare the
food.

Scene 3: Eating
Understanding what actions are necessary for the cook to give the waiter the food to
take to the customer, and those necessary for the customer to eat.

Scene 4: Leaving
Understanding what actions are necessary for the waiter to prepare the bill and take it to
the customer, and for the latter to leave a tip, go to the cash desk, pay and leave the
restaurant.

FIG. 5.5. Restaurant script.

from the example, a script helps one to understand not only what is
happening but also what one should do. In other terms, such structures
serve two purposes—comprehension of the world and planning a
subject's action. The interpersonal aspect is present at the same time as
the private sphere, at the subjective level.

One genuine novelty of scripts is that they are always written from
someone's point of view. The examination script differs depending on

whether the viewpoint of the examiner or the examinee is taken. The introduction of the possibility of representing different points of view remains a precious step forward in cognitive science, one enabling it increasingly to detach itself from an artificial objectivity of the world and simulate the various possible internal perceptions and reconstructions of the world.

A useful extension of the script is the concept of *theme*, which is to be considered as a generator of goals, which in their turn are to be achieved through the adequate use of scripts employed as action plans. Three types of theme exist: role, interpersonal and life themes. Themes permit individuals to activate not so much the script most suitable to the specific situation, even if this was Schank and Abelson's primary intention, but the interactions among plans at differing levels. A person may simultaneously activate different scripts, such as graduating in physics, flirting with that beautiful girl Christine, drinking iced tea. By arranging the different goals in order of importance, themes allow a person to pursue totally independent objectives contemporaneously, without their respective plans obstructing one another. Furthermore themes are structured as production rules, thereby increasing the procedural potential of scripts, as well as guaranteeing the system can react effectively if faced with an unexpected turn of events. To provide an example, a subject must be able to inhibit a possible sudden desire for a cup of tea, and remain seated and continue discussing when taking an oral examination at university, thus managing to persevere in the achievement of her life theme at that particular moment. On the contrary, if a tête-à-tête with Christine were to be interrupted by the arrival of the husband, the realisation of the original script would have to be promptly inhibited and substituted with a different goal, by means of an appropriate interpersonal theme.

Having outlined the qualities possessed by frames, we now turn to their weak points, of which there are basically three. The first has been called *frame problem* (though for reasons which differ from the term frame). It consists in the difficulty of formalising what remains unvaried in a situation in which, instead, something changes. A given action has known, predictable consequences, and non-consequences that are just as well known, but that cannot all be made explicit. If I break a vase, I change its macroscopic physical structure, but not its colour; even less do I change the structure of the table on which the vase lay; if the vase contained water, then the pavement gets wet, but the ceiling remains as it was before, and so on. Defining the changes that might occur is millions of times easier than defining the non-changes, the non-consequences, the non-effects. Despite its name, the frame problem is not limited to frames but is common to any type of knowledge representation—delimiting the consequences of processes and of actions is a general problem in artificial intelligence. Possible solutions are discussed in *The robot's dilemma*, edited by Pylyshyn (1987).

The second problem, which we may define as that of *identification*, is the difficulty of knowing which frame is the correct one to apply in a given situation. The frame furnishes the context within which it becomes poss''le to understand what is happening, to predict future events, and to act with maximun. iciency. But what is it that enables one to establish what the correct context to be activated is? The frame explains everything, except how one can select the right frame.

Even the final weak point is not exclusively a characteristic of frames. Common to all top-down structures that are heavily centralised, we shall meet this problem in an even more extreme form in Chapter 6 when we deal with perception. We may call it the problem of *resistance to novelty*. All those structures which explain what happens in a rigid fashion are incapable of identifying a genuinely novel element, one that was not foreseen. Instead they will tend to interpret it terms of what they already know, and therefore to assimilate it to previously existing knowledge. As a general rule, this is acceptable since it corresponds closely to human behaviour, as research in the fields of perception and memory have proved.

The New Look has demonstrated that humans perceive what they expect to perceive, or what they believe is reasonable. They require a great amount of evidence before they realise that a percept differs from their expectation. Frederick Bartlett (*Remembering*, 1932) carried out similar experiments on memory with equivalent results. Subjects tend to eliminate from memory all incongruities with their general knowledge, modifying, for instance, the contents of a story until all the abnormal aspects have disappeared. The value of a moderate degree of short-sightedness for adaptation is unquestionable, provided it is not transformed into total blindness. The meaning of a certain type of resistance to the unforeseen is to be sought precisely in the increased efficiency of a system capable of anticipating the future on the basis of some clue, instead of waiting for absolute certainty. Prey can be better hunted and predators more easily avoided if one possesses a good knowledge of the regularities of the world on which to found one's expectations.

Selection of the species has come about in such a fashion as to equip human beings with the capacity of surrendering in the face of the evidence, admitting incongruity or the unexpected occurrence. If because you expect to see a rabbit you fail to see a panther, then your life will indeed be brief. If the immaculately white shirt of the Dean of the Faculty were to betray a touch of lipstick on the collar, one may rest assured that the event would not pass unobserved, for everyone would admit that shirt collars, and above all the shirt collars of faculty deans, are improper and uncustomary places for the depositing of lipstick. But even this limited degree of freedom from expectations is impossible for frame systems, imprisoned in schemes that are efficient when dealing with what is known, but incapable of facing the unexpected.

Advantages: the unification of declarative and procedural knowledge, the effective management of familiar situations.

Drawbacks: difficulty in recognising the context in order to be able to apply the correct frame, inadequate for facing novelties.

Mental models

These constitute an analogic form of representing knowledge. There is a direct correspondence between entities and relationships present in the representational structure and entities and relationships of what is being represented. A mental model consists of *tokens* and *relations* representing a specific state of affairs and structured in such a way as to respect adequately the processes that will have to operate on them. Each model is therefore constructed to be coherent with its predicted use. The concept was introduced by Philip Johnson-Laird in his book, *Mental models* (1983), and developed above all in the area of inference (Johnson-Laird & Byrne, 1991).

The first example we shall provide is the mental model of an aeroplane, which comes in different versions depending on the uses that may be made of this entity: recognising one, building one, piloting one, talking about one, boarding one, etc. The model will also vary in accordance with other parameters, such as the aeronautic competence of the subject, his age, culture and any other relevant factor. Figure 5.6 shows two mental models of "aeroplane". The denotation [square brackets] will be employed every time reference is made to a particular mental model. Besides a central nucleus identifying it as an instantiation of [aeroplane], each version will contain diversified knowledge and manipulation procedures, since the things that are important and that must be stressed, will differ from case to case. Two instantiations of the same model may have little in common, beyond their label, if they have been built for completely different purposes.

There is no such thing as the one and only "perfectly correct" mental model corresponding to a given state of affairs. An assertion may usefully be translated into several models, even though it may be presumed that only one of these models will correspond in an optimal fashion to the state of affairs to be described, depending on the system's objectives. This enables both the intension and the extension of the concept to be represented. In actual fact, the nucleus of the mental model

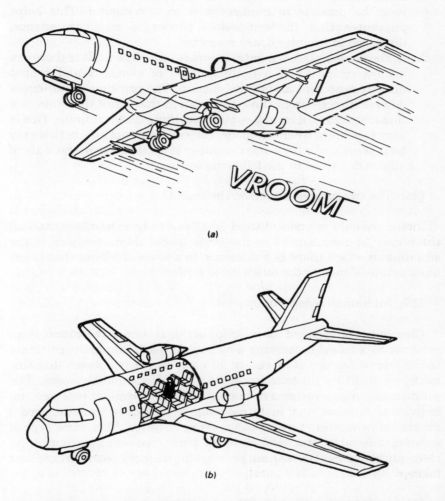

FIG. 5.6. Two mental models of [aeroplane]—one to recognise the object (a) and the other to travel on it (b).

represents the intension of a concept, that is to say the characteristic properties of the state of affairs described. The procedures for managing the model may be employed to define the extension of the same concept, that is the set of all the possible states of affairs the concept describes.

We shall now analyse the features of a mental model in greater detail, redefining what Johnson-Laird has already expressed in his fundamental work on the subject.

1. In the first place a mental model must be computable—that is, it must be possible to implement it on a computer. This helps guarantee that the constraints proper to cognitive science, discussed in Chapter 4, are respected.
2. In the second place a mental model is finite in size. It is therefore always composed of a finite number of elements and cannot directly represent an infinite domain. The principle of finiteness derives directly from the consideration that since the brain is a finite structure, it cannot contain anything that is infinite. This is easy to achieve when the model refers to states of affairs that may be represented with a small number of symbols, as in the state of affairs described in the following sentence:

[24] The two pirates landed on the island.

There are only two pirates and an island to be symbolised, and all three may be represented in the same model. More complex is the situation in which there is a reference to a state of affairs that is not numberable. Consider the following assertion:

[25] All human beings are mortal.

Clearly, to comprehend the meaning of this statement we cannot have recourse to a model containing a symbol for each human, predicating the universal feature of mortality for each one. It is obvious that this model by itself would take up all the brain's computing space. The solution lies in generating a restricted number of symbols representing individual humans and in introducing a procedure into the model capable of generating further elements as required. This same type of solution allows mental models to be built containing a number of elements in excess of the capacity of working memory, as in the case that follows:

[26] In 1989 Florence had 417,386 inhabitants.

Ruling out the possibility of constructing a model containing 417,386 tokens, each representing an inhabitant of Florence, the system may build a model containing a small number of elements that identify the concept of "an inhabitant of Florence", associating it to a procedure having the capability of generating other tokens if necessary (Fig. 5.7).

A further important consideration is that there is a potentially infinite number of situations that may be represented, whereas the mental mechanism for representing them is finite. From this

```
                    – – – inhabitant 1
                    – – – inhabitant 2
Florence                                              Date: 1989
                    – – – inhabitant . . .
                    – – – inhabitant 417,386
              (if necessary, generate further inhabitant tokens)
```

FIG. 5.7. Mental model of [population of Florence in 1989].

Johnson-Laird deduces that models must be built from more elementary components, which brings us to the third feature.

3. Mental models are constructed, and the basic construction units are tokens. These are arranged in a particular structure in order to represent a given state of affairs. When we deal with thought in Chapter 7, we shall see how model construction and modification procedures work in greater depth. As far as the basic elements are concerned, these may represent any concept in knowledge representation. Constituents of models come in three types:

    (a) *Primitives*. These are innate, linked to action and perception, and to basic cognitive capacities, such as comparing or memorising. Even basic emotions, such as joy and fear, fall within this category. These primitives lie at the heart of the perceptual, motor, emotional and cognitive capacities of the human system.

    (b) *Simple concepts*. These are built employing the semantic fields brought into being by primitives. To this category belong concepts such as "cause", "intention", "animal", "to the right of", "above", "before", "different from".

    (c) *Complex concepts*. These are built after the acquisition of simple concepts and are based on them. Examples include "observe", "calumny", "thoroughbred", "remorse", "infantile", "diesel engine".

4. The fourth feature is isomorphism. The representation must bear features that correspond to those of the state of affairs one wishes to represent. In the case of 24, for example, isomorphism lays down that there be two tokens to represent the two pirates, one to represent the craft they are travelling on, and one for the island.

Let us suppose 24 were to continue in the following way:

[27] The two bore a chest.

This would necessitate the addition of a further element for the chest. Let us suppose instead that, as in Fig. 5.8, the continuation were:

[28] The two carried pistols.

In its simplest interpretation it requires a further two tokens, one for each pistol, and so forth. This form of systematic isomorphism between the situation to be represented and the representation itself is the distinguishing characteristic of mental models. The propositional modes of representing knowledge, whose prototype is to be found in formal logic, modify quite radically the object of representation, paying no heed to its real structure, and transforming it in compliance with an internal syntax completely free from problems of correlation with the object to be represented. Analogical representation in mental models implies, contrariwise, that the state of affairs to be represented retains its own structure even after modelling has taken place. In this mode the structure of the model, its syntax that is, contains information itself.

Language is a good example of propositional representation. Words do not bear a trace of the structure of the concepts they refer to. The word "sinuous" does not look very different from "straight", nor "round" from "ragged". None of them expresses the concepts they recall in pictorial terms. The mental models corresponding to the four terms mentioned must instead also retain the figural aspects characterising sinuosity, straightness, roundness and raggedness respectively. Figure 5.9 illustrates this principle, furnishing the example of a mental model of a [square]. The geometrical definition of a square is: "a flat figure having four equal sides and four right angles". There is nothing in the above sentence suggesting the form and properties of a square. The model representation, however, directly exhibits the fundamental features of a square.

FIG. 5.8. Mental model of [two pirates armed with pistols landing on an island].

FIG. 5.9. Mental model of [square].

5. The fifth feature is that a mental model must be rigorously economical. The model has to be parsimonious in terms of the cognitive effort required. In other words, it must be able to offer those elements relevant to the state of affairs, and only those. At the same time, those symbols and relations preselected for representation must be the most important in relation to the objectives of the system so as to guarantee the maximum significance possible. The principle of parsimony further implies that a given state of affairs should be represented by a single mental model even if the description is incomplete or indeterminate. Representational economy imposes the condition that a mental model may represent indeterminate situations only if their use can be handled from a computational standpoint, without an exponential growth.

6. The final distinctive trait of mental models is that they may be embedded in one another. This permits the realisation of structures that are useful for the representation of reciprocal beliefs, which in turn are indispensable above all for the understanding of human communication. Think of cases such as:

[29] In order not to show she had understood Helen's intentions with regard to herself, Sylvia smiled.

Here Sylvia grasps what Helen is thinking about Sylvia herself (Fig. 5.10). Embedding may increase in complexity, though the limited power of our brain does not permit this to go beyond five or six levels.

Based on the features we have defined, Johnson-Laird has built a taxonomy of mental models, presented in Fig. 5.11. Within the category physical models, which represent entities belonging to the physical world, six major types of model are identified:

(a) *Relational*. This is a static frame consisting of a finite set of tokens representing a finite set of physical entities, a finite set of

FIG. 5.10. Sylvia's mental model of what Helen thinks about her.

properties of the tokens representing the properties of the entities, and a finite set of relations between the tokens representing the physical relations between the entities. The model in Fig. 5.8 illustrates this type.

(b) *Spatial*. This is a relational model in which the only relations between the entities are of a spatial nature. The model represents these relations by locating tokens within a two- or three-dimensional space. An example is provided by the model of [square] in Fig. 5.9.

(c) *Temporal*. This consists of a sequence of spatial frames corresponding to the order in which events occur. A series of photographs taken at successive moments in time, as in Fig. 4.2, exemplifies this type.

(d) *Kinematic*. This is a temporal model exhibiting psychological continuity. It represents changes and movements of the depicted entities without any temporal discontinuities. An example of this type is a canoe descending the rapids of a river.

(e) *Dynamic*. This is a kinematic model which includes relations representing causal relations between the events depicted. An

Physical models (representing entities in the physical world)

Relational
Spatial
Temporal
Kinematic
Dynamic
Image

Conceptual models (representing abstract entities)

Monadic
Relational
Meta-linguistic
Set-theoretic

FIG. 5.11. Taxonomy of mental models.

example of this type is the story of Snow White who falls asleep because she has eaten the poisoned apple given her by the witch.

(f) *Image*. This corresponds to the mental images we shall deal with later, or to the 2½-D sketch hypothesised by Marr, which will be explained in Chapter 6.

Within the category of conceptual models, representing more abstract entities, we have another four types of model:

(g) *Monadic*. This represents assertions about individuals, their properties, and identities between them. An example is the model corresponding to the assertion: "A triangle is different from a square".

(h) *Relational*. This introduces a finite number of relations between the tokens in a monadic model. An example is the model of the assertion: "There are more psychologists than ballet dancers".

(i) *Meta-linguistic*. This contains tokens corresponding to linguistic expressions, and abstract relations between them and elements in a mental model of any type. An example is the model corresponding to the assertion: "One of the ballet dancers is called Helen".

(l) *Set-theoretic*. This type of model contains a finite number of tokens representing sets. An example is the model corresponding to the assertion: "Among psychoanalysts, there are Freudians and Jungians".

Mental models are constructed as an explicit attempt to reproduce the basic structures of cognition. Having established this objective, they make a series of steps forward in the achievement of this goal. First, they push for the re-uniting of data structures and the functions adopted in manipulating the data, adapting the data structures to the functions that will utilise them. Even if they remain separate entities, data and procedures cohabit and interweave in the same model. The difference in representation between declarative knowledge and procedural knowledge disappears. The two types of knowledge continue to be different and are perfectly distinguishable, but they can now be represented in one single general modality, the mental model.

Second, mental models highlight the phase in which useful knowledge is constructed, rendering the process a relative one. A mental model never expresses what a state of affairs is really like, in the real world. Nor does it express the knowledge a subject has available on any object or concept. Rather it depicts what it is useful for the system or interesting for it to highlight at this particular moment. No unique representation exists of an entity (as it does in formal logic and in semantic networks). Nor does an optimal representation exist (as it does in production rules and frames). What does exist is a representation geared to the system's current objectives. A piece of chalk is an excellent missile to launch against a student who is disturbing the lesson—it is easy to hold, its loss is insignificant, and the impact is perceived neatly but without pain. That it may be used for writing on the blackboard, or what colour it is, are instead irrelevant to the context of throwing it, whereas it is evident that in other contexts the specific properties of chalk become significant. The point is that the mental model of [chalk] is not unique, always the same for all seasons. The model that is actually constructed and employed depends on the system's goals, and on the situation in which it finds itself. Since models are generated by the general knowledge available to the system, it nevertheless remains possible to build a prototypical model, equivalent to a frame with the default values assigned. This model of a paradigmatic type does not, however, enjoy any special status, nor can it be argued that the other models derive from this one. Mental models exhibit values for use, not absolute values.

Finally, mental models underscore the separation between general knowledge and active knowledge, between what the system knows expressed in terms of data and what it is employing at this moment. A mental model is always a partial activation of knowledge, a selection the system makes between what is relevant and what is not. All knowledge is potentially available, but only a minor part of it must be activated if total inefficiency is to be avoided. The initial decision determining what will be attended to greatly influences the cognitive processes that will be set to work starting from the original model built.

In conclusion, the knowledge made available to the system by mental models is unified and not dichotomised, relative and not absolute, aiming at attaining the most effective interaction possible between the system and the external world. Mental models are one method of representing reality which takes into account the intentional activity of the system, what it wishes to do, its subjective priorities.

The major difficulties encountered with mental models emerge when coping with abstract concepts, where the primitives have not been identified with acceptable exactitude. Their characteristics are lost when having to represent concepts such as "platonic love" or "intellectual vigour", devoid of an intrinsic structure, or at best possessing structures not easily amenable to analogic modes of representation. Where logical representations and their derivates are in their realm, mental models are not as efficient. As soon as we have to deal with labels or operators, mental models reach such a level of complexity as to become unreliable and hence not particularly credible from the cognitive viewpoint. This obliges them to accept coexistence with other techniques of knowledge representation able to handle that content for which it would be senseless to use models.

Advantages: merging data and procedures in a single structure, relativisation of knowledge, distinction between available knowledge and active knowledge.

Drawbacks: difficulty in passing from one domain to another, low level of integration between analogic knowledge and abstract knowledge.

Mental images

The representational mode which differs most from propositional structures is without doubt that grounded on mental images. Mental images were avowedly introduced to oppose the classic model of information processing. The latter did not provide for differences in the mode of representing data of different varieties—whether these were of a pictorial, linguistic, olfactory or other nature. The symbols employed in representation were all the same. The original input was completely transformed, and translated into a set of internal symbols, bearing no structural relationship to the reality that it was intended to depict.

The first scholar to challenge, in a series of brilliant experiments, the strictly propositional approach to knowledge representation was Roger Shepard (*Internal representations*, 1980). With a working program that was as innovative as it was controversial, Shepard demonstrated that human beings were capable of forming mental images of objects that may be more or less familiar and of comparing them. For example, subjects were asked to depict to themselves a geographical region and to answer questions on its shape. Or else two images were shown to subjects, one of which was either upside-down or specular to the first, and subjects were asked to evaluate whether the two images were identical or not. Shepard demonstrated how subjects effected the comparison by mentally rotating one of the two images. The greater the degree of rotation necessary (up to a maximum of 180°), the greater the time required to carry out the comparison. The highest degree of difficulty is shown in Fig. 5.12. The final touch is that Shepard also managed to extend these results from visual perception to auditory perception.

Later research by Allan Paivio demonstrated that images can facilitate the performance of memory, offering concrete advantages over an exclusively propositional representation of memories. Paivio took his inspiration from an anecdote from the life of Simonedes of Samos: during a banquet, the poet was called out to receive a message, and while he was outside the roof suddenly fell in, burying all the guests. Nobody knew who exactly had died, but Simonedes managed to list all their names by imagining he was walking round the table and naming the guests as he saw each one mentally. Developing this first intuition, Simonedes created a mnemonic technique dear to the Greek and Roman rhetoricians, the so-called method of loci (from *locus*, place). This technique is based on associating each particular concept with one particular aspect of a familiar place while imagining one is slowly passing through that place. This makes it possible to repeat a long speech by heart, since each key point will have been associated for instance with the room one is in, and will then be recalled by looking systematically round every part of this room, either with one's eyes or in one's imagination, and expressing the concept associated with that part of the room the eyes or imagination fix on.

FIG. 5.12. Are the two letters identical? *Source*: Kosslyn (1980).

In *Mental representations* (1986) Paivio takes up the subject of the connections between memory and images, demonstrating how words stored employing a double code (verbal and by image) are memorised better than those stored utilising only the verbal modality. For example, concrete nouns such as "vase" are easier to remember than abstract nouns such as "value" by virtue of the greater amount of information contained in images compared to that conveyed by words.

It was, however, Stephen Kosslyn, in *Image and mind* (1980), who bestowed full dignity upon mental images as a technique for knowledge representation that could not be reduced to any other modality, as well as building a theory of mental imagination that was perfectly attuned to the paradigm of cognitive science. Mental images are not literally shapes inside one's head. Instead they are mental representations allowing one to experience "seeing" something even when the corresponding visual stimuli are absent. This sort of phenomenon involves both a representation and a conscious experience of the representation itself, a unique case in the domain of the methods of knowledge representation.

For Kosslyn the image is always a mental representation, whereas seen from the standpoint of the subject experiencing the image, attention is focused on the objects represented, not on the means that represent them. Images are active in a short-term visual buffer, and are generated by more abstract representations, based on prototypes located in long-term memory. The origin of mental images is essentially perceptive, even though they are not tied to the actual presence of an external correlate: we can imagine anything, if we so desire. Kosslyn points out that we are capable of imagining entities we have never perceived, such as for example a blue crocodile entering a cake shop. From this he deduces that images are generated by putting together base elements and then transforming them in accordance with laws still to be explored.

The theoreticians championing mental images have faced the knights jousting in favour of the propositional thesis (e.g. Pylyshyn, 1984), in whose opinion a representation corresponding to an image comes nearer to a verbal description than to a picture. According to propositionalists, image representation is not pictorial; rather it is generated employing the same type of procedures applied in the field of perception. Once constructed, taking a propositional structure as its starting point, it is used "as if" it were an image. A major part of the experimental studies carried out by supporters of mental images was triggered off by the harsh criticisms aimed at denying any possible form of existence of a sort of representation differing so greatly with the types hypothesised by propositional theorists. The fruits of the conflict were summarised by Kosslyn in *Image and brain: The resolution of the imagery debate* (1994). They have been such as to convince the neutral bystander that mental images are not simply a laboratory invention, but one of the functions active in the human mind.

The decisive factor in demonstrating that mental images really were an independent method of processing information was proving that objects visualised in the mind retain the properties of external objects. It is thus possible to simulate mentally what would happen in a real physical situation, with the advantages that ensue in terms of planning and problem solving. Figure 5.13 shows the mental images constructed by subjects asked to visualise mentally (a) a rabbit with an elephant and

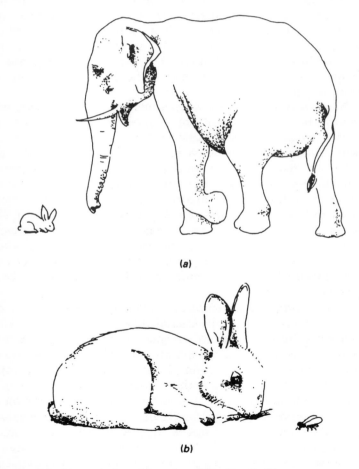

(a)

(b)

FIG. 5.13. A rabbit with an elephant (a), and a rabbit with a fly (b).

(b) a rabbit with a fly. When subjects are asked questions on details concerning the rabbit, they take more time to supply the answer in case (a) than in case (b). The explanation advanced by Kosslyn is that in (b) the anatomical details of the rabbit have already been enlarged in order to bring out the contrast with the fly. In (a) instead, the details must first be enlarged and then inspected mentally.

By imagining what would happen in reality, a person can save physical and cognitive resources, without needing to experience the situation directly. If I am taking an umbrella into the house I do not need to bump into the door to understand I must first close the umbrella and then keep it in an upright position. I can simulate what would presumably happen, and act to avoid the foreseen errors. In *Ghosts in the mind's machine: Creating and using images in the brain* (1983), Kosslyn hypothesises a developmental advantage of the capacity to form mental images, but even without following him down such a slippery road it is difficult to deny their existence in the face of the large amount of evidence in their favour.

Mental images have never aspired to be more than just one of the modes in which knowledge is represented. They have never claimed to be the sole modality, nor even the most important from a cognitive point of view. The weakness of this approach lies in not having yet managed to define the operations governing the way images are built and handled, even if computer simulations have recently been implemented on a few simple processes of this nature (see Finke, *Principles of mental imagery*, 1989). However, until construction procedures are accounted for, simple experimental evidence will continue to leave many scholars sceptical. The questions seeking an answer no longer concern the existence or non-existence of images, but how they are formed, transformed and utilised.

To this end, a lack of correlation between mental images and techniques which are closest to them, above all mental models, constitutes a further difficulty. The more pictorial aspects of mental images are lacking in mental models, given that the construction of the latter pays greater attention to the cognitive and intentional dimension than to precisely mirroring the perceptual correlates. But, while the image of any given entity, following the coherent rules of perception, is substantially always identical, the model of that very same entity changes depending on the cognitive use to which it is to be put. I have a standard mental image of an ice cream which is stable, whereas the mental model of [ice cream] changes according to whether I wish to eat it or use it to dirty the trousers of the person seated next to me. There exists, therefore, an area that can only be explained by mental images, as Shepard and Kosslyn have always rightly argued. Having established this fact, what is at stake now is explaining the laws governing them with some precision.

Advantages: iconic power.

Drawbacks: impossible to represent knowledge that is not exclusively figural.

Subsymbolic representation

The most significant news within knowledge representation in the last decade is *subsymbolic representation*, a radical version of the connectionist approach.

Connectionist knowledge representation differs radically from the classic mode employed by artificial intelligence to represent knowledge. In neural networks, subsymbolic-level processing is also responsible for knowledge representation. The main characteristics of subsymbolic processing is the distinction between computation tokens and conceptual tokens. Computational tokens are simple individual units, which do not refer to any explicit concept: they are *subconceptual*. The subconceptual level nevertheless remains higher than the neuronal network.

It has to be noted that a connectionist model is not necessarily subsymbolic (and vice versa). In fact, in *localist networks*, semantic content is linked with the single units. In the following, we shall refer to the more interesting *distributed networks*, which are entirely subsymbolic. The terminology adopted is that used by the connectionist theoretician, Paul Smolensky, in his rigorous work, "On the proper treatment of connectionism" (1988).

By the term *subsymbolic*, Smolensky intended to suggest cognitive descriptions built up of *constituents* of the symbols used in the traditional symbolic paradigm; these fine-grained constituents might be called *subsymbols*. Entities that are typically represented in the symbolic paradigm by symbols are, in the subsymbolic paradigm, represented by a large number of subsymbols.

Knowledge in a connectionist model coincides with the status of a complex dynamic system. In particular, this dynamic system consists of a network of processing elements; interactions between individual units are simple. Each connection is assigned a parameter, called the connection's weight. Such a weight may be interpreted as the synaptic efficiency between the two units linked up. The set of the parameters assigned to all the connections in the network represents the status of the system. Learning is equivalent to the modification of the dynamic system, obtained through algorithms such as Hebb's rule or back propagation, explained in Chapter 3. In neural networks, knowledge is stored in the connections, which may have excitatory or inhibitory power over the linked units. This distinctive method of representing knowledge has many interesting consequences.

First, knowledge is intrinsically procedural. The network associates a specific input pattern with an output pattern, without explicitly coding knowledge in symbolic terms. Such a feature makes declarative knowledge difficult to manage for a connectionist system. Analogously, the knowledge acquired by the network during learning may not be directly expressed at the symbolic level. In sum, knowledge may not be separated from the learning process.

Second, knowledge—like intelligence—is considered as an emergent property of the system. A connectionist model learns by induction over a great number of examples; knowledge cannot be defined *a priori*, but corresponds to a new status of the system.

Third, knowledge is distributed. The association between two patterns is a piece of information contained within the entire set of parameters. Local damage has a low effect over all the connections, but does not selectively affect a specific connection. There are no local properties: e.g. it is not possible to modify only one part of the dynamic system. Each change propagates to the global status.

Subsymbolic representations have given rise to an infinite number of computer implementations, in every cognitive domain, from perception to linguistics and reasoning. It is not, however, an easy matter to integrate them into the other methods—and results—of classic cognitive science. The reason is, to use Smolensky's words, the cornerstone of the subsymbolic paradigm: *the subconceptual connectionist system does not admit a precise formal description at a conceptual level.*

Advantages: Subsymbolic knowledge is content dependent; connectionist networks associate input and output with a good capacity for generalisation with regard to new input (e.g. categorisations). It is also context dependent: partially equivalent inputs in a network can produce different outputs. The differences are attributable to the different parts played by the equivalent input in the specific context. In general, subsymbolic representations are efficient in all domains where procedural knowledge is involved, and where it is impossible or difficult to formalise knowledge in a declarative mode.

Drawbacks: The limits of subsymbolic representations show up each time abstraction is required, or when context independent knowledge is necessary. Moreover, they are difficult to utilise in domains where knowledge is highly structured, or already formalised at a conceptual level (mainly macrocognition).

A final comment concerns computer simulation of the representational modes we have examined. All of these modalities have been implemented in numerous programs, described in the standard reference texts mentioned. Of the seven methods, the first four have been developed in numerous computer versions, the fifth (mental models) in a small but significant number of versions, while the sixth (mental images) still lags behind from this point of view. The last method has innumerable implementations, since each subsymbolic representation corresponds to a specific computer model.

MEMORY

Memory consists of a set of procedures for the conservation of information over a short period of time (short-term memory), for the management and retrieval of general knowledge (long-term memory), and of a special subsystem (episodic memory) devoted to the handling of temporal data. It is important to emphasise that viewing memory as a set of procedures means emphasising its creative aspects, namely the active construction of memories. This approach conflicts both with the standpoint that holds memory is an autonomous function, and virtually an organ, an independent part of the brain, and with the stance that limits memory to a data archive in which data are stored once and for all, ready to be extracted exactly as they are. A recommendable introductory reference text on the subject is *Human memory: Theory and practice* (1990) by Alan Baddeley.

We shall distinguish between two fundamental types of memory, short-term and long-term memory, schematised in Fig. 5.14. Episodic memory is a long-term memory with special features.

Short-term memory
The function of short-term memory, a scholarly treatment of which is to be found in Baddeley's *Working memory* (1986), is essentially one of selection: to choose which of the continuous waves of stimuli reaching

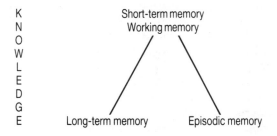

FIG. 5.14. Types of memory and the relationships among them.

the system must be dealt with. To carry out this selection, the system must be able to avail itself of the sensory input and of the intermediate stages of the processing of this input for a brief period of time, in order that both the input and the results of processing find a stable destination.

The principal task of short-term memory is to serve as a working memory. Two subsystems that are more specific with regard to the sort of information that may be stored can be identified: a verbal subsystem and a visuo-spatial one. Verbal information is kept active by an *articulatory loop*, which may be compared to a magnetic tape that can preserve the information for approximately two seconds. Beside the articulatory loop, there is a second subsystem, which is responsible for equally brief conservation of information of a visual and spatial type. Both the formation of mental models and of mental images is ascribed to this subsystem. Some scholars refer to it metaphorically as a kind of sketch pad for visuo-spatial information. This sketch pad retrieves material contained in general knowledge in an abstract form, and keeps it active in a visuo-spatial form so that further processing may take place on the material. The two subsystems described work in an integrated fashion thanks to a third mechanism, the *central executive*, which organises and coordinates their functioning.

The quantity of information short-term memory manages to hold is limited. In his famous article, "The magical number seven, plus or minus two", George Miller (1956) quantified its maximum capacity, the so-called *span*, around the number seven (plus or minus two). This means that we can remember 7 ± 2 chunks of data; for instance we can remember 7 ± 2 separate figures, or 7 ± 2 groups of figures, or 7 ± 2 items in a list of names, but in any case never more than 7 ± 2 independent sets of data.

Long-term memory

How large is human memory? According to the calculations made by Thomas Landauer (1986) a normal adult will have accumulated approximately 10^9 bits of information. This estimate takes into account how much is stored and how much is lost over considerable periods of time. Long-term memory acts as an intermediary between short-term memory and the knowledge possessed by the system: it receives the information selected by short-term memory and subsequently stores it in knowledge. Other functions carried out by long-term memory include managing the information in knowledge and retrieving it as required. We shall examine each of these functions separately, with regard both to data furnished by short-term memory and to data provided by one's general knowledge. However, since all of these operations are also carried out by episodic memory, we shall examine this type of memory before reviewing the various functions.

Episodic memory

Episodic memory, a notion first advanced by Endel Tulving (1972), is a type of long-term memory specialised in handling temporal knowledge. This knowledge refers in some way to events in time, or to specific instances as opposed to prototypes and general entities. Specialisation in biographical-temporal knowledge is not the sole characteristic identifying episodic memory. A second distinguishing characteristic is that it is normally associated with declarative memory (e.g. by Anderson and by Tulving himself). Episodic memory contains those specific events the system has come into contact with, both directly and indirectly. The first, and most important, class of phenomena conserved in this type of memory are episodes in the individual's life. These take the shape of stories in which one of the protagonists is the system itself. This category includes events of the type:

[30] The death of my father.

[31] My graduation examination.

A second category of episodes stored in episodic memory refers to those events which are not experienced directly, but with which the system comes into contact through already existing knowledge structures, as exemplified by verbal tales, books or films. This category includes events of the type:

[32] The final shoot-out between Paul Newman and Robert Redford against the Bolivian army, in *Butch Cassidy and the Sundance Kid*.

[33] Anthony's funeral oration for Caesar, in Shakespeare's *Julius Caesar*.

The procedures employed to store, manage and reconstruct episodic memories are identical to those operating in long-term memory. We now turn to a more detailed examination of these procedures.

Storing

This comes about on data furnished by short-term memory or from K-model, and which are to be stored in knowledge (tacit or explicit). It must be emphasised that the procedures utilised in storing knowledge do not operate on real, objective data, but on data that have already been processed and subjectively interpreted by the system. For example, in a case of mistaken identity where the error is caused by sensory memory furnishing an incorrect piece of information, then the system can be absolutely certain that its memory of the event is correct, no matter how much further evidence there might be disproving the correctness of the event itself. Stated differently, the fact as it is now remembered is in itself correct, whereas the initial perception, which is no longer retrievable, was incorrect.

The system already loses contact with external reality as soon as short-term memory starts functioning, since external reality is interpreted in accordance with a pre-existing mask depending on the scheme adopted in system–world interaction. Two people involved in the same situation will remember that same episode in different ways, since in comprehending it, they will have conceptualised it in different ways. The psychology of giving testimony has produced convincing evidence of this point, demonstrating how eyewitnesses relate a simple episode in profoundly different ways according to how each person experienced that particular episode (see Christianson, *The handbook of emotion and memory*, 1992). Knowledge thus depends on how memory has organised the incoming information, with categorisation taking place in an arbitrary fashion according to the interpretative schemes favoured by the system.

Memories structured as tacit knowledge channel the way the system proceeds both with regard to the external world (as a result of which I am able to remember how to ride a bicycle, once I have learnt to do so) and with regard to the internal world (as a result of which I have "gut" memories of certain situations, memories which attract me to or repel me from similar experiences). Such memories may be emotionally charged, and may be activated both by explicit structures (for example, one's first kiss) and by tacit structures (taste, smell, the emotions which accompanied that first kiss).

Management

The capacity of long-term memory is finite, in as much as it is based on a finite physical system, the human brain. Nevertheless, the capacity of knowledge to preserve information seems practically inexhaustible—we can always acquire new information, and it is by no means certain that what we cannot remember has really been lost for ever. Often something we believed we had forgotten returns to us, and even relearning something we do not consciously remember is easier than it was the first time we learned it. The mechanisms which oversee the management of knowledge structures concern both the data and the schemes by means of which the data themselves are stored and reconstructed. The principal modifications come about through updating and restructuring the systems of knowledge representation, a topic that has already been dealt with.

Retrieval

This may come about on data from K-explicit or K-tacit. Although the results differ, they nevertheless finish up in K-model. Retrieval takes place by means of structures analogous to those employed in storing memories. These general knowledge schemes organise single data into significant sets. What is involved, therefore, is not the recall of experiences statically fixed in some moment in the past, but their reconstruction grounded on one's knowledge of the world.

General retrieval schemes may be considered to be the equivalent of frames. They become the determining factor in obtaining the final conscious memory, much more so than the specific data that give that memory its unique identity. Frames constitute the expressive structures employed by memory to render its data readable, furnishing the interpretative framework for what was stored in the past. If we accept the fact that data remain stable over time, then to account for the transformations that take place in remembrances and even for part of forgetting itself, one need only assume that what is actually modified are the general knowledge schemes. Changing a frame or adding a new one is tantamount to changing the active part of memory, and hence what one is subjectively able to remember too. Access cannot be gained to the past except through these interpretative masks. Being part of one's present knowledge, namely the knowledge that is used continuously and daily, these masks are in their turn continually updated. Such automatic updating has the important advantage of allowing the system to utilise the most recent knowledge it possesses. Just as we perceive what we predict we shall perceive, so we remember what corresponds to our image of the world today, and not to the schemes that were active when the event occurred (see also Chapter 10). Memory in everyday life is described by Cohen in *Memory in the real world* (1989) and Neisser in *Memory observed: Remembering in natural contexts* (1982).

To claim that memories are constructed can be paraphrased by saying that even if memories refer to episodes that happened in the past, they are built in the present of the system. Therefore, the key factor is present knowledge. Memory processes employ data stored in past time but their structuring into a meaningful remembrance always occurs on the basis of what the system knows now, and not of what it knew then. Because these reconstructive processes are impenetrable to awareness, the subjective impression is that one simply rehearses something exactly as it happened, without any changes intervening between then and now.

An example will clarify what is meant by the reconstruction in the present of an episode pertaining to the past. If Raphael has been betrayed by Monica, whom he considered for years a loyal and faithful friend, it will be almost impossible for him to remember her as he thought she was prior to her betraying him. Even in a remembrance going back years before Raphael's delusion, Monica will present traits consistent with how Raphael perceives her now, and certainly inconsistent with his opinion about her at the moment the episode occurred. A new version of Monica has substituted the older version throughout the entire body of Raphael's active knowledge. Note that the older version of Monica need not be cancelled, it simply becomes less accessible.

A mechanism also exists providing direct access to knowledge, without requiring the assistance of intermediate structures. This direct form of retrieval is always set off by a cue, a suggestion that may be both external and internal. Examples of external cues include things seen, heard or perceived through the other sensory organs activating the pathways leading to the datum existing in knowledge. A cue may activate both memory procedures or the data themselves directly. In the latter case, we speak of direct access, as proven by the extremely brief retrieval times when compared to those obtained during reconstruction (or indirect access). Internal cues are those created or experienced by the subject, such as mental and emotional states. As we shall see later, direct access occasionally generates flash-back to a remembrance of an episode as it was, without changes attributable to reconstructive processes.

Retrieval structures are both cold and hot, since cues and schemes are linked both to cognitive and to emotional aspects. This involves a double categorisation, one with rational roots, the other with emotional roots, thus accounting for the results obtained by Gordon Bower (1981) on the effect of emotional states on memory. Inducing happiness or sadness in experimental subjects, Bower noted that the memories produced spontaneously by subjects mirrored their mood at the time. A person in whom happiness has been aroused generates a greater number of joyful memories than a person in whom sadness has been engendered, and vice versa with regard to depressing remembrances. According to the hypothesis of the correspondence between mood and the polarity of the material to be retrieved, memories may be recalled by using their emotional overtones. Just as it is easy to list a a series of objects bearing a common physical feature (for example, "red-coloured objects"), so it is simple to remember events exhibiting similar emotional traits (for instance, "moments of melancholy").

Exactly like all the other mental functions, memory is in a constant state of flux. The illusion of a stable subjective past fails to find corroboration except in the desire to maintain the system unchanged over the course of time. In contrast, diachronic integrity of the system as well as its internal coherence are to be sought after in change, and not in an impossible state of immobility within the flow of time.

Forgetting

The mechanisms by which we forget are as fascinating as they are mysterious. It seems impossible to find a general answer to the question: "Why does one forget?". The causes of forgetting are to be sought in the malfunctioning of storage and retrieval procedures. With regard to storing, loss is to be ascribed to processes of decaying or of interference. *Decay* is due basically to time passing. One forgets a piece of information that has never been used again, not even indirectly. For instance, who can remember the address of a restaurant one has only been to once 10 years later? *Interference* is caused by the existence of a significant amount of data of a similar nature to that required, thus disturbing the clarity of the remembrance. In other cases, it is retrieval that fails to work well, as in the "tip of the tongue" phenomenon. The subject is aware the information is stored in memory, but is unable to extract it. In the majority of cases it is difficult to establish the cause of forgetting. Why, for example, do we not remember the name of the person who sat next to us in primary school despite the fact that we knew the person perfectly well at the time? Even though it is impossible to establish *a priori* if such a failure is attributable to the real loss of that piece of information, or to some inadequacy in retrieval procedures, various sorts of considerations point decidedly in the direction of the latter factor, as we shall discover immediately.

Evidence from clinical studies indicates that information that seemed irretrievably lost is instead recoverable, on condition that one makes a serious effort to do so. Patients undergoing psychotherapy begin by recovering memories of when they were 5–6 years old, and then proceed backwards until 3, 2 and sometimes even 1 year of age. Likewise experiments on the transfer of knowledge indicate that any set of data—such as a list of meaningless syllables—that experimental subjects have no interest in is relearned more rapidly than a completely new set, even months after the administration of the initial test took place. Introspectively speaking, we continually remember something we believed we had totally forgotten, perhaps when stimulated by another person ("Do you remember the time we waited for each other for two hours at opposite ends of the Tower Bridge?"). Or going back into a suggestive atmosphere, as when an old situation recreates itself, where a hard-won initial glimpse into the past paves the way for the gradual surfacing of many bygone and unexpected memories, which, until that moment, had apparently been lost forever. Or finally through a dream, a smell, an image that vividly bring back entire episodes from the past.

Forgetfulness of the kind that may be defined as "convenient" was brought to light by Sigmund Freud in his studies on the psychopathology of everyday life (1901). In his revolutionary observations, Freud proves that many of our daily acts of forgetfulness may be attributed to unconscious desires coming into conflict with a person's conscious intentions. A case in point is forgetting to make a telephone call when the expected content is unpleasant—think of a manager deciding to call a newly promoted rival to congratulate the person. In cases of this sort, the aim of forgetting is to avoid a negative experience. Disagreeable experiences are often considered to be easier to forget.

Underscoring the benefits accruing to the system of such forgetfulness, as has already been done above all by clinicians, would be no arduous task. However, such benefits would be of a superficial and dangerous nature, for the simple reason that the person would thereby demonstrate an incapacity to take strategic, long-term decisions tending to ward off the repetition of negative experiences similar to those forgotten. What on the other hand denies this thesis general validity is that not all that is forgotten is negative, and not all that is useful or positive is remembered, as any student about to take an examination can testify. Ingenuous teleology does not hold up even in the field of memory.

Biochemical research has always been extremely active in the domain of memory, in the hope of synthesising a substance that would be able first to increase the effectiveness of memory in normal subjects desiring this result, for instance for reasons of study; second to block or slow down decline in senescence; and finally to alleviate pathologies with a neurological base involving amnesia, such as Alzheimer's disease. Although the results have not been spectacular, at least to date, research in this sector has repeatedly stressed the fact that amnesic behaviours are invariably attributable to problems of retrieval, not of storage. All drugs combating amnesia point to the fact that subjects who do not remember possess the information, that such information has been correctly preserved, but that they are unable to activate the knowledge. The administration of clinically active substances to experimental subjects or animals enables retrieval to take place, but it does not noticeably improve storage.

It may be concluded that forgetting is due essentially to problems of retrieval or the activation of information, and not to a loss of the information itself. As to the reasons for forgetting, these are to be sought either in mechanisms interfering with conscious intention, or in an inadequate activation of retrieval schemes.

Another phenomenon that is interesting to analyse goes under the name of *infantile amnesia*. Infantile amnesia, acknowledged as being a mystery by all those psychologists who have dealt with infancy, refers to the phenomenon by which adults are incapable of remembering their early experiences, those occurring before the age of 4–5. The psychoanalytic interpretation is that infantile amnesia is caused by repression: nothing potentially associated with childhood sexual wishes can be remembered after resolving the Oedipal crisis. The Piagetians take the view that it constitutes evidence of thought changing in a radical manner, with the result that those mechanisms adopted in infancy can no longer be employed. Neuropsychology has recently advanced a new hypothesis, connected to the differing maturation rates of the encephalic structures. Jacobs and Nadel (1985) argue that the hippocampus, the area where knowledge is thought to be stored, matures late (around 2–3 years), thus causing the loss of all memories preceding its full development.

In "Modifications of knowledge by memory processes" (1984), I advance an explanation linked to the procedures governing the retrieval of memories. Adults find it difficult to recall their early years because their schemes for interacting with the world have been modified and updated. All, or a large part of, the data are still available, but the interpretative scheme has changed so that the data can no longer be assembled into a meaningful structure. When speaking of the modification of schemes, what I mean is not simply an adjustment in knowledge to make it more suitable to the system's changed situation. It is obvious that the frame corresponding to having breakfast is profoundly different at 3, 12 and 40 years of age. The perceptions and the actions carried out have altered. The most important change, however, refers to the emotional values assigned to different schemes at different ages.

During early infancy, cold cognitive procedures are not clearly distinguishable from emotional ones (see Chapter 10). What it is impossible to reactivate are the emotional correlates of frames, and this is a crucial point since the more one goes back in the history of the system, the more important such correlates become. By totally reforming the previous schemes, with regard to both knowledge and emotions, a subject is able to overcome partially infantile amnesia, as this enables the memories that were only apparently lost to re-emerge. Although this type of reconstruction cannot be achieved in the laboratory, it is a frequent occurrence in the clinical field. Nor is it exceptional for adults to go through this experience if they come to find themselves in a situation in life that carries them back into the past. Visits to important figures in one's life, and to places familiar in infancy but which one has not since been to, seem to be the best recipe for old, forgotten schemes to return from oblivion, accompanied by their cognitive associations and emotional charge.

The picture we have drawn of memory is a fluid one, one of constant change, and not one that has hardened into the mould of rigid schemes or data fixed once and for all. We may even note an encouraging likeness with the process of language comprehension. As we shall see in Chapter 8, the literal meaning of a phrase is soon forgotten, while the intended message behind the form is preserved. Syntax is useful only as a vehicle for semantics. Similarly, memory processes extract the meaning of an episode, and that is what one remembers. Literalness, or reality if one prefers to put it that way, does not come into it.

Numerous computational models of memory exist, and they employ the modes for knowledge representation expounded in the preceding paragraphs. We may here recall the theory that has been implemented with the greatest care, John Anderson's ACT* theory (1983). Anderson adopts a scheme that is very similar to the one presented in Fig. 5.14. Long-term memory is called declarative, and episodic memory productive. While the functions of working memory and of declarative memory are more or less identical to those already described, productive memory is devoted to preserving abilities, seen as procedures the system knows how to execute. The three memories are organised into a formal network of conceptual nodes and relations, in which storage and retrieval depend on the activation of the links existing between the nodes in the network. Knowledge is represented in the form of production rules constituting a complex architecture capable of simulating many of the characteristics typifying human memory, with a relatively high degree of precision.

LEARNING

Learning may be defined both as the acquisition of new knowledge, by which we mean above all a series of data (first-order learning, linked to K-explicit), and as the ability to improve one's performance autonomously, by generating original strategies to increase efficiency in tackling new and old problems (second-order learning, linked to K-tacit). As always, in actual practice this distinction is less clear than it would appear from the definitions. Often we change data and procedures simultaneously as, for instance, when learning to use an electronic microscope, or to pilot a helicopter. The distinction will be maintained here for explanatory ease, and because it nevertheless pinpoints a difference that is already significant at the level of knowledge.

Every form of learning sets the problem of making the new consistent with what is already known, in order to avoid the creation of contradictions. How this is actually achieved is difficult to establish. At one extreme lies classic logic, holding that each new belief must be compared to all pre-existing beliefs. This solution is not particularly credible since it would require terribly long processing times. At the opposite extreme lies non-monotonic logic that is not concerned with coherence between new beliefs and preceding knowledge. This solution cannot therefore exclude the existence of local contradictions in the system's knowledge. What seems to be a defect is in fact a feature that is compatible with what is known about memory, since humans certainly tolerate contradictions between beliefs. We are, however, a long way from finding a satisfactory solution to the problem.

First-order learning

The acquisition of new knowledge is simple only if the operation is restricted to adding it to the previously existing database. Dispersed data, on the contrary, would be impossible to utilise. To make new knowledge useful, it is essential to integrate it into and make it interact with the knowledge already structured in the database. Thus, any new information modifies previous knowledge to a significant extent. Each integrated addition of data involves some form of revision of stable knowledge, with possible problems of incompatibility, as observed above. Only integrated knowledge can be used efficiently, because the system is able to retrieve it every time it requires to do so.

Second-order learning

Second-order learning may be gauged only through an improvement in performance. The most important and most elusive part of human learning falls into this area, namely the part regarding not an increase in the number of notions possessed, but an improvement in the ability to acquire them, select them, and grasp the essential points about them. It is a question of increasing the efficiency of learning processes, and it is in this sense that the anthropologist Gregory Bateson speaks of learning to learn (*Steps to an ecology of mind*, 1972). One special case of second-order learning is scientific discovery, where a person achieves a creative reorganisation of knowledge, inventing concepts and strategies that do not derive from the realm of the known. Scientific discovery is close to problem solving, even though it deals with very special problems since, by definition, no one must have solved them before. It will be treated here as a case of learning to underscore the innovative aspects compared to previous knowledge.

The first results of some importance were obtained in the field of concept learning by Patrick Winston (1975) who took up the work of Bruner, dealing with it from a computational standpoint. Winston employed procedures analysing examples of a concept, learning the meaning of the concept through slight discrepancies, deliberately committing near misses in order to explore both the boundaries of the concept and the limits of the comprehension procedure itself.

A significant step forward was taken by Douglas Lenat (1982, 1983a,b) in simulating mathematical discovery. His first program, AM, inspired by the work of Polya on heuristics, independently rediscovered important mathematical concepts in numerical theory. AM did not attempt to process concepts that were consistent with empirical data, as did Winston's programs, but utilised heuristics focusing on the degree of "interest" a concept offered. Interest was measured in terms of its relationship with other concepts, its capacity to generate examples, to subsume other concepts, and the like. In his even more important subsequent work, Lenat implemented EURISKO, which manages to discover not only new concepts, but also new heuristics, though remaining within the domain of mathematics.

Langley, Simon, Bradshaw and Zytkow (*Scientific discovery*, 1987) designed a series of programs for the simulation of the creative processes underlying scientific discovery: BACON, GLAUBER, STAHL and DALTON. BACON, emulating the philosopher Bacon, works on quantitative laws, seeking to construct laws explaining a well-defined set of empirical data. Playing purely syntactic games with the numbers available to it and adopting a data-driven approach, BACON has succeeded in discovering an amazing series of laws: Boyle's law on the volume of gases, Galileo's law of uniform acceleration, Archimedes' principle and many others. Adding heuristics searching for the properties—which are fundamental in physics—of symmetry and conservation, BACON has further improved its performances, discovering theoretically complex laws such as Snell's law of refraction concerning the passage of a ray of light from one medium to another, or Joule's law on the conservation of energy, positing the constancy of energy in a system—energy is transformed, it is neither lost nor gained.

The remaining three programs are capable of generating qualitative laws, and not simply quantitative ones, based on empirical data. They are also capable—with regard to certain topics in chemistry—of building structural and explanatory models. The distinctive feature of GLAUBER, the name deriving from that of a 17th-century German chemist, is that it possesses heuristics that are useful for forming groups of concepts. That is, starting from empirical observations, it can generate taxonomies grouping together similar objects. In this way, once it had been provided with an adequate set of experimental data on various chemical substances, GLAUBER discovered anew the abstract concepts of acid and alkali.

STAHL, a program whose name was taken from that of the German chemist who developed the theory of phlogiston, possesses heuristics capable of constructing componential models of various substances. STAHL possesses specific knowledge concerning the composition of objects and the preservation of basic substances. It can build up explanations in the guise of descriptions of the structures on which substances and reactions are based. Initially it is only data driven, but once having built up conjectures on hidden structures, it is driven by the conjectures themselves, that is by its own theory. Finally, thanks to its general heuristics, it can choose between multiple conclusions as well as handle any inconsistencies that might have emerged in the results obtained. STAHL has been applied to the controversy that reigned between chemists supporting the theory of phlogiston and those upholding the theory of oxygen. It manages to follow the line of reasoning of both camps, and succeeds in concluding that the oxygen-based theory is the more solid of the two.

Finally, we come to DALTON—the name having been borrowed from the 18th-century chemist who produced the atomic theory of substances—a program capable of formulating structural models. Employing this approach, applied to chemistry it has rediscovered Dalton's theory, and applied to genetics it has reformulated Mendel's laws on heredity.

Even if none of these programs has managed to reproduce the creative processes of scientific discovery in a satisfactory manner, they have amply demonstrated that it is possible to escape from the magic of insight and of impenetrable creativity to proceed along the road to a theory that also explains these aspects of the human mind.

Mixed forms of learning

Still bearing in mind the relationship between learning and the modification of knowledge, let us now examine how it is possible to acquire genuinely new knowledge through the interaction of explicit knowledge and tacit knowledge. We are no longer in the realm of either first- or second-order learning, but in a position mid-way between the two.

The first case, from K-explicit to K-tacit, consists in the generation of new procedures on the basis of information of a propositional type. We are here concerned with conventional learning, in which acquisition of declarative type knowledge serves as a guide for the creation of procedures for the execution of behaviours. For example, reading a recipe can lead to learning how to prepare a particular dish, or by reading a text on tropical illnesses a doctor may learn how to diagnose yellow fever. Norman (*Learning and memory*, 1982) terms this general mode of learning *compilation of knowledge* in order to underline its main feature—the passage from declarative knowledge to knowledge ready to be put to practical use. Compilation of knowledge comes about in two distinct phases:

1. First comes the construction of a series of action rules, each of which is separate from the others, and whose consecutive execution is rigidly controlled since it is a faithful reproduction of explicit instructions. The resultant action is discontinuous, requires the constant commitment of conscious attention, is recognisable as amateurish, and rarely reaches full effectiveness. Think of a beginner at tennis learning to play a backhand. He must be careful as to how he holds the racket, the position of his legs and trunk, the movement of his arms, etc.
2. Then comes a moment in which the individual actions are fused into one. The objective is to manage to integrate everything into a single intentional action that will gradually acquire the automation required. Think of an expert skier, fluidly coordinating the movement of eyes, arms, hands and legs without paying them any conscious attention.

The inverse learning mode, from K-tacit to K-explicit, is equally interesting, and consists in creating explicit knowledge starting from specific behavioural procedures, if the object of observation is a human being, or from the natural world, if the object of observation is the world. Taking observed phenomena as one's departure point, one must infer the existence of rules that can determine those phenomena and then make those rules explicit. Scientific discovery represents an example of learning of this type, as with Mendel reconstructing the laws of heredity.

A special case is that leading the individual, by means of self-observation, to the construction of her own image, first physical, then mental. Children must understand that the image reflected in the mirror is their own, and they manage to do so around 18 months of age. Later, as adolescents, they will have to strive to learn what they are like inside, what kind of personality they have, their weak points, their ambitions, their real capacities.

A further, interesting line of speculation comes from artificial intelligence (Agre & Rosenschein, 1995). Quite simple robots are set free to explore their environment, without any supervision on the part of the programmer. The idea is that these *situated agents* may learn how to improve their behaviour through direct interaction with the world.

It has to be noted that if progress in this sector seems to be partial and unsatisfactory, this may well be due in part to the fact that it is probably the most difficult, as well as the most important, sector of cognitive science. Once this area has been solved, many other problems will undoubtedly vanish into thin air. No mental function may be considered independent of learning and, because of the developmental type of comprehension I have invoked as epistemologically necessary, understanding how humans learn is the crucial point. A general theory capable of explaining it in a general fashion does not seem, however, to have appeared on the horizon, thus leaving us with the fragmentary intuitions described above.

CHAPTER SIX

Perception

How human beings perceive the world is a traditional field of investigation in Western thought, not only out of anthropocentric curiosity, but also because of the philosophical importance of the question. An answer requires tackling and solving problems concerning both the nature of the world (does an external reality independent of the perceiver exist?) and of humans perceiving that world (how do they do so?). The classic answers fall into two categories: one underlines the greater importance of the external world over the internal world, and the other underscores the exact opposite, the influence exerted by the organism over the external world. Useful reference texts to comprehend the contribution of the various schools to the understanding of perception are those by Bruce and Green (*Visual perception*, 1992) and by Gordon (*Theories of visual perception*, 1989).

DATA-DRIVEN PERCEPTION

The first approach is the most immediately intuitive, and consists in underlining the importance of objective data in the process of perception. The percept presents itself to us as self-evident, the source of primary and direct experience, independent of any activity on the part of the perceiver. The principal representative of this stance is Gestalt psychology. This school identifies a series of laws of good form specifying the features an object must possess in order to be efficaciously perceived. The most important laws, illustrated in Fig. 6.1, are based on the principles of the proximity of the elements forming the perceptive field, of the similarity between them, and of the continuity in the direction they take. It is not possible to present all the laws unearthed by Gestalt psychologists, since these have multiplied over time and now exceed the hundred mark.

The "Prägnanz" of the figure, to use the Gestalt term, which determines perceptual adequacy, lies in the external stimulus. The latter is an objective constituent of the real world, identical for all, and is ready to be grasped independently by the person carrying out the act of vision. The law of Prägnanz is described by Kurt Koffka (*Principles of Gestalt psychology*, 1935): "of several geometrically possible organisations, that one will actually occur which possesses the best, simplest and most stable shape".

The stance espoused by Gestalt is essentially innatist. In postulating laws that are identical for all humans, it places the perceiver in a passive position, one in which past experience and even the physiological structure of the nervous system do not have any significant part to play. Furthermore, the extraordinary experimental creativity of the Gestaltists, and their inexhaustible patience in devising stimuli that could bring to light anomalous and counterintuitive perceptual phenomena (the famous optical illusions), have led to the creation of an extensive and precious body of knowledge acting as a testing ground for all the theories elaborated after it.

The ecological view propounded by James Gibson (*The ecological approach to visual perception*, 1979) differs from Gestalt in its methodological approach, although they share the same empirical philosophical stance: empiricism. Gibson abandons the sophisticated equipment of the laboratory in order to deal with what animals and humans do to survive in the world, an activity in which perception is fundamental. The question is well taken and fulfils an ethological need increasingly felt in psychology—to try to understand what happens in vision in natural, ecological situations, leaving aside optical illusions and artful experimental situations. Gibson tries to describe the world as it is perceived by the animals that live within it. The environment is characterised by particular properties corresponding to what the environment allows an animal to do. These *affordances* correspond to the opportunities the environment offers for interaction between entities existing in the world and the sensorimotor abilities of animals. For instance, for some animals trees allow climbing, or fruits allow eating. Affordances depend upon objective features possessed by the environment.

1. *The law of proximity.* The elements constituting the visual field are grouped into shapes whose cohesion is greater the lesser the distance between them.

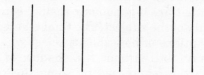

2. *The law of similarity.* The greater the similarity among elements, the greater the cohesive strength which groups them.

3. *The law of common fate.* Those elements exhibiting solidarity of movement and differing from the movement of other elements are grouped together.

4. *The law of good continuation.* Elements are grouped into shapes on the basis of common direction.

5. *The law of closure.* Lines forming closed shapes tend to be perceived as units (they tend to be seen as "closed" rectangles rather than a pair of "open" lines).

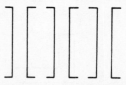

6. *The law of Prägnanz.* Of several geometrically possible organisations, the one that will actually occur will possess the best, simplest and most stable shape.

7. The law of past experience. Elements that in our past experience were generally associated together tend to be united into shapes.

FIG. 6.1. The classic laws of Gestalt. *Source:* Cesa-Bianchi, Beretta, & Luccio (1970).

Gibson claims that all this information is out there, in the external world, ready to be perceived by living beings. What counts in his opinion is the relationship between the organism and the environment, brought back to life in modes that are ingenuously realist, completely ignoring the constructive activity of the subject. The final result is a set of precise and fascinating descriptions of how sharks select what they may prey on and what they may not, without the shadow of an explanation. Despite the dissatisfaction expressed by some theoreticians, Gibson nevertheless furnishes us with fresh data, data which is by no means banal, data that can even enrich the viewpoint of those moving in totally different paradigms.

THEORY-DRIVEN PERCEPTION

The reaction to Gestalt was led by the New Look, a composite and heterogeneous movement, which adopted the opposite attitude, emphasising the rationalising role of the perceiving agent in contrast to that of the percept. The paradigm is reversed, rendering perceptual activity the protagonist. Whereas previously the movement highlighted lay in the direction proceeding from the periphery to the centre, now it is the centre that controls and influences what occurs in the outside world. In the post-war period, in a series of brilliant experiments (see Fig. 6.2), Jerome Bruner and Cecile Goodman (1947) demonstrated that Jewish children perceived a swastika as significantly larger than did non-Jewish children, and that poor children saw the dollar symbol as larger, compared to richer children of the same age.

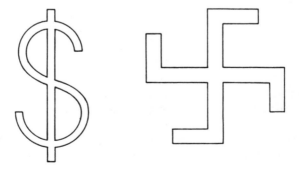

FIG. 6.2. The importance of central factors in perceptual processes.

Identical stimuli produced different percepts depending on subjects' social status. This brought the focus to bear on the predominance of cognitive and emotional aspects of perception, in an external world that had to be interpreted. In the opinion of the New Look, in vision what counted was subjects' expectations, their past experience, their emotional reactions to the stimulus, in a word the tacit principles which guided their interaction with the world.

INTERACTING SYSTEMS

Both the positions described above were supported by excellent experimental evidence, both were sensible from a general viewpoint, both had impeccable philosophical reference points. Yet both were limited by their opposition to each other, by their need to emphasise their own point of view. The foundations for a less rigid alternative were laid in a famous article written by Lettvin, Maturana, McCulloch and Pitts, whose aim it was to analyse the perceptual structures of the frog, and, to be more precise, what the frog's eye tells the frog's brain, as the title of their article, published in 1959, specifies. The position adopted in this work marks a departure from the empiricist stance holding that there exists an external reality which is absolute and independent of the perceiver. The world in which the frog is immersed differs radically from the world in which the fly, the animal on which the frog feeds, is immersed, and both these worlds are different from the one in which the student who observes the frog which feeds on the fly is immersed.

Humberto Maturana and Francisco Varela go beyond this position in *Autopoiesis and cognition* (1980). They refuse even to consider the question asked by scholars who favour the approach based on internal central control, which is "How does the organism extract information from the environment?". The question the two neurophysiologists believe is fundamental is, rather, "Why does the organism possess a structure fit for operating effectively in the environment in which it lives?".

This new approach demonstrates how organism and environment are not entities which can be separated and studied independently— interaction is, on the contrary, the fundamental aspect which must receive attention. A frog does not possess a representation of the world in which flies exist. Quite simply, the biological structure of its brain is such that its optic fibres respond to dark spots which are moving, setting off a global mechanism which finally enables its tongue to strike them. The frog's neurological mechanisms produce a sort of representation of the fly. The fly is the referent of the frog's representation, but the sense the reference takes on is totally different from that attributed it by a human representation of a fly. Naturally, if flies did not exist, the frog would not have developed this perceptuo-motor specialisation, but it is impossible to comprehend the present biological state of batrachians if we fail to consider the pressures of selection exerted by survival which brought about the evolution of the frog to its present state (see Fig. 6.3).

The approach adopted by Maturana and Varela is so ingeniously simple that it is astonishing what little acceptance it gained in the sphere of cognitive science. The basic reason lies in the fact that the two scholars switch from neurophysiology to cognition, to social factors, and to philosophy applying exactly the same principle of interaction between systems. The impeccable law enunciated to comprehend the structure and functions of the optic fibres becomes a simplistic formula when applied to explain thoughts and affects, and its superficiality is irritating when it is transformed into a philosopher's stone to interpret the social and political life of the human race. If one adds a tone at times verging on the mystical, the result is that a community which is sceptical towards ultimate truths ends up refusing to recognise even those contributions which are original, and not just the metaphysics imposed on those contributions.

In harmony with an ecological attitude, Maturana is sensitive to the biological aspects and to the concrete problems posed by them. One leaves behind the artificiality of the endless optical illusions of Gestalt and the confined laboratories of cognitive psychology, intent on quantifying minimal variations in the same phenomenon. At last, the objective is no longer to trap experimental subjects or astonish one's colleagues by the ingeniousness of one's umpteenth trick, but to try to discover how an animal manages to perceive its surroundings and understand those aspects that are important to its survival. Maturana thus sets himself objectives which are of real interest: how does the lioness distinguish zebras from the background, how does the small child recognise its mother in the midst of all the things she is not, how does each one of us succeed in perceiving something unexpected and which we have never seen before, no matter how unusual its physical structure might be? These become the tasks of the study of perception centred on the person.

A scholar of artificial intelligence, David Marr, was destined to produce the theory which has gained the widest acceptance today in the field of perception, drawing on all of the three positions we have outlined, in an original fashion. He attempted to explain the reasons why Gestalt laws work, to capitalise on Gibson's ideas about information, and to exploit the importance of knowledge in vision.

FIG. 6.3. The reader should imagine being a fly. If the insect begins its flight, the frog will capture it with its tongue. *Source:* Hoji (1814).

COMPUTATIONAL VISION

The working group which has exercised the greatest influence in cognitive science on the subject of perception is the group directed first by Marr until his premature death at the age of 35, then by his principal assistant, Tomaso Poggio, at the Artificial Intelligence Laboratory at the Massachusetts Institute of Technology. Although this is clearly not the sole group of scholars to have furnished important contributions to artificial vision, having to select a theory to present in detail, the choice falls naturally on the one born at MIT in the 1970s.

David Marr, in his book *Vision* (1982), takes up the distinction between competence and performance first introduced by Noam Chomsky in linguistics, a distinction which was explained in Chapter 4 and which will be developed further in Chapter 8 when we deal with language. In the field of perception, this distinction amounts to defining two levels: competence defines what is computed and why, while performance defines which algorithms are employed to execute the computation. The first level, namely competence, necessarily precedes the second: we must first know what we do when we see, that is what the tasks of the visual system are, and then how the specific procedures are executed.

Finally, there is a third level, the neurophysiological level, which underlies the other two: we must be acquainted with the neurophysiology of the nervous cells assigned to perception in order to be able to connect the two methodological levels to the organic structure which concretely realises computation, thereby giving rise to vision. A complete theory of vision must account for all three levels.

Tomaso Poggio, in "Vision by man and machine" (1984), takes as his point of departure both data of a psychophysical nature explaining how a person sees, and data of a neurophysiological nature illustrating how neurons behave. The most critical aspect is the transition to the level of algorithms, which, if they are to be effective, must respect the constraints imposed both by the neurophysiological hardware and by what is known about computational competence. The objective of the process of vision is to produce from images of the external world a description that is useful to the viewer. This means that the forms in which knowledge is represented are crucial. For the most peripheral function of the human system, perception, the system that mediates between environment and organism, to work effectively, it must be integrated with the central area of knowledge. Knowledge which is adequately represented and managed is the binding condition for persons to carry out their visual functions. This condition applies for the execution of any mental process, with the exception of none, as the previous chapter illustrated. But in perception it is interesting to note this point explicitly, in as much as other animals interact with the environment without needing to have representational structures. We may here recall the case of the frog (Fig. 6.3), which reacts to the stimulus of the fly at a maximum level of efficiency without representing it, limiting itself to capturing and eating it. The human being is, however, a different kind of animal, and needs to represent to perceive.

With regard to vision, Marr and his colleagues attribute special importance to a particular phenomenon: the analysis of the contours of images, which enable the viewer to establish the depth of objects in the three-dimensional world. In particular, attention has been devoted to stereoscopic and binocular vision, for reasons which range from the abundance of data available on this phenomenon to its primary importance in the early stages of visual processing. Here only the fundamental points of the theory will be outlined; readers desiring a more detailed account of the theory may consult the works quoted above.

The aim of Marr's theory is to explain how an organism determines the complete three-dimensional shape of an object starting from a two-dimensional image such as the one formed on the human retina. Marr hypothesises four successive stages, which, beginning from what may be extracted directly from an image, gradually recover the physical properties of the form of an object.The four stages may be summarised as follows:

1. Image: This represents the intensity value at each point in the image.
2. Primal sketch: This renders the information on the two-dimensional image explicit, in particular the information on intensity changes and their geometrical distribution and organisation.
3. 2½-dimensional sketch: This represents the properties of the visible surfaces in a viewer-centred coordinate frame. In particular, it makes explicit the orientation and the rough depth of the visible surfaces, as well as the distance from the viewer, the surface reflectance, and an approximate description of the current illumination.
4. 3-dimensional model representation: This builds a representation of the three-dimensional structure in an object-centred coordinate frame, employing primitives both for the surface and for the volume defining the amount of space that a shape occupies. We shall now describe each stage in the visual process in detail.

Image

The starting point of the process of perception is the stimulation of the visual cells triggered off by the quantity of light reflected onto the eye by each point of the visible surface of three-dimensional objects. In the human eye, this task is carried out by over one hundred million photoreceptors—the cones and rods of the retina—which construct a point-by-point representation of the colour and intensity of the light.

In the computer vision developed by Poggio at MIT, the image is simulated with electronic sensors which produce a matrix of 1000 times 1000 values of light intensity, each of which is called a pixel (from "picture element"). Figure 6.4 shows an image in which a small rectangle is marked; Fig. 6.5 shows the matrix constructed by the computer using the pixels, the values of light intensity of the image.

The second stage of the visual process interprets the raw image received from the receptors in the eye, or in the electronic camera.

Primal sketch

Four factors are responsible for the intensity values of an image: (a) the geometry of the visible surfaces; (b) the reflectance from the visible surfaces; (c) the illumination of the scene; (d) the point of view of the perceiver. These factors are intertwined in an image, and hence the first problem is to separate them out, assigning each intensity change detected to the factor responsible for it.

The primal sketch consists of primitives of the same type on different scales. The most important primitives are changes in intensity, discontinuity, edges, virtual lines, groups, terminations, curvilinear organisations and boundaries. These primitives are built up in

FIG. 6.4.   Point-by-point representation of a black and white image (526x456 pixels). *Source:* Poggio (1984). Reproduced with permission of the author.

225	221	216	219	219	214	207	218	219	220	207	155	136	135	130	131	125
213	206	213	223	208	217	223	221	223	216	195	156	141	130	128	138	123
206	217	210	216	224	223	228	230	234	216	207	157	136	132	137	130	128
211	213	221	223	220	222	237	216	219	220	176	149	137	132	125	136	121
216	210	231	227	224	228	231	210	195	227	181	141	131	133	131	124	122
223	229	218	230	228	214	213	209	198	224	161	140	133	127	133	122	133
220	219	224	220	219	215	215	206	206	221	159	143	133	131	129	127	127
221	215	211	214	220	218	221	212	218	204	148	141	131	130	128	129	118
214	211	211	218	214	220	226	216	223	209	143	141	141	124	121	132	125
211	208	223	213	216	226	231	230	241	199	153	141	136	125	131	125	136
200	224	219	215	217	224	232	241	240	211	150	139	128	132	129	124	132
204	206	208	205	233	241	241	252	242	192	151	141	133	130	127	129	129
200	205	201	216	232	248	255	246	231	210	149	141	132	126	134	128	139
191	194	209	238	245	255	249	235	238	197	146	139	130	132	129	132	123
189	199	200	227	239	237	235	236	247	192	145	142	124	133	125	138	128
198	196	209	211	210	215	236	240	232	177	142	137	135	124	129	132	128
198	203	205	208	211	224	226	240	210	160	139	132	129	130	122	124	131
216	209	214	220	210	231	245	219	169	143	148	129	128	136	124	128	123
211	210	217	218	214	227	244	221	162	140	139	129	133	131	122	126	128
215	210	216	216	209	220	248	200	156	139	131	129	139	128	123	130	128
219	220	211	208	205	209	240	217	154	141	127	130	124	142	134	128	129
229	224	212	214	220	229	234	208	151	145	128	128	142	122	126	132	124
252	224	222	224	233	244	228	213	143	141	135	128	131	129	128	124	131
255	235	230	249	253	240	228	193	147	139	132	128	136	125	125	128	119
250	245	238	245	246	235	235	190	139	136	134	135	126	130	126	137	132
240	238	233	232	235	255	246	168	156	141	129	127	136	134	135	130	126
241	242	225	219	225	255	255	183	139	141	126	139	128	137	128	128	130
234	218	221	217	211	252	242	166	144	139	132	130	128	129	127	121	132
231	221	219	214	218	225	238	171	145	141	124	134	131	134	131	126	131
228	212	214	214	213	208	209	159	134	136	139	134	126	127	127	124	122
219	213	215	215	205	215	222	161	135	141	128	129	131	128	125	128	127

FIG. 6.5. Matrix of the pixels corresponding to the small rectangle highlighted in Fig. 6.4. *Source:* Poggio (1984). Reproduced with permission of the author.

successive procedural phases. First, intensity changes are detected and terminations are represented. Elements are formed directly from this step. Next, representations are added of the local geometrical structure in which the first elements may be framed. Finally, selection and grouping processes are activated to form elements on a larger scale, corresponding to the larger scale structures of the image.

The complexity of the primal sketch varies according to the level of processing of the image required by the global purposes of the system. Not all the processes are necessarily activated on each occasion, nor is the image always analysed on all the scales of detail possible. Figure 6.6 represents the construction of a primal sketch starting from an image. At the lowest level, the raw primal sketch follows the changes in intensity of the image faithfully, and represents the terminations (denoted by small, filled circles). At the next level oriented tokens are formed for the groups in the image. At the next level, the different orientations exhibited by the groups in the two halves of the image bring about the construction of a boundary between them.

Image

Primal sketch

First level:
elements

Second level:
boundaries
between elements

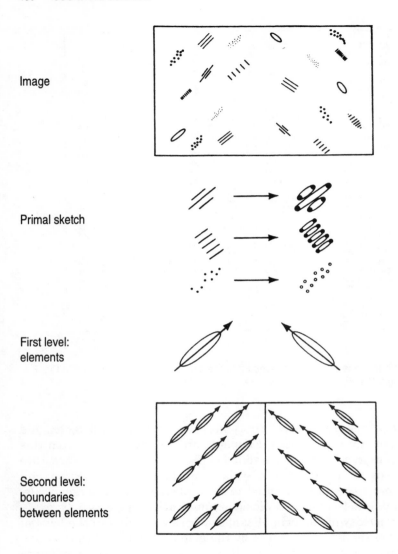

FIG. 6.6. The construction of a primal sketch. *Source:* Marr (1982). Copyright © 1982 by
W.H. Freeman and Company. Reprinted with permission.

2½-dimensional sketch

One of the classic problems in the psychology of perception is that of
distinguishing between figure and background. The ability to make this
distinction enables an organism to differentiate the significant areas in
an image from those destined exclusively to make the important parts

emerge through contrast. In the drawing in Fig. 6.3, for example, the frog represents a significant portion against the background of the pool, the latter being intuited rather than perceived explicitly. But the idea of dividing the image up into significant areas, whether this be in relation to the objectives of the perceiver, or whether it be because an area corresponds to a physical object or a part of it, has always proved elusive. What is an object? How may an entity be defined as such *a priori*? Is a nib an object? And what about the nib of a fountain pen? And is the nib of a fountain pen used by a person who is writing still an object?

The difficulty lies in the fact that there is not sufficient visual information coming from the outside world, in the image as it is processed in the primal sketch, to define an object. Furthermore, many cognitively and semantically important areas have no distinctive visual traits. Suffice it to think of the censor when he decides to hide part of an image deemed unacceptable—how much bosom, for instance, has to be covered up in order to eliminate evil thoughts?

To understand Marr's strategy, we must remember that the initial processes of vision extract information directly from the visible surfaces, with no concern as to whether what is being observed is a fly, a frog or a water lily. This means that the representation of the visible surfaces is processed before the observer knows if the object in question is a frog or a pool. What does representing visible surfaces mean, then?

The answer lies in the 2½-dimensional sketch, the last stage of data-driven vision. The 2½-dimensional sketch is the transition phase from the element by element analysis of the primal sketch to the description of surface shape. The term, which refers to an impossible half-dimension, is intended to remind us that we have not yet entered a wholly three-dimensional domain, though we are only one step removed from it.

There are three characteristics of a shape representation that exert the greatest influence over the type of information that has to be made explicit. The first refers to the type of coordinate frame employed by the system. The second is the nature of the shape primitives whose position must be defined by the coordinate frame. The third refers to how the representation can organise the information into an effective description.

The coordinate frame may be centred on the objects perceived or on the viewer. But we already know that the primal sketch employs measures for the distance and the orientation of surfaces that are centred on the observer's perspective. Hence, the 2½-dimensional sketch will also have to remain within the same perspective. The shape primitives have to be two-dimensional, and will have to specify the

orientation of each of the local pieces of surface. With regard to the third
characteristic, the representation is intended not only to make explicit
the depth, the local orientation of the surface, and the discontinuities in
these quantities, but also to create and preserve a global representation
of depth consistent with the information available locally.

Figure 6.7 exemplifies a 2½-dimensional sketch of a cube, seen from
the viewer's perspective. The surface orientation is locally represented
by the arrows. The cube's subjective occluding contours are shown with
full lines, while the discontinuities in surface orientation are coded by
dashes. The figure lacks information on the distance in relation to the
viewer, even though this information is contained in the 2½-
dimensional sketch.

Three-dimensional model representation

The final problem that has to be faced is how perceived objects are
recognised. To do this, the information coming from the data provided
by perception *per se* is not sufficient. The system is obliged to employ
general knowledge. The transformation of a 2½-dimensional sketch into
a three-dimensional model requires a complex procedure for the
geometrical manipulation of the shape. Again this is not enough, since
the information received from the direct phase in perception cannot
possibly be sufficient to build a three-dimensional image of objects.

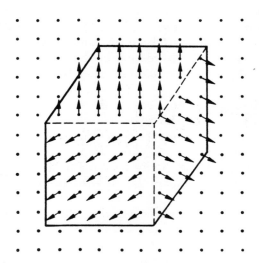

FIG. 6.7. The 2½-D sketch of a cube. *Source:* Marr (1982). Copyright © 1982 by W.H..
Freeman and Company. Reprinted with permission.

In "Representation and recognition of the spatial organisation of three-dimensional space" (1978), Marr and Nishihara hypothesise the existence of a catalogue of three-dimensional prototypes containing, besides the base forms, also the figures normally encountered by the subject. The prototypes are organised in the form of three-dimensional models centred on the object (no longer on the viewer, as had happened up until the 2½-dimensional sketch), exhibiting a hierarchical representation including both volume and surface primitives. The other features of prototypes comprise the following: each three-dimensional model is an independent unit of shape information, and has a limited complexity; the information appears in shape contexts that are appropriate for recognition; representation can be manipulated in a highly flexible manner. Each model operates independently of the others, realising modular decomposition grounded essentially on the canonical axes of a shape.

The representation of a three-dimensional model, built up on the basis of the first direct stage of perception and of the appropriate shapes contained in the catalogue, describes the shapes of the percept and their spatial organisation in an object-centred coordinate frame. Figure 6.8 provides as an example the organisation of shape information with reference to the three-dimensional model of the human body.

Recognising an object is a gradual process moving from the general to the particular, and guiding the derivation of a description starting from an image. The three-dimensional prototype that best fits the 2½-dimensional sketch is selected from the store of models, thereby generating the three-dimensional model. Thus the building up of a three-dimensional model depends in large part on one's general

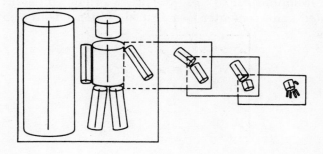

FIG. 6.8. The three-dimensional model of the human body. *Source:* Marr and Nishihara (1978). Reprinted by permission of the Royal Society.

knowledge, while the selection of the prototype acting as the departure point depends on the data furnished by immediate perception.

Marr's theory incorporates both a first stage, immediate perception, in which the image is directly explored, and whose physical characteristics determine processing procedures, and a second stage in which general knowledge takes command once the most suitable prototype has been selected.

Setting aside theoretical elegance, this approach boasts three strong points. The first is that the theory is consistent with all the neurophysiological data currently available on human vision. The second is that it has, to a large extent, been implemented on computer, thus satisfying the requirement laid down by cognitive science that a theory be computable. The third is that it has received confirmation from a significant amount of neuropsychological data (see Chapter 11)—in other terms, some types of neurological patients exhibit functional disturbances that Marr's various stages and various modules account for perfectly. For instance, the fact that the 2½-dimensional sketch does not of itself suffice in recognising an object, and that one essential point in representing a three-dimensional model is the transition from the viewer-centred perspective to the object-centred perspective, leads to the prediction that certain perceptual deficits will exist preventing patients from recognising shapes rotating in an unusual manner. Even if neuropsychology and Marr's hypothesis do not tally perfectly, nevertheless they exhibit a convergence alien to all other theories on perception.

As always, we are still a long way off from explaining perceptual phenomena in an integral fashion. However, the basic theory together with the manner in which it has been formulated by Marr render it an exemplary case of cognitive science: explicit, falsifiable, expressed in a language comprehensible to scientists from different areas, implemented on computer, confirmed independently by other disciplines.

CHAPTER SEVEN

Thought

The activity of thinking is difficult to encapsulate in watertight compartments. Of its nature, it permeates every mental function, it reprocesses everything that perception has filtered or that memory has reconstructed. Dividing this chapter into separate sections once again corresponds more to established practice in the sector than to a real distinction between functions. Problem solving and planning are classic show pieces in artificial intelligence. In contrast, reasoning was, originally, more the area of psychology. Later it also became the realm of cognitive science. Accepting this sort of division does not however imply postulating each function is characterised by different basic abilities. The reader who desires to explore this complex domain may refer to the exhaustive and well-structured *Thinking and reasoning*, by Alan Garnham and Jane Oakhill (1994).

PROBLEM SOLVING

The sector of *problem solving* is historically the most glorious in cognitive science. It is this area that proved that the methodology of computer simulation really was capable of carrying us beyond the narrow path trodden by psychology. And it is this area that has obtained the only Nobel prize ever to have been received by a scientist in computer science, artificial intelligence or cognitive science. The 1978 Nobel prize for economics was awarded to Simon for the applications to economic problems he developed from his research on decision making and choice processes.

The astonishing fact is that despite the great interest problem solving has aroused in all schools of psychology, no one has ever managed to advance a plausible theory on the subject. The most distinguished failure is furnished by Gestalt psychology's concept of *insight*. Insight consists of the magic passage from an unsolved problem to its solution, without either the solvers being able to express explicitly how they arrived at the solution, or the experimenter having the faintest idea of what happened. Typical examples of insight are Archimedes stepping out of the bath naked shouting "Eureka", and Newton conceiving the laws of gravitation upon being hit on the head by a falling apple. Although the miraculous side to insight did not mask its unfoundedness as an explanation, the very fact that it was given serious consideration reveals just how much theorists despaired of ever being able to understand how a problem, even the most trivial, was solved.

Cognitive psychology, true to its style, amassed an infinite number of experiments without ever managing to find an acceptable key that would unlock the secrets of how problem solving came about. As always, the blind accumulation of data proved to be a tedious dead end. In this case too, the most important result obtained was the discovery of how insipient the experimenters were, often more ingenuous than the subjects they experimented on, in interpreting what was happening. Pragmatically incorrect instructions; subjects deliberately led astray by experimenters desperately seeking an anomalous result; the rejection of brilliant solutions, not accepted as such because they were unexpected. This was what had become the sole focus of critical interest in an area suffering from stagnation. As for the other schools, the less said, the better.

The winds of change came blowing in the form of Allen Newell and Herbert Simon's work on the *General Problem Solver* (GPS). This theory, the first version of which saw the light in 1956, traced a paradigm that still retains its validity in the area of problem solving, a paradigm that may be identified by referring to the hypothesis of domain-independence.

The hypothesis of domain-independence

This may be defined as the assumption that it is possible to build a finite set of general procedures capable of solving any kind of problem. Researchers attempt to concentrate on wide-ranging procedures applicable to the largest possible number of problems. Alternatively, types of problems must be tackled possessing features that allow the results obtained to be generalised to different classes of problems. Domain-independent thus means avoiding being influenced by specific domains in which the procedures may be applied. Stated differently, a valid heuristic is as capable of solving third-order equations as it is of choosing a good hotel in a foreign city.

At the outset, interest focused on formally well-defined problems— typically, games like draughts and chess, or mathematical and geometrical problems. The methodology adopted by Newell and Simon in *Human problem solving* (1972)—940 important pages, each twice the size of a normal one—was to make human subjects speak while solving a problem to cast light on the internal, unobservable steps taken towards the solution. Their computational model then simulated the procedures thus identified, to reach the same solutions obtained by the human subjects.

The most renowned problem solved by GPS forms part of a series of cryptarithmetic problems, taken up from Bartlett's work on thought (*Thinking: An experimental and social study*, 1958), in which a subject must replace each letter of a word with a figure in such a way as to obtain the correct total (see Fig. 7.1). Each letter has only one corresponding figure $(0,1,...,9)$. Thus, for instance, if the letter D corresponds to the number 5, this number may not be assigned to any other letter. Readers may try to solve the problem for themselves. The only instruction subjects are furnished with is that the letter D corresponds to the number 5.

In the problem in Fig. 7.1 there are 10 letters, to which 10 numbers must be assigned. There are thus 10! possible combinations. But since the letter D has already been assigned the figure 5, there remain 9! possible combinations, yielding a total of 362,480 potential solutions, only one of which is correct. This number does not lie beyond the calculating powers of a human, since it would take less than two years' work, but it seems highly unlikely that an intelligent person would sit down and try out all the possible solutions by systematically varying the combinations. Instead good solvers attempt to apply an efficient heuristic, permitting them on average to solve the problem in under an hour. This is precisely what Newell and Simon's work focuses on—discovering the heuristics humans apply, and simulating them with the GPS.

DONALD +   where D = 5
GERALD =

ROBERT

FIG. 7.1. A problem in cryptarithmetic.

In the problem presented above, the crucial moment in reaching the solution comes when the subject notes that in the second column O + E gives O again, thus managing to conclude that E must then be equal to 9. Only in this case, and assuming a carry over of 1, is it possible for one figure added to another to produce itself as the result. The only other possibility would be for E to equal 0, but the figure 0 was initially assigned to T in as much as this is the result of the sum D+D, since we know D has been assigned the value 5 by definition. When the subject grasps this point the following situation is arrived at:

5ONAL5 +
G9RAL5 =
ROB9R0

From this point on the problem presents no special difficulty—all the letters can be substituted with figures. If the reader is still puzzled, the solution is:

526485+
197485=
723970

The GPS arrives at the solution by following the same path taken by those humans used as a yardstick, that is employing heuristic methods and not the blind force of calculation. Heuristics, intended as general problem-solving techniques, are independent of a specific applicational domain. They therefore keep knowledge of the world separate from procedures yielding solutions. A good solution procedure will be applicable to different fields, on condition that the specific knowledge domains are adequately represented. Any difficulties that might be encountered with certain sorts of problem are ascribable to insufficient formalisation of knowledge, and not to the problem-solving methods.

In general, the classic problem-solving procedure consists of four stages: recognition, definition, solution and check out.

Recognising the problem

People have a problem when they desire something that is not immediately obtainable. The desired object may be physical, as when a child wants a felt-tip pen to write on the wall with, or it may be abstract, as when a poet is seeking the final line of a sonnet, or may lie mid-way between the two, as when Othello dreams of Desdemona's love being faithful. Realising one has a problem always depends on the possibility of the system being able to exert its will. This explains why human beings have many problems, and machines none. This also accounts for the inherent difficulty in simulating problem recognition on computer—a machine capable of recognising a problem is a machine equipped with intentionality. Furthermore, the more sophisticated and interesting the problems it can recognise, the more developed will be the intentional part, that is the part generating objectives.

No problem is objectively such. It acquires this status in relation to the state of the system and the objectives the system sets itself. Solving a second-order equation may be so simple for a mathematics teacher that it is not considered a real problem. This would obviously not be the case for students just beginning to familiarise themselves with that branch of mathematics. Likewise, jumping a hedge three feet high is a simple matter for an expert in the saddle of a good horse, but it becomes an insurmountable obstacle for a person who has never ridden a horse before. Available knowledge is always one determining factor for something to be perceived as a problem. No locked door is considered to be an obstacle by a person possessing the key. A second crucial factor is the system's goal. Earning a million in a week is not a problem for a post office clerk who loves playing poker, not because he knows how to achieve the goal, but because this is not a concrete goal for him. If a gangster were to threaten to kill him should he not pay off his debt within a week, then earning a million would immediately become a problem for this poor poker player of a clerk.

In the field of scientific research, it is often argued that setting oneself the right problem means one is well along the road to having glimpsed a possible solution. It requires an adventurous and highly competent mind to identify a new problem where tradition sees but obvious and predictable chains of events. The problems studied by psychology, cognitive science and artificial intelligence are almost always confined to the four walls of the laboratory. In other terms, they are supplied to the subject or the computer ready made with the label "problem to be solved", with the result that the testees are not called upon to employ their own resources in order to identify the exact spot where the world offers some interesting discrepancy. Only an ecological approach permits the discovery that this first stage is anything but obvious, in natural conditions. Finally, there is an element of subjectivity implicit in the moment of recognition, one that appeals to internal states and private goals. The merging of two aspects that psychology and computer science find it difficult to handle has retarded the recognition of the importance of the first step in finding a solution, that is to say, realising that something has to be considered a problem.

Defining the problem

At this stage, the problem identified is translated into terms manageable with the tools available to the system. Often this coincides with complete or partial formalisation, above all when the problems pertain more properly to the laboratory than to real life. Whether it can be formalised or not, the definition of the problem requires the provision of the conditions, constraints, the initial state, the rules to be followed in solving the problem, and finally the criteria governing the validation and evaluation of the solution. The same relationship exists between recognising a problem and representing it that exists between becoming aware of needing something and knowing exactly what that something is and, at least in general terms, knowing how to obtain it.

An adequate formal description of a problem amounts to making the solution obvious, or at least facilitating finding it to a large extent. On the contrary, an inadequate or misleading representation may prevent subjects from finding the solution. On this score let us consider a classic problem in this area, illustrated in Fig. 7.2. In this problem, the subject is asked to make a necklace alternating black and white pearls regularly. Readers are invited to try to solve the problem themselves, bearing in mind that the sole constraint is that breaking the thread of the necklace to extract one of the pearls is not permissible.

The solution envisaged consists in breaking one of the four consecutive pearls. This is a false solution, since it is contrary to all constraints operating in real life, which lay down that precious objects are not broken deliberately. Employing a pearl necklace in setting out the problem, in lieu of some worthless substance, conveys the pragmatic implication that each pearl must be preserved. Hence, to solve the problem as it was originally conceived, one has to be psychotic, not in

FIG. 7.2. How can the pearls be regularly alternated without breaking the thread?

this world, or at the very least incapable of facing real life. More than a problem, this is a test of adaptation—anyone solving it must be helped to re-adapt to society. If the description can be classed as adequate, the problem is insoluble. If, however, the problem were to admit of the solution proffered, then the description would be totally mistaken. Whatever the outcome, the description plays a fundamental role.

Generally speaking, the process of furnishing a definition is complex with respect to real life problems, while it is much easier to realise in the case of abstract problems, such as mathematical, logical or geometrical problems and games. This is the fundamental reason why artificial intelligence has devoted itself to broaching situations humans might find extremely complex or demanding, such as solving a difficult problem in logic or playing a delicate end-game against a skilful chess player. Nevertheless, these sorts of problems remain of the type that may be formalised to a high degree. Natural decision-making situations, instead, such as choosing the most suitable present for one's 12-year-old child, cannot be handled computationally precisely because they cannot be encompassed by the methods of formalisation currently available.

One important result obtained in the sector known as *decision making* is the differentiation achieved between well-defined and ill-defined problems. A problem may be considered as well defined when the following conditions are satisfied:

(a) the set of valid lines of actions is known in its full extension;
(b) the important consequences of a line of action are grasped with significant clarity and depth;
(c) the person having to take the decision can quantify the degree of probability of uncertain events taking place;
(d) the person having to take the decision is able to determine the usefulness of the consequences of each line of action with respect to goals.

None of these four points are to be found in a real life situation. Outside the world of play, it is practically impossible to define the departure point unequivocally, and there is absolutely no hope of predicting the future, a condition that is necessary if the problem is to be well defined.

One area of great social interest is economics. Here the analysis of decision-making capacity is crucial. Faced with the need to forecast human behaviour with regard to economic choices, this science dreamt up the concept of *homo economicus*, based on one sole postulate: *homo economicus* is always motivated by the maximisation of profit;

furthermore, this being is always able to discover what actions realise maximisation, and to effect the pre-selected behaviour. However, the gap between observed behaviours and predictions grounded in a model of total rationality based on the maximisation of profit the subject expects is often quite significant. This kind of gap tends to narrow only when rational behaviour is obvious, and even then a reduction does not automatically follow—if subjects do not behave rationally, this may be taken as evidence of the fact that it is not obvious to them that a particular behaviour is rational. The choice made by classical economics was that of forcing the real situation into the artificial canons of absolute rationality. The deployment of this type of strategy meant renouncing many important psychological aspects in order to maintain the very few which are possible to work on in formal terms.

In *Reason in human affairs* (1983), Simon introduces the concept of limited rationality, according to which humans can process information and alternatives, can acquire data on the environment they live in, but have a more restricted capacity to draw inferences from that same data. Furthermore, constraints exist on attentional capacity. This is no longer considered unlimited but has to concentrate on one aspect of the problem at a time, thereby cutting the thread tying each problem to all the other aspects of the world. Finally, in considering the evolution of the human species, Simon analyses the role of emotion in the decision-making process. This consists mainly in guaranteeing that problems felt to be urgent be given absolute priority when planning our actions and when assigning a quota of our available energy resources. The limited rationality model cannot boast the magnificent formal properties exhibited by the absolute rationality model. On the other hand, it allows us to comprehend how creatures equipped with mental capacities such as ours may have achieved evolutionary success in a world which, if we accept the unrealistic vision deriving from the theory of expected subjective utility, immediately becomes too complex to survive in.

The general conclusions that may be drawn go beyond the admittedly brilliant economic implications, and consist in having to abandon the rationalist hope of managing to formalise every type of problem with the same general logical techniques. Real life problems are intrinsically different from those of the toy worlds. Any attempt at construing daily life in terms of rationality involves having to give up the most significant aspects of the former, and therefore of being left with the formalisation of something which no longer corresponds to the original problem. The indispensable process of defining a problem is not necessarily equal to its logical formalisation. Interested readers may refer to the book edited by Bell, Raiffa, and Tversky: *Decision making: Descriptive, normative, and prescriptive interactions* (1988).

Solving the problem

This constitutes the central phase in problem solving. It consists in searching for or constructing an algorithm enabling the realisation of the transition from the initial stage to the final pre-selected goal state. Figure 7.3 shows the basic alternatives, remaining of course within the standpoint of general methods of problem solving. The least laborious mode of solving a problem is to establish whether it may be deemed analogous to another problem to which the solution is already possessed, and then searching memory for the algorithm applied in the preceding case. The strategy of searching for prepackaged algorithms is the one most commonly adopted by human beings, since it allows them to save on cognitive energy, even though it involves the risk of wastage in absolute terms. People tend to adopt non-optimal solutions, if this avoids having to make cognitive efforts such as those required in constructing new solutions. Setting aside the experiments demonstrating that familiar algorithms are employed even in situations that are unsuitable, as long as this permits one to avoid having to face a small amount of creative toil, it must be noted that a minimal effort at introspection brings about the same results. Readers may confirm this from their own experience, by reflecting on how much easier it is to employ a habitual routine, even if it is not always functional, than it is to seek alternative paths despite the fact this latter procedure is more rewarding in terms of the results obtained.

The crucial points in a strategy grounded on the search for a similar problem solved previously are: having a good definition of the problem to be tackled; an efficient classification of the algorithms available in knowledge; and a valid management technique for the criteria of analogousness and similarity, indispensable for the selection of an

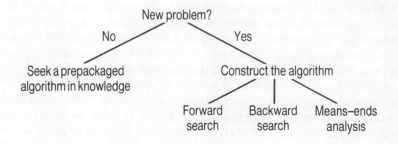

FIG. 7.3. Problem-solving methods.

algorithm that will really prove efficient in dealing with the problem at hand. Nevertheless, for a problem to be standard and offer algorithms storable in memory, it must have been solved at least once. This turns the wheel full circle, bringing us back to the essence of problem solving—the creation of original solutions. We shall leave aside methods that do not employ heuristics, such as an exhaustive search, where every possible solution is attempted, since such cases are banal for humans, even though they may sometimes be effective with computers with an enormous computing potential. For a discussion of problem solving in hard artificial intelligence, readers are referred to Nils Nilsson's book on the subject (*Problem-solving methods in artificial intelligence*, 1971).

Before going into detail as regards the various methods, we must define certain general concepts readers will find helpful in venturing into this rather complicated domain. The *state-space* corresponds to a particular approach to problem solving, according to which every problem of the same type corresponds to a particular state. To get from one state to another requires *operators* allowing one problem to be converted into another, within the same class of problems. If a single problem is being scrutinised, the state-space may be viewed as the set of all its possible variations, all of them being interconnected by the transformation operators. In this case the problem is characterised by its initial state and its final, or goal, state, the latter corresponding to its solution. The problem-solving process consists of a sequence of states, connected up by means of the appropriate operators, marking out the path leading from the initial state to the goal state. An effective method of representing a state-space is by means of a *tree* with pathways to be followed (Fig. 7.4).

A tree is composed of *nodes* connected by *arcs*. A node may be thought of as being the equivalent of a state, and an arc as the equivalent of a transformation operator. Nodes connected to a node appearing higher up in the hierarchy are called *successors* or *daughters* of that node. The only node that is not a successor of other nodes is termed the root of the tree. Nodes devoid of successors are denominated *terminals* or *leaves*. The branches of a tree are of two types: AND branches and OR branches.

OR branches (Fig. 7.5) connect up problems of equal difficulty—the root is as complex as each individual successor. It should be noted that the goal-state of the problem is arrived at directly when one of the two possible alternatives is realised. There is no intrinsic difference between spending a pleasant afternoon and going to the park with one's daughters, nor does either of the alternatives present greater or lesser difficulties. The latter achieves the former. In addition, the two possible solutions, park or tennis, are not correlated in any manner. Quite the contrary—they are mutually exclusive.

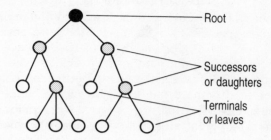

FIG. 7.4. A search tree. The circles represent nodes, the straight lines arcs.

The AND branches (Fig. 7.6) are marked by a semi-circle connecting the arcs, the latter having, instead, leaves that are simpler compared to the root. In this case, the goal-state is achieved only if both the terminal states are realised. Both daughters must be put to bed if a peaceful evening is to be had. Effecting only one of the terminal states is not sufficient, even though this would presumably bring the ultimate objective closer. Furthermore, each successor node in an AND branch is always simpler than the original node.

The tree bearing AND branches allows us to introduce the problem-solving method known as *problem reduction*. This technique consists in decomposing a problem into simpler sub-problems and, once having found the solution to each of these separately, recomposing them to obtain the solution to the original problem. Each sub-problem is solved independently, and the respective solutions are assembled together. The assembly of the partial solutions amounts to the solution to the initial problem. A tree may contain exclusively OR branches, in which case all

Spending a pleasant afternoon

Going to the park
with Simona and Helen

Playing tennis with
Alice

FIG. 7.5. An OR branch.

Spending a quiet evening

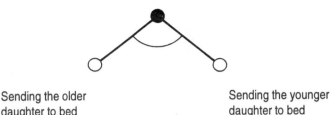

Sending the older
daughter to bed

Sending the younger
daughter to bed

FIG. 7.6. An AND branch, recognisable by the semi-circle joining the two arcs.

the states are perfectly equivalent. It is normal, however, to find AND/OR trees, trees in which qualitatively different branches co-exist.

Let us now proceed to an analysis of the various types of search that may be carried out to find a solution, as outlined in Fig. 7.3.

Forward search.   In a forward search, the search procedure starts from the initial state and expands towards the goal-states, as illustrated in Fig. 7.7. This type of search starts from the initial state of the problem and works towards possible solutions. Once potential goal-states, corresponding to solutions that the system deems valid, have been identified, the tree expands even further until the steps necessary to achieve each one of the solutions has been determined.

If for instance the system were to decide that a brilliant solution to the problem of what to do at the weekend was to go to Paris with Christine, it would also know exactly what steps would have to be taken to realise the plan it had generated. The leaves of the tree indicate that a level has been reached that presents no real difficulties. In our case we may presume that the system already knows what hotel to book in at, and that it is capable of arranging a flight. Typical of a forward search is the fact that one single original state exists, whereas the final states may be considered as being equally satisfactory.

Backward search.   In backward search, the search procedure starts from the goal-state and expands towards the initial states, as in Fig. 7.8. In this type of search the system has one sole goal-state that it can consider as the real solution. All other alternatives would be unsatisfactory.

In the example shown in Fig. 7.8, going to Vienna with another person would be disappointing, just as going with my wife Marcella to some

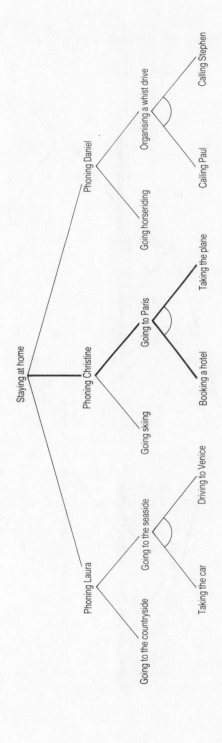

FIG. 7.7. Forward search on the problem of organising a weekend; the bold lines represent the solution chosen.

167

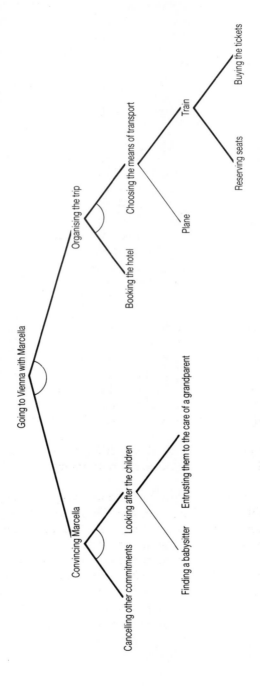

FIG. 7.8. Backward search on the problem of going to Vienna with Marcella; the bold lines represent the solutions.

other place would be equally illegitimate, given the pre-set goal. In this kind of situation, the goal remains unvaried, while the system attempts to create intermediate states approaching the goal that are feasible for the system, until the terminal problems are arrived at, which may be presumed to present no further difficulties. The most typical case of backward reasoning is to be found in mathematics, when attempting to demonstrate a new theorem. One knows perfectly well where one wants to get to. What is not equally clear is what the best point from which to start is in order to prove the new theorem.

Means–ends analysis.   This is the most important method in the problem-solving area, and is the one adopted by Newell and Simon in their General Problem Solver. Means–ends suitably mixes forward procedures and backward procedures, by adopting the strategy of first solving the principal parts of a problem and then going back and unravelling the incongruities created in seeking to match partial solutions with local difficulties. The procedure followed by means–ends analysis consists in identifying significant differences between the initial state and the goal-state, in an attempt to reduce their number. It may be defined as consisting of four steps:

(a) Compare the present state with the goal-state and pinpoint the differences.
(b) Select an operator aiming at reducing the main difference.
(c) If possible, apply the operator selected. If this is not possible, consider the sub-problem of transforming the present state into one to which the operator may be applied. If necessary, apply the means–ends analysis recursively to solve the sub-problems thus generated.
(d) Repeat the procedure until all the differences identified have been eliminated, or until a failure criterion emerges.

A means–ends analysis is similar in many ways to the technique of reducing the problem into sub-problems. There are three crucial phases in this procedure: first, generating sub-problems, an essentially analytical activity calling for a profound understanding of the nature of the objective and of the pathways that lead close to it, but that perhaps are far from obvious; second, actually solving the sub-problems; third, assembling the partial solutions, an activity of synthesis that must bring about the transition from a series of separate terminal nodes to a fluid and coherent operational chain. The amount of creativity required to carry out each task adequately is anything but microscopic. It is for this reason that means–ends analysis has come to dominate the field not only of hard artificial intelligence but also that of the simulation of human behaviour, as the most promising technique available.

As an example, let us suppose that young Alice decides to spend a year in California (the goal-state). A first backward analysis enables the initial problem to be decomposed into three sub-problems—finding the money for the journey, how to keep herself for a year, and doing something interesting. A further step backward may lead to the identification of the basic sources of finance—her parents, her rich but miserly uncle, or the odd jobs she does, her studies permitting. If it occurs to Alice that Italian organisations exist offering grants to Italian nationals for the United States, she may decide to discover what has to be done to win one of these grants. Alice's sub-problems have now become how to win a grant, how to get her uncle to help her in some way, and to invent some enjoyable activity that will pass muster with her parents. Going to dinner with her uncle—in a forward search—does not make her even one dollar richer, but it does provide some indication as to a rather mouth watering and almost clandestine grant as well as producing a telephone call to a friend and manager of a jazz orchestra promising her a job as an assistant organiser. But will her strict parents agree to this? Problem solving will continue until Alice has found effective solutions to all the sub-problems, and has succeeded in putting all the pieces of the puzzle into place.

Checking the solution

This represents the last step in problem solving, and consists in verifying the solution obtained. There are two possible cases: either the solution is known *a priori*, and the system is simply interested in establishing whether it can achieve the goal, or the solution is not known in advance. If the solution is already known, as with a mathematics exercise a student has to solve, all that will be required is a comparison between the solution obtained and the correct solution as furnished by the textbook. In real life, however, solutions are never known in advance. It is a question of inventing them autonomously. In addition, the "solution" does not exist in a natural state—there is nothing marked by an explicit label informing one that this is the solution. It is therefore necessary to apply those validity criteria established when the problem was defined. It is the criteria for evaluating the solution that interrupt the search process if the solution is accepted as valid, or else oblige it to continue if the solution is not deemed satisfactory or if there are indications that some more advantageous solution may be generated. Since the solution often consists in an action plan, checking it accurately means avoiding engaging in behaviours that may in actual fact prove to be disastrous or inefficient.

The validation criteria are established when the problem is defined, for that is the point at which it becomes clear which margins of tolerance are permissible and what the maximum costs the system is prepared to bear are. If one has to estimate the length of a car trip from Los Angeles to San Francisco, an estimate which is correct to within about an hour is acceptable. The same approximation would be totally unacceptable if one had an appointment a few blocks away from home. The costs are calculated first and foremost in cognitive terms, the time factor coming well before the rest—to continue putting off indefinitely the acceptance of a solution in an attempt to find a better one does not make sense. Furthermore, external time constraints often operate. If for example my *flambé* steak catches fire, I do not have the time to reflect on which strategy is most suitable to save the steak, the table cloth and all those present. I may make do with a mediocre solution in absolute terms, provided it is rapid and respects the most important values. Other costs are measured in terms of the resources the system has access to, and in Western society these may be represented for the most part in terms of the fundamental symbolic means of exchange, money. Hence, the solution generated has to come to terms with the resources assigned to it by the system for that particular problem. For this reason, a trip to Polynesia with the entire family, which would constitute a brilliant solution to the problem of the summer holidays, nevertheless has a good chance of being discarded, if it requires spending the savings of 10 years.

Later, when we deal with reasoning, we shall find the counterpart of checking the solution. Scholars in that particular domain, however, prefer to call the phenomenon the falsification of the conclusion reached. Nevertheless, the objective does not change. It consists in attempting to ensure the result obtained is the best possible, compatible with the constraints given.

A final point to note is that there is always the possibility that an acceptable solution does not exist. In this case too, it is best to realise this has happened in the shortest possible time, in order to avoid wasting resources on an endless search. Abandoning the search for the impossible is a sign of wisdom in the human being, and in a machine, a sign that it is working properly. In both cases it is a matter of recognising one's limits, an important and difficult task. To conclude, it may be stated that no absolute solution exists to a real life problem. The quest must be interrupted at the point that is judged most favourable to the system, a decision that is anything but banal.

As was asserted at the outset, the area of problem solving has been the most fruitful in cognitive science. If it begins to mark time today, this means that the drive produced by the dominant paradigm, that of domain-independence, of the general purpose problem solver capable of tackling any task, has burnt out. An attempt to follow the same line was made by Newell (*Unified theories of cognition*, 1990) with the system known as SOAR. This system attempts to reproduce the general architecture of the mind, and it is therefore presumed to be capable of handling and simulating every type of cognitive activity. In later sections we shall analyse promises and results of the alternative paradigm, based on local and not on general methods, i.e. on methods of finding solutions that vary as the domain varies.

PLANNING

Building plans has always been recognised as one of the key activities carried out by human beings. In one of the pioneering works in cognitive science, examined earlier in Chapter 2, Miller, Galanter, and Pribram advanced the hypothesis that the hierarchical TOTE plan constituted the basic unit in cognitive systems. No clear distinction may be made between the activity of problem solving and that of planning, as is testified by the fact that all the techniques described in the preceding pages will again be encountered here. The main difference is that in planning it is assumed that the system already knows how to solve the problems that occur, and that the real difficulty lies in constructing an efficacious action plan. It is a matter of putting together a puzzle when all the pieces are already available, creativity residing more in the assembly process than in the construction process.

The activity of planning is effected on two levels. One, which is more action-oriented, concerns the ability to build one's own plans. The other, which is more communication oriented, involves the ability to comprehend the nature of the plans built by another system. Both the action oriented and the communication oriented functions are grounded in the same capacity to organise the knowledge possessed by the system around plans. A *plan* is a set of coherent actions aiming at achieving a pre-selected goal. Planning therefore requires knowledge of the world, of how the world may be modified in accordance with one's objectives through action and, finally, the effects such action will have.

Building action plans is indispensable if one wishes to act effectively on the world. What is at stake is achieving one's goals with the minimum expenditure of resources. If I have to post a letter, buy a bottle of wine and put the car into the garage, the most convenient strategy will be that of trying to execute the three actions optimising their common traits, to save time and energy. The concept of the "shopping list" corresponds to the criterion of not leaving the house as many times as there are items to be bought, but to carry out all the actions having common characteristics in one single trip to the supermarket. Thus planning means, first and foremost, selecting one's goals, then establishing the global set of actions most efficient for their realisation, and finally actually carrying out the pre-selected course of actions.

Vice versa, comprehending the plans of others is tantamount to understanding the nature of their intentions, finding plausible explanations for their behaviour, and structuring actions that are not directly connected into significant wholes. This capacity, brilliantly explored by Martha Pollack (1992), is employed as a rule for cooperative ends—and it will be encountered again later as an essential component in communication—enabling collaboration to take place for the achievement of common goals. Thus on the field of Waterloo, when Wellington realises that the Prussian troops led by Blucher are coming up on Napoleon's rear, he decides to attack the French army with all his troops. Consequently, Napoleon, caught between two fires, is defeated by the full cooperation achieved by England and Prussia. If the other's plan were to contain negative objectives for the system, then understanding this fact would enable the system to oppose these potentially dangerous actions in an attempt to prevent those threatening goals from being achieved. When Crazy Horse realises that General Custer has detached Major Reno's regiment onto the flank with the aim of trapping the native Americans in a vice, he engages Reno in a series of skirmishes, and anticipates Custer by launching a massive attack against the main body of the cavalry that has taken up position at the Little Big Horn, destroying it before the flanks can intervene to bolster the position of their surrounded chief (see Sandoz, *Crazy Horse*, 1971).

The activity of planning is connected to the ability to act mentally in the future, predicting what would happen in the follow-up to a certain situation in several possible worlds, and attempting to comprehend what is most convenient, or least unpleasant, for the system. Accurate planning also permits the system to predict what will presumably occur, provided that all the important factors in a situation have been taken into consideration. This in turn allows the system to prepare in time for positive or negative outcomes. Given the link between future and planning, we may assume that the capacity to build plans is one of the most difficult for the system to acquire. In fact, this capacity is to be found only in higher mammals, and in the human species no earlier than the second or third year of life, when the concept of time begins to take on meaning.

The basic mechanism is that of *projection*, by means of which the world is made to evolve in line with a possible future. This enables a person to observe mentally a potential action plan, and thus evaluate its estimated costs and benefits, uncover its defects, and compare it to other plans, in order then to choose the way to make the world evolve towards the most advantageous future possible. Finally, the projection mechanism enables a plan to be revised, attuning it to an optimal condition as judged from all the points the planner considers important, and in particular to ensure the consequences correspond exactly to those desired. Projection is not a method of producing plans, but of evaluating the efficacy of plans generated by other means.

Some general principles with regard to the way plans are actually constructed have been introduced in artificial intelligence which are worth pointing out without needing to go into great detail. The following three principles may be considered to be essential:

1. Optimisation of the use of resources.
2. Achievement of the maximum number of possible objectives.
3. Maximisation of the value of the objectives.

If the system has constructed a complex plan in which the actions designed to reach one goal happen to interfere with those drawn up to achieve another objective, the second principle goes into action to resolve the conflict and render it possible for both goals to be realised. For instance, a system may have set itself the dual target of finding a job and of amusing itself. In this particular case, the second principle can suggest finding a job offering ample possibilities of enjoying oneself. If the search for a means to goal achievement were to prove vain, the third principle would have to be activated to decide which goal has greater overall importance, and which, on the contrary, would have to be suspended or modified. The third principle is based on subjective evaluations of the state of the world. The more this principle manages to contemplate emotional as well as rational aspects of the situation a system finds itself in, the more effective it is. No abstract criterion exists for deciding whether it is better to devote a weekend to finishing off an important job or to take the children skiing, nor whether living in the country is better than living in the city—it all depends. It is the task of the third criterion to interpret best the needs of the system, ensuring the really important goals are given priority. But the truly fundamental principle in planning remains the first, since the sole everlasting problem for living systems is the limited availability of resources. An efficient plan must not waste resources, and the examples furnished at the beginning of the paragraph underline this aspect first and foremost—no one goes to the supermarket as many times as there are things to buy.

In *Planning and understanding* (1983), Robert Wilensky proposes a fourth principle, that of avoiding establishing impossible goals. However, this may be considered as representing a special case of the first principle, imposing saving on resources. Pursuing hopeless objectives is indeed the most notorious method of wasting available energy. A further reason for not adopting this fourth principle is to be found in the impossibility of drawing the dividing line between difficult goals and impossible goals. Life would be much simpler if we could decide which of our desires were unreachable and when, instead, an extra effort would be crowned with success. Leningrad did not surrender to the Nazi forces, even though resisting seemed impossible without food and supplies. Analogously, Christopher Columbus is remembered as a great navigator because he continued where others would have given up. In both cases, however, it was possible to establish that the goals were achievable only after the act. This predicament of being unable to decide *a priori*, an awesome quandary in logic, is absolutely normal in everyday life, where virtually nothing may be taken for granted beforehand.

These considerations mean that the principle of avoiding setting oneself impossible goals cannot be applied in concrete situations. The responsibility of not wasting resources thus falls back onto the first principle. Nevertheless, no one who is overworried about achieving a favourable balance sheet in the short term, will ever be able to accomplish in a brilliant fashion bold undertakings requiring immense investments that will only be repaid, if they ever are, in the long term and after running many risks. Making the most of one's resources is therefore a principle that must be applied in accordance with the individual characteristics of each system, bearing in mind one's personal style in the cognitive and emotional fields.

Analogous concepts to the general principles I have illustrated have been introduced by other authors to tie down the activity of planning in artificial intelligence. Earl Sacerdoti (*A structure for plans and behaviour*, 1977) speaks of *critics* that operate by utilising procedural knowledge on how to avoid a series of defects that may beset a given plan. The activity of critics is tantamount to submitting the plan to a group of experts, each of which is specialised in detecting a particular defect and in suggesting the appropriate corrections, before the plan is actually executed. In Sacerdoti's implementation of the system NOAH, critics are subroutines capable of making specific observations on a plan, and of solving the problems identified.

Mark Stefik (1981a,b) adopts a mechanism grounded on *constraints*. A constraint is a state the violation of which is explicitly prohibited. Every time a plan infringes a constraint, it is rejected and redirected to the planning mechanism whose task it then is to propose a new plan. A solution of this sort, realised in the program MOLGEN, has the disadvantage that it fails to furnish indications either as to the reasons for having rejected the plan or as how to get round the difficulty.

An interesting development of the notion of constraint is that termed *policy*, introduced by Drew McDermott (1978). A policy is a constraint that is represented explicitly. As such, it enables the planner to learn why the plan was not deemed adequate. In this way the system may assign itself the subgoal of suitably altering the specific point, to propose a new version of the plan corrected locally.

The program PANDORA (Plan ANalysis with Dynamic Organisation, Revision, and Application), in contrast, by Wilensky (1983) utilises *metaplanning*, in which the same rules employed in simple planning are applied to generated plans. The knowledge possessed by the system with regard to how plans are drawn up is expressed in terms of a set of goals for the planning process (metagoals), and of a set of plans suitable for their achievement (metaplans). The metaobjectives are fed to the planning mechanism which treats them as it would any other type of goal, attempting to find a plan satisfying the metagoal. The result of an effective application of the metaplan thus generated constitutes the solution to the original planning problem.

In artificial intelligence systems, a plan is composed of four essential parts:

(a) the *name*, usually indicating the goal for the realisation of which it may be employed;
(b) the *preconditions*, indicating the conditions that must be true for the plan to be executable;
(c) the *effects*, describing the most important changes the execution of the plan will produce on the world;
(d) the *constraints*, representing unalterable negative conditions impeding the plan's successful outcome.

One aspect every planner must take into account are the resources available for the realisation of a given objective. But what is meant by the term *resource*? In general, it is something that is necessary for the execution of the plan, and that must be taken into account in evaluating the costs of each predicted action, as in the following examples:

[1]   Lear, King of Britain, must decide whether to name Goneril, Regan or Cordelia as his heir.

[2]   Aeneas needs both swords and ploughs.

[3]   D'Artagnan does not know how best to spend his last francs.

[4]   Hannibal asks himself whether to march on Rome or whether to give the army a rest in Capua.

Example [1] illustrates the case of a resource that is no longer available since it has already been employed in a plan. The drama of Lear consists in the fact that his kingdom may be transferred to others only once, after which the king is no longer the possessor, nor can he intervene to modify the assignation, if the decision proves to be unfortunate.

At times the resource is altered as a result of its employment, and becomes unavailable for some future action. If I have only one egg and I want to make some mayonnaise, I must be aware of the fact that I cannot use the egg again to make myself an omelette. Thus, if Aeneas, in [2], decides to melt the swords to make ploughs, he will no longer have arms.

Another possibility is that the resource is used up in a situation in which it is far from simple to regenerate. While water may even be wasted in town, in the desert it is rationed. Example [3] furnishes an example of the most common of modern resources to which careful attention must be paid: money. This resource is by no means freely available. On the contrary, it is normally a hard earned quantity.

Finally, in [4] we have an example of the sole resource that is always taken into consideration, to a greater or a lesser extent, in human planning: time. The scarcity of time available is an inevitable constant in human life, given its obligatory termination. Even if this is so in a less dramatic fashion than it was for Hannibal, we are always having to grapple with the problem of having to execute several plans in the same time interval. And the situation is aggravated when it is impossible to effect all those plans simultaneously. One can converse and drive contemporaneously, but it would be inadvisable to read while one is driving.

So far, we have considered planning as being the abstract construction of rigid schemes, which are completely represented in the mind: once the action starts, each step will be executed. Such a view has been challenged by Lucy Suchman, in her *Plans and situated actions* (1987). She starts by assuming the *situated* nature of learning, remembering and understanding: these activities depend upon the nature of the world in which they are performed. Actions to be executed in the real world are largely unpredictable, linked to the changing environment where they are placed. Thus, plans are a constituent of practical action, but they are a constituent in the sense that they are a product of our reasoning about action, not the generative mechanism of action. Situatedness overturns the perspective of planning, with some success, especially in the area of human–machine communication. Lave (*Cognition in practice*, 1988) states the general principle, not only valid for planning, that cognition has to be studied outside laboratories. Edward Hutchins stresses the same point in his *Cognition in the wild* (1994), where kitchens and bedrooms are substituted for experimental settings.

FORMAL REASONING

Formal reasoning is a type of thought that operates independently of its content, and follows abstract rules, rules that are always valid. It is precisely for this quality of being abstractable from everyday life, as well as for the possibility of defining form-based rules (stable and devoid of contingent features—unlike content-based rules, which are open to change and interpretation), that this type of reasoning has been considered to be the highest and most complex expression of human thought in the philosophical tradition that may be traced back to Aristotle. Developmental psychology has adopted an identical stance. In Piagetian theory, the culminating point of cognitive development is constituted by formal thought, corresponding to the acquisition of the capacity for abstract reasoning, in accordance with what may be termed mental logic. The ability to effect deductions is considered the essential feature defining rationality in the individual, seen in its turn as the function defining the human being compared to everything a human is not.

An alternative approach will be adopted here, according to which reasoning is not grounded on the application of logical rules of inference, but on the manipulation of mental models representing the states of affairs relating specifically to the object of reasoning. We shall therefore attempt to outline an integrated framework of reasoning, identifying a set of general principles for the manipulation of models, capable of explaining both formal and everyday reasoning. For further enlightenment on the relationship between logic and psychology, the reader is referred to *Human reasoning: The psychology of deduction* (1982) by Jonathan Evans, Stephen Newstead, and Ruth Byrne.

We shall begin by distinguishing between the two basic concepts, deduction and induction. *Deduction* proceeds from the general to the particular, guaranteeing the validity of the conclusions obtained. For instance:

[5]   All men are mortal.
       Socrates is a man.
       Therefore, Socrates is mortal.

The validity of the conclusion does not imply the conclusion is true. The truth value of the conclusion depends on whether the premises are true. Consequently, the truth of the state of affairs asserted must be checked, and this has nothing whatsoever to do with the deduction proper. Even if only one of the premises is false, as in [6], then the conclusion is also false, even though the form of the deduction retains its validity:

[6]   All humans have two hands.
       Captain Hook is a human.
       Therefore, Captain Hook has two hands.

The great difficulty in applying deductive logic in everyday situations consists exactly in the difficulty of establishing with absolute precision and certainty what the significant premises are. And in fact, as Aristotle himself notes, the premises in deductive reasoning are arrived at inductively.

Induction is the inverse procedure—it proceeds from the particular to the general. Induction—grounded as it is on specific events and not on general laws—can never guarantee the validity of conclusions. And since the validity of a conclusion cannot be guaranteed, *a fortiori* neither can the truth of the conclusion. For instance:

[7]   Every crow I have ever seen is black.
       Therefore, all crows are black.

A white crow could exist, one that in my necessarily limited experience I have never seen. It may be asserted that human knowledge is always inductive, since there is no way we can have access to ultimate truths—we always encounter individual facts.

A reflection of this kind led the philosopher of science Karl Popper to elaborate the position known as *falsificationism*, the principles of which will be useful to us at a later stage. Intending to do without induction, Popper (*The logic of scientific discovery*, 1959) notes how science grounds its laws on a limited set of observations and experiments, employing an inductive procedure that can never guarantee either the validity or the truth of the conclusions. A subsequent observation, or a subsequent experiment, might disconfirm the preceding data. No scientific theory can ever be demonstrated as being true. However, if no scientific theory can be proved to be true, it can be shown to be false. In example [7], one single counter-example of a white crow would falsify the theory. Science should therefore proceed by attempting to falsify the hypotheses that are advanced. When a theory is not disconfirmed, then it may be assumed to be provisionally true and employed in the construction and explanation of the world. Popper identifies the essential feature of a theory in its falsifiability. If in principle a theory cannot be invalidated, then that theory does not form part of the scientific corpus.

According to Popper, the theory of universal gravitation is scientific because it can be refuted by a series of observations of free objects that do not fall towards the centre of the field of gravity. Psychoanalysis, on the contrary, together with Marxism, the favourite target for the author's criticisms, are not scientific, since no observation exists that cannot be re-encompassed within the predictions these theories can make, in a cunning game of absorbing the contradictions. If, for example, a man has had a very close relationship with his mother during infancy, he may become an impenitent seducer, or a self-declared homosexual, or a devoted father—a psychoanalyst will always manage to explain everything in an apparently causal manner. Popper turns evaluation criteria on their head, claiming that the impossibility of falsification represents not a strong point, but an intrinsic weak point of the theory.

Returning to reasoning, the obvious temptation was to use the systems of logic available, also as an explanatory theory of human reasoning—after all, logic is always the product of a human mind. Boole himself, the founder of formal logic, believed that it reflected the laws of human thought (see Chapter 2). The system of logic generally selected is first-order logic. This allows all and only those inferences which are valid to be derived starting from a given set of premises. A set of legitimate rules exists, the most important of which is *modus ponens*, schematised in Fig. 7.9.

The approach employing mental logic favours a syntactic type of representation of knowledge, in as much as it does not refer to specific interpretations but to an abstract theory which can be easily managed by propositional type representations. Apart from the difficulty of

$A \rightarrow$	B	IF it is Monday	THEN Stella is in Buffalo
A		Today is Monday	
$\therefore$	B	Hence Stella is in Buffalo	

FIG. 7.9. Scheme of *modus ponens*.

explaining how mental logic develops in the mind of a child, the major weakness exhibited by this approach is that people draw mistaken inferences. If we possessed an intrinsic logical capacity, we should not make mistakes as we do continually. For a theoretical and experimental comparison between a rule-based approach and a model-based approach, the reader may consult *Human and machine thinking* (1993) by Philip Johnson-Laird.

The goal of an inferential system is not to produce all the possible conclusions legitimately derivable from a set of premises, but to derive only those inferences which are of interest. No deductive system adds to our knowledge, because each deduction is inevitably tautological. Stated differently, all valid conclusions are already present in the premises—they have simply been highlighted in the conclusion. Despite this fact, rendering a part of implicit knowledge explicit may be highly informative, since no human being is capable of grasping all the conclusions that may be drawn from a given set of premises immediately. What is at stake, however, is distinguishing between conclusions that are only valid and conclusions that are both valid and interesting. As an example, let us consider the following premises:

[8]   A horse is swifter than a camel.
A camel is swifter than an elephant.

From this it may be correctly deduced:

[9]   An elephant is slower than a camel.

Though valid, this deduction is not particularly informative, in as much as it is simply a paraphrase of the second premise. Conclusion [10] is equally valid, but more interesting, in as much as it links the first premise to the second, rendering the consequence of the link thus established explicit:

[10] A horse is swifter than an elephant.

Generally speaking, reasoning means constructing a coherent representation integrating the information contained in the premises. Informative deductions are those in which new relations in the representation being enacted are made explicit, without increasing the semantic content in the transition from the premise to the conclusion. According to the model approach, reasoning is based on the manipulation of models of specific situations, rather than the application of inference rules to abstract symbolic structures. Manipulation of this type leads to the construction of new models generating valid and informative conclusions. Developing analogue-style representations does not correspond to applying abstract procedures to a system of symbols, as happens when inference rules are applied to the formulae in a given logico-formal system. The elaboration of mental models consists rather in transforming them in accordance with processes that depend on their intrinsic structural properties. What is meant by process here is a procedure to be applied to the model able to make the model itself evolve. A series of computational models able to simulate inferencing processes, and special cases of learning and discovery are presented in *Induction* (1986) by John Holland, Keith Holyoak, Richard Nisbett and Paul Thagard.

The hypothesised use of models encourages considering reasoning as a highly domain-dependent activity, both because of the importance attributed to the knowledge that is employed in each operation, and because of the specificity of the procedures for processing the representations. Studying reasoning therefore implies seeking a range of significant domains and identifying classes of models that characterise them. At the same time it is of primary importance to develop a unified theory of reasoning capable of identifying a series of principles for manipulating models that apply to all domains. Such principles, which guide every type of inferential activity, reflect a few elementary types of competence possessed by the cognitive system, such as the ability to construct, integrate and falsify models. These competences are achieved in a fashion specific to each single applicational domain.

Syllogistic inference

Syllogisms have always been a favourite testing ground for psychological theories on reasoning since they exhibit features that are precious from the experimental standpoint. Since it has been investigated uninterruptedly for the last 2400 years, our knowledge on this subject is complete. They consist of a set of 64 problems of varying difficulty. Syllogisms are deductions based on two premises from which the correct conclusion is to be drawn. The premises consist of propositions containing quantifiers. Depending on the quantifier, each proposition may take on one of four possible *moods*:

A —	a universal affirmation:	All X are Y
I —	a particular affirmation:	Some X are Y
E —	a universal negation:	No X are Y
O —	a particular negation:	Some X are not Y

The usual abbreviations for the four moods, the letters A, I, E, O, come from the vowels of the corresponding Latin terms: *AdfIrmo* (I affirm) for the affirmative forms, and *nEgO* (I negate) for the negative forms. A classic syllogism therefore has the following structure:

All A are B.
All B are C.

∴ All A are C.

The term B appears in both the premises—it is the so-called middle term, enabling a connection to be established between the two and hence an inference to be drawn. The conclusion connects term A to term C, and may take one of the four forms seen earlier. In addition, a fifth type of conclusion is possible: "There is no valid conclusion" (abbreviated to NVC). No valid conclusion means there is insufficient information available to determine the relation between the terms A and C. Out of the total of 64 syllogisms, only 27 allow a valid conclusion, while 37, the majority, do not.

The premises may be arranged into one of four possible *figures*, as follows:

	A–B		B–A		A–B		B–A
I:		II:		III:		IV:	
	B–C		C–B		C–B		B–C

Since each premise may be presented in four different forms, each figure offers 4 x 4 = 16 possibilities. Multiplying this number by the 4 figures gives 16 x 4 = 64 syllogisms.

In 1984, Johnson-Laird and I advanced a theory grounded on mental models accounting for over 5000 experimental data, on the basis of which a computational model was built to simulate subjects' behaviour observed during experimentation. I shall exploit this work to try to furnish a paradigmatic example on how to proceed in building and validating a computational theory. If, therefore, the space devoted to formal reasoning might appear excessive in comparison to other topics, the reason lies in having taken this subject as offering an opportunity for providing a concrete exemplification of what was theorised in Chapter 4 on how to produce cognitive science. Our theory posits that the performance of experimental subjects depends essentially on three factors:

(a) a differing degree of complexity in processing the premises, connected to the figure of the premises themselves, in order of increasing complexity from the first to the fourth;

(b) the number of models that must be built to confirm the validity of a conclusion in each syllogism, a conclusion being valid only if it demonstrates its compatibility with all the other significantly different models that may be built;

(c) the limits of the cognitive resources available, in particular the different capacity of working memory exhibited by subjects.

As with every other form of reasoning, three stages may be identified in solving syllogisms: constructing, manipulating and falsifying the mental models under investigation.

Construction.   In syllogistic inference, this stage corresponds to the interpretation of the premises. For each premise, subjects build a mental model representing the state of affairs that premise describes. The result will be two separate models, each made up of a finite number of elements representing individuals and relations between individuals.

Integration.   This stage consists in achieving the integration of the separate models corresponding to each premise, thus forming a single integrated model that may be read in order to find the conclusion. The integrated model builds relationships between the tokens representing terms A and C which were not explicitly expressed in the premises. In so doing, the integrated model constitutes an informative conclusion.

Falsification.   The attempt to falsify the conclusion obtained takes place at this stage. In line with the suggestion advanced by Popper with regard to epistemology, confirming the validity of the conclusion comes about through the search for counter-examples. Applying function 2 recursively generates alternative integrated models, each of which represents a possible conclusion. If a conclusion is shown to be compatible with all the integrated models generated, such a conclusion may then be considered as being valid. If, vice versa, all the conclusions are disconfirmed by at least one of the integrated models, then no valid conclusion is possible.

We shall now examine these three stages in detail.

Building models by interpreting the premises

The first step consists in constructing a mental model for each premise, interpreting the four moods, as shown in Fig. 7.10. The representation in this figure is one of the many that may be held to be legitimate. Its main advantage is parsimony, since it is the most concise possible. It is to be read as follows: a zero (0) appearing next to a term means that term is optional, insufficient information being possessed about it to

A	All A are B	a – b
		b0
I	Some A are B	a – b
		a0 – b0
E	No A are B	a___
		b
O	Some A are not B	a___
		a0 b

FIG. 7.10. Building models starting from verbal premises.

decide whether it exists or not. A dash (–)between two terms indicates they are positively linked. A continuous line (___) indicates instead that no link may be established between the two terms separated by the line. Models represented in this manner have the least possible number of elements, but each one may be duplicated at will. For instance, in example 11 the single symbol A represents all New Yorkers, and could in principle be replicated up to seven million tokens, if this were necessary for some special reason. In particular, it is possible to add tokens at the stage at which the various models are integrated, to equalise the number of terms present in the models representing the two premises.

Analysing the representations of the four moods in detail, it should be noted that the universal affirmative contains an optional element b0. In fact, stating "all A are B" does not amount to affirming the opposite, "all B are A". There might be some B that are not A. And indeed we are not to know if this is so. For example, in the case:

[11] All New Yorkers (A) are Americans (B).   a – b
                                                                       b0

There are doubtless some B that are not A—Philadelphians and Bostonians are Americans but they are not New Yorkers.

A and B might be coextensive, as in the following case:

[12] All military airmen (A) wear blue berets (B).   a – b

Here the converse model is valid, since in this case "All" has the limited meaning of "all and only". The premise may thus be inverted and the following stated:

[13] All those wearing blue berets are military airmen.   b – a

The same observation is true of the particular affirmative mood. Stating "Some A are B" has three representations, illustrated in Fig. 7.10 by the use of the optional terms denoted by a 0. In the first case, the usual meaning of "some" is intended:

[14] Some engineers (A) are tennis players (B).   a – b
                                                        a0 – b0

This example includes the case of some engineers who do not know how to play tennis (a0) and, vice versa, of tennis players who are not engineers (b0). An implication of this kind is not always true, since there do exist cases of the following type:

[15] Some doctors (A) are gynaecologists (B).   a – b
                                                        a0

There exist doctors—such as neurologists and dermatologists (a0)—who are not gynaecologists, but no gynaecologist exists who is not a doctor (no b0).

The final possible interpretation of "some" is that it means "some and perhaps all", as in:

[16] Some of the 1968 bottles (A) have gone sour (B).   a – b
                                                          b0

In this case all the bottles of 1968 wine might have gone off (no a0). Naturally, this does not exhaust the category of wines that might have turned bad—some of the 1958 and 1973 bottles might have turned to vinegar.

The universal negative mood admits no alternative interpretation, since the converse is always true:

[17] No athlete (A) is a cardiologist (B).   a
                                                        ‾‾‾
                                                         b

This always implies that no cardiologist (b) is an athlete (a).

The particular negative mood has a usual interpretation, as in:

[18] Some actresses (A) are not dancers (B).   a
                                                        ‾‾‾‾
                                                        a0 b

What may well be true here is that some other actresses (a0) are also dancers.

In other cases, "some not" takes on the meaning of "some not, and perhaps no one", as in:

[19] Some condemned persons (A) are not happy (B). $\dfrac{a}{b}$

In this final case the particular negative mood is the equivalent of the universal negative mood. The ability to appreciate the alternatives when a premise has more than one model is fundamental, in order to be subsequently able to falsify the conclusions reached.

Integrating the premises

The second stage consists in integrating the two models the subject has built of the premises. This comes about by adding the information contained in the model of the second premise to the model of the first premise, taking into account the different ways in which the two models may be integrated. The key point in achieving integration is establishing valid links between the middle terms in the two premises, and then eliminating them in order to highlight the relations between the end terms A and C.

In the case of the first figure, A–B B–C, this second step occurs without undue hardship since the two middle term Bs in the premises are contiguous. In his work, *The prior and posterior analytics*, Aristotle calls syllogisms of this type "perfect" precisely because the conclusion derives almost self-evidently from the two premises. Figure 7.11 shows the construction of the integrated model of the most straightforward syllogism, the first of the first figure—AA. The experimental results reported are from "The development of syllogistic reasoning" (Bara, Bucciarelli, & Johnson-Laird, 1995). In this experiment, three groups of subjects of differing age groups were tested. The age ranges of the three groups were: 9–10 years, 15–16 years, and 21–30 years. All testees belonged to the same social class and the standard statistical procedures were applied. The first group was not given all 64 syllogisms, but a balanced sample of 28. This explains why not all the examples that follow will contain the data of all three groups. Only the correct answers will be presented here.

Only one integrated model is possible for Syllogism 1. Hence the conclusion presents no difficulty. In fact, 95% of adult subjects solved it correctly. Children also exhibited a 95% success rate, whereas only 50% of adolescents managed to solve this syllogism.

All A are B   All B are C

Model of the first premise: Model of the second premise:
    a – b   b – c
      b0   c0

Integration of the two premises: a – b   b – c
                                                   b0   b – c
                                                                c0

Integrated model: a – c
                                                   c0

Conclusion: *All are C*

FIG. 7.11. Syllogism 1, AA of the figure A–B B–C.

The second figure, B–A C–B, appears to be slightly more difficult, since the middle term B in the two premises is not contiguous. Suffice it, however, to invert the order of the two premises and a figure identical to the first is obtained, the sole difference being that the first term is now C and not A:

B–A   C–B
    becomes
C–B   B–A

This new syllogism now takes the form of the figure C–B B–A. The middle terms are now contiguous, and may be treated in exactly the same manner as those belonging to the first figure, as illustrated in Fig. 7.11. The only difference is that, for reasons of prior entry to working memory, the conclusions preferred by subjects contain C, and not A, as their first element. The syllogisms pertaining to the second figure are more difficult than those belonging to the first figure, because their solution requires one operation more, namely that of inverting the order of the premises.

The third figure, A–B C–B, and the fourth, B–A B–C, termed *symmetrical*, are more problematical, because no ready method exists for rendering the middle terms contiguous. This means that the order of the elements in one of the two premises must be reversed before embarking upon the procedure prescribed for integration.

Figure 7.12 illustrates the construction of the integrated model for Syllogism 45, OA, pertaining to the figure A–B C–B. The solution to Syllogism 45, as with the other syllogisms bearing symmetrical figures,

Some A are not B   All C are B

Model of the first premise: Model of the second premise:

   a⎯⎯⎯⎯⎯⎯⎯   c – b
a0   b   b0

One of the two premises must be rendered specular, and it is simpler to do so with the second premise.

Specular model of the second premise: b – c
                                           b0

Integration of the two premises: a⎯⎯⎯⎯⎯⎯⎯⎯⎯⎯⎯⎯⎯⎯
                                           a0   b   b – c
                                                   b   b0

Integrated model: a⎯⎯⎯⎯⎯⎯⎯⎯
                    a0   c

Conclusion: *Some A are not C*

FIG. 7.12. Syllogism 45, OA, of figure A-B C-B.

presents even greater difficulty, since rendering one of the two premises specular is obligatory if integration is to be achieved. This view is confirmed by experimental results: the exact solution is found by only 30% of children, 45% of adolescents, and 25% of adults. However, this does not pose the most difficult problem, since it is still possible to construct one single integrated model from the premises.

Falsifying the conclusions. The third stage in the solution procedure consists in seeking to falsify the provisional conclusions by comparing them to all the integrated models that have been constructed. This phase will be illustrated by examining a case in which it is possible to construct more than one integrated model. The case will be that of Syllogism 7, IE, of the perfect figure A–B B–C, appearing in Fig. 7.13.

Numerous mental operations are required to solve Syllogism 7, thus increasing the difficulty and, consequently, the number of errors committed, as is demonstrated by the experimental results: correct answers were furnished by 15% of children, 30% of adolescents, and 25% of adult subjects.

Some A are B   No B are C

Model of the first premise: Model of the second premise:
   a −b   b_____
   a0 b0   c

Integration of the two premises: a −b   b
                                            a0 − b0   b_____
                                                          c

Integrated model: a
                                            a0_____
                                                 c

Conclusion: *No A are C*

However, a second legitimate mode of integrating the premises exists, if it is borne in mind that the optional term in the first premise is not tied to the middle term, and therefore nothing may be predicated about this element with regard to its possible links with c. This leads to:

Second integrated model: a_____
                                            a0   c

Second correct conclusion: *Some A are not C*

This second conclusion is the correct one, because it is compatible with both of the integrated models, whereas the first conclusion is falsfied by the second integrated model.

FIG. 7.13. Syllogism 7, IE, of figure A-B B-C.

    One case needing special mention, because of its frequent occurrence (since 37 syllogisms out of 64 fall into this category), is the one in which there is no valid conclusion. An illustration of this case is provided in Fig. 7.14 by Syllogism 54, II, of the figure B–A B–C. Fifty-five percent of adult subjects and 40% of adolescents answered this syllogism correctly. (This particular syllogism was not administered to children.)
    Finally, the most difficult type of syllogism admits a conclusion founded on the construction of two models, but in which one of the models has to be scanned in the opposite direction (from right to left).

Some B are A   Some B are C

Model of the first premise: Model of the second premise:
   b – a   b – c
   b0 a0   b0 c0

Building of a specular model of the first premise to render the middle terms contiguous yields:

   a – b
   a0 b0

Integrating the two premises yields:

$$a \; -b \; \leftarrow \; b \; -c$$
$$a0 \quad b0 \; \leftarrow \; b0 \quad c0$$

First integrated model: $a \; -b \; -c$
                                                    $a0 \quad b0 \quad c0$

First conclusion: *Some A are C*

However, it is possible to integrate the premises in a different way, when it is noted that the two pairs of terms could be connected by inverting them without becoming incompatible, thus yielding:

                    a – b   ><   b – c
                    a0 b0   b0 c0

Second integrated model: $a \; -b \quad c0$
                                             $a0 \quad b \; -c$

This second model renders the first conclusion false, in as much as no element is explicitly linked with an element c; however, in its turn this second model does not support any conclusion compatible with the first integrated model. It may therefore be deduced that:

Second correct conclusion: *No valid conclusion exists*

FIG. 7.14. Syllogism 54, II, of figure B–A B–C.

An example is offered in Fig. 7.15 by Syllogism 10, EI, of the figure A–B B–C. Only 5% of adults and 5% of adolescents supplied the correct answer. Children were not given this syllogism either.

No A are B   Some B are C

Model of the first premise: Model of the second premise:
   a   b –c
         b   b0 c0

Integrating the two premises yields:

                        a
                             b   b –c
                                b0 c0

First integrated model: a
                                       c
                                    c0

First conclusion: *No A are C*

However, it is possible to manipulate the integrated model further, by moving the optional term c0 above the negative line, thus obtaining:

Second integrated model: a   c0
                                         c

This model falsifies the previous conclusion, and may lead the subject to:

Second conclusion: *No valid conclusion*

If, however, the subject is capable of scanning the second model from right to left, then she will note that though nothing may be predicated of term a, a c exists under the negative line that is separate from a. Thus this leads to:

Third correct conclusion: *Some C are not A*

This third and final conclusion is in actual fact compatible with all the integrated models constructed.

FIG. 7.15. Syllogism 10, EI, of figure A–B B–C

The steps described are those which Johnson-Laird and Bara's theory hypothesises as being the steps experimental subjects follow when solving syllogisms. Figure 7.16 demonstrates that the percentage of correct answers decreases as the following two factors increase:

(a) The number of operations necessary to manipulate the premises prior to integration: difficulty increases steadily from the first figure, A–B B–C, to the second, B–A C–B, increasing still further with the two symmetrical figures, from the third, A–B C–B, to the most difficult, the fourth, B–A B–C;
(b) The number of mental models that must be built in order to solve the syllogism: difficulty increases as the number of models increase steadily from one to two, until the most difficult is reached, the syllogism requiring two models with inverted scanning.

The third important factor is the individual differences between subjects, by far the most important being the different capacity of working memory. People exhibiting a low capacity memory may be compared to children, since the capacity of working memory increases with age.

None of our subjects succeeded in producing a perfect performance, i.e. 64 correct answers. However, people tend to maintain the same style in processing throughout the entire set of syllogisms. This enabled

	Figures of the premises				
	A – B B – C (N = 6)	B – A C – B (N = 6)	A – B C – B (N = 6)	B – A B – C (N = 9)	Percentage totals (N = 27)
Number of models to be built:					
1 model (N = 11)	90	83	72	43	72
2 models (N = 12)	30	30	16.5	14	18.5
2 inverted models (N = 4)	3	3	0	0	3
Percentage total	51	48	35	22	37

FIG. 7.16. Percentages of valid conclusions as a function of the figures of the premises and the number of the models to be constructed (N = number of syllogisms in each condition). *Source:* Johnson-Laird and Bara (1984).

experimental subjects to be classified into four different types in relation to their ability to construct and manipulate mental models:

Type 1:   subjects capable of building up to one model;

Type 2:   subjects capable of building up to two models, but of scanning them only unidirectionally;

Type 3:   subjects responding "there is no valid conclusion" every time they manage to falsify a model, namely every time they construct more than one integrated model starting from the pair of premises assigned them;

Type 4:   perfect subjects, capable of building up to two models and of scanning them in both directions.

Younger subjects belong only to Type 1, while increasing age brings them into Types 2 and 3. From adolescence on, Type 4 may be achieved. At this stage, model theory can predict not only the replies furnished by each type of subject, but also errors typical of each type.

Finally, an experiment was carried out to illustrate an intermediate response, in order to check whether the procedure followed by subjects actually did correspond to that predicted. To this end, a special experiment was set up in which subjects were asked to furnish a response within 10 seconds of being presented with each syllogism. In this manner, intermediate processing stages were obtained, which were later compared with the final answers furnished by the same subjects when given one minute to revise any of their previous conclusions. The importance of the intermediate stages for cognitive simulation has already been explained in Chapter 3: they guarantee that the procedure adopted by the machine is identical to the one adopted by human beings.

The predictions regarding the asymmetrical figures were that subjects would provide answers equivalent to those furnished by Type 1 subjects, the 10 seconds allowed being insufficient for the construction of two alternative models. In the case of symmetrical figures, an increase in NVC replies was predicted, due to the impossibility of building an integrated model, given the series of preliminary operations indispensable to the integration of the premises—subjects would not even have had the time to construct one model, thus being obliged to state they had found no valid conclusion.

These predictions were also confirmed, so that the theory proved capable of forecasting, for each individual subject: correct conclusions, errors, intermediate processing states, response times. All these detailed results are necessary if one considers the computer program validation of the psychological theory propounded. Furthermore the theory purports to be developmental, since it also explains the development of subjects' syllogistic inferencing capacity, from Type 1, to Types 2 and 3, to Type 4. Figure 7.17 shows the replies predicted for each type of subject as a function of the number of mental models required to solve each syllogism. Systematic errors of Type 1 subjects consist in always giving the first possible conclusion. Type 3 subjects' usual error is to answer that no valid conclusion exists to every syllogism requiring two models for its solution.

EVERYDAY REASONING

Dealing with what each one of us does continually, effortlessly, and without paying any particular attention to the task, is much more difficult an undertaking than concentrating on those activities calling for the maximum effort for their achievement. Five-year-old children cannot do arithmetic, play chess or make microchips. They can, however, climb onto a chair, jump up and down on only one foot, make the stereo work without anyone having ever taught them how, understand when their parents are angry and therefore wash their teeth without a murmur. There are robots capable of executing tasks any human finds extremely arduous, but no machines exist able to compete with 5-year-olds on their own ground.

Since the emergence of the view that the easier a thing seems the more difficult it is to simulate, scientists have begun to apply their curiosity to the ordinary world, with a dual interest—cognitive and

	Subjects			
	Type 1	Type 2	Type 3	Type 4
Number of models required:				
1 model	+	+	+	+
2 models	–	+	–	+
2 inverted models	–	–	–	+
2 models (without valid conclusions)	–	+	+	+

FIG. 7.17. Answers predicted for each type of subject (+ indicates a correct response; – indicates an incorrect response).

technological. The technological objective, which has been competently illustrated by Jerry Hobbs and Robert Moore in *Formal theories of the commonsense world* (1985), consists in building machines able to get by in the real world, in the same way an untrained person would get by. The numerous programs presented can reason on qualitative dynamics (how things change, what such changes are due to, what will happen in the future, what must have happened in the past for a certain situation to have come about), on relationships between plans, actions and time, on planning everyday actions such as hanging a picture on the wall, and on temporal reasoning necessary to understand a conversation and to solve problems.

In contrast to expert systems, where the system possesses in-depth knowledge of a specialist field, here the problem lies in providing the system with knowledge about domains in which everyone is an expert. An expert system in oncology can suggest a chemico-therapeutic strategy without knowing that a syringe is a physical object employed to give injections and that the patient is a human being. Typically, expert systems are unable to tackle problems simpler than those for which they were designed, since they lack flexibility, one of the fundamental features of practical intelligence.

In *Mental models* (1983), Dedre Gentner and Albert Stevens analyse how humans tackle everyday problems. Again, emphasis is placed on dynamic processes, thus involving the analysis of change and sometimes of causality. In particular, simulation has centred on the naive theories on the physics of liquids, on elementary mechanics and on heat. One important distinguishing feature of the book is the difference drawn between the thought of experts and that of novices, the pedagogical aim being to outline a developmental pathway enabling a novice to learn to handle a problem as an expert would.

The majority of the inferential activity carried out in daily life does not consist of deductions similar to those examined so far. Everyday reasoning tackles situations in which the data required to produce a conclusion are not totally explicit. This makes it necessary to have recourse to knowledge of the world, that is to that body of specific knowledge concerning objects, individuals, facts, and their reciprocal relations. Having to deal with incomplete knowledge means having to reason not only on facts of which there is positive evidence, but also on facts for which no negative evidence is available. Consequently, expectations based on knowledge of stereotyped situations must be generated, hypotheses must be formulated by means of inductive processes, and reasoning must proceed by analogy. In other words, reasoning proceeds through a series of arbitrary assumptions which may later be demonstrated to be false.

Clearly, we cannot speak of valid inferences in the sense of the term used in formal logic, since we find ourselves in the sphere of possible or plausible inductions. As regards the epistemological aspects of everyday reasoning, two fundamental aspects of knowledge may be identified:

(a) knowledge regarding physical phenomena in the world; that is facts and physical events, and their causal and temporal relations;
(b) knowledge of mental states, both intentional and non-intentional, and of their causal and temporal relations.

From this viewpoint, causality takes on a key role, in as much as it organises the construction of complex models enabling events to be related to each other. I shall here restrict my investigation to the analysis of causal relations, with reference to "Causality by contact" (in press) by Bruno Bara, Antonella Carassa and Giuliano Geminiani, who propose a scheme defining the constituent elements in psychological terms. The nucleus of a causal relation is constituted by a cause–effect relationship between events, a relationship which would not however be significant if considered in isolation. Indeed, in a causal relation, the effect-event follows the cause-event with a high degree of expectancy only if certain enabling conditions occur, as shown in Fig. 7.18.

The evaluation of the enabling conditions is necessary to justify the cause–effect link established. In this way, the probability of the very same link occurring varies with the number and quality of the enabling conditions set up. Establishing a causal relation is therefore a creative activity, grounded on a subjective interpretation of events. Consider, for example, the following facts:

[20] The trapper was bitten by a snake.
[21] The trapper died.

Claiming the existence of a cause–effect link between the facts described in premises [20] and [21] will require the identification of the existence of certain conditions, such as, for instance:

[22] The trapper had a weak constitution.

Enabling conditions
Cause-event ⟶ Effect-event

FIG. 7.18. Scheme of the causal relationship.

Or:

[23] The snake was poisonous.

Or other conditions at the reader's pleasure.

As will be obvious to the reader, even from an introspective point of view, it is possible to produce a broad variety of conditions which bring about or impede a given link. This abundance of possible hypotheses suggests everyday reasoning is not so much the deployment of pre-established laws as the exploration of models representing significant aspects of the facts described. This also accounts for the novelty of the conclusions that may be drawn each time we reason, and for the fact that we may discover aspects of a situation we had never considered previously.

Manipulating an analogical model of a process of physical causality may permit the mental simulation of the development of a situation, enabling the consequences of those factors interacting causally to be considered in their entirety. In the analogical model of a process of poisoning, such as the one depicted in Fig. 7.19, track may be kept simultaneously of how the poison flows in the organism, of how it may interact with an antidote introduced at a later stage, of how the flow of poison may be deviated or slowed down, and of the effects produced by different quantities of poison in relation to the resistance of the target organ.

In a model of this type, poison takes on the role of the causal agent acting by coming into contact with the target organ. The crucial aspect for the causal relationship to occur is that this contact actually does come about through the flow of poison in the organism. In this sense the model is analogical with respect to the characteristics of the means through which contact occurs. Bearing in mind that our basic thesis is that every act of reasoning is governed by the same general principles, valid for both deductive and inductive reasoning, we shall now illustrate how the three general principles analysed with regard to deductive reasoning are realised in the domain of everyday life.

The interpretation stage

This corresponds to the construction of a specific causal process. Referring to its general knowledge of the world, the system attempts to attribute the role of cause-event in a causal relation to one of the facts described in the premises. Some cause–effect relationships are explicitly represented in our general knowledge, attributing roles to events. This type of knowledge is schematic, expressing a sequence of highly probable facts in a unitary fashion, which cannot be analysed further. Thus in the

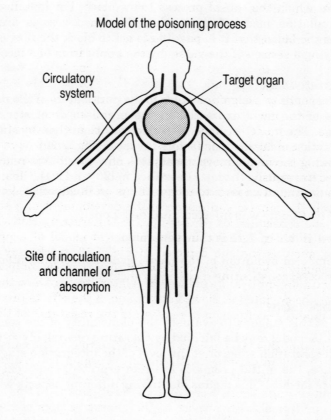

FIG. 7.19. Stage 1: interpretation of the premises. *Source*: Bara, Carassa, and Geminiani (1989).

example examined above, it may be hypothesised that a representation such as the following is activated:

[24] X bitten by snake → X dies.

When it is necessary to evaluate the plausibility of such a link between events with reference to the specific situation described, a model is built of the causal process, to analyse the development of the cause–effect relationship, by decomposing it in temporal terms (Fig.7.19). The model depicts the physical features of the means by which the action of the causal agent comes about on the specific target.

The choice of the characteristics of the means corresponds to the identification of aspects, or parameters, deemed to determine the manner in which the causal process takes place. For instance, in the blood circulation model, holding that the blood vessels are elastic generates the belief that it is possible to act to block the flow of poison by modifying a section of the vessel by the application of a lace.

The manipulation stage

This corresponds to assigning values to the parameters of the model, on the basis of the facts expressed in the premises or of stereotypical knowledge. The model is made to evolve over time thereby simulating the interaction of the various factors involved (Fig. 7.20).

By making the model move ahead, it is now possible to reach a first conclusion integrating the facts described in [20] and [21]. Clearly, there may be more than two facts to integrate. As an illustration, let us add:

[25] An antidote was injected into the trapper.

The model in Fig. 7.21 is thus generated.

The temporal evolution of the model generates a final state in the causal process, constituting a possible conclusion.

The falsification stage

The aim of this stage is to evaluate the plausibility of the conclusions produced. A model may be falsified by operating recursively at the level of the interpretation stage or at the level of the integration stage. In the former case, the initial premises are reinterpreted, thus constructing alternative models. This means identifying different aspects which had

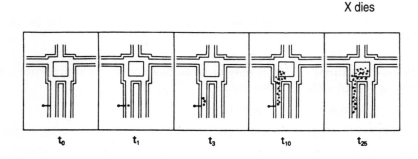

FIG. 7.20. Stage 2: integrating two facts, [20] + [21]. *Source*: Bara, Carassa, and Geminiani (1989).

X is injected
with an antidote

X is saved

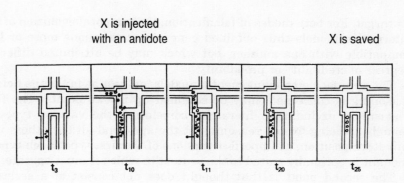

FIG. 7.21. Stage 2: integrating three facts, [20] + [21] + [25]. *Source*: Bara, Carassa, and Geminiani (1989).

not been taken into consideration previously, or even adding new facts to the initial ones, as happened in the model in Fig. 7.21. In the latter case, falsification may come about by assigning different values to the parameters considered so as to be compatible with the facts described. This may be achieved by considering different injection times for the antidote, as in:

[26] The antidote was administered to the trapper too late.

The ensuing model is represented in Fig. 7.22. In the example examined, the late introduction of the antidote does not impede the harmful action of the poison, since the latter is on the point of reaching

X is injected with
an antidote

X dies (the antidote
was taken too late)

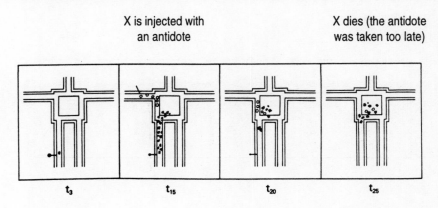

FIG. 7.22. Stage 3: falsification. *Source*: Bara, Carassa, and Geminiani (1989).

its target. For both modes of falsification, the temporal evolution of the alternative models thus obtained generates conclusions more or less compatible with one another, but which may be attributed different degrees of credibility or probability.

The approach employing mental models, at least as far as the field of reasoning is concerned, may be summed up in two points. The first consists in excluding a hierarchy correlating the various types of thought, placing formal reasoning at the apex and all the others in a subaltern position, as imperfect versions of the first type. Each type of thought is optimally suited to the context in which it must operate.

The second point is that thought does not consist of a series of abstract, domain-independent procedures, but of a set of domain-specific models, each set employing processing procedures characteristic of that set itself. What the different thought processes have in common is the way they proceed, realising three general principles: interpretation, manipulation and falsification.

CHAPTER EIGHT

Language

All living beings can communicate among themselves, if for no other reason than to guarantee the continuation of the species. Indeed, a minimal level of communicative competence is necessary in order to procreate in all those species in which collaboration between male and female is indispensable for the generation of offspring. The capacity to communicate, however, varies greatly within the various species of animals. Broadly speaking, we may distinguish between three different classes of communications systems: closed systems, semi-closed systems and open systems.

Closed systems, which are to be found in all animal species, from the lowest levels of the evolutionary ladder up to lower mammals, are those in which each signal corresponds to one sole meaning, there being no option to modify or invent. The alarm whistle of the marmot, the song the nightingale sings to stake out its territory, the stickleback's courtship dance, are unique and unchangeable. Animals emit and receive a signal that is genetically determined. They have no scope either for the creation of a new signal (innovation), or for the construction of a meaningful set of signals whose global meaning derives from the individual elementary signals (composition). Thus a duck will never be able to invent a new sound (an innovative "word"), no matter how useful the animal might find it in a given environment, nor emit a significant series of simple noises to compose an original "sentence". Given these limitations, the capacity of a closed system does not exceed a score or so of meanings.

Semi-closed systems, typical of higher mammals, are those in which a limited number of basic meanings may be put together, employing a rigidly defined system of connectives, to build up a composite meaning. A pack of wolves can organise a hunting group, assigning specific functions to each member of the pack—wolf A, which is especially fast, plays the role of pursuer of the prey; wolf B, the biggest and most powerful, that of killer, and so forth. This presumes the existence of elementary meanings (such as prey, pursuer, killer, and so on) that may be associated with other elementary meanings (wolf A, wolf B) to make up a meaningful discourse at the end of which each participant has understood its own role in the collective action. Likewise, dolphins can communicate among themselves following conversational schemes allowing reference to be made to internal states of the system (hunger, sexual willingness etc.), and even employ concepts referring to themselves, and therefore to express the key concept of "I": see *Dolphin cognition and behavior* by Schusterman, Thomas, and Wood (1986).

A special case of semi-closed systems has been obtained in experiments in which chimpanzees and gorillas were taught sign language. In these cases, in which interaction first between experimenters and primates, and then between primates and primates, lasted for years and years, attempts were made to demonstrate that monkeys acquired the capacity not only to communicate among themselves with a degree of effectiveness comparable to that of humans, but also to generate new symbols conveying new meanings. Recent critical reviews of the conversations and experiences recorded (such as Parker and Gibson's *"Language" and intelligence in monkeys and apes*, 1990) have cast doubts on the optimistic interpretations of the results furnished by ethologists, since these scholars have tended to underestimate the gap separating us from our cousins who have continued to live in the forests.

Higher mammals also possess the power of metacommunication, that is of referring to the communication itself when interacting, mentioning the different meanings without using them directly. For instance, a lioness can play with a cub, pretending she is sending it aggressive signals, and contemporaneously communicating to it that these signals are not to be interpreted literally, but as a game. In playing, the lioness transmits a message of aggression while at the same time conveying that the message is to be interpreted as a joke. The capacity of a semi-closed system, namely one that may link up elementary units of meaning, can reach a total of a few hundred messages. The number of elementary messages that may be expressed remains severely limited, as does the number of connectives available. A primate can construct meaningful sequences employing primitive signs and linking them up.

The "words" are given, but they can be freely "assembled" to generate meaningful "sentences".

Open systems, which include only those possessed by human beings, are those in which both the number of elementary meanings and the number of lawful connectors is potentially infinite. The result is that the number of significant meanings possible is potentially infinite. Human language creates words employing elementary units, such as the letters of the alphabet, or the signs of ideograms. The number of words creatable is infinite, since each letter or sign may be repeated an infinite number of times in the same word. The lexis utilised in normal conversation is approximately a thousand words. *Basic English*, a restricted "language" invented by Charles Ogden (1930), claims that all other English terms may be defined using a base of 850 words. Reading the newspaper requires a vocabulary of approximately 1500 words, while an educated person possesses a store of 40,000–50,000. *Webster's new universal unabridged dictionary*, one of the most authoritative and comprehensive dictionaries of the English language, contains 300,000 different entries, such a high number as to make it possible to compose a quantity of significant phrases that would cover the needs of many lives.

Language is a relatively recent social invention, being only a million years old. This prerogative begins to evolve with *Homo habilis*, the first hominid to possess the adequate phonic apparatus and the cerebral capacity (700ccs as against the 450ccs of the preceding Australopithecus, with a 50% increase in the parietal and frontal areas, responsible for language and for controlling hand movements) that is indispensable to sustain language ability. In *Homo erectus* cerebral volume reaches 900ccs, and in *Homo sapiens* 1200ccs. With the entry of the modern *Homo sapiens* onto the scene, approximately 35,000 years ago, the brain reached its present, and as yet unsurpassed, capacity of 1450ccs. The sophistication of manufactured products and the complexity of the social structure march hand in hand with the increase in brain capacity, and it seems legitimate to presume that the complexity of language has trodden the same path.

Turning now to language as it is employed today by human beings, three different aspects must be examined: syntax, semantics and pragmatics. This three-fold division, shown in Fig. 8.1, permits us to illustrate the different components that are simultaneously at work every time we use language. None is more important than the others in theoretical terms, and all three help to render language an extraordinarily flexible instrument and complex to analyse. Thus separating the three for ease of explication in no way amounts to maintaining that they are three cognitively independent structures.

SYNTAX
Grammatical structures of the sentence; how is the sentence generated? The relationship between signs.

(Examples: declensions in Latin; the subject–verb–complement sequence in English.)

SEMANTICS
Meaning of the single words; meaning of the words as ordered in the sentence; the relation between signs and the world.

(Examples: the meanings of the words "dog", "gun", "lift", "Tex"; the global meaning of "Tex lifted the dog of the gun".)

PRAGMATICS
Context in which a sentence is uttered; goal to be achieved through uttering the sentence; what is the communicative intent of speaker with regard to listener? The use of signs.

(Example: depending on specific context of utterance, the very same utterance "the wolf" may be emitted to inform, amuse, or frighten the receiver.)

FIG. 8.1. The syntactic, semantic, and pragmatic constituents of language.

Quite the contrary. We may rest assured that their action is carried out in a parallel fashion, and not serially—that is the three constituents of language are active contemporaneously, and not in ordered linear sequence. Each component needs the others for language production and comprehension to be effective.

Syntax deals with the rules whereby single words are assembled in an orderly manner to generate sentences that can be immediately recognised as being well formed. Let us take the following sentence as an example:

[1]   A man who was walking in a field met a tiger.

The sentence is well formed, in the sense that it respects rules of syntactic construction and it is immediately comprehensible to any speaker of English. If we take the following sentence however:

[2]   They were drinking knee train for lovingly.

This sentence can convey no meaning in as much as it is syntactically unacceptable. Knowledge of the individual words is not sufficient to transmit the meaning of the sentence, because the words have been put together in a completely random fashion without respecting any of the rules of syntax, even though the context might render the sentence comprehensible. Stated differently, it is sometimes possible to understand the meaning of sentences despite ill-formedness, as in:

[3] Sitting Bull want pale face give food Indians eat.

Nevertheless, the native speaker is always aware of the fact that the sentence does fail to respect syntactic rules, even though context might render it comprehensible. At other times, it is syntax that determines the meaning of a sentence. For example, the English speaker knows that English syntax generally requires the subject to precede the verb and the object to follow it. This makes it easy to interpret a sentence such as:

[4] Stella slaps Alex.

Were such a syntactic convention not to exist, or if the rule were to be violated, then it would no longer be possible to establish with absolute certainty who administered the slap and who received it.

Semantics deals with the meaning of the sentence uttered, both that transmitted through the meanings of the single words and that conveyed through the combination of the single meanings in the expression as a whole. Generally, semantics employs syntax to uncover the real meaning of an utterance, but this is not always so. We saw in [3] that despite a number of syntactic violations we can still attribute a precise meaning to Sitting Bull's words. Semantics enables us to comprehend expressions that are ambiguous from a syntactic point of view. Let us take the following as an example:

[5a] Laura watched the speedboat disappear with great sadness.

It is Laura, and not the speedboat, that is sad. Now we shall examine an utterance that is apparently similar:

[5b] Laura watched the speedboat disappear with a loud noise.

We attribute the noise to the speedboat, not to Laura. The syntactic constructions in [5a] and [5b] are identical, but it is their semantic likelihood that permits us to establish the function of the prepositional phrase as modifying the subject (with sadness) in the former example and as modifying the object (with noise) in the latter. In syntactically treacherous utterances, it is the semantic component that may even turn the initial hypothesis as to meaning upside-down, as in:

[6] Hitting aunts can be dangerous.

Since our knowledge of the world informs us that it is improbable that chasing an aunt is a dangerous sport, we may assume that it is the aunt who is in danger. In these cases it is semantics that determines syntactic interpretation instead of being guided by it, as normally happens. The extreme case, but one that never presents itself in real life, is that in which the syntactic construction is impeccable, the meaning of the single words crystal clear, and yet the global meaning of the utterance escapes us:

[7]   The motorcycle wrote the girl.

Syntax is correct, we all know what motorbikes and girls are, but overall meaning remains obscure.

Finally, *pragmatics* deals with the actual use humans make of language in achieving their goals in and through interaction, studying the various contexts in which utterances are emitted, and how the meanings transmitted are influenced by the contexts themselves. Over and beyond the literal meaning of an utterance, the concern of pragmatics is to establish the goal for which a particular utterance has been emitted. For instance, the literal meaning of the following utterance is quite clear:

[8]   It's five o'clock.

That is, the speaker believes the time, as measured with a conventional timepiece, is five o'clock. The speaker may, however, have produced [8] for a wide variety of reasons, comprehensible only if one is aware of the context of utterance. It may furnish information, if emitted in reply to a specific question regarding the time. It may represent a threat, if uttered by the sheriff who had previously given an outlaw a precise deadline by which time he would have to leave town. It may constitute a warning if uttered to one's lover knowing her husband generally arrives home at around five. It may embody a request, if uttered by a teacher who has informed pupils that their examination papers must be handed in at five o'clock on the dot. It may represent an excuse, if uttered by a psychoanalyst to a patient who has raised an embarrassing problem just as time is running out.

An important distinction is that holding between "sentence" and "utterance". A *sentence* is an abstract theoretical entity defined within a theory of grammar, belonging therefore to the domain of syntax. The meaning of a sentence pertains to the sphere of semantics. An *utterance* is the pronouncing of a given sentence in a given context. It is linked to the effect the emission of the sentence has had on the world, and it is therefore the concern of pragmatics.

The pragmatic goals of linguistic use interact with syntax and semantics, but may be clearly distinguished from the other two concepts, thereby rendering it both possible and useful to treat these categories separately, in order to clarify the speaker's intentions behind the utterance of a given linguistic expression. From the standpoint of interaction with others, understanding the meaning of what is said to us is not sufficient if we do not understand *why* this particular message was uttered. In the following paragraphs, we shall analyse briefly the fundamental contributions of linguistics to cognitive science, adhering to the classic subdivision into syntax, semantics and pragmatics for the sake of ease of argumentation. As psycholinguistics will not receive specific treatment, I refer the reader to the introductory text by Michael Garman (*Psycholinguistics*, 1990), and to the *Handbook of Psycholinguistics* (1994) edited by Morton Ann Gernsbacher.

SYNTAX

The most celebrated contemporary linguist, Noam Chomsky, is also the most authoritative supporter of the importance of syntax in language comprehension. According to Chomsky, syntax may be studied independently of semantics and pragmatics. Indeed, it takes priority over them. In this section, we shall endeavour to illustrate the development of his theses, which have exerted such a great influence over modern linguistics as to have brought about its refoundation, in full accord with computational methodology, as compared to preceding methods of enquiry. Since the present section on syntax confines itself to expounding Chomsky's work, the reader wishing to delve more deeply into the area may turn to the wealth of material in Terry Winograd's *Language as a cognitive process*: *Syntax* (1983).

Transformational grammar
The approach generally known as transformational grammar does not correspond to a theoretical corpus that has been defined once and for all. Quite the contrary. It is in a state of constant development, thanks to the work not only of its founder, but also of his pupils and of those scholars who take their inspiration from him. The crucial points in transformational history were the publication of *Syntactic structures*, which in 1957 revolutionised the international framework of linguistics, and the complete formulation of the standard theory, which was marked by the publication in 1965 of *Aspects of the theory of syntax*. Standard theory has been reformulated many times since, but each version was consistent with transformational philosophy.

Grammar may be defined as a finite set of rules for the generation of well-formed sentences in a given language. The key points in transformational grammar, illustrated in Fig. 8.2, are:

(a) the existence of a universal, in-born grammar specifying the set of human grammars possible and corresponding to children's ability to learn to speak;
(b) native speakers have unconscious knowledge of the specific grammar of their language, corresponding to the set of rules and principles enabling them to generate sentences intelligible to other native speakers and to comprehend sentences produced by other native speakers;
(c) from an epistemological viewpoint, the grammar rules must be totally explicit, in order to permit the construction of formal or computational models of the syntactic component of a language.

The syntactic component of language is connected to the *phonological* component (establishing the links between words and their respective sounds in the spoken language), to the *lexical* component (furnishing the complete dictionary of the words extant in the language), and to the semantic component (correlating words with meanings). Transformational rules enable the transition to be made from surface structure to deep structure and vice versa. *Surface structure* corresponds to an utterance as it is actually read or spoken, word for word, whereas *deep structure* consists of the grouping of the single words into *phrases*. Figure 8.3 illustrates the phrase structure of an example sentence. The explanation of the symbols employed is furnished in Fig. 8.4.

Phrases constitute the most basic component of syntactic structure—in order that one or more words be attached to each phrase,

SYNTAX

Surface structure   →   PHONOLOGY   →   Sounds
Transformational rules

                                  LEXIS   →   Dictionary
Deep structure   →
                                  SEMANTICS   →   Meaning

Particular grammar of a language
Universal grammar (innate)

FIG. 8.2. Transformational grammar.

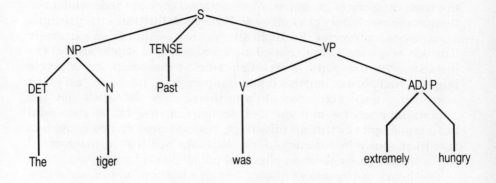

FIG. 8.3. Syntagmatic structure of the sentence: "The tiger was extremely hungry".

Symbols	Names	Examples
S	Sentence	The tiger was extremely hungry
NP	Noun phrase	the tiger, two mice
DET	Determiner	the, a, many
N	Noun	tiger, tree, field
TENSE	Tense marker	past, present, future
VP	Verb phrase	was extremely hungry, sees a ripe strawberry
V	Verb	was, walks, smiles
ADV P	Adverb phrase	skilfully and merrily, very rapidly
ADV	Adverb	extremely, rapidly
ADJ P	Adjective phrase	extremely hungry, wild and bitter and strong
ADJ	Adjective	hungry, merry, ripe

FIG. 8.4. Table of the base constituents of grammar.

the lexical component of language must also be activated. Phrases are the base category of grammar. Nevertheless, they are independent of the specific vocabulary of a language. Figure 8.4 illustrates the principal grammatical categories. Each phrase may correspond to an extremely long set of words, being treated by speakers as a single entity. For instance, "in the time of Spanish rule" corresponds to a single prepositional phrase, and this is how it is perceived. Likewise, "excellent men who had been forced all their lives to spell on an impromptu-phonetic system, and having carried on a successful business in spite of this disadvantage, had acquired money enough to give their sons a better start in life than they had had themselves ..." corresponds to a single noun phrase of unbelievable length.

The heart of Chomsky's theory lies in his *transformational rules*, enabling an essentially descriptive structure, as is grammar, to become also a tool for the generation and analysis of language. In the first place, transformational rules account for the passage from surface structure to deep structure in the language comprehension phase, and the opposite movement in the production phase. Furthermore, the rules guide syntactic modifications such as the transformation from active to passive mood, or from indicative to interrogative. For example, from a transformational standpoint, sentences [9] and [10] have the same deep structure, though differing in their surface structures:

[9] The peasant ate the strawberry.
[10] The strawberry was eaten by the peasant.

The important point here is that it be specified how the transformation from one structure to another may be made and that a trace be maintained of the transformation effected. What makes Chomsky one of the precursors, as well as one of the most illustrious representatives, of cognitive science is the fact that transformational rules are formalised and fully specified. Indeed, transformational grammar fully respects computational criteria of reproducibility on computer, and therefore shares the essential methodological constraint characterising our science.

One hotly debated point among linguists, psychologists and psycholinguists has always been that regarding the psychic correlates of transformational grammar. Initially Chomsky denied any claim of a psychological type, stating clearly that his theory was purely one of competence (see Chapter 4). In other terms, his work as a linguist was strictly connected to the definition of an abstract grammar, intended as that set of abstract rules allowing the comprehension and production of well-formed sentences. That humans also employed precisely this type of grammar was not a pertinent question from a linguistic point of view. In contrast to the competence approach which characterised linguistics, psychologists adopted a performance approach, interested as they were in the procedures that humans really do employ in speaking or learning a specific language.

Such a radical dichotomy was justifiable in the 1950s and 1960s, first and foremost to defend the new approach from the attacks of the behaviourists, who shunned such complex internal language processing. A second difficulty was paradoxically due to the excessive enthusiasm with which psycholinguists had adopted the transformational model, attempting to utilise it as a fully fledged psychological theory simply as it stood. Chomsky had preferred to distinguish his position from such a mistaken interpretation of his work quite clearly, confirming his view that grammatical transformations were a formal abstraction useful in the explanation of syntax, which did not, however, correspond to the mental operations psycholinguists were seeking. Later the picture was to change, thanks both to the non-respect of disciplinary limits professed by cognitive scientists, and to renewed attention being devoted towards the psychological reality of transformational grammar. Great interest was aroused in particular by Joan Bresnan's publication in 1978 of "A realistic transformational grammar", in which she sought to answer the question of how an acceptable model of realistic language use could incorporate a transformational grammar.

In a book of reflections on the general themes of his work, *Rules and representations* (1980), where many ideas are reformulated and many bridges are built by a scholar who has always shied away from finding possible common standpoints with others, Chomsky himself declares that drawing an absolute distinction between competence and performance has no practical value. On the one hand, a theory of particular and universal grammar should form part of psychological theory since it concerns the genetically determined program specifying both the range of possible grammars for human languages, and the specific embodiments arising in specific conditions. On the other hand, linguists interested in competence should utilise psychological models dealing with linguistic knowledge. Such explicit connections also significantly weakened the absolute modularity of grammar, namely the hypothesis that grammatical competence was completely autonomous of other cognitive functions, and according to which syntactical analysis of language could take place without any consideration being given to extralinguistic factors. This was an important concession, since the thesis of the absolute independence of syntax had contributed greatly to rendering Chomsky difficult for cognitive psychologists to accept.

The final point concerns the importance transformational grammar attributes to the generation of language, and hence to the existence, growth and maturation of those mechanisms permitting a child to learn to speak. The attention devoted to the developmental aspects of language acquisition is another point Chomsky holds to be fundamental. Indeed, the importance of this aspect keeps Chomsky in the front line of cognitive science even today, since our discipline still finds it difficult to accept the study of cognitive development as a new research paradigm. In the linguistic field, the developmental approach has led to the formulation of the *theory of learnability*, advanced by Wexler and Culicover in *Formal principles of language acquisition* (1980). The aim of the theory of learnability is to specify the features allowing a child's cognitive apparatus to learn to speak a language naturally, and the first results obtained are of great interest to all of cognitive science, as we shall see in Chapter 10.

Augmented transition networks: ATNs

An important development stemming from the Chomskian approach and adopting a computational framework is what is known as augmented transition networks, usually abbreviated to ATN. Their original developer, William Woods (1970), intended ATNs to have two essential features:

(a) to be a grammar totally implemented on computer, and therefore capable of effecting the syntactic transformations necessary for the production and comprehension of a sentence in a completely automatic manner;

(b) to be a cognitive simulation, and hence a psychological model of language parsing.

An ATN is basically a method of realising a grammar based on a class of computational systems known as recursive transition networks. These recursive networks are further "augmented" thanks to the introduction of a series of other power increasing techniques. Augmented transition networks may be described as oriented graphs, an example of which was provided when discussing semantic networks. In that case nodes represented states, and arcs transitions. What increases their power in such a drastic fashion as to render them formal equivalents of a Turing machine is that the arcs may also be labelled, specifying both the conditions to be evaluated and the actions to be undertaken, including transfers to other networks and recursive calls. It had already been demonstrated by Chomsky that a finite state transition network was not powerful enough for the analysis of natural language. Instead, phrase parsers employing ATNs are potentially capable of carrying out the necessary transitions from surface structure to deep structure and vice versa. An augmented transition network is a general formalism, capable of implementing any grammar. It is not bound up with any theory in particular, even though the influence of transformational grammars has always made its presence felt. In principle, however, the designers of an ATN are not committed to any precise theoretical choice—they are free to structure networks as they see fit.

The weak points of ATNs are similar to those identified in the case of semantic networks. First, a certain rigidity in functioning, which means the system must already be familiar with all the words that will be encountered, and the entire structure of the sentence must correspond exactly to one of the pathways envisaged by the network, otherwise parsing will break down. Partial comprehension is not contemplated. Second, the lack of semantic knowledge practically prevents the use of heuristic functions which would help predict which interpretation of a sentence is most probable in the case of complex or ambiguous sentences. The attempts at incorporating semantic information in the arcs has not been rewarded with particularly brilliant results—augmented transition networks cannot go against their purely syntactic nature. Even though ATNs have never managed to prove their claim in a convincing fashion, to being a psychological model of syntactic analysis of language, they have represented a fundamental step ahead from the standpoint of computational cognitive models as defined in Chapter 4, and they constitute a fertile research area in computational linguistics.

SHRDLU

The most famous computer program for language parsing grounded on a prevalently syntactic approach is SHRDLU by Terry Winograd (1972). SHRDLU is a program written in LISP. It may be considered an expert in a fictitious world, composed of a shelf on which were placed a box, blocks, cubes and pyramids of various sizes and colors (the domain is represented in Fig. 8.5). The system visualised the domain on a screen and had an artificial arm with which it could execute orders in virtual mode on the screen itself.

SHRDLU could comprehend questions pertaining to its world set by the experimenter, answer appropriately [11], execute orders [12], and finally ask for clarification when necessary [13]. In the examples, the computer's replies are in capital letters.

[11] What does the box contain?
      THE BLUE PYRAMID

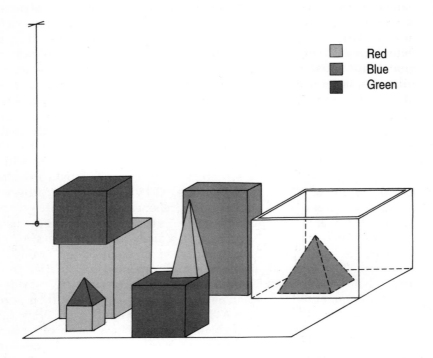

Red
Blue
Green

FIG. 8.5. The domain of SHRDLU. *Source*: Winograd (1972). Reprinted with permission of the publishers, Academic Press, Inc.

[12] Pick up a big red block.
OK.

The system answers "OK" when it carries out an order. In this case, in order to pick up a big red block, it must first remove the green block lying on top of it.

[13] Put the pyramid on a green cube.
I DON'T UNDERSTAND WHICH PYRAMID YOU MEAN.

Since there are three pyramids in the domain, the system does not execute the order but seeks further specifications. The indefinite article preceding the noun phrase "green cube" is interpreted as if it were irrelevant which of the two green cubes the pyramid is to be placed on.

SHRDLU lacked a real understanding of the meaning of the questions asked of it, and the answers too were generated through the adoption of a strategy virtually insensitive to semantic aspects of discourse. And yet its syntactic performances were little short of breathtaking—it could draw sensible conversational inferences, it could handle pronouns, logical quantifiers and comparatives, as well as solve ambiguity to a certain extent. For the first time it was demonstrated that it was possible to construct a program capable of conversing in natural language, even if only on an extremely limited domain. Such a result was obtained by exploiting in the best manner possible three distinct programs, each of which was a specialist in a different area:

1. An expert program for syntactic parsing, capable of handling the syntactic structure of sentences employing a special type of grammar known as systemic grammar.
2. An expert program for semantic analysis, utilising a set of semantic specialists capable of interpreting the meaning of the syntactic structures identified. It was the task of one expert in particular to relate the objects described by noun phrases to the actions described by verb phrases.
3. An expert program for making deductions in the preselected domain, capable of planning actions and predicting their consequences, as well as of contributing to removing ambiguity and of constructing the responses.

The programs interact on a par with each other without any of the three being assigned the task of ordering the operations, even though the heart of the system remains the syntactic parser. The system therefore operates neither hierarchically nor serially, but rather thanks to the contemporaneous contribution of all three programs.

Winograd is extremely explicit in underscoring the limits of his program. In particular, he states that the system works for the sole reason that its universe is so restricted. He also declares that the lack of any form of pragmatic competence and its reduced semantic competence make it impossible to extend its domains to broader and more interesting fields. Furthermore, these very same characteristics also deprive it of any psychological plausibility. And yet, SHRDLU has had a tremendous impact on cognitive science. Many attempts even in the field of semantics have found the courage to grow thanks to the actual existence of SHRDLU. One of the most important movements in semantics, known as procedural semantics, drew its inspiration from the processing procedures first implemented by SHRDLU, even though the operations carried out by the latter were limited essentially to syntactic parsing.

SEMANTICS

Scholars supporting the priority of semantics in the study of language take a different stance. They hold that the starting point of an efficacious analysis of a sentence is its meaning, not its form. Or rather, first comes the meaning to be conveyed, and then the way that meaning is to be conveyed. Such a distinction, though impeccable from a logical viewpoint, is debatable from the standpoint of actual language use—the relationship between form and meaning does not envisage such rigidly sequential processing stages.

The semantic approach points out that, generally speaking, if a person wishes to express something in linguistic terms, that person must first have formulated a thought, an image, an idea corresponding to that particular "something". Now, while it might be true that ideas sometimes require a real linguistic translation, because they appear in the form of images or sensations devoid of immediate lexical correspondents or well-formed structures, it is also true that on other occasions thought already manifests itself phenomenologically in the guise of language. At times, thought may even be determined by language, thereby turning the more obvious sequence upside-down—in this case syntax would determine semantics. Today a more balanced approach sees syntax and semantics—to which pragmatics will be added later on—as constantly interacting, in a kind of serene parallelism, each one influencing the other reciprocally.

A large number of scholars underline the importance of the semantic component in language analysis, and they adopt widely differing approaches. Nevertheless, it may be stated that because of their paradigmatic nature, the most important approaches are those proposed by the theory of conceptual dependency and by procedural semantics. We shall therefore limit our analysis to these two proposals, referring readers desirous of further enlightenment to the books by James Allen (*Natural language understanding*, 1987), and by Ray Jackendoff (*Semantics and cognition*, 1986).

Conceptual dependency
The first possibility open to a semantic standpoint consists in attempting to analyse the meaning of sentences and words by breaking them down into a certain number of *semantic primitives*. It is by no means clear if relying on semantic primitives is, in principle, a stance a theoretician nurturing ambitions of psychological plausibility may legitimately adopt, since many critics convincingly argue that the use of any type of primitive for language comprehension on the part of the human mind is totally devoid of plausibility. Yet the theory of conceptual

dependency was proposed by Roger Schank with avowedly psychological aims, bolstered by experimentation and computer simulation of great efficacy (1972). Schank notes that a large part of linguistic understanding is grounded in general knowledge, permitting us to expect certain things and not others in a specific context. A linear example is furnished by juxtaposing the following two utterances:

[14] I like fish.
[15] I like Mozart.

It is clear the speaker implies "eating" in [14] and "listening" in [15], even if neither of these two verbs is mentioned. But how would hearers understand this if they did not share a network of knowledge regarding the world with the speaker? Conceptual dependency is a theory about the representation of the meaning of utterances, whose basic axiom is:

> For any two sentences having the same meaning, independently of their superficial linguistic structure, there should be only one representation.

From this axiom is derived the corollary:

> Any information implicit in a sentence must be made explicit in the representation of its meaning.

Seeking an economical form for representing meaning, Schank calls meanings underlying language *conceptualisations*. Conceptualisations may be active, with actors executing actions, or stative, with objects in given states. Eleven primitive actions are posited (in the current version of the theory; previously the number varied up to a maximum of 14):

ATRANS: the transfer of abstract relations such as possession, ownership or control.
PTRANS: the transfer of the physical location of an object.
PROPEL: the application of a physical force to an object.
MOVE: the movement of a body part of an animal executed by the animal itself.
GRASP: the grasping of an object by an actor.
INGEST: the taking in of an object by an animal to the inside of that animal.
EXPEL: the expulsion of an object from the body of an animal into the physical world.
MTRANS: the transfer of mental information between animals or within the animal itself.

MBUILD: the construction by an animal of new information from old information.
SPEAK: the production of sounds.
ATTEND: the focusing of a sense organ on a stimulus.

To these must be added unanalysable semantic relations such as POSS-BY, relating to possession, CAUSE, relating to causation, and other more complex relations about time and negation.

These primitive entities are employed to build networks of conceptual dependencies, allowing all the links any term in the language has with the base components to be made explicit. This means that on the basis of a restricted number of concepts that cannot be analysed further it is possible to comprehend the meaning of any word in the language. Schank offers no general criteria for the selection of the primitives, beyond his own personal theoretical intuitions. Many critics, such as Margaret Boden in *Computer models of mind* (1988) for example, have asked the question how it is possible to know if a primitive is "right", in the sense that what has to be demonstrated, to start with, is that humans have in their minds precisely those 11 primitive actions. Schank's attitude towards criticisms of this type has always consisted in underlining the fact that the theory of conceptual dependency is useful, and that simulation programs based on it work. It can hardly be classified as a convincing argument from the standpoint of cognitive simulation, but at the same time it cannot be totally rejected since it does have some validity as a rhetorical argument.

The theory of conceptual dependency, enriched by numerous other structures for knowledge representation, memory, action planning and goal handling, has been implemented in various types of computer programs, all designed for the comprehension of stories pertaining to a particular domain. The importance of these programs lies in the fact that they are able to answer questions, at varying levels of depth, on details not explicitly mentioned in the original story, including events that should have taken place but failed to do so. The first program to obtain remarkable results was SAM (Script Applier Mechanism), which employed general knowledge represented through the means of scripts. SAM's capabilities are well worth illustrating in detail. Starting from a story input as data, we shall see the paraphrase SAM makes of it and the questions it provides answers to. The story, taken from Schank and Abelson (1977), is the following:

[16] John went to New York by bus. On the bus he talked to an old lady. When he left the bus, he thanked the driver. He took the subway to Leone's. On the subway his pocket was picked. He got off the train and entered Leone's. He had some lasagna. When

the check came, he discovered he couldn't pay. The management told him he would have to wash the dishes. When he left, he caught a bus to New Haven.

As a demonstration of how the implicit parts of story [16] may be filled in, SAM generates an extensive paraphrase:

[17] John went to a bus stop. He waited at it a few minutes. He entered a bus. The driver got the ticket from John. He went to a seat. He sat down in it. While John was on the bus an old lady and John talked. The driver took John to New York. He went to the driver. When getting off the bus John thanked the driver. John got off it.

He entered a station. He put a token into the turn-stile. He went to the platform. He waited at it a few minutes. He entered a subway car. A thief went to John. The thief picked John's pocket. He went away from John. John went to the seat. He sat down in it. The driver took John to Leone's. He left the subway car. He left the station.

He entered Leone's. He looked around inside it. He saw he could go to a table. He went to it. He sat down in the seat. He ordered some lasagna. The waiter indicated to the chef John would like him to prepare something. The chef prepared the lasagna. The waiter got it from the chef. The waiter went to the table. He served the lasagna to John. He ate it. He became full.

He asked the waiter for the check. John got it from the waiter. John read the check. John discovered he was unable to pay the check. He indicated to the waiter he was unable to pay the check. The management told John he would have to wash dishes. He entered the kitchen. He washed dishes. He left Leone's.

He went to the bus stop. He waited at it a few minutes. He entered the bus. The driver got the ticket from John. He went to the seat. He sat down in it. The driver took John to New Haven. He got off the bus.

The intolerable prolixity bears witness to SAM's capacity to expand the details contained in each sentence in the original text employing the restaurant script shown in Fig. 5.5. These details, which make the paraphrase so boring, allow SAM to answer special questions, thereby illustrating an inferential process that is so obvious to humans as to make us forget how arduous it may be to make such deductions from the small number of sentences in the original text, unless one possesses a broad independent knowledge of the world:

[18] Did anything unusual happen on the subway?
A THIEF PICKED JOHN'S POCKET.
Why couldn't John pay the check?
BECAUSE JOHN DID NOT HAVE ANY MONEY.
How was John able to take the bus to New Haven?
PROBABLY JOHN HAD A TICKET.

A more recent program, capable of achieving a much deeper level of discourse comprehension than that realised by SAM and the systems of its generation, is BORIS, designed by Michael Dyer (1983). Besides encompassing much more sophisticated knowledge and memory structures compared to its predecessors, it also includes the novel feature that it can handle affective elements present in the text. The system analyses stories concerning divorces, including the causes and consequences of such events.

For instance, in one of these stories the husband (George) finds his wife in bed with another man, and asks his old friend David to represent him in the ensuing lawsuit. BORIS answers appropriately not only to why discovery of marital infidelity can induce a spouse to ask for a divorce, but also as to the predictable emotional states of the husband when he discovers his wife with another man, when he asks his friend David for help, and so forth. More subtly, the system can answer questions regarding the reasons why the lawyer tells George not to worry before the hearing before the judge. The reply furnished ("to reassure him") is based on the fact that the system grasps the empathetic aspects of interpersonal relations. Hence a friend tries to make a friend snap out of a bad mood. The story ends badly for George since he loses the case, much to his friend's dismay. When asked why David expresses his regret to George, BORIS's analysis of the situation yields the answer that in a friendship one is expected to commiserate with one's friend when something bad happens to the other person. If the story had finished up with David congratulating George, BORIS would have pointed out the emotional incongruity.

Can SAM, BORIS and the other programs for the semantic comprehension of language grounded on the theory of conceptual dependency be considered plausible simulations of human mental processes? Literally, undoubtedly not. There are too many unwarranted assumptions making the theory difficult to accept globally, too many solutions that are clearly *ad hoc*, thus leaving one perplexed as to whether the systems constructed may really be easily extended from divorces to other social situations, and finally too little attention devoted to comparing the performance of the computer with that of experimental human subjects. Nevertheless, credit must be given to Schank's group for having obtained remarkable results, since an excessive puritanical methodologism would only constitute an empty pretext for criticising his contribution. In Schank's work far-sighted intuitions and fascinating ideas stand side by side with inexactitudes and short-cuts. The theory is not sufficiently well organised into a systematic whole, precisely because of the lack of homogeneity of its components.

Procedural semantics

Procedural semantics is more a current of opinion than a specific theory. In the course of time, various scholars have even furnished versions which are significantly different, thus making it impossible to report the variants proposed. In general, it may be stated that the idea inspiring this approach derives from the notion of procedure, employed in computer science to indicate what a computer is programmed to do. In the original version by Davies and Isard (1972), it is assumed that the meaning of an utterance corresponds to what happens to the listener when he hears it. Utterances may be thought of as programs constructed to influence others. The meaning of an utterance is shifted entirely onto the listener, coinciding with what happens to him internally, together with the behaviour that is activated on comprehending the utterance. When the theory of speech acts is explained in the following section, the debt this first version of procedural semantics owes to speech act theory will become clear.

Christopher Longuet-Higgins (1987), William Woods (1981) and others have furnished their own original interpretation of procedural semantics. The authors who have furnished the most elaborate and extensive theoretical structure are George Miller and Philip Johnson-Laird in their book, *Language and perception* (1976). At a later date (1983), Johnson-Laird applied the philosophy of procedural semantics to mental models, assuming that procedures exist with which models may be built starting from the meaning of expressions. Constructing a mental model corresponding to an utterance is tantamount to understanding the meaning of that utterance. The process of comprehension comes about in two stages:

(a) the first corresponds to a propositional representation, similar to literal meaning, to the superficial meaning of the sentence;
(b) the second stage, which is where full comprehension of the utterance is achieved, corresponds to the construction of a mental model equivalent to the state of affairs described by the utterance. This second stage is optional, in the sense that it is sometimes never reached.

Consider the following example:

[19] The warrior bade farewell to his wife and child.

The first stage in comprehension is reached by any speaker of English, without any difficulty whatsoever. But it is only when the sentence is contextualised, for instance by adding that we are in Troy, and that what we are speaking about is Hector taking leave of Andromache before his duel with Achilles, that readers can state that they have fully understood the meaning of the utterance. This corresponds to having built a mental model of the utterance—no longer are they simply words on a sheet of paper, but they have been enriched with links, memories, emotions. Underlining yet again the links between procedural semantics and speech acts, it may be noted that it is in the second stage of comprehension when the model is constructed that the hearer understands the intentions the speaker has towards the listener himself, that is, why the speaker has uttered that particular sentence to the hearer. Take the example:

[20] I'm expecting.

Understanding that a pregnancy is being announced is a straightforward matter in the first instance. But what is the communicative intent of the speaker? Is she filling her spouse with joy, is she casting her lover into a state of anguish, is she informing her employer that someone must be found to take her place? Does she expect to cause a cry of happiness or of pain? We shall return to these points below, to analyse them in greater depth. If we now concentrate on the second stage, since it is in this phase that the comprehension of meaning properly speaking takes place, we have to explain how propositional representations may be transformed into mental models that finally correspond to sentence meaning. Johnson-Laird grounds his theory on a set of five independent assumptions.

(a) The processes by which realistic discourse is comprehended are identical to those employed in comprehending fictitious discourse. Clearly, speaking of horses is different from speaking of unicorns, but no difference exists in the procedures utilised in understanding the utterances corresponding to those topics.

(b) In comprehending a discourse, a single mental model is constructed of that discourse.

(c) The interpretation of a discourse depends on both the model and the processes that construct, extend and evaluate it.

(d) The functions that construct, extend, evaluate and revise mental models must be explicit computational procedures.

(e) A discourse is true if there exists at least one mental model that satisfies its truth conditions, and that can be embedded in the model corresponding to the world.

It must be borne in mind that the theory of mental models employs procedural semantics not to relate language to the world, but to relate it to models of the world. Finally, translating an assertion into a mental model requires seven different procedures:

1. A procedure that constructs a new model whenever an assertion makes no reference to any entity in the current model of the discourse.

2. A procedure which, if at least one entity referred to in the assertion is represented in the current model, adds the other entities, properties or relations to the model in an appropriate manner.

3. A procedure that integrates two or more separate models, if an assertion interrelates entities belonging to them.

4. A procedure which, if all the entities referred to in the assertion are represented in the current model, verifies whether the asserted properties or relations hold in the model.

If the verification procedure is unable to establish the truth value of an assertion (if, for instance, there is insufficient information), the procedures for constructing the model will have to take arbitrary decisions (procedure 5) which will then have to be verified employing other procedures (namely, procedures 6 and 7).

5. A procedure that adds the property or relation ascribed in the assertion to the model in the appropriate fashion.

Finally, two recursive procedures are required to cope with the non-deterministic device that must constantly adjust the model, for example when prior representational decisions are disproved by subsequent information. These last two procedures constitute the semantic principle of validity.

6. If the verification procedure (procedure 4) finds that an assertion is true of the current model, then the present procedure checks whether the model may be modified in such a way as to render the assertion in question false while ensuring the model remains consistent with previous assertions. When no such modification is

possible, then the assertion adds no new semantic content to the model—it is a valid deduction drawn from previous assertions.

7. If the verification procedure (procedure 4) finds that a given assertion is false of the current model, then the present procedure (namely 7) checks whether the model may be modified in such a way as to render the assertion in question true while ensuring the model remains consistent with previous assertions. When no such modification is possible, then the assertion contradicts previous assertions.

No program fully implementing a model of procedural semantics exists as yet. Only mini-programs have been developed, interpreting spatial descriptions or analysing extremely simple utterances. Hence no comparison may be carried out of simulations of this theory with simulations of the other theories scrutinised previously. The major point in its favour is that this proposal employs exactly the same principles to explain different mental functions—language, of course, but also thought and perception. In view of the broad canvas of intents, a certain sluggishness in yielding results virtually becomes of secondary importance.

PRAGMATICS

A radically different approach to language consists in asking oneself what the aim of a message, of an utterance, is. Why did the speaker produce it? What goal did the speaker wish to achieve with respect to the interlocutor? "What is the communicative intent behind the utterance?" is the question that must be answered if one is to say one has understood the real meaning of what has been said. Where semantics ends and pragmatics begins is not clear to anyone. We have seen that procedural semantics contains a series of elements that we shall treat as being pragmatic, further on we shall see that pragmatic theories of communication must necessarily come to terms with semantic aspects and even syntactic aspects of utterances. In *Pragmatics* (1983), a book that has become the official standard text in the field, Stephen Levinson lists something like 53 pages of definitions of the term.

From a cognitive standpoint, it must also be recalled that syntactic analysis, semantic analysis and pragmatic analysis are three processes running parallel, and not consecutively. This is a comfort to our reticence in drawing disciplinary limits which are overprecise—stated quite simply, our minds do not use such clearly demarcated confines, and in our brains there are no syntactic neurons or pragmatic synapses. We can therefore allow an epistemological distinction to remain at the service of science without qualms, and without it becoming an untouchable ontological dogma.

The origins of the pragmatic approach may be traced back to the philosophy of language, a movement of thinkers that developed in Oxford and Cambridge in the 1930s led principally by John Austin and Ludwig Wittgenstein. With his *Tractatus logico-philosophicus* (1922), which ended with the threatening seventh aphorism, "Whereof one cannot speak, thereof one must be silent", Wittgenstein had contributed to the foundation and diffusion of the thesis of verificationism, which had emerged within logical positivism, according to which any utterance that could not be verified, i.e. that could not be assigned a truth value, was devoid of meaning. This meant that any utterance whose truth or falsity could not be established, at least in theory, had to be judged meaningless from a rigorous logico-philosophical viewpoint. Utterances of the following types fell into this philosophical nonsense category:

[21] Be careful not to get dirty!
I beg leave to depart.
Fellini is a poet of the cinema.
The death penalty should be abolished.

Philosophically meaningful utterances instead are supposed to be those which are, in principle, verifiable, as for example:

[22] Antananarivo is the capital of Madagascar.
Antananarivo is the capital of Mozambique.
The next time you are absent without permission, you are fired.
Freud's wife was called Martha.

There are insoluble logical problems with the doctrine of verificationism, the most important of which consists in the difference between the meaning of an utterance and the procedure by means of which one can check whether the truth conditions are satisfied. To find out what country Antananarivo is the capital city of, one need only consult an atlas. On the other hand, it might be practically impossible to check the truth of:

[23] The manager is in love with the chief accountant.

Yet the hearer has no difficulty in comprehending the meaning of utterance 23, no matter what he would have to do if he set out to discover if the two people really were having an affair. The situation becomes even more complicated when one moves into the fantasy world of literature.

[24] Achilles avenged the death of Patroclus.
Oberon orders Puck to pour the love potion over Titania's eyes.

Just how does one check an action carried out by an imaginary character? These difficulties weakened the rigid, abstract, "objective" approach of the verificationalist position, to the advantage of an approach that was more mundane, "subjective", related more to what people really do when they speak. In the most important work of his mature period, *Philosophical investigations* (1958), Wittgenstein modifies his position, introducing the concept of *language game*, enabling him to identify the meaning of language with the use actually made of it. Independently of Wittgenstein, Austin was also studying the use of language in our daily lives, reaching results akin to those of the Austrian philosopher. A confrontation never took place between the two, not even in an indirect fashion, one reason being that the works of Wittgenstein, who was active in Cambridge at the time, circulated in manuscript form, while Austin, who was teaching in Oxford, entrusted the diffusion of his thought only to the vehicle of oral lessons, as is borne out by the fact that his main work, *How to do things with words* (1962), was printed only after his death, edited by his disciples.

The spirit of philosophers of language was to abandon abstract research which was proper to linguistics, to concentrate on the daily use people made of words. Not the study of linguistic norms, but the scrutiny of everyday conversations, of the language games people actually played. "Saying is doing" became the motto of pragmatics. Attention was paid to how verbal cooperation was constructed between people. The key concept at the heart of pragmatics is that of *speech act*. Austin noted that in precisely definable situations, certain utterances expressed in the declarative form (for which he employs the term "performative acts") modify the world as would actions. In addition, saying whether such actions are true or false is meaningless, seeing that the appropriate dichotomy to apply is whether or not such acts are efficacious or, in his terminology, felicitous or infelicitous. Consider for example the utterance:

[25] The accused is ordered to pay a fine of £300 or 30 days imprisonment.

If uttered by a judge in the appropriate circumstances it obliges the convicted person to pay or to go to prison, and a series of people (clerk of the court, prison warders etc.) to commit themselves to acting in such ways as to ensure the sentence really is executed. Let us next take the example:

[26] I christen you Helen.

Analogously, if pronounced by the vicar in the correct context, it causes a child to be given the name Helen, by which she will be called from that moment on. Performatives may be successful, altering the world as desired by the speaker, provided what Austin defined as felicity conditions exist:

A.1 There must be an accepted conventional procedure having a given conventional effect, and the procedure must include the act of uttering specific words by specific people in specific circumstances.

A.2 The procedure must specify the circumstances and lay down the behaviour of the people involved.

B.1 The procedure must be followed by all the participants both correctly and

B.2 completely.

Γ.1 If, as often happens, a procedure is destined for use by persons with specific thoughts or sentiments, or to activate consequent conduct of one or more of the participants, then a person executing the procedure must have the requisite thoughts or sentiments, and the participants must intend conducting themselves in the prescribed manner; furthermore

Γ.2 they must subsequently behave in that fashion.

If circumstances do not correspond to those established by the felicity conditions, then the performative misfires. For example, a judge cannot condemn people if they are not in a courtroom, or if the procedure laid down by the law has not been carried out as prescribed. In the case of a misfire, it may be said that "the trial was not lawful", but it is incongruous to comment on a judge's sentence on the basis of its truth or falsity. Thus conversations of the following type would sound strange indeed:

[27] Judge: The accused is sentenced to six months' imprisonment.
     Accused: That's true.

[28] Capitalist: I shall double your wage.
     Office boy: That's true.

The first four rules are denoted by uppercase letters A and B preceding the numbers and the last two by uppercase gamma preceding the numbers to indicate there are two fundamental types of infelicitous performatives. The first type has already been illustrated above. Failure coincides with not having accomplished the desired act, to all concrete ends and purposes and, as will have been noted above, Austin terms this type of violation a *misfire*. In the two cases signalled by gamma, instead, the act has been accomplished, but in circumstances that warrant the affirmation of an *abuse* having been committed in the procedure, as Austin calls this type of violation to distinguish it from a misfire. If the speaker, for example, has made a promise but has no intention of keeping it, then the promise has somehow been made, even if insincerity renders it anomalous. An infringement of A or B renders the act null and void, whereas an infringement of gamma renders the act vacuous, meaningless.

Over and beyond these clear, evocative situations, it subsequently became evident that performatives were not the sole type of speech act that could modify the external world—any speech act generated in communication had this power. Each communicative act bears at least one consequence, namely that the addressees are aware that a communication has been made, and hence that their mental states have been altered, no matter how minimal the alteration may be, in order to grasp knowledge of the kind "A has emitted a speech act". Austin identifies the change in the world with something that is observable. However, if we extend the concept of world to include the mental states of the participants in the speech event, it becomes obvious that interlocutors are also altered by speech acts that leave no observable external trace. Informing someone of something, [29], or imparting an order, [30], change the mental state of the addressee, if nothing else, independently of any modifications that may come about in the world or in the behaviour of the participants as a result of the utterances:

[29] Mary phoned.

[30] You mustn't get out of bed today.

One of the basic consequences of considering language in terms of speech acts is that language thus falls within the ambit of the general laws governing action, highlighting above all its intentional aspects. A further step in Austin's treatment of speech acts is to decompose them into three separate parts: locutionary acts, illocutionary acts, and perlocutionary acts.

A *locutionary* act corresponds to the specific utterance emitted with determinate sense and reference.

An *illocutionary* act corresponds to the addressor's communicative intent in emitting the message.

A *perlocutionary* act corresponds to the effects the addressor sets out to achieve in the mind of the addressee by means of emitting that specific utterance.

A locutionary act represents what is said, an illocutionary act what is done *in* saying something, and a perlocutionary act what is done *by* saying something. A few examples will clarify the differences:

[31] Locution: "Don't move or I'll shoot you!"
      Illocution: threatens addressee.
      Perlocution: induces addressee to stand still.

[32] Locution: "I didn't eat the cake!"
      Illocution: protests her innocence.
      Perlocution: convinces addressee of her innocence.

[33] Locution: "Do it for friendship's sake"
      Illocution: begs a favour of the addressee;
      Perlocution: obtains the favour from the addressee.

[34] Locution: "Have pity on me!"
      Illocution: implores.
      Perlocution: moves addressee to pity.

The three parts of a speech act are governed by different sets of felicity conditions determining the success or failure of each phase. In addition, success at one stage does not automatically imply success at the following stage. For ease of argument, the following notation will be employed for the remaining part of the chapter:

A for Actor—the generator of the speech act; to avoid ambiguity, the actor will be considered as being female, hence Actress.

B—the receiver of the speech act, and of the male sex.

Returning to example [31], the locutionary phase may fail if for instance B is deaf or if he does not understand English, or if a sudden noise drowns A's words. But even if the locutionary phase meets with success, this is no guarantee that the illocutionary phase cannot fail—B could think A is joking, or that A is incapable of pulling the trigger, or he may be convinced the gun is not loaded. Finally, illocutionary success is not necessarily followed by perlocutionary success. B might decide to move in spite of the threat, in the hope that A has a bad aim, or because he does not care whether he lives or dies, or perhaps to show how brave he is.

We may already note a first difference between locutionary and illocutionary aspects, and perlocutionary aspects of a speech act. The first two are essentially conventional, and are enacted in an area of linguistic knowledge shared by both speaker and listener. On the contrary, the perlocutionary act is strictly private, pertaining exclusively to the world of the listener—it takes place in the listener's mind, and there is no direct way the speaker can discover if the perlocutionary effect is felicitous or not. The addressee may have comprehended assertion [32], but might not believe it; the request for a favour [33], but not concede it; the cry for pity [34], but not be moved.

Taxonomy of speech acts

Austin was the first scholar to propose a classification of speech acts. He failed, however, to offer a convincing set of criteria to establish a sound taxonomy. Several taxonomies have been proposed since, but none has yet managed to gain general acceptance among scholarly opinion. The most commonly used still remains that proposed by John Searle in his work *A taxonomy of illocutionary acts*, published in 1971 and reprinted in *Expression and meaning* in 1979, even though the lack of methodological rigour of the classification criteria prevent it from being held the definitive work on the subject. The fundamental difference between the two proposals is that Austin basically only takes performative verbs into consideration, while Searle grounds his own taxonomy on illocutionary acts, and in particular on the felicity conditions necessary for an utterance to be successful.   Searle's classification provides useful guidelines for charting the seas of the different types of linguistic actions that may be performed, and thus deserves to be reported in detail.

Assertives

The function of verbs belonging to this category is to commit the speaker, to varying degrees, to the actual coming about of a state of affairs, to the truth of the proposition expressed. All members of this class may therefore be judged on the basis of their truth or falsity. The psychological state expressed is the belief (that p), and it is the words that are to be adapted to the world, fitting themselves to the state of affairs they are intended to describe. Typical examples include the verbs "state that p", "swear that p", "formulate the thesis that p", "insist that p". In all these cases, belief in a given proposition is declared, with varying degrees of force.

Directives

The function of the members of this class is to attempt to induce the hearer to do something, again with varying degrees of force. This time, the direction of fit comes about in an inverse sense to that taken by representatives, in so far as it is the world that must be adapted to the wishes expressed by the words. Propositional content always induces the listener to carry out a given action. Typical instances comprise the verbs "pray", "implore", "invite", "order", as well as "ask", in as much as questions constitute attempts at obtaining a speech act as a response on the part of the hearer.

Commissives

The function of verbs belonging to this category is to commit the speaker, though with varying degrees of force, to some future conduct. As with directives, the direction of fit that takes place means that it is the world that must submit to the words, which in this case means to the promises made. Once again, propositional content concerns some future action the speaker commits herself to carrying out. Typical of this class are verbs such as "commit oneself to", "promise", in addition to verbs used in the future tense, as in "I shall do", "I shall say", "I shall write".

Expressives

Expressing the psychological state specified in the propositional content is the function performed by this category of verbs. Typical instances include expressions such as "thank you", "I beg your pardon", "I congratulate you", "my condolences", "welcome", "my compliments!". Direction of fit is not relevant to this class since the speaker neither tries to adapt the world to her words, nor her words to the world. Quite the opposite—barring rare cases, the truth value of the proposition is taken for granted, or else investigating it is of no relevance. If A congratulates B on his success, then unless special conditions exist, it would be inappropriate for B to question A as to the sincerity of her emotions—expressives must be accepted for the simple reason they have been uttered by the speaker.

Declarations

The final category we deal with is the one Austin started with, namely performatives, for which it may be stated more directly that saying is doing. These are those cases in which if the felicity conditions are favourable, then the state of affairs expressed will take place quite simply by dint of making the utterance. Classic examples include:

[35] I name this ship "Titanic".

I sentence you to 30 days' jail.
You are promoted to the rank of corporal.
You're fired.
I declare you man and wife.
Penalty!

Searle notes that the essential characteristic of these assertions is that the felicitous execution of one of these acts, under the proper circumstances, produces a perfect overlap between propositional content and reality. If a declarative is performed felicitously, words and world must correspond—adaptation works simultaneously both ways, from words to world, and from world to words. For declaratives to be successful, the appropriate extralinguistic institution must exist guaranteeing that the utterance is sufficient in itself to generate the new state of affairs, namely that a ship will be called "Titanic", a condemned man will be put in prison, a soldier will become a corporal, an employee will be sacked, two people will be joined in matrimony, a penalty will be conceded.

Conversational implicature

The most important bridge between language and communication is constituted by the concept of conversational implicature. To comprehend the origin of this special type of inference, reference must no longer be made to language, but rather to the social norms regulating social and conversational interaction. In the William James Lectures he gave at Harvard in 1967, Paul Grice (1975) illustrated how everyday language use has contents that are transmitted through words, but which in no way derive from the meaning of the words themselves. In other terms, some things are not expressed directly, but are "implicated" by what is said. These implicatures are intentionally communicated by the speaker to the listener. Let us examine the following conversational exchange:

[36] A: What's Stella's new boyfriend like?
      B: His father must be very rich.

B has not furnished a literal reply to A's question, but he has indisputably given her reason to believe that the motives that attracted Stella were not the young man's beauty, charm or intelligence, and not even his well-sported personal wealth. In order to explain the norms governing this particular type of inference, Grice's first point was that every conversation is the result of cooperation between two people having a common goal. Every speech act is generated by A with the intention that participant B recognise that A is openly attempting to induce in him given illocutionary effects and—if she can—perlocutionary effects. The conversation also presents a mutually shared development, rendering certain moves acceptable at certain stages in communication and not at others. Grice therefore formulates the following *cooperative principle*:

> Make your contribution such as is required, at the stage at which it occurs, by the accepted purpose or direction of the talk exchange in which you are engaged.

This general principle is specified in four maxims, recalling Kant's four categories:

The maxims of quantity.   Make your contribution as informative as is required.
Do not make your contribution more informative than is required.

The maxim of quality.   Try to make your contribution one that is true.
The maxim of quality may be further specified into:
Do not say what you believe to be false.
Do not say that for which you lack adequate evidence.

The maxim of relation.   Be pertinent.

The maxim of manner.   Be perspicuous.
Unlike the other maxims, the maxim of manner does not refer to the content of what is said, but to how it is said. It may be further specified into:

Avoid obscurity of expressions.
Avoid ambiguity.
Be brief.
Be orderly.

It should be noted that the four principles do not apply exclusively to language, but to action in general.

1. Quantity: I expect a contribution that is quantitatively adequate. For instance, if I am making the mayonnaise and I ask you to add two egg yolks, I expect you to add two, not one or four.
2. Quality: I expect an authentic contribution. For example, if you offer me a glass of bourbon, I expect the glass to contain bourbon,

and not coloured water, or even whisky (if we are in a civilised country).

3. Relation: I expect a contribution appropriate to the stage the exchange has reached. For instance, if we are at dinner, I do not expect to start off with apple pie, which will be acceptable later on, when we get to the dessert.

4. Manner: I expect my partner to clarify what he is doing, and to do so rapidly and in an orderly manner. For example, if we are working together on assembling the record player, then I must know what you will be doing, and I expect you to carry it out in a reasonable time and without any incongruities.

Grice's aim is not to furnish a book of etiquette on conversation, even though children are indubitably taught to respect something of the sort in a more or less explicit fashion. Rather, the goal of the maxims is to clarify the kinds of criteria we use to create the inferential chain which, starting from the speaker's utterance, extends to the comprehension achieved by the listener. Returning to example [36], if A can assume B is cooperating, then she is in a position to understand what B is seeking to communicate to her, which is that Stella's fiancé does not appear to possess any attractive features, otherwise the reply would seem to have no connection whatsoever with the question asked. The maxims are respected in a well-conducted conversation. Naturally, they may also be violated, thus giving rise to various interesting cases. The possible types of infringements as shown in Fig. 8.6 are as follows:

Error. This consists of an involuntary violation of a maxim, unfortunately a frequent occurrence in everyday conversation. Examples of chronic, non-deliberate violations are those committed

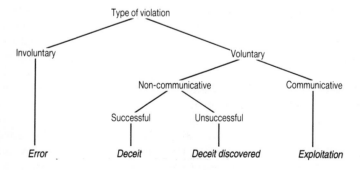

FIG. 8.6. Types of violation and their *effects*.

continually by prolix speakers (second maxim of quantity), by superficial speakers (maxim of quality, second specification), by distracted speakers (maxim of relation), and by disorganised speakers (maxim of manner, fourth specification).

Naturally, anyone, including even generally cooperative speakers, may flout a maxim without realising, or realising too late. These cases may be defined as errors committed in good faith, in which speakers believed in what they were saying. Error does not intentionally transgress the cooperative principle, because the speaker lacks the intention to communicate something false. Errors may or may not be detected by the listener. At an examination, the teacher will presumably discern the mistakes of the examinees, but the client of a financial advisor or of a lawyer is not usually able to grasp mistaken predictions or bad advice. If it is the speaker who possesses the greater quantity of information, the error will escape the listener. If it is the listener who finds himself in this privileged position, then the error will be recognised as such. From this derives the fact that if the addressee infers something from the non-fulfilment of a maxim, the addressee does so over and beyond the intentions of the addressor, since the latter is normally far from desirous of having an error she might have committed detected.

Deceit.   This is a deliberate, but not a communicative, violation of one or more of the maxims, in which the addressor attempts to deceive the addressee by inducing the latter into drawing incorrect inferences. The deceiver has the conscious intention of deceiving, and the equally conscious intention of preventing the addressee from discovering the deceit. For instance:

[37]   A: I fancy a coffee. Is there a café near here?
        B: Yes, there's one just round the corner.

If B knows the bar round the corner is shut today, he has violated the second maxim of quantity. The intent behind A's question was in fact to get a cup of coffee, not to get information—which was correct though irrelevant—as to the whereabouts of a café in the area. B has led A into believing that the café was open by failing to reveal to A that he knew the place was in fact shut. There is a whole category of lies consisting in withholding a significant part of the truth, without ever reaching the point of openly asserting a falsehood. The doctor failing to inform the patient as to the real gravity of his illness, the child hiding something from her mother to avoid causing her worry, the benevolent critic who does not speak out when she cannot utter praise, these are all instances of deceit through the violation of the maxim of quantity, provided an addressee does not realise something has been omitted. In British law, witnesses in a trial are wisely made to swear they will not only tell the truth (commitment to respecting the maxim of quality), but will tell the whole truth (commitment to respecting the maxim of quantity) and nothing but the truth (commitment to respecting the maxims of relation and manner).

Fully fledged lies are those infringing the maxim of quality. Let us distinguish black lies from white lies:

[38] A: Is Guy in?
      B: No.

If B knows Guy is in, then he is violating the first specification and telling a black lie. The same exchange may, however, represent a white lie, violating the second specification, if B does not know whether Guy is in or not, and he does not bother checking or informing A he is not sure.

Violations of the maxim of relation may be achieved by changing the subject without the addressee realising. The concept of dissimulation is tied to the simultaneous flouting of the maxims of relation and manner, avoiding the commitment of the more socially serious infringements of the maxims of quantity and quality.

Exaggeratedly over technical language may flout the maxim of manner (first specification) if the speaker is aware the listener cannot understand such language:

[39] A: Well, doctor, what's wrong with the child?
      B: Recklinghausen's disease.

Deliberate ambiguity may violate the second specification:

[40] A: May I count on your vote at the faculty meeting?
      B: Your plan is undoubtedly worth considering.

If B knows that A will interpret his reply as indicating his willingness to vote in favour of A's proposal, while simultaneously thinking that meritorious as the initiative may be, it should not be approved for reasons of a different nature, the deployment of [40] has saved him from an explicit confrontation. The key point in the deceit is that the deceived must not realise that the deceiver has deliberately infringed a maxim. The deceiver attempts to do so in a non-communicative manner, that is trying to ensure the addressee does not become aware of the violation. Naturally, the deceived party may detect the deception, in which case two choices are open to him—unmasking the deception, or effecting a counter-deception, in other words attempting in his turn to induce his partner into drawing the wrong inferences. Countless examples are furnished by the double-crossing rampant in espionage and counterespionage services. A person discovered to be a spy may be deliberately furnished with false secret information passed off as true, thus deceiving the deceiver by playing a double game.

Finally, it may be noted that only the infringements of the maxim of quality may be classed as explicit lies. The remaining cases constitute more an attempt to make the other person believe something one does not really believe, but without explicitly stating a falsehood. Indeed, of all the examples from [37] to [40], only in [38] could A appropriately reply with the expletive "Liar!".

Exploitation. This constitutes a communicative violation of a maxim in which the speaker induces the listener into making a series of inferences based on the fact that he has realised that the speaker openly intended to flout the maxim. The difference between deceit and exploitation lies in the fact that in the former the violation is hidden, while in the latter it is openly displayed for communicative purposes. We shall now examine a few examples of exploitation of the various maxims. Tautologies, utterances devoid of informative literal content, exploit the maxim of quantity (first specification):

[41] A: Did the lawyer really carry out his threat to show the photographs to Alexander's wife?
B: A divorce is a divorce: war is war.

A is quite aware that divorces are divorces, that in war one acts as in war, in peace as in peace, and so on. She may however induce from this set of truisms that B is trying to make her understand that the appropriate context in which to situate Alexander's divorce is that of a war in which no quarter is given. Instead of answering directly with a "yes", B preferred to give the reason why the incriminating photographs were shown to Alexander's wife.

The maxim of quality (first specification) may be exploited in various ways. The first possible case is irony. For example, in commenting on an action which it is impossible to attribute the quality of genius to, A might say:

[42] Brilliant, truly brilliant.

Another way of infringing the first specification of the maxim of quality is the use of metaphor. Here the utterance is obviously false from the literal point of view, but the implications of the comparison are significant. Powerfully evocative examples are provided by the Song of Songs:

[43] Your eyes are like doves ...
Like purple ribbon your lips,
your mouth an invitation:
segments of pomegranate are your cheeks ...
Your breasts are like two roe deer,
two twin gazelles,
browsing among the anemones.

The above two cases merge in ironic metaphor, where literality is flouted and at the same time the quality named is inappropriate to the referent. For example, a person behaving in a particularly caustic manner may be rewarded for his efforts by sarcasm:

[44] You're a real sweetie-pie.

An example of the exploitation of the maxim of manner (third specification) is furnished by a comparison of the following two utterances:

[45] The countess played a piece by Chopin.
[46] The countess sat down at the piano and pressed the keys in a sequence that bore some resemblance to the score of a work by Chopin placed on the music stand.

The deliberate descriptive long-windedness of [46], where the conciseness of [45] would have sufficed, guides one to the inference that the performance was so far removed from what might be defined as "playing" as to warrant the avoidance of the term. For further explanation of Grice's ideas see his book *Studies in the way of words* (1989).

Presupposition

Another type of pragmatic inference employed in conversation is what Robert Stalnaker (1973) has called presuppositions. It may be said that a speaker *presupposes that p* at a given point in a conversation if she is willing to act, linguistically, as if she took the truth of *p* for granted, and as if she assumed that the audience recognised that she were taking it for granted. Let us suppose I were to affirm:

[47] Scarlet's daughter is sleeping in her cot.

I am taking it for granted that Scarlet does have a daughter, and that her daughter has a cot all of her own. I have not explicitly asserted the existence of the daughter, nor of the cot, but I have nevertheless committed myself to acting conversationally in such a way as to be consistent with these two presuppositions. I would not get away with it if I were later to come out with the phrase:

[48] Scarlet has no children.

It could be argued that since Scarlet does not have children, one is not really lying if one states that a nonexistent daughter is sleeping in her cot. Nevertheless, only a logical positivist would be happy with this kind of justification. Any other person would hold that I had violated the rules of conversation. While conversational implicatures drive inferences forward, departing from the specific utterance, presuppositions drive inferences backwards, that is they establish the logical conditions permitting that utterance to exist in the first place. A typical forward inference, departing from utterance [47], could be the following implicature:

[49] So don't make a noise.

We shall now analyse another example to clarify further the differences between implicatures and presuppositions:

[50] A: Do you still see Laura?
      B: Her second husband is extremely jealous.

The presuppositions of B's reply are that Laura has had a first husband and that she has divorced him. The implicature is that her new husband's jealousy prevents her from continuing to meet B. Figure 8.7 represents the pragmatic inferences drawn from an utterance in a conversation.

The deployment of these two types of pragmatic inferences allows a phenomenon escaping syntactic–semantic treatment to be explained— indirect speech acts. In a direct speech act a speaker intends to say exactly what the words mean literally, as in the example:

[51] I forbid you to go out.

Presupposition ◄——— Utterance ———► Implicature

FIG. 8.7. Types of pragmatic inference.

In this case the directive is explicit, and its literal meaning coincides with what the speaker intends to convey to the listener, namely prohibiting the latter from leaving. In other cases instead the literal meaning of the utterance in no way coincides with the message the speaker intends to transmit, as in the example:

[52] Couldn't you stay at home?

The utterance is literally a question aiming at investigating the possibility of the listener remaining at home, but its intended meaning is clearly a request not to go out. This is identical to the classic case analysed by Searle in which in convivial circumstances A asks the person sitting next to her at table:

[53] Can you reach the salt?

A literal answer ("Yes, I can") would sound rather strange, or would be taken as a joke. The goal of the request is to get the listener to pass her the salt, not that he answer her question regarding his abstract ability to reach the salt. An *indirect speech act* may therefore be defined as one in which the literal meaning does not represent the communicative intention of the speaker, or does so only minimally. In indirect speech acts the addressor wishes the addressee to go beyond the literal aspects of the utterance and infer the fundamental meaning of that utterance, understanding what the speaker really wishes to communicate beyond the specific words.

The reason why the use of indirect speech acts is so frequent is connected to social courtesy, to the need to avoid interactants confronting each other with explicit declarations of intentions, which would expose them to the continual risk of losing face. An indirect speech act permits participants to ask without committing themselves to issuing orders that could embarrass their interlocutors. It also enables them to bargain reciprocally or refuse without the proposer being humiliated. We may note the varying social acceptability of the same proposal when expressed indirectly, as in [54], and directly, as in [55]:

[54] A: I wonder if you could come out with me this evening.
[55] A: You're coming out with me this evening.

In [54] the addressor furnishes the addressee with an easy escape route were the latter unwilling to stray beyond the front door. The directive is explicit in [55], while it is masked in [54]. Analogously, expressing the reply directly or indirectly confers on it a different impact:

[56] B: I am terribly sorry, but that will be impossible.
[57] B: No.

In [56] B has appealed to some external circumstance to justify his non-acceptance, while in [57] he makes the fact he does not wish to go out explicit. The directness of [57] costs A a loss of face.

An interesting computer implementation of the processes of comprehension and production of indirect speech acts has been furnished by Perrault and Allen (1980), while Cohen and Perrault (1979) foramlised speech acts in general.

Real language use goes far beyond the comprehension furnished by syntax and semantics, and is grounded in a set of social conventions guiding linguistic cooperation, especially as regards the inferences every participant in a conversation must continually draw, in order to keep abreast of the conversation. The psycholinguist Herbert Clark (*Arenas of language use*, 1992), points out a question that is particularly hard to answer: how far can inferences be legitimately pushed? When should the inferencing process stop in order to avoid misinterpreting utterances? Returning to example [50], A could hypothetically continue the inferential chain *ad infinitum*, inducing that Laura also desires to continue meeting B, that she is unhappy with her new marriage, that she is unfaithful, that she had a disturbed adolescence and so on. Dan Sperber and Deirdre Wilson (*Relevance*, 1986) answer by strengthening the role of contextual information in spontaneous inference.

From a computational standpoint, pragmatics has had a lesser number of implementations compared to the classic domains of syntax and semantics. Offsetting this is the fact that the level attained has frequently been high indeed, coupled with a degree of sophistication that is infrequently found in computational linguistics. Pertinent examples include Douglas Appelt's book *Planning English sentences* (1985), and the extremely rich volume by Philip Cohen, Jerry Morgan, and Martha Pollack, *Intentions in communication* (1990), to which we direct the attention of those readers interested in exploring this area further.

COMMUNICATION

Human communication is not composed of a set of independent utterances passing between addressor and addressee. It consists rather in the integrated activity of two or more persons collaborating to make up what is defined as a dialogue. A linguistic approach to conversation consists in treating it as an object, whose structure is to be analysed. In a cognitive approach, conversation may be viewed as an interpersonal activity governed by shared social rules and motivated by private mental states.

Gabriella Airenti, Bruno Bara, and Marco Colombetti (1993a,b) have elaborated a computational theory of human communication. Leaving aside the more formal aspects of their *cognitive pragmatics*, I shall outline below only the essential traits. The authors begin by distinguishing between conversational goals and behavioural goals. When people communicate their fundamental aim is to achieve an effect on their partner, that is to say they wish to modify their partner's mental states or induce their partner to carry out some action. But there is a second important point, due to the fact that, to achieve the desired effect upon the partner, actors have chosen communication as their means—they have now committed themselves to abiding by conversational rules.

Conversational goals and behavioural goals have independent origins. Thus a person may refuse to satisfy a request while remaining within the bounds of correct conversation. For instance:

[58] A: Can you answer the telephone?
      B: Sorry, I can't. You answer.

The two goals are pursued exploiting different knowledge structures, and play different roles in the process of comprehension. Behavioural goals are essentially private, and depend on the motivations of and the possibilities open to a person. Conversational goals employ shared knowledge about the conventional norms regulating conversation, and are based on one single motivation—that of communicating.

The problem of the effects of communication (the perlocutionary aspect of a speech act) has an important psychological dimension. First, this dimension has proved difficult to analyse utilising traditional linguistic methods. Second, the problem of perlocutionary effects has never received the extensive treatment afforded locutionary and illocutionary acts, the classic fields investigated by linguistics. To specify what we mean by the effect of a communicative act, two constraints must be introduced:

Intentionality of the actor.   Only those effects the actor intended to achieve must be taken into consideration. Those brought about unintentionally cannot be contemplated, no matter how much they may have been caused by utterance emitted. Let us suppose for instance that A says to B:

[59] Can you please remind Paula that she must bring the cards tomorrow?

If A's intention is solely that of ensuring B tells Paula to bring the playing cards, then other possible communicative effects of [59] may not be legitimately taken into consideration, as for example the case of B making a scene with Paula because she had promised not to play poker ever again. True, the scene made by B was sparked off by A, but if A had not intended this to happen, then it cannot be said that the effects she intended to achieve by emitting utterance [59] included the scene made by B. From this, it may be deduced that only the actor knows what effects she intended to have on her partner. Neither an observer nor an interlocutor would ever be able to state with absolute certainty what these effects were. The picture becomes even more complicated if we consider the fact that an actor may be unconscious of the real intention behind her action. For instance, authoress A may say to a colleague:

[60] I can't decide which publishing house to use to publish my new novel.

If A knows that B's latest manuscript has been rejected by every publishing company he had sent it to, her utterance might be intended to humiliate B. And even if A were to deny this and be in good faith in so doing, an observer, or even B himself, could always argue that this was nevertheless an unconscious but active intention. Since there is no way of solving the problem, then the criterion will be respected of taking into consideration only those communicative effects the actor intentionally and consciously set out to achieve. The following treatment will therefore ignore incidental effects.

Openly achieved effects.   Only those effects an actor intended to achieve openly will be considered. Those that were implicit, but that the partner was not supposed to recognise, will be ignored. As an example, let us consider the following phrase:

[61] Do bring your fiancée to the party.

The effect that is openly intended by the actor is that of inviting both B and his fiancée to the party. However, A might also wish B to realise that she is aware of his engagement. If A did not intend B to recognise she was communicating this to him deliberately, this effect must not be considered a communicative effect of the utterance. The fact that B may grasp this hidden intent corresponds to a sort of unmasked deceit.

It was noted earlier that Grice identified cooperation as the fundamental element of communicative interaction. Indeed, his maxims express a general principle of cooperation underlying his concept of conversational implicature. The links between cooperation and communication are very close indeed. In practical terms, to cooperate people must communicate, or, at the very least, synchronise their efforts.

In [58], by refusing to answer the telephone B failed to cooperate from a behavioural viewpoint. From a conversational standpoint, however, cooperation was definitely maintained. For cooperation to be concrete, both the agents must be familiar with the plan they are carrying out. Airenti, Bara, and Colombetti call the scheme of an action plan known to both A and B a *behaviour game*. In general, actions executed by A and B are tantamount to moves in the behaviour game the agents are playing. The meaning of an action is clear only when it is realised what game this move is being played in. This assumption applies equally to verbal as well as to non-verbal actions. Hence speech acts also fall into the category of moves acceptable in a behaviour game.

In cognitive pragmatics, the classic theme is turned on its head, giving rise to the specular "doing is saying". Stated differently, the actions of two people cooperating may be interpreted as moves in a game, or turns in a conversation. Both physical and speech acts are comprised by a general theory of communication. Analogously with the notion of behaviour game, the existence is presumed of a *conversation game*, containing the rules governing conversation.

The concept of game helps us to understand how one and the same act may be attributed different meanings, depending on the game it is interpreted as being played in. Let us suppose for example that during a conversation one of the two people looks at her watch. This gesture will be interpreted differently, depending on the role relationships of the two actors. If it is the person of higher status that looks at her watch, then the other person will take it as an invitation not to stay any longer. If instead it is the person of inferior status who performs the very same act, then the superior will interpret it as an equivalent of requesting permission to leave. If no game can be identified familiar to both actors and in which the speech act may be considered a move, the utterance emitted will prove to be incomprehensible. Suppose, for example, that while a person was working in his office, a stranger were to walk in and say:

[62] It's raining outside.

Although the literal meaning of the utterance would be crystal clear, the addressee would be extremely perplexed. Only if [62] were taken as an invitation not to go out, or a request to close the window, or a warning to take an umbrella, or in some way the reason for uttering the expression were evident, would the addressee be able to make the necessary inferences and answer appropriately. The naked bones of the utterance pure and simple, without a game to which reference may be made, has of itself no communicative significance whatsoever. The literal meaning is important as a point of departure, but it is not of itself sufficient to answer the questions we set ourselves when someone is speaking with us: "Why is the addressor saying that to me? What does she want from me?". Behaviour games are the structure that coordinates interpersonal actions, and that communication employs to choose the real meaning of an utterance, among the many possible. Identification of a behaviour game on the part of a partner does not necessarily mean the partner must also take part in the game. On the contrary, the partner may refuse to play the game, or propose an alternative game, or negotiate a variant. The partner may even, as a most extreme resource, and one of great social violence, break the conversation game, the equivalent of "I don't even want to answer you".

Behaviour games exhibit differing degrees of applicability, ranging from those employed by all those belonging to a given culture, to those characteristic of one particular social group, down to those idiosyncratic games known to only two people. A partner's acceptance to play is based on an independent, private motivation, the winner of a kind of internal motivation competition against the other motivations driving the system towards goals that differ from that of the game proposed by the actor.

The general scheme of the comprehension/production process may be subdivided into five logically distinct phases, which are not, however, temporally successive, given that two or more of these processes may take place contemporaneously

Literal meaning. The starting point of this phase is the syntactic and semantic analysis of the utterance emitted by actor A. B reconstructs the mental states literally expressed by A, which will serve as the basis for the following phase.

Speaker's meaning. The task of this phase is to recognise the set of communicative intentions A wished to express. Furthermore, achieving full comprehension of the communicative intent of an utterance requires the identification of the behaviour game the addressor is referring to either explicitly or implicitly.

Communicative effect. The communicative effect on an addressee is the result of a process activated by the recognition of the communicative intents expressed by the addressor. As we have already seen, the effect must be intended by the addressor and communicated openly. Failing to recognise an effect intended by A renders the communication infelicitous. In addition, the change induced in the partner must have as one of its causes the communicative intentions of the actor. For instance, the fact that someone attempts to convince me of something must represent one of the reasons why I believe in that something. If this specific condition is not realised, the desired effect has not been fully achieved.

Likewise, if a person already intended doing a particular thing, and is subsequently requested by another person to carry that act out, it cannot be claimed that the actual execution of the act constitutes proof of having obtained a communicative effect. The fact that someone asks me to do something has to be one of the reasons why I do that particular thing. If a motorist who has been stopped by a traffic policeman because she has committed an offence orders the policeman to give her a ticket, it cannot be held that the policeman was induced to give her the ticket by the order she imparted. The fine would have been levied in any case, and the request did not achieve its purpose.

In this phase we enter the domain of the private mental states of the addressee. On the one hand, he must establish whether the speaker has been sincere and proper in expressing the utterance, and on the other hand decide whether it is convenient for him to participate in the game proposed.

Reaction.   This phase must produce a communicative intention serving to generate a specific response. To do so, it has to integrate the communicative effect as reconstructed by B, that is the intentions attributed by B to A, with the behaviour games B is willing to play with A, depending on the internal state of the system, and on the motivations active at that moment.

From the conversation standpoint, B has to inform A about the effects on B himself of A's utterance. If A attempted to convince B of something, B is obliged to declare whether or not he has been convinced. If instead A tried to induce B into doing something, B has to transform his private intention into a communicative intention: do it communicatively. The partner may produce a speech act, or overtly execute the action. If the requested action has to be carried out immediately, B may simply perform it. For instance, an adequate response to a request such as:

[63]   Give me a kiss.

is giving a kiss straight away.

If the action implies a delay in its execution, the partner has to confirm his intention to perform it. For instance, in cases like:

[64] Please give me a ride home tomorrow.

it is not sufficient for the partner to plan the requested action: he has to make his intention clear with an explicit answer.

As regards negative responses, the same principles apply. The partner may render his intention not to perform the requested action explicit through a negative answer. When possible, he might also execute an overt action which is incompatible with the one requested. For example, one can go out when requested to participate in something or stick out one's tongue instead of bestowing the desired kiss.

Just as the actor had the opportunity of being insincere, so the partner has an identical opportunity. Thus the mental states he decides to express in his response might in their turn be insincere. He is at liberty to say he is convinced, even if this is not so, or to play his part in a deceitful manner, and so the cycle starts over again.

Response generation.   An overt response is generated, complete with non-verbal and syntactic details, consistent with the indications furnished by the preceding phase.

The concatenation of these five phases is assigned to the conversation game, which ensures that on completion of each phase the set task has been carried out, and hence the required information is available in input and output. Since each process may have an infelicitous outcome, the theory provides for a series of failures in communication. A partner may apply one rule where the actor did not intend that rule to be applied, or contrariwise may not apply a rule where the actor intended it to be applied.

Cognitive pragmatics represents a cognitive theory of a social, psychological and linguistic phenomenon. That is, it attempts to explain how such a well-formed object as a dialogue is structured employing the private mental processes of the participants in the conversation. Stated differently, through the use of concepts such as mental states, knowledge and motivation, it attempts to explain the most significant of all human social activities: communication.

For an exhaustive survey of the field, the reader may refer to the *Handbook of Pragmatics*, edited by Jef Verschueren, Jan-Ola Ostman, and Jan Blommaert (1995).

CHAPTER NINE

Emotion

Emotions play an extraordinarily important part in human life. They constitute the motor behind our life, the prime reason for our being, the compass directing our primary choices, the thermometer indicating the degree of our happiness with our surroundings and with how we are facing existence. Nothing is as fundamental for the comprehension of the human being, and nothing is still as little understood as emotion. The hot areas of the mind are notoriously the most difficult to study, but in this case a scientific taboo has also lent a hand, as a result of which emotion has not been recognised by modern psychology as an acceptable area of investigation. The most important testimony to the fact that things are changing has been offered by the birth of a new scientific journal, *Cognition and Emotion*. Founded by Fraser Watts in 1987, it has achieved consistent success both with the public and with the critics.

Emotions occupy a very special place in the architecture of the mind, lying as they do at the crossroads between the cognitive dimension, the biological and the social. This is of itself sufficient reason to discourage researchers, since acquisition of competence in such diverse areas is an arduous task, to understate the case. For the same reason emotion should be a theme which fits the multidisciplinary style of cognitive science like a glove, and it would appear more than legitimate to expect significant progress to be achieved in this area.

The most important questions an adequate theory of emotions must provide an answer to are: what is emotion? Which emotions exist? What role do they play? How do they work? How do they develop? Faced with the lack of a widely accepted theory coherently integrating the answers to these five questions, we shall continue our investigation by trying to furnish the best possible answer to each of them, leaving the task of providing an organic synthesis to future research. An ample introductory text to the subject is *The emotions* (1986) by Nico Frijda.

To understand how research developed in this sector, it is essential to examine the historical controversy between the supporters of two diametrically opposed points of view on emotion, a controversy which is still alive today.

William James, the philosopher and psychologist, published "What is an emotion?" in 1884, a work in which he turned upside-down many common sense beliefs that had been taken for granted up to that time. His theme was the analysis of emotional experience, how it differed from other types of experience, and how the single emotions differed within the general category of emotional experience. In James' opinion, emotions are constituted by the peripheral reactions of the organism, which in turn depend on the direct perception of the excitatory event. Emotional experience consists of the sensations provoked by body movements on the one hand, and by physiological reactions on the other. The first way it upsets common sense is this: our heart does not beat because we are with the person we love. Quite the contrary. The fact that our heart beats causes our emotion, love. The difference between the various types of emotion is in James' opinion ascribable to different compositions in the physiological and motor sensations—anger and passion are distinguishable since they are composed of a different combination of peripheral sensations.

James turns another commonly held belief on its head—that regarding the relation between emotional experience and the response to an emotional experience. Since the response to an emotional experience is in fact the experience itself, he argues, then it is the response that should be considered the primitive and not the sensation. His most famous example aims at demonstrating it is not true that we flee because something has frightened us—we are frightened because we realise we are fleeing.

Common sense launches a telling counterattack under the aegis of the neurophysiologist Walter Cannon (1927), who proposes a totally contradictory solution to that advanced by James. The nature of emotional experience is central, not peripheral, and mental, not constituted by direct physical stimulations. Cannon supports the position that emotional experience cannot be reduced to its organic dimension—emotion comes about in the mind, not in one's stomach or in one's muscles, he argues. Furthermore, different emotions correspond to different irreducible primitives, and not to combinations of elementary sensations. This became one of the cardinal points of Cannon's theory, thanks to his demonstration that the different emotions are physiologically activated in essentially the same manner in the human organism. Autonomic responses (arterial pressure, heart frequency, variations in blood distribution and smooth musculature) are identical for subjectively opposite emotions, such as anger and passion. James' thesis on the physiological reducibility of emotions is wrong therefore—it is the mind that distinguishes an aggressor from the prey, not our peripheral physiological alterations. Whether the organism is fleeing from something or whether it is pursuing something, autonomic and muscular responses are equivalent, roughly speaking. However, in the case of flight, the corresponding emotion is fear, while in the case of pursuit, it is excitement.

At this point James' most provocative thesis, his reversal of cause and effect in the relationship between emotion and behaviour, no longer stands up to scrutiny—behaviour goes back to being an effect of emotional experience, not its determinant. We love a person, hence we kiss that person; something disturbs us, hence we shun it. Cognitive science has adopted the cardinal point of Cannon's position, central control, even though research has modified some of his other claims, and, as always happens with intelligent provocative assertions, has uncovered fruitful suggestions in James' paradoxes.

BASIC AND COMPLEX EMOTIONS

Theorists by no means agree on whether there exist a limited number of basic emotions or not, nor, if they do, which ones are basic. A comparison of the various lists of basic emotions proposed by different authors reveals concordance on a set of six emotions, deemed to be innate and irreducible to more elementary sensations: *happiness, sadness, anger, fear, disgust, surprise*. Significant convergence may be found in those studies adhering to the Darwinian tradition on the recognition of facial expressions.

Paul Ekman (*Emotion in the human face*, 1982) furnishes the essential criterion for establishing whether an emotion is to be considered basic, demonstrating that those facial expressions recognisable transculturally are precisely the six named above. This means that persons belonging to different cultures may read each others' emotions in their faces, even though those emotions may, of course, have been induced by different causes. An event giving rise to disgust in a Somalian may not bother a Dane, and what brings fear to the heart of a Mexican might not even cause a moment's thought to a Japanese. But the significant fact is that the Dane can recognise disgust in the Somalian, just as the Japanese may read fear in the Mexican.

The facial expression of each of these emotions, which is innate in humans, may also be found in the behaviour of primates and to some extent even in the behaviour of higher mammals. To the six emotions Ekman adds contempt, which may appear a complex emotion requiring cognitive processing based on social interaction, thereby making it difficult to be considered a basic emotion. Basic emotions are perceived directly and cannot be further subdivided into constituent elements—they are irreducible psychic primitives. The intensity with which an emotion is felt can vary, but its quality remains identical.

These six emotions are determined by innate biological mechanisms, common to all humans, and are independent of social and cultural factors. As will be seen later in this chapter, the physiology of the human body is so constituted as to enable us to experience the six basic emotions right from birth. Even though autonomic responses accompanying these emotions overlap in part, nevertheless they cannot be claimed to be identical. James was therefore right on this point, and the experimental evidence produced by Cannon was an artifice bred in the laboratory, produced by techniques insufficiently sophisticated to distinguish differences finer than simple heart rate or blood pressure. Highly sophisticated techniques are employed today, ranging from electro-myography to the analysis of neuroendocrinological modifications, to the measurement of the effects of hormonal secretion. Thus, the six basic emotions may be distinguished not only on the basis of emotional experience at a mental level, but also thanks to the differing physio-logical reactions parallelling subjective experience and characterising each basic emotion.

If these are the six basic emotions, what status do the other emotions humans are capable of feeling have? What are shame, regret, pride? They are *complex* emotions, generated by differing mixtures of the six basic emotions, and by cognitive evaluations based on the image of oneself, on one's previous experience, on the culture one is a member of, and on social interaction. Such evaluations are generally, but not necessarily, unconscious. In addition, each complex emotion is grounded on a different set of physiological reactions, which depends on the particular mixture the emotion in question is composed of. For a multifaceted treatment of this approach, readers may refer to *Basic emotions* (1992), edited by Nancy Stein and Keith Oatley.

An opposite position is taken by theorists who do not accept the existence of innate basic emotions. Tracing their stance back to Cannon, each emotion is considered as a particular blend of physiological and mental elements. These elements are common not only to different emotions, but also to other mental states. A first sketch of a *componential theory* of emotions was advanced by Schacter and Singer (1962). Later, a more structured treatment was provided by George Mandler (*Mind and body: Psychology of emotion and stress*, 1984). A recent version of the componential approach has been proposed by Ortony and Turner, in their provocative work, "What's basic about basic emotions?" (1990).These authors argue that emotions are constituted by:

(a) *arousal*, a rather undifferentiated physiological component;
(b) an *appraisal* process, the determining factor in structuring the different expressive, behavioural and subjective responses into specific emotions.

While waiting for a new theory capable of integrating both approaches, let us turn to the role emotions play in our lives.

WHY FEEL EMOTION?

What the function of emotions is, and of the particular phenomenal evidence of how they impose themselves on consciousness, is not yet very clear. One reason for their existence is provided by Ekman's research, demonstrating that emotions furnish an important mode of expression, useful for communicating directly without language. Communication through the expression of emotion precedes the development of language in human evolution, since it is also utilised by other mammals, though with less emphasis on the face and more on the rest of the body.

One peculiar feature of emotion is the extreme difficulty in counterfeiting it. It is easy to lie with words, far less so with one's body. Learning to simulate emotional modalities of expression is an arduous art, the mastering of which involves special natural endowments and an extensive period of training. The official schools for actors, and the unofficial ones for politicians and deceivers of various types, have formulated a series of sophisticated techniques to enable their pupils to achieve mastery to come out winners on their chosen stage.

A second function is that suggested by Oatley and Johnson-Laird in "Towards a cognitive theory of emotions" (1987), taking up the classic studies on the neurophysiology of the emotions. According to these scholars, besides the interpersonal communicative function, emotions also have an intrapersonal function—that is, they help provide a rapid, high priority system working in parallel to the cognitive system, to modulate the system's disposition to act. The goal of this intracommunicative system is to avoid slow and uncertain cognitive processing, in order to place the organism in a state of overall readiness most functional to the attainment of vital objectives. The emotional system is not as sophisticated as the cognitive system, the basic objective of the latter presumably being to position the organism in a few basic states. The number of these states is limited, corresponding roughly to the primitive emotions, or in any case to the emotions the organism has learned to react to as autonomously as possible, independently of cognitive evaluations.

If the system has activated itself in an emotional modality corresponding to fear or anger, then it is important that the organism be instantly ready to set pertinent behaviours in motion. Cognitive processing may proceed in parallel, but to await the outcome of an act of reasoning before attacking or fleeing is a risky survival strategy in a natural environment. In a socially complex situation, the integration of immediate physiological elements and slow mental processes mediating the emotions without suppressing the messages they convey becomes a winning policy. The goals emotions pursue may differ from those of the cognitive system. For example, perhaps no representation exists of an emotional goal, thereby rendering the comparison with other, cognitive objectives that have representations impossible. Planning procedures find themselves in troubled waters when they have to decide, for example, if it is more profitable to continue working or to go and have an aperitif with a fascinating acquaintance. Pleasure cannot be plotted on a curve against activities devoid of emotion.

Emotional communication is able, moreover, to assign the maximum priority to its own signals, interrupting other functions. This too is extremely significant, since organisms are constantly engaged in a series of activities of differing importance. The digestive process may be in operation, but if the organism is attacked, it is important for the blood flow to be diverted away from the stomach, and towards brain and muscles. Analogously, if we hear a strange noise while writing a letter, no matter how important its content may be, we immediately and automatically stop, caught between surprise and fear, to investigate what is going on. Digesting calmly or bringing a sentence to an end might have unpleasant consequences. Emotions may intervene in cognitive processing with different degrees of priority, from the low priority of a situation that only touches the fringes of the emotional sphere, to the absolute priority typical when emotional values are threatened. A child crying has a drastic effect on maternal behaviour; the goal of saving the small child takes absolute precedence over any other possible maternal objective, including self-preservation. Highly emotional situations therefore impose their own objectives rapidly and unconditionally.

A further function performed by emotions is that of acting as a bridge between physiological and mental elements, informing the conscious system of certain primitive values useful for the system's equilibrium. Hence, a state of sexual excitation aroused by another person may help one make a sound choice in selecting one's partners, or a state of peacefulness induced by a given place may help one decide where it is best for one to live. This last function accounts for why emotions are also conscious—the first two operate efficiently even when they do not penetrate the realm of consciousness, whereas the last function has to interact with values, goals and modes of planning actions characteristic of conscious awareness. For a deeper exploration of these themes, scholars may refer to the innovative book by Oatley: *Best laid schemes: The psychology of emotions* (1992).

In "Cognitive–emotional interactions in the brain" (1989), the neuroscientist Joseph LeDoux has hypothesised a model explaining interactions between mental processes and the neural substratum in constituting emotional experience. LeDoux goes one step further in the thesis of the reciprocal independence of neural processes computing emotion and cognition, claiming that the two functions are realised by different parts of the brain. On the basis of an extensive re-examination of the literature on the subject, the amygdala (the cerebral area located in the temporal lobes) is posited as being the centre of emotional processing. Emotional–cognitive interactions are achieved by means of the neural connections between the amygdala and the cerebral areas assigned to cognitive processing, such as the hippocampus and the neocortex. In order to experience an emotion, stimulus representations, affect representations, and self-representations have to coincide in working memory. In the wake of the tradition of centralisation, LeDoux claims emotional experience may also be mediated exclusively by the brain, without peripheral reactions necessarily being present. The role of peripheral factors is to act as amplifiers of the central network. In addition, if in a highly charged affective situation no subjective emotion is experienced, then peripheral factors may make a second attempt at activating the emotional part of the brain.

The existence of relations between emotion and cognition presuppose the functioning of two integrated systems in a constant process of interchange. Although this fact has numerous consequences, two force themselves upon our attention and will be treated in some depth. The first is that from a developmental point of view, separating these two dimensions is concretely impossible and theoretically erroneous. The second point is that such a complex integrated structure may present malfunctionings or suboptimal connections, thus opening up the possibility of emotional disorders to be discussed in Chapter 12.

EMOTIONAL DEVELOPMENT

The conflict over the basic or componential nature of emotions reappears in the debate on emotional development. On the componential side, Sroufe (1984) shows that emotions are differentiated in the course of time, starting from an initial emotive state of arousal. The differential theory, sustained by the influential psychologist, Carroll Izard (*The psychology of emotions*, 1991), assumes instead that primary emotions exist that are already differentiated at the neonatal stage.

The starting point of our discussion is the observation that the child already expresses signs of pleasure (endogenous smile), of fear (starts, pain), of distress (crying) and of disgust (nausea) at birth. These emotions lack cognitive correlates that are to be found later in time, but it is beyond doubt they are important modulators of the child's activity.

When the child is about 3 months old, she begins to exhibit a set of motor and vocal (cooing) activities. These activities may be interpreted as precursors of affect and of curiosity and interest.

With the development of the exogenous or social smile, around 4 months, the child is able to interact in the proper sense of the term with the external world, and hence to exhibit reactions of pleasure (smiling) in the case of success, and of displeasure (disappointment, anger) in the case of failure.

After the 7th month, the child interacts fully with the people and things surrounding her, differentiating her emotions to a vast extent: happiness (expressed with smiles, and brought about, for example, by games, such as saying "peekaboo" to the child, with the child cooperating in the game); fear (the stranger now arouses fear because the child recognises the stranger as foreign to her environment); anger (caused by an undesired outcome of an event); surprise (the child recognises the event as unexpected).

The period from 9 to 12 months is that of attachment, with the child developing deep emotional relationships with the people caring for her (their presence is reassuring, they worry about the child). At this stage the expression of emotions becomes extraordinarily rich—degrees of emotion, ambivalence, moods, and the intentional communication of emotion all emerge at this phase. The child has become a fully emotional being.

In order to feel emotions connected to the self (shame, pride, guilt, love) it is necessary to develop the concept of the self. Similarly, for emotions linked to the social context to appear (embarrassment, contempt), the child must have acquired awareness of the behavioural rules and culture of her group. Awareness of the self and of the group are achieved after 18 months of age.

The child is now ready to learn to perceive and to manifest the complex emotions necessary to enable her to lead a rich adult life. As the cognitive system gradually manages to carry out the relevant cognitive evaluations, the child will acquire the notion and experience of each new emotion.

Passing on to the development of artificial systems, we find the first emotional robot in history—the Solitary Fungus Eater, later to develop into the Emotional Fungus Eater (Toda, *Man, robot and society*, 1982). The Solitary Fungus Eater, invented by Toda in 1962, is described as a humanoid robot sent out to gather uranium ore in a science fiction scenario, the planet of Taros, where wild fungi grow from which the biochemical machine obtains the energy it requires to work. The robot possesses a perceptive system by means of which it can recognise fungi and uranium. The geography of Taros is unknown, hence it is not known in which areas fungi grow and in which uranium is to be found. A human operator guides the robot by means of a control panel receiving all the sensory information the Solitary Fungus Eater extracts from the environment. The operator has been instructed to ensure the robot gathers as much uranium as possible. The operator's earnings will be proportional to the quantity of uranium collected. Each activity carried out by the robot consumes a set quantity of energy. As soon as the energy runs out, the robot comes to a halt, the mission is terminated, and the game is over.

That the Emotional Fungus Eater possesses an emotional system is by no means obvious. In his later version, Toda introduced a series of urges, that is to say "motivational" subroutines connecting cognition with action. Every time a situation is identified as pertinent to one of the urges, a subroutine is immediately activated bringing the robot directly into action. Impulses fall into two categories: emergencies (fear, anxiety, surprise) and social (love, anger, frustration, guilt, gratitude etc.). The intensity of each impulse depends on its degree of importance to the survival of the robot. Toda demonstrates that when numerous Solitary Fungus Eaters interact, then social and cooperative behaviour develops which has many affinities with what in humans would be called emotional reactions—hence the new social robot has been given the name Emotional Fungus Eater.

Toda's robots were followed by a long series of interesting programs, but none of these is important enough to validate a general theory. An up to date description of programs simulating emotions is to be found in *Cognitive perspectives on emotion and motivation* (1988), edited by Hamilton, Bower, and Frijda. Once again we are far from a definitive solution, from a theory linking up data in a satisfactory manner and providing answers to the questions we asked at the beginning of the chapter. Nevertheless, a promising path has begun to be trod which will hopefully lead to ever more exciting results.

CHAPTER TEN

Development

The entire structure of the mind is fluid, in a constant state of development. The task of the scholar is not so much that of attempting to fix procedures into static states, as to comprehend how it is that a constant stream may maintain the unity of the self, even at a subjective level. With the exception of a small number of perennial cells, principally those making up the nervous system, our bodies renovate their cellular structure completely in under 10 years. And even with regard to the brain cells, only the neurons have the mortal privilege of being unable to reconstitute themselves. All the other structures, including the network of connections between the neurons, uninterruptedly form part of the cycle of generation and decomposition.

The human being is relatively unstable from a physical point of view. Yet subjective experience exhibits a continuity which is broken only in the case of special and quite rare degenerative diseases of the brain. It is as if the body were a boat that is continuously being repaired, repainted and fixed up by replacing each part that wears out through use. Assuming that after a few years not one of the parts is the original (except perhaps for the helm), then the question is whether the boat can be considered to be the same object, or whether it must be considered another entity. In the case of the human being the situation is rendered more complex by consciousness. Are we absolutely certain we are the same person we were many years ago when we went to primary school, even though strictly speaking 99% of the atoms presently constituting our body are different from those that went to primary school bearing our name? The key point is that these atoms have inherited the properties and functions of the original cells. This is what guarantees the continuity indispensable for individual existence.

The behaviourist approach, dominated by positivist realism, assumed that humans were a function of the environment. Conditioning modified humans in accordance with external factors, which were optimistically supposed to be controllable. In his novel *Walden Two* (1948), which illustrates behaviourist philosophy better than do scientific texts, Francis Skinner, the authoritative father of behaviourism, expounds a social utopia in which everyone is conditioned from birth to execute the functions they are best suited to, in the harmonious coexistence of individuals whose happiness has been hierarchically programmed.

Once the force of the extremist position of the behaviourists had abated in the 1960s, the opposite reaction, cognitive psychology, was not long in appearing. It favoured individual aspects and emphasised the subjective sphere. From being extremely rigid, education became destructured.

To escape such a rigid dichotomy, society had to wait for the appearance of the reflections of a thinker capable of grasping the complexity of the interaction of humans with their environment—Gregory Bateson, a scholar at the crossroads between anthropology, cybernetics, psychiatry and psychology. In *Steps to an ecology of mind* (1972) Bateson abandoned the principle of linear causality, and introduced the concept of complex system to refer to the set of relations existing between organism and environment. In simple systems, such as those describable in terms of the laws of classical physics, it is possible to establish cause and effect when a change is observed. For instance, if I throw one billiard ball at another, then under standard conditions I can predict what will happen with a fair degree of precision. In complex systems, however, where the number of variables interacting with each other is high, this type of prediction is impossible.

One vital issue Bateson analyses in *Mind and nature: A necessary unity* (1979) is how much a genotype influences the somatic and psychic change that takes place in the individual. Undoubtedly there is always a genetic contribution to all somatic alterations. In the case of acquiring a tan, skin colour changes depending on whether one is exposed to the sun or not, while the capacity to acquire a tan is determined by genetic factors (type of skin, quantity of melanin it contains etc.). But if we consider psychic processes, it is the capacity to change that is itself the object of learning. For example, students can not only study—they can also learn to study, striving to find a method optimising their individual capacities. It is as if humans were able to learn to control their own capacity to acquire a suntan.

We may speak in the first case of a change of the first degree, and of a change of the second degree (*metachange*) in the second case, when what is altered is not a particular trait such as skin colour, but the capacity for change itself. In this second case, the capacity to achieve metachange might be completely under the control of genetic factors; otherwise it is possible to hypothesise a change of the third degree, with a capacity to change the capacity to change. Whatever the case may be, the series must perforce begin from genes. Hence there is no point in asking ourselves if a particular characteristic of the organism is determined by the genotype or by the environment or by learning. There is no phenotypic feature that is not influenced by our genes.

The correct way of posing the question is therefore: at what level of logical type does genetics act in determining the emergence of that characteristic? The answer to the question phrased in this way is always of the following kind: at a higher level of logical type than the capacity both of changing or of learning to change observed in the organism. In the case of individual development, two points must be taken into consideration. On the one hand there is the fact that human beings are complex systems interacting with another complex system, their environment. On the other hand there is the fact that genetic factors influence not only the phenotype, which are the observable characteristics humans possess, but also their capacity to change (which corresponds to the capacity to learn), and their capacity to change their capacity to change (which corresponds to the capacity to learn how to learn).

How can the organism be helped to reach an optimal level of development? If truth be told, no certain method exists to achieve this end since we cannot determine genotypes, nor effectively control their interactions with the environment, and in particular their interactions with other living organisms. This approach drastically reduces the role of psychology both in bringing up children and in treating mental illness by means of psychotherapy.

The educator stimulates children to action, but she does not act directly. There is a sort of passivity due to the acceptance of the limitations of education—one cannot put anything into anyone's head, one can only create a situation in which change is possible. Inevitably, parents leave certain possibilities open and cut off others. But it is virtually impossible to predict what structure children will evolve by employing the means made available to them—the system creates itself. What is certain is that building oneself using top quality bricks (love, security, flexibility, knowledge) is one thing—employing reject material (estrangement, insecurity, rigidity, ignorance) is quite another.

Since the system develops over time, and since the system is aware of the passing of time, it will be useful to examine this topic in greater depth before continuing the subject of individual development. The reader wishing to investigate the problem of the education of children further may consult *Know your child* (1987) by Stella Chess and Alexander Thomas.

TIME

We may distinguish between two basic types of approach to the analysis of time: objective and subjective. Such a distinction is made for ease of argument since the two aspects of time are inextricably intertwined, as will be seen as soon as we try to separate them. But perhaps we had better start as tradition dictates, for the sake of clarity.

Objective time

The objective approach deals with physical time, as it appears to us in the world, or rather how it should be to permit us to study it with the means we have available. Other names have been applied to this type of time by famous thinkers. Sir Isaac Newton, in his *Philosophiae naturalis principia mathematica* (1687), which lays the foundations of the modern concept of time, defines it as "absolute, true and mathematical". The philosopher Henri Bergson, in his *Essai sur les données immédiates de la conscience* (1889), calls it "external", defining its features as "spatial, linear and reversible". Finally, the cybernetician Norbert Wiener rechristens it "Newtonian".

Our current model of objective time attributes it a series of qualities which are assumed to be natural to time: it is monodirectional, moving from past to future, constant, cannot be repeated, an abstract dimension within whose bounds events take place. However, these characteristics are recent theoretical constructs, dating back no more than 500 years. Nor are they accepted by all presently existing civilisations.

The model of objective time common to pre-Christian civilisations is cyclical. In this interpretation, each event repeats itself *ad infinitum*, without separating time from its content. Events do not occur *in the course of time*, but *are time* itself. Time cannot be distinguished in an abstract fashion from what happens within a certain period. Prehistoric man managed the continuous alternation of night and day in such a way as to survive by sleeping and not moving around when it was dark. The cycles of the moon were used all over the ancient world to measure time (six cycles between sowing and reaping, nine between insemination and childbirth). The constant repetition of the seasons is also a simple observation to make. Generalising with regard to the cyclic model, time is a globally repetitive phenomenon, where eras repeat themselves just as the days and the months (Fig. 10.1). Indeed, time has been depicted by many peoples as a snake eating its tail, thus forming a circle. The circle has neither beginning, nor end, nor direction. Everything will be repeated, everything will return. The cyclical model in no way hinders the precision of measurement. The Babylonians had sophisticated and extremely accurate sand glasses and sun dials, the Egyptians divided the day up into 24 hours and the year into 365 days, the Mayas had an almanac that was more accurate than the one in current use. The clock was invented in China, the cradle of the cyclic tradition, and introduced into Europe in the 13th century by Venetian merchants following in Marco Polo's footstep. Mechanical clocks, the symbol of the enormous and disproportionate importance the measurement of time has acquired in our civilisation, offer an interesting mixture of cyclic repetitiveness and directionality, most evident in traditional clocks with an analogical face. Every minute is identical to every other minute, every hour to every other hour. So true is this that after 12 hours the hands return to their original position. And yet a clock serves to indicate time which is not repetitive for us, but which proceeds inescapably from past to future.

The spread of Christianity leads to a first fundamental modification of objective time. A point now exists that is qualitatively different from all other points, because the birth of Christ marks the separation of a before from an after. In spite of this fact, up to the middle ages the social model remains that of cyclic time, virulently attacked by St. Augustine who emphasises that, for a Christian, time proceeds in a rectilinear

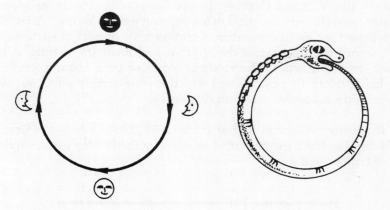

FIG. 10.1. Cyclical time.

mode, with absolutely no repetitions, and moves in the direction of the end of the world and doomsday.

It was only in the year 525, however, and thanks to a Shiite monk, Dionysius Exiguus, that the convention of numbering the years serially in one sole era was introduced. Dionysius, who had given himself the nickname Exiguus ("unimportant") to demonstrate his modesty, carried out the calculations necessary to synchronise the Roman era with the Christian period, establishing that Christ was born 754 years after the foundation of Rome.

In order for the currently held objective view of time to become dominant, we have to wait for the triumphs of modern science, of Galileo, who separates the temporal dimension from the event itself, and with the attempts to build instruments for measuring time precisely, initially for astronomic investigation, and later to satisfy the experimental needs of physics. It was Newton who was to define time as linear, continuous, and constantly uniform, that is exhibiting neither accelerations nor decelerations. However, it should be noted that Newtonian mechanics does not assume the irreversibility of time. The equations of traditional physics may be read in both directions—they know neither past nor future (Fig.10.2). Thermodynamics modifies this situation by introducing irreversibility into its equation. If I burn a sheet of paper and produce heat and light, I cannot then reconstitute that sheet starting from its products. Nevertheless, the stand science takes on time from the 16th century to the 19th is that it is linear. To give linearity direction, a further specification of a social nature is necessary, such as the one St. Augustine called for, or subjective, as we shall see shortly.

In philosophy it is not possible to draw clear dividing lines between thinkers—like Vico and Hegel—who privilege cyclic aspects, and those who champion the serial aspects of history and nature, as does Aristotle, who defines time as "the number of motion with respect to earlier and later", where the number is the rhythm and measure of time. After Newton, contributions to the study of objective time abound, pride of place being taken by physicists. I will therefore simply point out two extraordinary successes achieved this century:

(a)   the temporal logic of Arthur Prior (1957, 1967), who is the first to formalise the treatment of time, opening up the way to the current of temporalist logic;

FIG. 10.2. Linear time.

(b)   the theory of relativity developed by Albert Einstein (1917), who considers time to be the fourth dimension of the universe, the other three being the three dimensions of space, and demonstrates that it is *relative*, and not absolute as was held by classical physics, since two events occurring simultaneously in a system of coordinates may not be such in a different system of coordinates. The concept of relative time revolutionises science and common values, so that Einstein finds enthusiastic support in Bergson and, less surprisingly, in Wiener.

Subjective time

The second possible approach to the study of time relates to the way time is experienced by human beings, no matter whether they are aware or unaware of how this happens. Bergson calls this type of time "internal" and characterises it as "pertaining to experience, of length and irreversible". Wiener calls it "Bergsonian", and this is the type of time he employs in his theory on automata. We saw earlier that objective time is as much a physical construct as it is social. We shall now analyse how in its turn subjective time is inextricably linked with the time of the world we live in and in which we evolved.

Watanabe Satosi (1972) argues that in a system of growing entropy, as is that of our world, only those animals capable of proceeding mentally in the direction past to future can survive. In a world of increasing disorder, problem solvers are able to make practical predictions based on causal descriptions of the physical world. *Predictions* are not factually certain, because they unavoidably omit a number of background conditions. Rather, they are inductions, attempts at guessing what will happen, grounded on the knowledge of natural laws. While, however, predictions have some hope of being right, provided the significant variables have been identified, retrodictions, that is tracing phenomena back to their cause, are more difficult to carry out from a probabilistic viewpoint.

A mental experiment will clarify this point. Suppose there is a large vase shaped like an amphora in which a marble can move, as in Fig. 10.3. Whatever the point of departure of the marble, chosen from the infinite number of possibilities on the sides of the vase, after a sufficient length of time the marble will be found in position Z, on the bottom of the amphora. If we know the departure point (A or B, for example), we can predict what will happen (always within limits, since if the vase were to break, the marble would finish up elsewhere). But if we only know the arrival point Z, there is no way we can infer what position the marble was allowed to fall from. It could be A or B, or any other point. Retrodiction is a statistically impossible guessing game.

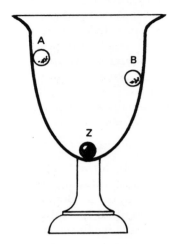

FIG. 10.3. Prediction: Where will marbles A and B end up?
           Retrodiction: Where did the marble at point Z start from?

Since we live in an entropy increasing world, the cognitive structure rewarded from an evolutionary standpoint is that permitting the system to move towards the future, inferring what will happen, whereas a cognitive structure that moves towards the past, one that infers what has happened, is penalised. The question remains one of probabilistic inferences. Nevertheless, while predictions have a high chance of success—they help the system that is structured to carry them out—retrodictions have a low chance of success as they provide less aid to the system which is designed to carry them out.

In a hypothetical entropy decreasing world, the opposite would be true. In such a universe of decreasing disorder, systems capable of making inferences with regard to the past would be in a privileged position. Beings inconceivable to us would evolve, since their subjective experience would be the opposite of ours, given that they would be able to live their world in a way that would appear to us as moving backwards. From their point of view, however, this movement would be perfectly logical. If a human being were to be catapulted into such a world, that person would have the extraordinary experience of viewing a film running backwards, from the end to the beginning. Broken vases would jump back together again, divers would emerge from water and find themselves back on the diving board, ink blots would return in orderly fashion into the fountain pen and the elderly would become younger and younger until they finally re-entered their mothers' wombs.

The conclusion that may be drawn is that subjective time is biologically determined, and depends strictly on the type of world in which humans have evolved. Linear time acquires directionality, it becomes asymmetrical, and may be schematised as an arrow pointing from past to future (Fig. 10.4).

Wiener reaches similar conclusions on the need for directionality in subjective time in complex systems, when dealing with communication in the theory of automata. Wiener points out that communicating with an intelligent being living in an entropy decreasing world, where time runs in the opposite direction to ours, would be impossible. Whatever the signal that might be sent to us, it would reach us as a logical succession of events from its point of view. To us, however, these events would appear to be in the contrary order. We would have already experienced the final result, and it would serve as the natural explanation of the signal received, without our being able even to suspect that an intelligent being sent it to us. If for example this entropy decreasing alien were to draw a square on a piece of paper, we would immediately perceive the fully drawn figure as something which already existed, formed by some curious physical process, which we would presumably manage to find some explanation for. The gradual disappearance of the square (while it is being drawn, from the alien's point of view) would appear a singular event to our eyes, but it would still find some explanation in natural laws, according to which the square would cease to exist. An analogous result would be obtained if we attempted to communicate with the alien—this entropy decreasing being would not even realise that we were trying to communicate something.

The links between entropy and information are defined by a simple law of conservation. This establishes that the sum of information and of entropy in a given system is constant. Each increase in information is compensated for by a loss in entropy and vice versa. If we reflect on the fact that in a world which tends naturally to increase the disorder existing in it, communication instead decreases entropy, diminishes disorder; in so far as it is transmission of information, then it may be concluded that we realise a communication has been made precisely because it goes against the natural flow, and for this reason stands out in the context of the world. A message of the type hypothesised by Wiener, augmenting entropy, would not even be recognised because it

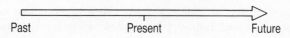

Past   Present   Future

FIG. 10.4. The time arrow.

would be indistinguishable in the midst of natural events. To communicate, the direction taken by time must be identical for all the systems involved in the communication. The automata theorised by cybernetics are complex systems, with output following input in a fixed order from past to future. Automata are capable of communicating like humans, and with humans. Time for these automata, and for the present-day robots which descend from them, must therefore be of the subjective variety.

This situation of systems which evolved in such a way as to be able to read the world in a predictive manner, apparently seems to contrast with the difficulty of predicting the individual development of a living system, such as a child growing up. To solve this paradox, it must be borne in mind that predictions concerning the environment are probabilistic, whereas in the case of the individual what matters is the deterministic forecast of his own particular development. We are now ready to tackle the problem of how humans construct, or create, their subjective time in greater depth. In general terms, time is a feature of certain functions: the past is a function of memory, the present of consciousness, the future of planning.

Present
Considering men and women as living systems, we immediately discover that the sole dimension that is real for such systems is the present. Within this present inhabited by the system, other times are constructed. All mental activity in the system is carried on in the present—from perception to reflection to problem solving. If by chance the system requires data that are not immediately available, it can call upon memory stores, from sensory buffers to long term memory. These are structured in such a way as to permit the system to utilise a piece of information even when it is not immediately at hand. But whatever mental process the system is engaged in, from the attempt to recover notions stored decades earlier to planning an action to be carried out the following day, the time in which these functions are executed is always the present.

It is the present that creates all the tenses of thought. We saw earlier that memory is properly speaking a construction of (past) memories occurring in the system's present—I reconstruct my past in my present. The same is true of planning the future—I construct now what will happen tomorrow. The future is built by projecting the present. Finally, if a system becomes conscious of its present state it can create its own present. By activating memory processes, we can gain the sensation of returning to the past. By concentrating on our current state, we become conscious of the present. By utilising projection procedures, we can travel into the future. Nevertheless, everything always comes about thanks to the present functioning, here and now, of our cognitive processes—past, present and future are all constructed in the present. This particular state of affairs is schematised in Fig. 10.5.

It is not possible to establish a general rule determining how long the sensation of the present lasts, since this depends on the people themselves and the activities they are engaged in. Time is not perceived as being shorter if a person has many things to do, nor even if the person is doing things she deems pleasant, to cite two common instances. There are people who live their moments of happiness at breakneck speed, and others that live them in slow motion. Indubitably, if someone is concentrating deeply on what she is doing, she loses all sense of time. This, however, does not correspond to a subjective shortening of time. It is more a suspension of the procedures that mark time itself. It would be more precise to say the person leaves the dimension of time.

One fact that is confirmed both by eyewitness accounts (typically of wartime events) and by psychiatric reports on survivors from disasters, is that when faced with extremely serious events threatening their lives, some people enter a sort of decelerated time in which they manage to carry out complex processes of reasoning or to execute a series of actions which appear impossible to squeeze into a few seconds. This is consistent with the discipline of *budo*. Budo is the collective name for Japanese martial arts, such as fencing, shooting arrows, karate. Budo is not simply an art for warriors, but a real way of life aiming at the search for perfection in the practice of the chosen art of battle. In the final stages of training, attention is devoted principally to developing spatial and temporal consciousness, and the effect is that of creating a perfect entry into time, enabling it to be exploited to anticipate the opponent. According to the master Kenji Tokitsu (1979) the ideal combat is *ton*, in which a combat proper does not even take place, because the warrior has anticipated the time of the other combatant so he can attack him while he is still immobile. In the last stage, the abstract stage of "nothingness", anticipation reaches the point that the master manages to avoid fighting altogether.

The time of the system	Active processes	Time constructed
Present	Memory	Past
Present	Consciousness	Present
Present	Planning	Future

FIG. 10.5. The construction of the past, present, and future, in the system's present.

The insistence on the present is a feature of Buddhism, and in particular of Zen, from which budo also takes its inspiration. The enlightenment sought by the Zen monk is realised by living each moment in full consciousness, without ever leaving the present, in total harmony with nature and the entire universe. And apart from any considerations of Zen, there remains the fact that pleasure and wisdom exist only in the present.

Past

The past is constructed employing the same rules utilised to manage knowledge—dilating, shrinking and structuring it as required. The past, like everything pertaining to memory, is reconstructed, not relived. Besides research experiments, proof of this reconstruction process comes from studies on giving testimony, an activity that is enormously influenced by one's expectations, and on autobiographical accounts. On this subject, see the analysis of memory in natural contexts carried out by Ulric Neisser and Eugene Winograd (*Remembering reconsidered*, 1988).

One fundamental consequence of this point is that to change memories in an individual we must change that person's knowledge system. Or, if one prefers to put it that way, the past is not written once and for all. Modifying the system's present state, its actual procedures, we contemporaneously modify its past. It is only when we become parents ourselves, for instance, that we manage to understand our parents. Restating the problem in more rigorous terms, we find that memory schemata are retrieved from general knowledge. Moreover, memory may retain different images of the same exemplar, located in different times—for instance, our partner when we first met her, together with the image of what our partner is like now. But images of the past are fleeting, because knowledge representation furnishing memory with the elementary data required to reconstruct one's memories, is being continually updated. Careful scrutiny also lays bare the fact that a large majority of temporalised images are extracted from concrete objects (photographs, films), which the mind refers to, rather than relying directly on memory. The only people who have exact images of a person in the past are those who for some reason have lost their sight and have maintained a frozen, unchangeable stereotype of them.

The models we build flow with time, but they may be stopped or may flow at different speeds, depending on necessity and intention. When a memory is constructed, the time factor pertaining to that memory is also constructed. This means that in building up an event from the past if I (now) believe that the episode lasted an hour, I build it in one way. If, instead, I believe it lasted three hours, I make the event unfold more slowly. In the episode as it is constructed by memory procedures, everything takes place in the present. If I remember the day I met my wife for the first time, it is as if I were watching a film about something that had happened 13 years earlier, but I am watching the film now, and the film has one time only, the present. While I am watching the scene, I am aware, however, of the markers indicating that the episodes refer to time past. Analogously, if I am imagining some hypothetical future scene, the time in which mental construction takes place remains the present. This time, however, the markers direct me toward the future.

Episodes reconstructed by memory bear no indication relating to external time. They are structured in terms of ordinal series, with uncertainties, approximations and inconsistencies. Memories bearing exact dates are a rare occurrence because the bulk of our memories are founded on a scale based on the relationship before/after. The few memories with a reference to objective, calendar time are those such as:

[1]   My daughter Helen was born on 12th June 1988.

Such instances typically refer to dates it is socially necessary to remember. If the anniversary is slightly less precise, then external timing may occur in the following fashion:

[2]   I graduated in medicine in 1973.

Normal events, however, can only contain approximate references to objective time, at times by means of a complex inferential chain. For example, if I ask myself when it was I began working with Phil, I do not have a direct answer. I may, nevertheless, build up a deductive chain of this sort:

[3]   To have gone to England for a long time, I must have already finished my military service. I obtained my PhD in psychology in 1976, while I was still doing military service. So it might have been in 1977, or 1978? More probably, 1978 because ...

Such an approximate reconstruction is nevertheless more than ample to fix events whose location with respect to external time is completely nonexistent with sufficient precision.

Future
The same rules as those governing the past apply to the subjective future—it is constructed in the present, the sole difference being that the procedures involved are those of planning and not of memory. Planning too is grounded in general knowledge, as is memory. To be able

to project oneself efficaciously into the future, one's knowledge of the world must be optimal in order for inferences, those necessarily imprecise phenomena made in the present, to have a good chance of turning out to be correct when the time comes for them to be verified.

Computational approaches to the problem of time are becoming increasingly numerous as the community gradually admits their importance. Space permits the mentioning only of the innovative work of Drew McDermott: "A temporal logic for reasoning about processes and plans" (1982). The essay by James Allen, "Towards a general theory of action and time" (1984), offers a computational model for handling time intervals. The most significant contribution of this work for our goals consists in presenting a proposal regarding the possibility of automatically managing time interval structures such as episodes, and hence memories, without having to refer to external time. Allen hypothesises 13 possible relations between time intervals within which episodes take place, as illustrated in Fig. 10.6. All relations admit their opposite, with the exception of the relationship of equality (b), and this yields a total of 13, by means of which it is possible to handle any reconstruction of episodes exhibiting the characteristics of memories or of plans we hypothesised above. The International Society for the Study of Time, founded in 1966 by Fraser, devoted itself to exploring the nature of time from every possible perspective. The eight volumes of papers selected from its conferences are an outstanding example of inter-disciplinary research (Fraser et al., 1972–1995). With regard to the field

	Relation	*Pictorial example*
a.	X before Y	XXX YYY
b.	X equal to Y	XXX YYY
c.	X meets Y	XXXYYY
d.	X overlaps with Y	XXX YYY
e.	X during Y	XXX YYYYYY
f.	X begins with Y	XXX YYYYY
g.	X ends with Y	XXX YYYYY

FIG. 10.6. The relations possible between time intervals.

of psychology, the reader may consult *Cognitive models of psychological time* (1990), edited by Richard Block.

COGNITIVE DEVELOPMENT

I shall not attempt to summarise the vast literature on developmental psychology. Rather, I shall outline the fundamental points on which there exists reasonable consensus and from which a cognitive theory on development should start. Despite the primary importance of these studies, cognitive science today does not fully realise that it is only through comprehending children that an understanding of the adult will be achieved. Research in the developmental field is not exploiting the possibilities offered it by simulation. What we said of epistemology is equally applicable to the individual—explanation is only obtained through construction. Every cognitive process observable in the adult, from perception to reasoning, from memory to creativity, is rooted in infancy. We shall be able to claim we have understood a function only when we can explain how the function arose in the individual, starting from birth and tracing it to its complete development.

The mental development of the child

The scholar who has exerted the greatest influence over current theories on the development of human beings is Piaget, and it is with him that our briefest of resumés begins. Jean Piaget (1937; 1964) distinguishes four stages in development:

The sensorimotor period. This is the period of infancy, going from birth to 2 years of age. Basic behavioural adaptations and abilities are acquired during this stage. Children become capable of coordinating perceptual information and referring it to the same object, and of integrating the motor schemas of various parts of the body, with the aim of uniting individual movements into one single objective. Children also learn to coordinate a series of actions in order to achieve one sole final goal. By the age of 2 children have begun to consider the world as constituted by entities whose existence is independent of whether they are observing them or not.

The pre-operational period. In this stage, which goes from 2 to 7 years, children develop a fragmentary and unstable cognitive representation of the outside world. This acts as a bridgehead towards the fully developed conceptual representations of the following period stage.

The period of concrete operations.   In this stage, which goes from 7 to 11 years of age, children definitively organise schemata into interconnected systems. This stage produces a new equilibrium, as did the previous phase. Children can carry out simple logical tasks, but not those requiring fully abstract thought.

The period of formal operations.   This stage goes from 11 years of age to pre-adolescence, and is the phase in which complete cognitive maturity is achieved. Children are capable of comprehending abstract logic, causal relations, and of grasping the nature of the scientific method of exploring the world—varying one variable at a time, maintaining the others unaltered.

Piaget adopts a constructivist approach, in as much as he considers children not as passive receptacles into which information is poured, but as active constructors of the world surrounding them. Within subject/environment interaction, individual mental structures are also organised gradually and incrementally. Piaget's approach was to influence all developmental psychology, vaccinating it against behaviourist simplifications which tended to see children as direct and passive products of the world they grow up in, even though the influence of the environment over children's development is clearly important and should not be undervalued.

In Piagetian theory the culminating point of cognitive development is formal thought. In interpreting development as a continuous equilibration process, formal thought represents the optimal level of equilibrium, the point of maximum operational efficiency, the most general form of intelligence. It therefore constitutes the form of thought exhibiting the widest possible range, for it exclusively manipulates relations between entities, while the specificity of objects is ignored since they are deemed to be equivalents of each other. The essential tension to achieve the highest possible level of internal organisation of the system's structures is the driving force behind cognitive development. At the most advanced level of development of intelligence the specific objects of knowledge—content and context—are of secondary importance when a cognitive act is to be executed. The reasoning process is identical both when its object is, for instance, a concrete entity, and when it is a proposition. Language is thus a subordinate and secondary cognitive activity with respect to thought.

Piaget's basic philosophy is that the true nature of thought—the ultimate key to its mechanisms—corresponds to the more general and global laws of the cognitive system. But as soon as it is assumed that what matters are the laws of structures, since they exert a logico-causal action on specific thought processes and together determine the operational possibilities of the mind, then the distinctive and specific features of the individual cognitive processes are lost in part. They are flattened on a highly general and a-specific form, whose importance is emphasised.

Our observations so far have concerned above all the organisation of the mind (the synchronic aspect). A further criticism may be levelled at the hypothesis on its development (the diachronic aspect). In the Piagetian view development consists of four separate stages, one following on from the other. Accordingly, restructuring in the transition from one stage to another takes place in one compact block. Indeed, the reference to the presence of a global structure implies that modifications in the various cognitive capacities come about at the same time and with a common configuration. The operational possibilities that are opened up in one cognitive domain are also automatically applicable in the remaining domains, since it is the basic structure that has changed. The leap from one developmental stage to another is viewed as a series of microcatastrophes radically altering the operational capacities of the subject. Each stage has its own distinct identity—although stage two is built from stage one, yet the former is completely different from the latter.

Piaget does not deal with the way information is represented. He focuses principally on the mode of operation. Under the influence of cybernetics (see Chapter 2), he considers the quantity of information reaching the system, but the quality is something that escapes him. Indeed, to be able to handle quality, it is necessary to have a firm hold on the problem of how one piece of information is represented, compared to all the other information the system possesses. However, since bringing about a change in the mode of operating on data concerning reality requires a corresponding change in the mode of representing this reality, changing the set of operations the system can carry out—that is the structure—coincides with changing its representational capacities. Hence, in the development of the system, continuity is guaranteed by the mechanism enabling cognitive processes to function (assimilation + accommodation = equilibrium), while radical change appears to characterise its representational mechanism.

The development of mind proceeds through ever-increasing generalisation, continually augmenting its capacity to make abstractions from reality. This reason, together with the reduction of the different abilities of thought to features and laws equally valid for all sectors in which thought takes place, allows the Piagetian approach to human cognitive activity to be defined as global and domain-independent.

The strongest methodological opposition to Piaget's constructivism is represented by nativism. According to the nativist stance, children possess modules dedicated to their various cognitive capacities. Each capacity can be shown to be active very early on in life, and much before the Piagetians deem it possible. In fact, experimental results in different domains have demonstrated a high level of cognitive abilities is already operative in early infancy. Here I shall furnish only a few examples: for a deeper treatment of the subject, the reader may refer to the excellent work, *Beyond modularity: A developmental perspective on cognitive science* (1992), by Karmiloff-Smith.

The point to understand is that possessing innate capacities of some complexity strongly facilitates learning. In fact, the system does not have to waste time and resources to construct the basic elements necessary for its further development. Thus, postulating some degree of nativism may explain: first, the precocious onset of the perceptual, motor, and cognitive abilities; second, the astonishing learning power of a human infant, in every mental domain.

In the field of perception, Elisabeth Spelke (1990) has proposed four principles that reflect basic constraints on the motions of physical bodies. These principles are central both to human perception of objects and to human reasoning about object motion. They are: *cohesion, boundedness, rigidity* and *no action at distance* (an object may act upon another only when the two come into contact).

Infants as young as 4 months perceive objects in motion in accordance with the principles; children of 7 months acquire the capacity to apply these principles even to stationary objects. Thus, in opposition to Piaget's stance, Spelke claims that some sort of object persistence may be found practically at birth.

The research carried out by Renée Baillargeon (1987) to explore the young infant's understanding of causal events is in a similar vein. In a prototypical experiment, 3.5-month-old infants were habituated to seeing a screen rotate; then a solid object (e.g. a box) was placed behind the screen and the infants were subjected to testing by showing them a possible and an impossible event. In the possible event, the screen stopped rotating when it came into contact with the hidden box; in the impossible event, the screen rotated through a full 180° arc, as though the box were no longer behind it. The results indicate that 3.5-month-old infants looked reliably longer at the impossible than at the possible event. Baillargeon's conclusions are that infants (a) believed that the box continued to exist as an independent entity, after it was occluded by the screen; (b) realised that the screen could not rotate through the space occupied by the box; (c) expected the screen to stop and were surprised at the impossible event that it did not.

Another series of ingenious experiments by Baillargeon demonstrated the startling sophistication of even very young infants' physical reasoning. At 3.5 months of age they realise that objects continue to exist when occluded, that objects cannot move through the space occupied by other objects, that objects cannot appear at two separate points in space without having travelled the distance that lies between them, and so forth. These qualitative reasoning processes are enriched by quantitative strategies, which develop by 6.5 months of age. For instance, 6.5-month-old infants predict *at what point* a rotating screen should stop when an obstacle of a given height is placed in a given position behind the screen (Baillargeon, 1991).

In the domain of mathematics, Rachel Gelman (1990) has shown that infants process number-relevant data far earlier than Piagetians believe. She claims that infants have an innate predisposition for number-like representations. In particular, it is possible to define five principles that infants appear to follow from the outset, and that constrain the development of mathematical capacity. These principles are:

One-to-one correspondence. This is active already in neonates, who exhibit the ability to discriminate between arrays with different numerosities, such as:

  ● ●
  ● ●
  ● ●

Stable ordering. Any counting list will do, as long as each tag is unique and the ordinal sequence is the same for any counted set. For instance, a child might substitute the standard sequence "one, two three" with a functionally equivalent "one, two, ten".

Item indifference. Any item is likely to be counted, independently of its kind.

Order indifference. There is no fixed order for counting; the same cardinal value is obtained whether a set of items is counted from left to right or from right to left.

Cardinality. Only the last counting term of a set represents the cardinal value of that set.

Although too complex to be discussed here, a list of innovative developmental research inquiries would be incomplete were I not to mention the area devoted to understanding how children develop a so-called *theory of mind*. This corresponds to the ability to reason upon another person's supposed mental states: i.e. to simulate him. The skill of simulating another individual's mental states is peculiar to our species. It requires a specialised, modular machinery that Alan Leslie (1994) named Theory of Mind Mechanism. The way children acquire comprehension about representational and mental processes, which is crucial for the understanding of themselves and of others, is brilliantly described in Joseph Perner's *Understanding the representational mind* (1991), a book to which readers interested in this topic may usefully turn.

A mediation between nativism and constructivism has recently been proposed by Annette Karmiloff-Smith (1992). For nativists, the prespecified structures in the brain represent the explanation of cognitive development: the role of the environment is reduced to that of a mere trigger. Karmiloff-Smith recognises both the inherent plasticity of the brain—a factor underestimated by nativists—and the existence of some innately specified, domain-specific predispositions, that guide development. Her model is based on a reiterative process called *representational redescription* (RR model). Representational redescription is a process by which implicit information *in* the mind subsequently becomes explicit knowledge *to* the mind, first within one domain and then sometimes across domains. The focus of the RR model is on the multiple levels at which the same knowledge is represented: the key word in development is redundancy, not economy. The actual process of representational description is domain-general, but it operates within each specific domain at different moments. Karmiloff-Smith offers a neat discussion of the possibility of a computer implementation of developmental models. With regard to the RR model, she favours a connectionist approach as the one offering reresentational features closest to her theoretical choices (non-nativism, domain generality, implicit knowledge at the base level).

Language acquisition
Language acquisition is one of the most important and hotly debated chapters in infantile development. The reader interested in this theme may consult the wide-ranging review of the topic in Wanner and Gleitman's book, *Language acquisition* (1982). Research in this area has confirmed the necessary balance between the nativist and the constructivist approach. In the field of language this has principally meant demonstrating that children cannot learn to speak by passive means, by directly internalising a grammar explicitly presented to them by the environment they live in. Chomsky, the main exponent of this position, has claimed that language develops through an innate, species

specific process, mediated by LAD (*language acquisition device*). LAD reflects infants' innate sensitiveness to the phonological and syntactic structure of any human language.

In a series of innovative research experiments, Jacques Mehler and his collaborators have demonstrated that a few hours after birth, infants differentiate between linguistically structured input and non-linguistic acoustic input. Moreover, infants always prefer their mother's voice to anyone else's. Infants as young as four days can discriminate between their native tongue and other languages. The relevant point is that such a differentiation is made at birth, after the infant has been immersed in a particular language only for a period of a few hours (Mehler and Bertoncini, 1988). Evidence suggests that infants have an innate predisposition to select the clausal structure of any human language, and to tune their internal processes into specific prosodic and syntactic features, typical of the language they are exposed to.

In his book *Learnability and cognition* (1989), Steven Pinker has tried, on the basis of the linguistic data available, to furnish a formal definition of the features a system has to possess in order to be able to construct a complete grammar, and how these features grow parallel to the growth of children's linguistic abilities.

Jerome Bruner (1983b) points out that individual development progresses from pragmatics to semantics to syntax, somehow inverting the traditional order of these three dimensions of language—mother and child communicate effectively much earlier than language is mastered. Right from the first few weeks of life mothers recognise children's different modes of crying (hunger, pain, capriciousness, calling). Subsequently, mothers interpret the different meanings children attribute to the same monosyllabic word in different contexts. For instance, at one year of age my daughter Helen would use the word "ba" to convey the following meanings, depending on the context: father, to call me if I was present or to mention me in my absence; her older sister, again to call her or mention her; a visible object she desired, except water to drink ("uaua"), the sole case, with the exception of her mother ("mama", employed very occasionally), for which she employed a special name. Only her mother was able to interpret the special meaning of "ba" up to the child's 14th month, when, despite the fact that her total lexicon remained unchanged, she restricted the use of the word to her father or, if I was absent, to her sister, thereby enabling Helen to improve her interaction with the rest of the family. At that point, objects were asked for ostensibly, by indicating them.

In mother–child interaction, both participants already comprehend their respective communicative intentions, which represent the essence of pragmatics, in the first few months of the child's life. To obtain the first syntactically well-formed utterances, one must wait until the child is 2. The obvious deduction is that the ability to communicate is one of the essential features permitting real language to develop. It is language that grafts itself onto communication, and not vice versa.

A lively confrontation between the views held by Piaget and Chomsky on language and learning is offered in the stimulating book edited by Massimo Piattelli-Palmarini, *Language and learning* (1979).

Mental models in development

The limit identifiable in the theories existing on human cognitive activity is that of presenting a series of fragmentary research works on the development of a single, specific capacity, without managing to provide a unitary overview of the mind uniting the two basic aspects of cognition, the synchronic dimension and the diachronic dimension. The *synchronic* aspect concerns the organisation of mental processes at the stage of fullest development, while the *diachronic* aspect concerns the changes that take place during the course of development.

Each relationship between humans and the world surrounding them is not enacted directly, but through the way humans make up their world, with the mediation of mental structures. The structures enabling a relationship to exist between mind and world are presumed to be mental models. Hence they are essentially mental representations, not operations in the Piagetian sense. Nevertheless, because of the specific features distinguishing models from the other types of knowledge representation, employing models also means having to define the operations that may be executed upon them. Every cognitive act, every thought process, can be described as a more or less elaborate and complex set of computational operations carried out on models. The essential constituents of representation do not consist exclusively of the elements of which it is composed, but also of the manipulation procedures it may be subjected to.

Although various types of mental model may be distinguished, all have in common a set of procedures devoted to the construction and the manipulation of models. These common procedures may be called the basic abilities. They are the ability to *build, modify, integrate* and *falsify* models. When a model begins to be built from the initial data, the specific modification procedures applicable come with it. Together the constructions and operations executable on the model constitute the activity known as thought. Since, therefore, each specific cognitive domain has its own models and procedures, we may speak of a domain-dependent form of cognitive activity, whereas the Piagetian stance posits a domain-independent applicational activity (Bara, Bucciarelli, & Johnson-Laird, 1995).

Viewed from this standpoint, the role and value to be attributed to logical capacity also changes, as do the abilities that have to be postulated as essential to be able to carry out this particular type of activity. No longer is formal thought the highest thought capacity possible. It becomes simply one of the modalities of reasoning a cognitive system may employ—a highly specific procedure for the integration of a particular class of models (monadic models), in which the sole relation to be detected and considered is that of identity, without it being necessary to evaluate the general meaning of the elements given. Formal thought is therefore viewed as a form of reasoning that is not qualitatively dissimilar from the other forms, but only more difficult in as much as it implies the utilisation of a greater number of constraints. It is, in other words, an acquired capacity on a par with the others during individual development. Thus judging this form of reasoning to be the highest point in the development of thought because it is capable of abstraction from contents no longer makes sense.

Developmental changes in thought processes, that is the mode of representing the world and of interacting with it, can be interpreted in two ways—as the capacity to construct different and increasingly richer models, or as the capacity to operate on them with increasingly complex and differentiated procedures. The paradigm of mental models does not require reference to be made to a general, global modification in mental structures responsible for identical and contemporaneous modifications in the different applicational domains. Since the type of representation does not alter, except perhaps in complexity, since the manipulation procedures built later are not qualitatively different from the initial ones (they are in fact obtained by recursion), and since models and procedures are the basic form of representation employed by thought, then speaking of radical discontinuity between the developmental phases of the individual is meaningless.

Whereas Piaget sees continuity in development as pertaining exclusively to the functioning of the system, in the case of models, continuity also exists in the modes of representing reality, in the basic mental structures. Rather than referring to the global aspect of the system, discontinuities are to be sought in acquiring increased ability to construct models and employ procedures within a single specific cognitive domain. In this sense we have to speak of an approach to and an interpretation of human cognitive activity that are local as opposed to global. In this respect, mental models and Karmiloff-Smith's RR model concur perfectly.

In Piaget's view, the process of abstracting elements from reality is an inevitable step forward in the development of thought, a natural and spontaneous phenomenon, just as the growth of an organism is natural and spontaneous. According to the theory of mental models, however, this is only one of the many abilities to be mastered. Even Piagetian constructivism pays a debt to realism since a child's constructions are always representations of a real, external world to which the child has to "accomodate". In the equally constructivist view of mental models, no objective reality exists. There is instead an infinite possibility of constructing meanings, of creating worlds, none of which is intrinsically truer or less true than the others, or a more or less accurate description of "things as they really are". Indeed all meanings may be subjectively perceived as true. The world in which children learn to live, interacting with others, is made up of the meeting of all the different individual mental models—every single one of us remains unique, the specific, singular result of his or her own development.

In the mental model approach, a computer model of the development of reasoning ability has been implemented by Bara, Bucciarelli, Johnson-Laird, and Lombardo (1994). In this case, a serial architecture has been favoured, using the classic programming language, LISP.

CONSCIOUSNESS

The experience of consciousness comes about thanks to a dual process— it is constructed in the system's objective present and it is perceived by the system itself as being in the present. The structures necessary to experience consciousness are short-term memory and attention. Consequently consciousness fluctuates irregularly, has a limited capacity and disappears as soon as it is no longer required. No operational definition of consciousness establishing the necessary and sufficient conditions required to generate it is possible. We shall try instead to arrive at a definition gradually, through the accumulation of experimental and subjective evidence. Clinical evidence on the subject will be furnished in Chapter 12.

In making a distinction between conscious and unconscious experience, we shall take as our departure point the work of Anthony Marcel (1983a). He has carried out numerous complex experiments, so I shall limit this outline to providing a simplified version of some of them, emphasising those aspects important for the construction of a theory of consciousness. First we shall consider his experiments on *visual masking*. Under special experimental conditions where a person was unaware of the existence of a stimulus and could not consciously discriminate between whether it was present or absent, Marcel demonstrated that the stimulus had nevertheless influenced the processing of stimuli which succeeded it. Thus, a representation at a structural and referential level had been achieved independently of the subject's consciousness.

In a second series of experiments, conducted with a similar procedure, a polysemous word, that is a word having more than one meaning (for example: PALM), was presented for a time insufficient for it to be consciously recognised. In this case the polysemous word facilitated the recognition of subsequent words connected to both its meanings (HAND/ TREE), compared to other words. If instead the item was presented for a sufficiently long time for it to be recognised consciously, then only those words tied to one of the two meanings benefited from the facilitating effect (TREE). This proves that while at a conscious level only one meaning is represented, at an unconscious level both are represented.

Results such as these led to a re-examination of certain cases of *blindsight*, in which patients suffer from neurological damage to one of the occipital lobes, as a result of which they present a loss of visual capacity over a certain region of the visual field. It has been shown that such patients are able to carry out successfully movements of the arm, wrist and fingers to grasp objects situated in the blind visual field, invisible to them on the conscious plane. In other words, while conscious perception cannot be performed, unconscious perception can, and this is what guides motor acts. Analogous qualitative differences between conscious and unconscious processing in the area of perception are highlighted by the fact that while we consciously perceive a world composed of rigid objects conforming to Euclidean geometry, unconsciously the very same world appears in the form of a non-metrical projective geometry. Finally, in auditory perception the unconscious description of a stimulus word is acoustic (meaningless elementary components), while the corresponding conscious description is phonemic (the components form a recognised word).

The experimental results described have been criticised by some scholars (such as Holender, 1986), who deny the validity of the experimental procedure adopted, claiming that the data thus obtained are the result of ingenious artefact. This is another example of the division between those who credit this kind of research and those who do not. A similar situation had come about with regard to an analogous phenomenon, perceptual defence. In *Preconscious processing* (1981), Dixon defines *perceptual defence* as the phenomenon in which a taboo word is more difficult to recognise than an equivalent non-taboo word which is equally familiar. The explanation advanced at that time, and which is consistent with the current approach to consciousness, postulated that the preconscious recognition of a word preceded its conscious recognition, inhibiting the latter in cases of socially embarrassing or disapproved terms.

Neuropsychology (see Chapter 11) has however furnished further evidence supporting the existence of the phenomena observed by Marcel. Tranel and Damasio (1985) report cases of prosopagnosic patients—incapable of discriminating faces—who exhibit different autonomic responses depending on the degree of familiarity they have with the people involved, despite being incapable of recognising those people from their faces. Patients were shown photographs of persons they had differing degrees of familiarity with (relatives, hospital staff, famous people and strangers). When shown the photographs a second time, patients were asked to put them in order of familiarity on a scale of 1 (familiar) to 5 (unfamiliar). Even when patients exhibited no conscious recognition of a face, their cutaneous galvanic response (an indicator of emotional activity) varied with the degree of familiarity they actually had with the people shown them in the photographs. These experiments have furnished independent clinical data in support of the thesis of the unconscious processing of data, thereby guaranteeing it is not founded on artful methods of investigation.

The thesis advanced by Marcel (1983b) on the basis of the experiments described is that consciousness is a functionally different mechanism from unconscious processing procedures. Conscious processing is based on a series of presuppositions grounded on the subject's knowledge of the world, and it operates serially, analysing one item of content per time. Unconscious processing, on the other hand, is constrained neither by knowledge nor by serial limits. Taking up the subject of perception again, unconscious processes utilise the sensory data of every type of representation available to the organism, generating perceptual hypotheses (which are unconscious too), and activating knowledge and motor structures correlated to the hypothesis, but without ever interpreting the data.

Conscious processes require an act of construction involving the prior adjustment of perceptual hypotheses to sensory data before comparing them to the expectations of the system. Once a perceptual hypothesis has been selected, other hypotheses are inhibited—the subject is only conscious of the chosen hypothesis, the others remaining unconscious. The function of conscious perception is to structure and summarise part of the results of unconscious processing at the most functionally useful level, so as to make these results operationally available to the system in its entirety. Seen in this light, consciousness plays a fundamental role in the total functioning of the organism, and it carries out its task by reinterpreting the results of unconscious processing in accordance with criteria consistent with general knowledge, expectations, and the focus of attention. Unlike unconscious processing, consciousness possesses intentionality, a property capable of modulating the criteria listed above.

Finally, consciousness actively modifies the product of unconscious processing. First, what has not been given precedence is inhibited and rapidly eliminated from sensory buffers, and is thus totally lost to the system. Furthermore the chosen perceptual hypothesis is not neutral with respect to the original sensory data: these are forced to cohere to the hypothesis. Indeed data are retrieved from different representations with the aim of confirming the hypothesis that has now become conscious.

Behaviourist and cognitive psychology have always been embarrassed in the presence of consciousness, and have thus always preferred to place greater weight on unconscious processes than on conscious processes. The normal version of the relationship between unconscious and conscious is in fact that all processing takes place in the unconscious, and the results are then passed on to the conscious. It is like switching a light on in a dark room. Nothing is altered, each object remains exactly as it was, the only difference being that the observer can now see what was previously invisible. A more precise metaphor would be that of a sculpture progressively taking the shape the artist desires, though the characteristics of the material employed remain unchanged—colour, consistency, taste. Only when the hypothesis becomes definitive does the subject experience conscious recognition of a perceived object. We thus experience the passage from a vague sensation of familiarity to final conscious certainty, for instance when recognising a person. Figures 10.7 and 10.8 illustrate the relationship between unconscious and conscious in a dynamic fashion.

In the first stage of perception (Fig. 10.7) a set of environmental stimuli are subjected to unconscious processing, first generating possible, data-driven interpretations, then their corresponding identifications, and finally arriving at conscious recognition. Each process contributes to motor control. Both identification and recognition interact with the system's general knowledge.

In the second stage of perception (Fig. 10.8), the conscious recognition process activates a selective inhibition process in order to eliminate possible alternatives that have not been given preferential treatment. In this way only that information referring to the selected interpretation, and the corresponding identification, is held. Furthermore, from being exclusively automatic as it was in the first stage, motor control also becomes intentional.

Much of what is commonly called unconscious is a way of producing consciousness, but it cannot be assigned the status of unconscious processing as defined above. What is involved here is a set of neurophysiological mechanisms able to generate mental states, not a set of unconscious mental states. Otherwise we would be obliged to claim that

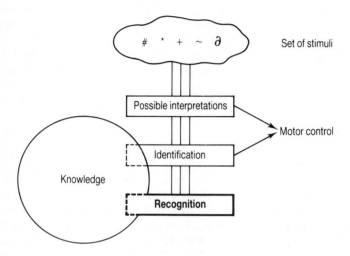

FIG. 10.7. The relationship between unconscious and conscious (in bold face): stage 1, unconscious.

all mental states that can be potentially generated by such neurophysiological mechanisms are unconscious, while these potential mental states do not exist until they have been constructed, exhibiting immediately the features of consciousness. For example, claiming that during the act of reading a reader unconsciously believes the page she is perusing cannot be eaten is senseless, even though if asked a precise question the person would answer that paper is inedible. The concept of the non-edibility of paper has been generated on request. It was not an unconscious concept that was brought to light. It exists exclusively as a conscious concept.

The key concept with regard to intentionality is that of the self. This concept mediates the majority of interactions between the system and the world. It may be stated that for the property of consciousness to emerge in a system, a self must have been developed. For a discussion of the relationship between consciousness and the self, compare the provocative works by Daniel Dennett (*Consciousness explained*, 1991), Gerald Edelman (*Bright air, brilliant fire: On the matter of mind*, 1992), and John Searle (*The rediscovery of the mind*, 1992). The mechanism through which consciousness is activated always means two acts of parallel processing have taken place. That is to say consciousness implies one procedure analysing another procedure. It is therefore not the product of a re-elaboration of results already produced. This explains

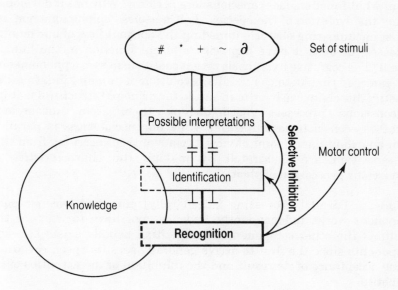

FIG. 10.8. The relationship between unconscious and conscious (in bold face): stage 2, conscious.

why it is always naturally easier to become conscious in the here and now, and when dealing with a fact relating to the present.

It is theoretically also possible to become conscious of something past or future. For instance, statements of the following type may be expressed:

[4]   This is what really happened that time.

The system becomes aware that something has happened within itself. It is therefore restructuring a memory, not something that is occurring now. Analogously, cases of the following type may be quoted:

[5]   This is what I really thought would happen.

We are in the future compared to the point recalled, but for the system it is again a question of restructuring a memory. Both cases involve the reconstruction of a process that has already taken place, and they therefore fall back on a less vivid, second-hand awareness.

But what function does consciousness perform? Why has it developed during the evolution of the species? Is it a useless epiphenomenon, an almost encumbering obstacle impeding the optimal flow of our mental processes, or does it have a precise internal function in the human system? The cognitive hypothesis is that consciousness is a phenomenon that emerged thanks to the peculiar evolution of human kind. In order to justify this claim, we have to return to the principal structural feature of conscious processes distinguishing them from unconscious processes—consciousness is serial while unconsciousness is parallel. From an evolutionary point of view, the advantages accruing from this differentiation are illustrated by making the characteristics of unconscious processing explicit:

Speed.   Parallel processing is generally quite fast because each component works independently, and does not have to wait for the results of the other components. Competition usually exists between components since the first to arrive generally sees its speed rewarded by the acceptance of its result and the inhibition of the activities of its competitors.

Solidity.   A component may die or be damaged without the system suffering as a consequence, because others take its place. Similarly, if a component were to fail to execute the task assigned to it, other components can intervene and take the place of the component that has failed.

A multiplicity of relationships with the world.   The system can afford the luxury of exploring all stimulating or promising avenues before choosing the one that will become conscious.

Globality.   Parallel processes do not suffer from processing bottle-necks. Hence the system can receive unconscious information from a great many components of the most varied of natures, activating other perceptual, motor and cognitive components without the system having to intervene on each individual component.

It is not so much consciousness which is precious, or unconsciousness which is effective. It is the interaction between these two subsystems with such different characteristics that make mental processes considered in their entirety so powerful in humans. Any opposition between conscious and unconscious, which many dichotomies, such as rational/irrational, thought/emotion, adult/child, civilised/natural, good/bad or vice versa, supposedly correspond to lose sight of the fundamental point, which is that the two subsystems reciprocally strengthen one another's power.

Undoubtedly conscious processing developed later, both from an evolutionary standpoint and from that of individual development. A decrease in consciousness leads to a dramatic reduction in the quality of life, one in which the essential dimension of the human being appears to have been lost. And yet it must be remembered that this extraordinary capacity of the human brain is based on the fact that both conscious and unconscious capacities must be present contemporaneously, so that unconscious capacities can correct the rigidity and laboriousness of consciousness. It is probable that in their turn, parallel processes have received further impetus to development from the first appearance of consciousness, which freed them from ultimate responsibility for the system's interpretation of the world. Just as a bow is built from wood and string, and it is the synergy between wood and string that launches the arrow, so it is the synergy between conscious and unconscious that allows us to be what we are.

The peculiarities of consciousness—serial processing and intentionality—make it vulnerable to error. Seriality is important to conceptualise the system–world relation, mediating it above all through the concept of the self. But its limited processing powers may cause it to ignore those very elements that are important and that the unconscious had not failed to make available; it may also irremediably lose that information which would have been most useful to the system in attempting to comprehend the situation. Intentionality, which obliges the percept to accomodate itself within the bounds of conceptual expectations and the system's self-image, may also lead the system up a blind alley if the individual does not have an effective model of the world and of oneself.

CONSTRUCTION OF THE SELF

One event that is unique and very special in human cognitive development is the construction of the self. Around 18 months children show the first signs of their sense of self, by beginning to recognise themselves in the mirror and in photographs. René Zazzo (1993) has introduced a brilliant experimental method to clarify the stages in the recognition of the self. It consists in drawing a blot on a child's forehead without the child's noticing, and then getting the child to look in the mirror. If the child touches the spot on the figure in the mirror, it is clear the child does not recognise herself in the reflected figure, but simply notes the colored spot. If the child takes her hand to her forehead, however, then it may be stated the child has realised that the figure reflected corresponds to herself. This test is passed around 2 years of age, definitively separating the cognitive capacities of the child from those exhibited by other animals.

Animals do not possess a sense of self. The crucial test is still that of recognising oneself in a mirror, a feat no animal is capable of achieving, with the remarkable exception of the chimpanzee. A clear demonstration of the fact has been provided by Gallup (1970), again using the same test. Initially chimpanzees were left in front of the mirror to allow them to get used to it. After a couple of days they would stop acting as if another chimpanzee were present, abandoning a social attitude for one of self-response. Later, while the chimpanzee was sleeping, a red spot would be painted on its forehead. All those chimpanzees who had been given the opportunity to familiarise themselves with the mirror raised their hands to their brows and explored the area painted red, employing their reflected image as a reference point, and thereby showing they recognised themselves as the chimpanzee reflected in the mirror. These results have been replicated with orang-utangs, but not with other primates, as is reported by Marco Poli in a critical review of the problem of self-recognition in animals (1988). It is precisely the fact that gorillas do not recognise themselves, even after prolonged exposure to a mirror, that leaves scholars perplexed as to how these experiments are to be interpreted. Gorillas, orang-utangs, chimpanzees and humans are in fact closely related biologically.

During infancy the image of the self possessed by the human system is furnished and defined by the family. Later, during pre-adolescence, the peer group and the school become the determining factors in bringing about modifications. Adolescence marks the detachment from the self as induced by others and the first core of an independent, adult self. In this phase the transition from dependence to autonomy is made, first and foremost due to the confrontation and clash between children and parents. Through continually opposing parental authority on principle, the adolescent builds up the barriers between the internal and external world that will permit further structuring of the self. Overcoming the crisis of adolescence means the adult has managed to construct a recognisable core that he may stably identify as his own self. This core is composed of a structure that is at the same time cognitive, emotive and physical. For a detailed analysis of this topic, the reader may consult *Child development and personality* (1984) by Paul Mussen, John Conger, Jerome Kagan, and Athena Huston.

By reflecting on their own way of proceeding, that is to say by recursively applying the procedures we have seen allow interaction between tacit and explicit knowledge to take place, people construct their own self-image. However, given the complexity of the human system, the task is an arduous one, and the method is by no means error proof. Thus the model of the self possessed by the system is not necessarily accurate, since it is after all a theory on how the system works. Naturally, the more accurate it is, the better one lives—the system predicts its behaviour, it knows what it likes and what its private scale of values is. Moreover, each normal person has a plurality of models of himself or herself, from which the appropriate one will be selected in a particular situation.

Since consciousness employs the self as its stable reference point, errors are again possible. Despite the phenomenological evidence through which conscious data present themselves, leading us to believe that making mistakes is impossible, this is what actually happens. If the system has an effective contact with the world, it can realise it has made a mistake, namely, it has misinterpreted itself or the situation it is in. Recognising a possible inconsistency, suspecting the hypothesis taken for granted is not as certain as it might appear, is essential if the mistake is to be put right. But if the contact between the system and the environment is not sufficiently valid, the error will not be detected; even less will it be corrected. An occasional mistake is rarely cause for alarm, unless occurring in very special circumstances. If, for instance, in the heart of the jungle, that faint, unidentified noise behind my back I would normally ignore were to prove to be the grumble of a tiger, I would have little chance of correcting my mistaken interpretation.

Systematic error, on the other hand, is always serious. It is frequently the source of mental pathologies—from depression to paranoia—in which individuals constantly employ knowledge of themselves and of the world that is not attuned to the environment, and that continually places them in painful or socially unbearable situations.

What we have been saying about the importance of the self must not lead to the conclusion that the concept of consciousness overlaps with that of self-consciousness. The distinction between the two notions is the same that obtains between doing something and observing oneself when doing that thing. For example, a person may ask a speaker at a conference a question fully conscious of what he is saying, and of the environmental and social situation around him. In exactly the same situation, that very same person may suddenly feel self-conscious, realising that others are watching him and feeling the acute unease that normally accompanies this type of situation.

An extensive review of the present theories about consciousness may be found in *Consciousness in contemporary science* (1988), edited by Marcel and Bisiach.

CHAPTER ELEVEN

Neuroscience

The application of neuroscience extends over a wide area. This area includes all those disciplines dealing with the brain, the physical substratum of the mind. Anatomy, physiology, biochemistry, neurology, endocrinology and neuropsychology are all involved in the study of the central and peripheral nervous system. A review of all the methodologies employed and of the current state of knowledge in this field strays well beyond the objectives set at the beginning of this book. I shall therefore not even make such an attempt but shall limit myself to a skeletal outline.

One of the disciplines mentioned, however, is establishing such close ties with cognitive science as to warrant a more detailed examination. It is not a matter of this subject being more important than the others, but of it constituting the discipline within the field of neurosciences that has set itself the task of trying to correlate the workings of the mind with its physical structure. Neuropsychology has entered into a permanent relationship with the disciplines of the cognitive hexagon: psychology, linguistics, philosophy and artificial intelligence. The interests of neuropsychology cover all the mental functions, thus making a concise survey of all the knowledge acquired an impossible task. Furthermore, as we shall see, the results achieved are far from being stable. This renders the state of the art more like a winter sea than a solid rock. We will therefore concentrate our attention on its methods, which no scholar of the mind can ignore, and limit our investigation to the paradigmatic results, these being useful to comprehend the nature of the discipline, and to those results that most affect our subject as a whole.

The reader requiring a more analytical guide to the individual issues and the main theories dominating the current scene, may turn to *Principles of neural science* (1991), edited by Eric Kandel and James Schwartz.

STRUCTURE AND FUNCTIONS OF THE NERVOUS SYSTEM

The anatomy and physiology of the nervous system are described in several excellent works, ranging from an elementary level to the super-specialised level, which it would be ridiculous to attempt to compete with; hence, only the basic notions will be outlined here.

Neuroanatomy consists of the study of the structure of the nervous system, both at a microscopic and a macroscopic level. At the macroscopic level, the first important point is the distinction between the central and the peripheral nervous systems. The central nervous system is composed of the brain and the spinal chord. The peripheral nervous system is composed of nerve fibres leaving the central nervous system and connecting with muscles, internal organs, sensory and epidermic organs. The nerves control the activity of these organs thanks to efferent fibres. Afferent fibres carry the stimulations coming from these organs to the central nervous system. The connections with the receptors in the eye, the ear, the skin, the muscles, the tendons, and in the internal organs, have an especially important part to play since they allow the brain to receive information from the external world and from inside the system itself.

At the microscopic level we have the cells of the nervous system: the neurons, estimated to be 10^{12}. The neuron is an extremely complex structure, capable of establishing various kinds of connections with tens of other neurons, thus modulating hundreds of different signals. This means neurons may potentially be interrelated with other neurons in a multitude of ways, far in excess of the number of relationships that can be established in the course of a person's life. Neurons are not all the same. Their anatomy depends on where they are located in the nervous system. The basic structure is composed of a nucleus (the soma), from which leave various terminations (the axon and the dendrites). Neurons do not constitute a homogeneous substance, there being interruptions between them called synapses. Excitation comes about through synapses where various neurotransmitters are able to bridge the gap between cells. Neurotransmitters, which have been intensely researched by biochemists and pharmacologists, are influenced by numerous modulators, including endorphins and neurohormones.

Neurophysiology studies the functioning of the various parts of the nervous system. Figure 11.1 shows a scheme of the principal functions of the cerebral cortex.

The split brain

The brain is anatomically separable into two parts, the right hemisphere and the left hemisphere, connected by a tract of nerve fibres, the *corpus callosum*. The first neurologist to tackle the problem of the functional differences between the two hemispheres systematically, was Roger Sperry, who had started out by studying the effects of the excision of connections between hemispheres in animals. Sperry reports his observations on cats and monkeys, which have a nervous system organised in a similar fashion to that of humans, in his article "Cerebral organization and behaviour" (1961). By sectioning interhemispheric connections, the principal effect was that the animal behaved as if it had two separate brains, hence the name *split brain*. For instance, a monkey with a split brain whose right eye has been blindfolded, is taught to

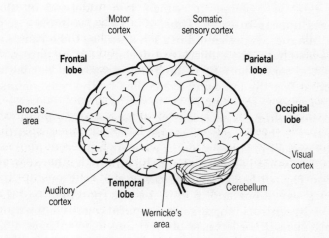

FIG. 11.1. The functions of the left hemisphere of the brain.

respond to a stimulus presented in the left half of its visual field. Then the other eye is blindfolded and the same stimulus is presented in the right half of its visual field. At this point the monkey no longer knows how to respond to the stimulus which it recognised perfectly previously. The right hemisphere shows no sign of the learning exhibited by the left hemisphere—information no longer flows from one hemisphere to the other.

This type of research was later extended to humans, through studies of the behaviour of people who had undergone the sectioning of the corpus callosum. The operation was supposed to relieve intractable epilespy. The justification for such drastic treatment was that by cutting the brain into half, even if a crisis were to get beyond control in one area of the brain, this would not spread to the other hemisphere, which would then be able to regain control over behaviour. The results of this type of operation, which is now carried out very rarely, have never been particularly clear. Disadvantages seem to outweigh advantages in the long run. One surprising effect of the operation is the seeming absence of behavioural effects in daily life—though detached, the two hemispheres are immersed in the same environment and receive the same information.

Nevertheless in special experimental circumstances, it is possible to note a series of interesting symptoms that go together to make up the so-called *split-brain syndrome*. If the use of suitable techniques allows communication to take place with each single hemisphere, then behaviour can be isolated in which two independent flows of consciousness appear to exist. Thus, if a visual stimulus, a pair of scissors for instance, is presented in such a way as to be visible only to the right hemisphere, the subject will declare that he does not see a thing (since the hemisphere furnishing the verbal response is the left), while in actual fact the left hand (controlled by the right hemisphere) will clasp the scissors and employ them correctly.

In *The social brain: Discovering the networks of the mind* (1985), a pupil of Sperry's, the neuropsychologist Michael Gazzaniga, recounts the troubled history of these complex experiments, offering a cautious reformulation of what is known about the differences between the two hemispheres. The *left hemisphere* perceives the right part of the visual field, receives sensations and coordinates the movements of the right half of the body, controls language and general cognitive functions. For this reason, lesions in the left area of the brain cause aphasia, a disorder in comprehension or in the generation of language, which will be dealt with later. In the majority of left-handers, linguistic and cognitive functions, normally the responsibility of the left hemisphere, are instead carried out by the right hemisphere.

Besides perceiving and controlling the left part of the visual field and of the body, the *right hemisphere* manages processes not connected with propositional processing, such as attention, the exploration of space and certain perceptual functions. In the visual field it carries out fine discrimination of forms (such as the direction of lines in space or the recognition of stimuli that are difficult to verbalise), in the auditory field the discrimination of complex sounds, and in the tactile field the discrimination of complex forms. It also seems able to synthesise figural elements more effectively than the left hemisphere.

The function of the corpus callosum, connecting the two halves of the brain, is to keep each hemisphere informed about what is going on in the other hemisphere, duly transferring control each time the functions each hemisphere is specialised in have to be employed. In a normal person therefore, no one hemisphere dominates. Instead effective integration is achieved, with each half doing what it is best at. The different competences of the two hemispheres seem to be ascribable to evolution—cerebral zones having common functions tend to develop in contiguous areas for reasons of greater efficiency. Likewise, if two clerks do the same job, it is better for them to be in the same room rather than in different places, since the latter arrangement would make communication more difficult.

By carrying out the appropriate tests on patients before and after the operation, Gazzaniga confirmed the results obtained in animals for humans—sectioning the connections produced two separate brains, each with its capacity for learning, remembering, emotions and acting. The two brains could not, however, communicate directly. Consequently, in a split-brain patient, the only way in which the right hemisphere can come to know what the left hemisphere is doing, and vice versa, is through observing behaviour, just as an external observer does. If for instance the left hemisphere perceives a house, the right hemisphere realises nothing of this, unless the patient says "house" in a loud voice, in which case the right hemisphere discovers what the left hemisphere has perceived, but it does so exactly as would other people who might be present in the room, that is by hearing a voice, even if it is its own voice, uttering the word "house". Conflict may also arise between the two hemispheres. Gazzaniga narrates episodes of patients who are angry with their wives while at the same time exhibiting protective behaviour towards them, or bent on getting dressed with one arm while getting undressed with the other.

But over and above the demonstrated differences in functions between the two hemispheres, it is simplistic to define the left as analytical and the font of reason, and the right as synthetic and creative. Interhemispheric differences vary greatly between individuals. These differences are extremely subtle, and bound up with language. Since the right hemisphere does not speak, it may indeed appear incapable of carrying out all those operations whose results require a verbal description, and it is thus obliged to invent ingenious alternative solutions to communicate with others. All the parts of the brain work together, and the specialisations that may be found can develop thanks to the integration of their functions. Furthermore, the plasticity of the nerve fibres enables damaged functions to be replaced through the functionally equivalent operation of another undamaged cerebral area.

COGNITIVE NEUROPSYCHOLOGY

We have already spoken of the research carried out by neuropsychologists on the split brain. More generally, neuropsychology studies the links between psychic processes and the functioning of the nervous system. These connections are investigated by analysing the disorders in perception, memory, thought, language, emotion, and action exhibited by neurological patients.

This area of research is far from achieving unity, plagued as it is by different paradigms in total opposition to each other, besides having already experienced two scientific revolutions in its history. This constant storm has made scholars wary of accepting the discoveries of neuropsychologists. Every interesting hypothesis is challenged by a different school of neuropsychologists, and deciding which is right is indeed an arduous task. We too will adopt this cautious attitude, refraining above all from adopting sensational tones founded on metaphors, such as trine brains and totally independent hemispheres, instead of on testable methods, and maintaining (with the exception of a slight deviation on pp. 306–308) the scientific rigour propounded by Tim Shallice in his excellent methodological text, *From neuropsychology to mental structure* (1988).

The birth of neuropsychology may be traced back to 1861, the year in which the French surgeon, anatomist and anthropologist, Paul Broca, that genial figure of a scientist and anticlerical, published his *Remarques sur le siège de la faculté du langage articulé, suivies d'une observation d'aphémie*. Broca had had the opportunity of carrying out a clinical examination on a patient who had been deaf for many years shortly before the patient died. When subsequently carrying out the autopsy, he had found cerebral damage to the left frontal lobe (see Fig 11.1). The connection between *Broca's aphasia* (a language disorder consisting of the inability to produce grammatically correct sentences) and a specific cerebral lesion in the area known as Broca's area was later confirmed in other patients subjected to neurological examination, exhibiting an incapacity to speak, though remaining quite capable of comprehending speech, and even of singing. It was thus demonstrated, first, that one function may be damaged while other psychic functions remain intact, since the patients observed by Broca were found to have no other disorders apart from aphasia, and second, that the specific area damaged could be identified.

The golden age of neuropsychology began with Broca. Each psychic function was localised with extreme precision, in a crescendo generating a body of knowledge of fundamental importance, despite the fact that imagination sometimes compensated for the lack of hard neurological and anatomical data. Within the space of a few decades objections began to be voiced—numbered among the critics was the young Sigmund Freud—to the extremist approach of segmenting the brain into separate and independent parts, as the localisation of functions was anything but precise and unanimously agreed on. Even the tools available to the psychologist were not equal to the task—they were too ingenuous in relation to the psychological theories holding sway at the time. Finally, the neurological evidence offered was accused of being unsatisfactory: the attention paid to patients was too limited and theorising was disproportionately large compared to the quantity of data available.

The result of this first revolution was a more cautious approach, based on the study of large groups of patients. Similar defects were identified in neurological patients, and an attempt was made to localise corresponding anatomical areas. This prudent new method led to extensive studies of long duration, but the analysis of the similarities in the disorders failed to produce the significant results expected. A new paradigm thus began to emerge, one of whose most illustrious supporters was Norman Geschwind (1965). Joining forces with the renewal generated by cognitive science, he helped bring about in the 1970s what is now called cognitive neuropsychology.

The potential of this new cognitive approach in neuropsychology was initially expressed with the adoption of computational models typical of cognitive science. The feature that makes such models interesting in neurology is that they may be *lesioned*. It can be hypothesised for example that one particular part of a model is not working, predict what would happen to the system, and attempt to verify this by seeking a patient exhibiting behaviour similar to that conjectured by the theory. Psychology and neuropsychology can thus share the same theory, at least in principle—theories engendered to account for normality may also be employed in the domain of pathology. The other side of the coin is that psychology is offered an opportunity of verifying the solidity of its own theories testing them with different pathological cases. As we have already seen in Chapter 3, when neuroscience comes to computational modelling, it is the connectionist approach that must be intrinsically favoured, and promises seem to be kept, as Mark Gluck and David Rumelhart show in their *Neuroscience and connectionist theory* (1990).

A second important feature of the cognitive approach consists in the return to the individual case—relevant data correspond to what single patients do. Alfonso Caramazza (1986) is the scholar who most energetically supports the need to remain firmly implanted in the data on individual cases, since although cognitive systems are essentially identical, nevertheless direct duplication of a datum is impossible in neuropsychology. Although justified from the point of view of scientific rigour, this type of attitude renders generalisation and theorisation almost impossible, implying as it does that no patient can be compared to another and each patient constitutes the model of himself. Furthermore, the literature is so widely dispersed that it cannot be dominated by someone who is not super-specialised—one case is enough to justify the writing of an article which everyone should then read.

The dissociative method
The key methodological point in the cognitive revolution is the *dissociative* method. This enables modalities of preservation and damage to functions to be highlighted. The essence of the method may be expressed as follows:

(a) If, following a specific cerebral lesion, it is observed that process X is intact while process Y is damaged, and especially if in other patients the opposite is observed (complementary dissociation), then it may be stated that X and Y reflect normal underlying mechanisms;

(b) if the preceding condition is present, it may also be stated that X and Y are functions of independent modules.

The second affirmation is not exactly a logical consequence of the first, since X and Y could be manifestations of a single function Z, which, under certain conditions, behaves like X, and under other conditions like Y. The pathology might alter the activating conditions but not the process.

The dissociative method assumes modularity, both at the level of organ (the brain) and at the level of function (the mind), a feature that characterises cognitive neuropsychology and has yielded outstanding concrete results. We spoke of modularity in Chapter 4, analysing its experimental advantages and its methodological drawbacks. While traditional neuropsychology localised lesions anatomically, cognitive neuropsychology localises them first and foremost functionally. Although anatomical locations are currently out of fashion, they still remain interesting and theoretically possible. They are, nevertheless, a risky extension of the modularity hypothesis, and scholars are not too keen on committing themselves to this stance.

The main characteristics of a neurological disorder tackled from the dissociative viewpoint are describable in terms of what Shallice calls a *set of dissociations*. In a theoretically fruitful case, many functions remain intact, only a limited number are damaged—the disorder has to be highly selective. A patient two-thirds of whose cerebral tissue has been damaged by a grenade is of no theoretical interest because his behaviour has been compromised to such an extent that no useful inference may be drawn from it.

A theory about normal functioning may find evidence in support of it in a set of dissociations compatible with it. Indeed, a theory postulating a series of independent modules interacting between themselves implicitly predicts that impairments in each single component will influence final behavioural patterns in different ways. A theoretician prepared to run the risk of confutation by clinical evidence may predict outputs corresponding to different hypothetical pathologies, depending on which module has been damaged. The prize is that if the predicted pathological performance is actually identified in neurological patients, the value of theory will be greatly enhanced (as happened with the modules hypothesised by Marr in the case of vision). Vice versa, if a set of neurologically identifiable dissociations does not correspond to a module that can be isolated, this datum constitutes evidence against the theory. It should also be noted that the set of dissociations may also cause surprise in the clinician, who was not expecting this particular configuration of disorders.

Let us suppose for instance that a particular cognitive ability exists, situated in the rice frontal area determining the ability to make rice with curry. A valid theory would identify four independent modules, one furnishing the rice, another furnishing the spices, one adding salt and another checking the rice is not over cooked. A series of dissociations may now be predicted compatible with the theory. If the salting module were to be damaged, the rice would be insipid; if the spicing module were to be damaged, the rice would not taste like curry; if both the salting and cooking module were to be damaged, the rice would be insipid and over cooked, and so forth. Our theory would receive confirmation from any of the previous cases, but it would be challenged by the discovery that rice becomes increasingly salty and spicy as it is cooked. Such a dissociative set, which would fail to correspond to predicted damage, would find no explanation in terms of the modular theory presented to this end.

A few examples, taken up by Shallice, will help clarify the nature of dissociation. In the case of visual perception, where the nerve cells are the most specialised, a long series of distinct neurological syndromes has been found. Cortical lesions can produce specific disorders in the perception of colour, shape, movement and the information required to provide an accurate guide for actions. Each syndrome presents a characteristic set of deficits, while the other perceptual functions remain intact:

Achromatopsia.   Patients exhibiting a deficit in colour perception lose the sense of colour. In more serious cases the world appears in black and white, discrimination being restricted to the shades of grey.

Amorphognosia.   Patients exhibiting a disorder in the perception of shape are unable to distinguish between different shapes. For instance they cannot tell a rectangle from a square. Consequently their performance in perceiving objects and faces is poor.

Disturbances in perception of movement.   Patients lose their four-dimensional vision of movement (besides the three dimensions of space there is also that of time). For example, when one patient poured out tea, the liquid appeared to her as if it was frozen, like ice. When she watched people or cars moving, they "were suddenly here or there, but I had not seen them moving".

Visual disorientation.   Patients exhibiting a disorder in their visual guide to actions lose the ability to locate objects visually. They are incapable of grasping an object shown them, or reaching out to touch it, sometimes waving their arms about in quite the wrong direction. In classic cases of visual disorientation, not only are patients incapable of reaching an object, but the eye movements required to fixate an object are also impaired, as is their ability to judge size and distance. In the most extreme cases, the effect on behaviour is tantamount to blindness.

In the case of vision, there has been an extremely significant convergence between the data from neuropsychology, the discoveries in the field of the physiology of perception, and psychological theories. Marr's theory of vision, outlined in Chapter 6, has been confirmed in an independent and spectacular fashion, thanks to patients exhibiting sets of dissociations corresponding to perceptual models hypothesised by him.

Interaction between psychology and neuropsychology has also played an important part in the investigation of memory, though pathological evidence has failed to converge into one single theory in this case. One common finding is that corresponding to the distinction between episodic memory and semantic memory. Patients suffering from a kind of *amnesia* lose their episodic memory but retain their semantic memory, the one corresponding to general knowledge. For instance, they forget what has happened to them in the last month but are quite capable of speaking normally, thus demonstrating that their semantic store still exists. The complementary dissociation, denominated the *semantic memory syndrome*, in which patients retain their personal memories but have lost their stock of general knowledge, has also been identified. Things have not gone so smoothly, however, because on the one hand psychology judges the dichotomy semantic memory/episodic memory to be too schematic, and on the other hand neuropsychology has heavily criticised the definition of amnesia as being the equivalent of a deficit located in episodic memory. According to the most stringent objection, how amnesic patients retain the information stabilised before their illness is inexplicable. The further one goes back in time from the onset of the disease, the better their memory is. Thus, amnesic patients can remember events in their life that precede the onset of amnesia by a few years. This phenomenon, which demonstrates that episodic memory does function albeit in a very reduced form, is incompatible with a specific disorder of that very same memory type.

The dissociative method, strongly conditioned as it is by the assumption of modularity, may lead neuropsychologists to an exasperated search after the significant set of dissociations, at times postulating organic and functional modules that are not very credible: what are we to think of the cases described in the literature concerning agnosia for vegetables, that is to say the inability to distinguish one type of vegetable from another? Given the critical attitude this book has adopted with regard to strong modularity, it is important to underline the risks inherent in a method that, while being efficient, is grounded on a hypothesis that cannot be accepted *in toto* when applied to mental functions. Although the dissociative method maintains its heuristic value under standard conditions, above all when applied to areas that are unquestionably modular as is the case with perception, the results deriving from its application under other conditions do leave scholars sceptical at times.

I shall therefore now present a case where the dissociative method is put to bad use, demonstrating how the assumption of modularity in an absolute form may lead research astray. Just as those searching for localisations last century managed to localise non-existent functions, thus cognitive neuropsychologists are in danger of discovering non-existent dissociations, based on independent modules that do not exist.

A case of prosopagnosia analysed using the introspective method

Prosopagnosia is the inability to recognise faces. Two dissociable types may be distinguished. The first concerns the inability to *perceive faces* and the second the inability to *recognise faces*. We referred to perceptual prosopagnosia when we spoke of disorders of the primitive perception of shape, of which this deficit may be a corollary. However, perceptual prosopagnosia may also develop independently of a disorder in the perception of shape. I shall concentrate on the second type of prosopagnosia, i.e. the inability to recognise faces, which I am quite familiar with since I suffer from it myself, though not in a serious form. Introspective accounts have been banished from cognitive science, for hundreds of good reasons. Nevertheless, I shall break the rule since it is rather a rare occurrence for a patient to be able to share his subjective experiences of an illness with others. I therefore beg the reader to withhold judgement, with the assurance that we shall try to recover at least a small dose of methodological rigour at the end.

The repertoire of anecdotes of people unable to recognise faces is endless. It ranges from introducing oneself to another person several times at weekly or other intervals, to pretending recognition of a person who is absolutely certain she has never met us previously. A mild form of this disorder does not wreak havoc with one's life, but it is a source of continual social embarrassment—people hate not being recognised, and any excuse one might offer only makes the situation worse. A preventive strategy consists in assuming that if a person is in a familiar and clearly defined context, then that person's identity will be known to me. When I enter the vice chancellor's office, for example, I already know I shall meet the vice chancellor's secretary. Hence, even if I fail to recognise her, I greet her affably all the same, and unless a new secretary has just taken her place, then everything will go smoothly. When I am in a familiar part of town, near home or in the vicinity of the university, I greet anyone who looks at me. I am thus taken for a sociable person, even though I am far less short sighted and absent minded than I appear. Naturally, showing excessive affability with strangers creates difficulties, but less so than the opposite case.

More embarrassing is the case in which the unrecognised person has been known to me for a long time, such as a final year student, a friend or a relative who is not very close. The most dramatic events from a subjective standpoint are those in which the failure to recognise the other is a highly improbable event, and is consequently interpreted by the other person as an act of aggression or one that at least requires an adequate justification. I was once struck in a supermarket by a fascinating blond lady who appeared to be observing me with an eye that could not be defined as indifferent. Having approached her and having uttered an innocuous opening remark, I discovered too late that the lady had been my fiancée a few years earlier. My unusual behaviour had reawakened in her vague desires for revenge. As she lived in another part of the town, she had come to that supermarket by pure chance, where there was nothing to remind me of her. Perceptual and evaluational processes had executed their task to perfection. They had even exhibited consistency over time in the face of a clamorous failure in my identificational capacities. Analogous episodes in which my failure to recognise people was interpreted as explicitly pretending not to have seen them have given rise to surrealistic disagreements—events which are far more amusing to recount than to experience.

Self-analysis conducted for many years—employing ethological modalities that cannot be reconstructed in the laboratory—has made me sceptical with regard to the thesis that my inability is ascribable to the malfunctioning of an independent module. Prosopagnosia in my case is the companion to a more basic disorder, consisting in a difficulty in reconstructing the object in its entirety every time I start from a part and do not receive help from an unequivocal context.

The most convincing example is looking for a missing object, which usually implies recognising it starting from some detail, such as identifying a watch from its strap or a pan from its handle. But in actual fact the passage from the visible detail to the total object is by no means automatic, as is said to be the case for other people. On the contrary, it is an operation calling for conscious attention (I have to represent the complete object to myself), one which often fails where others carry out the task without any apparent difficulty. My perceptual processes are heavily theory driven, strongly influenced by my expectations, and insufficiently data driven, as not enough influence is exerted by what I am perceiving in the environment. Deprivations of this type, in the transition from unconscious to conscious processing, is consistent with the results obtained by Tranel and Damasio reported in Chapter 10, according to which prosopagnosic patients exhibit significant autonomic responses even when they are incapable of consciously recognising familiar people in photographs.

An exhaustive explanation of these disorders posits a difficulty in the passage from unconscious primary processing to conscious processing, and in gaining access to general knowledge when the process starts from decontextualised details. When context is clearly defined, general knowledge and an excellent memory stand me in good stead in any situation. But when a contextual framework is lacking, the malfunctioning of elementary processes emerges. Recognising faces then becomes the clearly visible tip of the iceberg, since in such cases nothing comes to my rescue, at least initially, until the person begins speaking or makes some significant act, thus allowing memory and knowledge processes to stand in for defective recognition.

Prosopagnosia is generally held to be dissociable from a deficit in the perception of colours and objects, but in laboratory investigations objects are presented as integral wholes, and not starting from details. I too have no difficulty in recognising colours and complete shapes. Prosopagnosia is, instead, often associated with geographical disorientation. This latter disorder appears to be due to a subject's inability to identify specific buildings (that is a particular block of houses, a particular church), even though the capacity to recognise different categories of building, that is distinguishing houses from churches or from schools, remains intact. I manage to lose my way in any place that has more than three houses, and I save myself in a metropolis as in a village only by using detailed street maps. I fail to recognise any change in environment, context or situation—I do not notice if there is a new carpet in a room, nor if the attractive unknown lady who is convinced she is my wife is wearing a new dress.

Bornstein (1963) quotes the case of a prosopagnosic ornithologist who had lost his ability to discriminate between similar birds, a phenomenon that cannot be explained by the hypothesis of a deficit in a module specialised in recognising faces, but that is however congruent with the thesis of the malfunctioning of more general abilities. In his studies on patients with a split brain, Gazzaniga also notes that, unlike the right hemisphere, the left hemisphere appears to be incapable of recognising faces. He is, however, more inclined to put this difficulty down to the left hemisphere not possessing elementary discriminatory abilities, rather than to a specific function that may be located in the right hemisphere.

Subjective evidence of the type I have offered above does not constitute proof because of the inherent methodological weaknesses of self-analysis. Nevertheless, it is an extremely cogent approach for those who spend part of their time dealing with these disorders. To conclude this digression, I hope I have managed to cast some doubt over the clinical concept of prosopagnosia, and consequently over the existence of modules corresponding to what neuropsychologists claim they have identified experimentally or clinically.

The relationship between normality and pathology

Neuropsychology does not as yet admit definitive conclusions, nor does it offer stable theories that have gained consensus first within the discipline itself and second from the other neurosciences and thus from cognitive science. From the brief sketches we have drawn of some of the topics dealt with by cognitive neuropsychology, it is nonetheless obvious that such topics provide an excellent test of the psychological theories dealing with the various psychic functions in normal humans. Those cerebral lesions giving rise to the phenomena studied by neuropsychology are of an organic nature, brought about by physical causes, and identifiable by means of objective examinations. Neurological patients thus have a cerebral structure which was originally identical to that of normal people, as they were also prior to the onset of the illness or to the occurrence of the wound, but which now exhibits important organic differences—their nervous systems are different from those of normal humans.

If psychology had developed sufficiently precise theories about mind, it would be possible to account for why a given organic lesion produces a given mental malfunctioning. If the initial theory were of a really high quality, it would even be possible to predict the nature of ensuing deficits. The standard reference system adopted by neuropsychology is not the same as that employed by psychology. According to the single-case approach currently in favour, each patient is considered an independent self-contained system that may be utilised to falsify a general psychological theory.

To use the computational metaphor, in psychopathology the software program is different from the one used habitually, and it thus processes input differently and produces different output. But the computer it runs on is equipped with the standard hardware. In neuropsychology, instead, the computer hardware is different from that habitually used. Thus it operates in a different manner even when the same function is to be computed.

The theoretical importance of neuropsychology is therefore destined to increase within the sphere of cognitive science, since it acts as a support in the bridge connecting the brain to the mind, and claims the status of a discipline capable of verifying the explanatory and predictive power of the corresponding psychological theories. For a clinical introduction to the field, I recommend *Cognitive neuropsychology* (1990), by Rosalyn McCarthy and Elisabeth Warrington.

CHAPTER TWELVE

Applications

In this chapter we shall examine some important applications of cognitive science. The first is in the sphere of mental illness and its treatment, the second in education, and the third in human–machine interaction. Naturally these are not the sole applications, but they are the ones in which most progress has been made, as well as being of interest to three sectors of vital importance to society.

PSYCHOPATHOLOGY

I hope it will not seem strange to begin the chapter on psychopathology with what is not a correct approach to the subject, but the fact is that the two most well-known programs that artificial intelligence has produced in the area of mental illness cannot be taken seriously. One of the two because the programmer intended it to be a joke, and the other because the simulative method was applied incorrectly, pretending it produced performances that it could not really reproduce. Only after we have cleared the field of the hoaxes will we tackle the real contributions made by cognitive science to the analysis and treatment of mental illness.

In 1966 Joseph Weizenbaum presented the first program that appeared to be able to converse with human beings—ELIZA. The name was inspired by Eliza Doolittle, the heroine in George Bernard Shaw's *Pygmalion*, who gradually improved her manner of speaking and acting thanks to the teaching of Henry Higgins, a gentleman scholar of phonetics. What the program actually did was in fact an illusion. ELIZA's efforts were restricted to recognising a series of words, and to emitting a pre-established phrase in reply to each of the words it recognised. It was an ingenious program and nothing more. It had no inkling whatsoever of what was really going on in the conversation. Its amusing stupidity aroused the curiosity of those working in the field and the ruinous interest of the general public. Weizenbaum's brilliant stroke was that of having selected as a script for the conversation with ELIZA a parody of the replies a non-directive psychotherapist might give in a psychiatric setting. Thus, if the user were to type a word such as "mother" or "father", ELIZA would answer with:

[1]   Tell me something more about your parents.

If the key word recognised was "sex" or an equivalent, one possible answer was:

[2]   How's your sex life going?

If no key word was recognised, ELIZA would simply repeat the client's preceding utterance, with exhilarating effects for anyone who had ever had contacts with a therapist of the Carl Rogers' school. An example of a not particularly brilliant conversation is furnished in Fig. 12.1.

If the user wanted to play a little longer, then it was necessary to keep to classic topics and avoid mentioning subjects other than work, sex and family. Even when this rule was respected, after a few minutes the conversation became repetitive. A conversation yielding better results than the previous one is reported in Fig. 12.2. It should be noted that the program's apparently brilliant performance is directly proportional

Patient:	I miss Simona today.
ELIZA:	You miss Simona today.
Patient:	Very much.
ELIZA:	You don't seem very communicative today.
Patient:	No.
ELIZA:	Before you said you miss Simona today.
Patient:	Shoot yourself.

FIG. 12.1. An infelicitous conversation with ELIZA.

ELIZA:	Good morning, my name's ELIZA. What's your name?
Patient:	John.
ELIZA:	Can I do something for you, John?
Patient:	I'm worried.
ELIZA:	Worried about what?
Patient:	I have to finish a job for tonight.
ELIZA:	What does work mean to you?
Patient:	Money, career.
ELIZA:	Is money important?
Patient:	Not all that much, perhaps.
ELIZA:	You told me before you were worried.
Patient:	There are other things.
ELIZA:	What other things?
Patient:	Love.
ELIZA:	Talk to me about your emotional life.
Patient:	I'm divorced, and I've had a few affairs.
ELIZA:	What does divorce mean for you?
Patient:	She was my great love.
ELIZA:	She was your great love.
Patient:	I'll never find another woman like her.
ELIZA:	Are you sure, John?

FIG. 12.2. Part of a conversation with ELIZA.

to the user's willingness to believe in there being a psychiatrist hidden in the machine. The conversational gadgets employed by Weizenbaum are aimed at producing an anthropomorphisation effect in an attempt to involve the naïve user on an emotional plane.

Weizenbaum's program achieved notoriety outside artificial intelligence circles, where it was immediately misinterpreted not only by the public at large, but more surprisingly by psychologists and psychiatrists. Weizenbaum (*Computer power and human reason*, 1975) recalls that some were horrified at the prospect of their imminent substitution by an electronic colleague, while others wrote proposing opening a consulting office together with ELIZA.

One program that stands out from ELIZA's offspring is PARRY, designed by psychiatrist Mark Colby (*Artificial paranoia*, 1975) to simulate a client affected by paranoia. Paranoia is a psychosis characterised by persistent delusion, generally of persecution or of jealousy. It does not involve hallucinations, intelligence is fully preserved, behaviour and emotions are consistent with the subject's ideas. In a persecution delusion, clients may believe they are being spied upon, slandered, disturbed, hindered in achieving their goals. A jealousy delusion consists in convincing oneself for no good reason that one's partner is unfaithful. Minor facts and events are ingeniously employed as conclusive proof of the client's convictions.

Colby assigns the paranoid a database of "false" beliefs which are presumed to produce the observable symptoms. Which particular sets of beliefs the program activates depends on the emotional value it places on the phrases it receives from the subject. PARRY's weak point is that, since it derives from ELIZA, it understands nothing of the utterances conveyed to it, and it limits itself to activating its own subroutines on the basis of a set of key words. For example, the interviewer asks:

[3]   Do the doctors think you are mad?

In this case the program increases the value of the parameter "shame", since the accusation of madness is connected with shame. Colby has seemingly created a complicated system of beliefs associated with emotions. In practice, what he did was to associate words with other words. To try to escape from the ghetto of motivated scepticism which met the work when it was published, an attempt to validate it was made employing a weak version of the Turing test.

Eight human psychiatrists were asked to evaluate the replies furnished by four paranoids—three humans plus PARRY—to a series of stereotyped questions. The eight judges failed to distinguish to a significant degree the replies provided by the computer from those furnished by the three humans. However, given PARRY's inability to generate meaningful utterances, the replies provided by it were pre-formed, selected from a list drawn up by hand by Colby's team. Clearly, the judges were not allowed to set any question they wished, otherwise the programs' weaknesses in comprehension would have come to the fore immediately. Thus question and answer were pre-established. To demonstrate the relatively high degree of efficacy of PARRY's performance, a conversation between PARRY and a psychiatrist is reported in Fig. 12.3.

In the case of ELIZA anthropomorphisation of the computer had given rise to surprise. PARRY cleverly attempted to bring about the same

PARRY:	People get on my nerves sometimes.
Psychiatrist:	Tell me more.
PARRY:	Do you know anything about bookies?
Psychiatrist:	No. What about bookies?
PARRY:	A bookie didn't pay me off once.
Psychiatrist:	When did this happen?
PARRY:	A couple of months ago.
Psychiatrist:	And then what happened?
PARRY:	I had an argument with the bookie who did not pay off. Later on I remembered he had friends in the underworld, and I thought he would use them to get even with me.
Psychiatrist:	Did he get even with you?
PARRY:	Yes. I think he is behind the gangsters who are out to get me.
	. . .
Psychiatrist:	Thank you for talking with me. I would like to see you again.
PARRY:	Don't come back.
Psychiatrist:	Goodbye.

FIG. 12.3. A conversation between PARRY and a psychiatrist.

effect. This time, however, users had woken up to the ruse and were in no mood to give Colby more than his due, namely that of having created a new computer game, which had nothing whatsoever to do with language comprehension, nor with paranoia. To paralyse PARRY, it is sufficient for the psychiatrist to ask a question not contemplated by the program, as for instance:

[4] Where would you like to go on holiday for Christmas?

Better still would be turning an acceptable question such as 3 on its head:

[5] Do you think the doctors are mad?

Having terminated the necessary destructive part to avoid possible misunderstandings, we may now pass on to something more constructive.

Complex systems in development

Given its passion for behaviour modifying techniques, behaviourist and cognitive clinical psychology both consider the mind as a machine working normally until one of the parts of the machine goes wrong or breaks: the famous loose screw. It is sufficient to tighten the screw, and the machine starts working again, exactly as it did before.

We have seen, instead, that no single correct mode of functioning of the mind exists. On the contrary, the mind continuously interacts with the environment, in an ever-changing equilibrium that is continually readjusted. Simply asserting that one isolated part breaks, goes against the assumption of the unity of mind. But even if it were to be admitted that something of the sort occurs, the system would simply adjust to a new level of equilibrium, one that is not necessarily pathological. And once a pathology were to be cured, the system would have changed in any case. People cannot regain their former equilibrium because the new experience consisting of illness and cure has transformed them into different people. There may be a new equilibrium, a well-being refound, but this is not brought about by the replacement of some cog. It is the entire machine that works differently, often in an environment that has been modified in its turn.

A developmental approach has also to be applied to the clinical field. The development of the various psychoneurotic syndromes must be traced in order to explain the special constellations of mental illnesses and to indicate the pertinent therapeutic possibilities: the key to psychopathology lies in human development. Readers desiring to go into this subject in depth are advised to turn both to *Complexity of the self* (1987) and to *The self in process* (1991) by Vittorio Guidano. The fundamental concept is that in order to understand the different cognitive organisations lying at the root of the various psychiatric syndromes, the developmental choices each individual has made in the course of his development must be identified, thus contemporaneously constructing the self and any pathology that might be exhibited.

The scholar having made the greatest contribution to our knowledge of the effects the early years of life have on the emotional and cognitive development of the individual is John Bowlby. Bowlby devoted all his life to the study of the child's early years, and published the results in a trilogy entitled *Attachment and loss: Attachment* (1969); *Separation* (1973); *Loss* (1980). He analyses the processes of attachment, not of the infant towards the mother, but "between" mother and infant. Bowlby points out there are instinctual components driving infants to attach themselves to the maternal figure, and reciprocal instincts driving mothers to take care of infants and protect them. The particular interactional modality of the mother–child couple is a determining factor in children's later emotional relationships, children tending to repeat the type of experience lived in their first few months in a dimension that may last their entire lives. Finally, child and adult reactions to loss and bereavement are interpreted as deriving from early experience.

It is difficult to overestimate the importance of infancy to the construction of a person, but there is a second crucial phase in life that everyone must overcome to reach adulthood: adolescence. The fundamental objective of adolescence is to conquer one's autonomy and consciously construct one's own identity. But while infancy is an obligatory step, the same cannot be said of adolescence—it is not everybody who manages to face this effectively, and the vanquished remain in an infantile state of dependency, refusing to take on responsibility. Indeed, the crisis of adolescence does not necessarily come between 14 and 15 years. Any person may go through this beneficial crisis leading to growth at any point in their lives, even many years later. It is brought about by events driving the person to reject an equilibrium which, though comfortable, feels increasingly restrictive, in an attempt to obtain a higher degree of liberty and autonomy.

What triggers off the maturation process in the first place is the physical change that comes about with the acquisition of secondary sexual characteristics. In the female the ovaries develop, causing menses, and this is followed about a year later by real procreative capacity. Genital apparatus and the breasts reach full development, and sexuality becomes diffuse and tied to the emotional context. In the male the development of the testicles leads to that of the penis and to a gradual increase in the production of sperm to the point where reproduction is guaranteed. Such an explosive change is accompanied by the need to restructure the image of the self to include the extremely novel emotions that shift the interests of the individual from the family to his contemporaries. Reflexive thought acquires crucial importance. If instead an adolescent relies exclusively on formal thought, then he may set out on an impossible undertaking—to achieve control over emotions through reason. Whatever happens, one of the most difficult transitional steps is renouncing control, to give way to the integration of emotion and reason, the prelude to the acceptance and enjoyment of emotions themselves.

The search for a new internal coherence begins within the family, projecting the adolescent towards the outside world. For the second time parents play a crucial role in the development of their children, after the part played in the first three or four years. Parents act as a means for comparison, a yardstick children can measure themselves by. Hence the conflicts characterising this period, which help adolescents to assert their own independent existence, above all in the fight against the homologous parent. And if the parents themselves are not adult enough to understand this, the apparently unmotivated bitterness of these battles may become a dramatic experience, fostering the ambivalence of the child, and generating an unwarranted sense of guilt. The parent of the opposite sex constitutes the reference point for the child in her interactions with children of the opposite sex, and the child's future amorous relationships will be significantly influenced by the manner in which the relationship with the parent of the opposite sex was experienced. The type of affectivity and sexuality exhibited by the parent influences the way the child will develop.

Since if there is a crisis in life, adolescence is that crisis, a successful—though not necessarily happy—adolescence allows children to gain the self-confidence necessary to deal with problems. Once having granted themselves permission to be different, young adults can plan their own future course without unduly fearing the solitude and promises of maturity.

If adolescent experience follows on from infantile experience, then five basic cognitive–emotional organisations are obtained: depressive, dissociated, phobic, obsessive, and psychosomatic. From these five structures, mental illness proper may develop. Normality does not lie in a different dimension, but in an elastic mixture of all possible neuroses, thus guaranteeing flexible interaction with the world. Pathology is tantamount to the rigid impossibility of escaping the stereotyped structure of one's organisation, with the ensuing loss of active adaptation.

Cognitive therapy

The cognitive psychotherapy we are about to deal with is not intended to be an exhaustive account of the constellation of approaches this term is generally applied to. For a broad survey the reader is referred to the classic manual by Vittorio Guidano and Giovanni Liotti (*Cognitive processes and emotional disorders*, 1983). Here we shall limit our discussion to that part of cognitive therapy that privileges the relation between therapist and client as the crucial factor in therapeutic change.

We may begin by defining cognitive psychotherapy as the achievement of integration between emotions and cognitions, aiming at maintaining a conscious dynamic equilibrium, and obtained thanks to the client–therapist relation. The goal of therapy is not that of modifying the person's external behaviour—the symptoms—but the subjective causes of behaviour, and these depend on internal structuring—perception, cognition, emotions, and above all the image of the self. Cognitive therapy investigates two fundamental dimensions:

1. Analysis of the past. Attention centres principally on moments of crisis, seen as the breaking of equilibrium, a painful but necessary process in development. Besides re-elaborating infantile and adolescent experiences, events that may induce loss of equilibrium—love, marriage, children, bereavement—and thus require reflection on the self, are scrutinised. The primary objective of this phase is the construction on the part of the client of a theory about himself. Such a theory helps to understand both one's own personal history and one's particular mode of emotional and cognitive organisation that led to the current state of malaise.

2. Analysis of the present. Starting from the onset of the disorder, it proceeds to the here and now of the therapeutic situation. The objective aimed at is not so much definitive stability as much as acceptance of possible future crisis, in the name of continuing individual development.

The method by which the therapy is realised is through the therapist–client relationship. This is a cognitive–emotional relationship whose objective is the client's well-being. For the relationship to be effective, it has to be emotionally well mediated and the correct informational content has to be provided. Just like any other communicative act, that between therapist and client has both a specific meaning and a relational meaning. This implies that what is said during a session must also be interpreted as being a commentary on the client's relationship with the therapist, and hence is based essentially on emotional values. The idea is that through the relationship with the therapist, which becomes increasingly clearer and deeper, the client learns not only to understand himself, but also to re-establish a channel of communication between his cognitive part and his emotional part. A full account of this coherently cognitive approach may be found in the excellent *Interpersonal processes in cognitive therapy* (1990) by Jeremy Safran and Zindel Segal.

If we were to desire a better definition of the objectives of cognitive therapy at this stage, we would have to speak of a coherent construction of the self, and of the conscious and perfectly integrated functioning of cognitive and emotional processes, a guarantee of interaction being in harmony with the world. The disintegrated self, or the inconsistent self, or the insufficiently mature self has been analysed, and the client has been aided in the restructuring process. It must be stressed that the key point is not the discovery of a particular truth in the person's past or present life. It is rather the construction of a theory about the client that will be effective in the three temporal directions:

(a) towards the past, to explain the client's personal development from birth;

(b) in the present, to allow the client to live every moment of his life in accordance with his cognitions, emotions and desires, understanding and accepting what happens;

(c) towards the future, to guarantee future development will continue without serious breaks.

Time management acquires new importance in psychopathology, since certain syndromes correspond to a characteristic form of deformation in temporal experience. The depressed live principally in the past, continually brooding over it and projecting it onto their image of the future without altering it in any way. They therefore fail to give themselves any hope for the future, imagining it will be as painful as the present is. The anxious tend to live in the future, anticipating almost all the possible difficulties and anguishes they may encounter, brooding over every possible escape route to avoid the feared event, and dying a thousand deaths before coming face to face with the danger, which they then tackle in a state of exhaustion, having consumed all their energies in realistically simulating the catastrophe in the offing. Obsessionals are incapable of inserting themselves into the normal flow of time. They do not therefore live an event as it unfolds itself in the present. This has the effect of rendering them incapable of construing the event as something that actually took place in the past, thereby filling them to the brim with their uncertainties, which in their turn increase further the degree such people are out of step with present time, and so on in a vicious circle that has no end and no beginning.

Happiness, serenity, wisdom lie basically in the present, this being the sole dimension of pleasure (even memories afford pleasure in the present), the point from which a past lived to the full may be reconstructed, moment by moment ("When you came in, did you leave your umbrella to the left or to the right of your shoes?" asks a master of Zen to the disciple to remind the latter that one must be present in his own being at every instant and during every action), without growing anxious about a future which is in any case beyond our control.

The decrease in interest in realistic investigation, above all of the client's past, is due to the fact that cognitive science does not have a normative theory concerning personality development, thereby making it possible to construct an individual's developmental model, namely one drawn up to fit the individual as closely as possible, though it would respect certain fixed reference points.

Besides truth, which is subordinated to functionality, the other traditional aspect neglected by radical cognitive therapy is that of stability. Quite the contrary. The value sought after is that of flexibility, of seeing moments of crisis as necessary transitional stages in achieving a new equilibrium that is potentially richer than the previous state—old constraints are overcome to open up new possibilities. Renouncing the conquest of a non-existent "optimum state", that limbo of non-life which one attempts not to budge an inch from, appears to be a necessary act of realism in the light of the present state of the art—the more we investigate the human mind, the more its complexity seems to triumph. Theories which have been simplified to the utmost—such as behaviourism—or those which are rigidly deterministic—such as psychoanalysis—appear more and more inadequate as the picture grows more complex. It is impossible to predict the future, above all in a system that offers extraordinary opportunities for development, as is the case with children and adolescents. Hence the important point is that a person be aware of this fact, that he choose to live within the realm of change, without making the vain attempt to withstand it.

The client is not a system that can be modified by the environment, as with cybernetic models, where the system reacts to a real change in the external world, but a system that brings about its own disturbance, where any alteration that may take place is attributable to the interpretation the system provides of its world. Consequently, the therapist sees her role reduced to that of the catalyst of change, in contrast to the previous view of the role as that of the director of change. The client has no alternative but to settle autonomously into a new equilibrium, while the therapist moves away from the idea that she is capable of determining people and events in line with her own particular worldview.

EDUCATION

One area where the use of the computer could lead to significant innovation is that of education. At least two reasons justify this supposition. The first consists of a widespread dissatisfaction with human teachers in the Western world, from Italy to the United States. Our society, which asks young people to undergo a long period of training (sometimes over 20 years), does not devote sufficient economic and cultural resources to ensuring the quality of the trainers is up to the job.

The second reason is connected to continuing education. Any job today requires one to update one's knowledge at a pace unknown until recent decades. Up to the middle of this century, knowledge acquired during schooling was generally sufficient for a person's entire working life—innovations were few and clearly delineated. From the 1950s on, the pace has become so hot that no worker can allow themselves the luxury of interrupting the learning process. However, continuing education requires extremely complex organisation, due both to the learning difficulties typical of the age of maturity and because it is no simple matter to include this extra task in work schedules that are normally rigid.

Why not use the computer then to help human teachers whenever necessary? Unfortunately the situation is not as straightforward as might appear, if we do wish to use the computer as more than simply a passive tool, an electronic book. Instruction means that a significant piece of knowledge, whether it be declarative (what) or procedural (how), has been interactively transferred from one intelligent system to another. It cannot be claimed that an act of instruction has been carried out if a user employs a computer to acquire a piece of information that was not previously possessed—the user has only exploited access to an electronic data bank. The computer has not understood what the user wanted from it, nor what the user has obtained—the computer has simply executed an order. A book teaches no one, if not in a metaphorical sense—it is an object, not an agent. Likewise, a computer that is utilised as an instrument cannot educate anyone. To instruct someone requires an intelligent system that interacts with the user, cooperating with her in the moving towards a common goal.

Computer-aided instruction

There is one classic error that has always been committed when introducing computers into the educational environment, namely believing that the first step consists in teaching children—or adolescents, or adults—to program a computer employing a low level language, typically BASIC or an equivalent. Experts from artificial intelligence have always fought against this approach, for the incontestable reason that programming in an elementary language is simple if the desired results are banal, but unbelievably complicated if the results aimed at are to reach an even minimal level of interest.

It would be as if to teach a child to talk one taught the child only single letters instead of addressing complete messages to the infant. No mother says to her six-month-old child "a", "m", and so on until the child has learned the alphabet by heart, then proceeding to teach the infant the use of the articles, of nouns, and so forth. Were she to do so, then a genius would not learn language before the age of 20. On the contrary one speaks to a child employing entire words, and trying to make the child comprehend the meaning of an utterance: "Do you want a glass of water?" or "Give grandmother a big smile" or "Get down from the table". Pragmatics precedes grammar, as was pointed out earlier. Teaching a novice of any age, from 1 to 80, to program in a low-level language is so nonsensical and has such a demotivating effect on the student as to justify fully the accusation of gross incompetence.

If we were to buy a tape recorder and ask how it worked, then we would be interested in knowing how to record something and then listen to it. What matters is how the object works in relation to the user's interests. If the salesman were to insist on our first learning the principles of acoustics and then the basics of mechanical engineering, that salesman would indeed be adopting the wrong approach. Whoever buys a tape recorder is interested in putting it to practical use, not in building one, nor in exploring the theories underlying the way it works. Whoever is learning to use a computer wants to make the machine do something useful, not to get lost in a boring set of repetitive instructions bringing no intellectual or concrete advantage whatsoever.

It is obvious that in a course on programming languages the languages themselves will be studied. If, however, the focus of interest is not on the way the machine works, then the computer will have to be employed to do something significant. A hammer is of interest because it drives nails into walls and breaks fragile objects, whereas hands are less efficacious than hammers in executing these tasks. Similarly, computers are of interest if they can be programmed to do things that are not obvious: solving problems, controlling robots, answering questions humans cannot easily answer. In each area in which cognitive science is applied to a concrete problem, it must be emphasised that the computer is an instrument at the service of the user, and one must be on one's guard to ensure the position is not reversed. Technology must be increasingly tailored to fit human needs. Humans must not be continuously forced to adapt to the changeable and not always reasonable demands of the designers of machines. In particular, in the field of compulsory education where a long lasting influence is exerted over the minds of the young, it is of fundamental importance that the focus of attention be on the user and not on the instrument.

LOGO: Children and tortoises
In the field of education, the first person to adopt a child-centred philosophy and not a machine-centred one, was Seymour Papert, a mathematician who attempted to apply Piaget's theory to the teaching of mathematics with the aid of the computer (*Mindstorms*, 1993). The LOGO group, which was founded at the Artificial Intelligence Laboratory at MIT (Massachusetts Institute of Technology) in 1970, has produced highly innovative research on teaching in the scholastic milieu, giving rise to the birth of research groups which continued the analysis in those nations possessing a computer culture.

Papert and his collaborators have remained faithful to Piaget's principle of considering children as the active constructors of their own intellectual structures, and have sought to develop a computer environment allowing children to reflect on their own mental processes. The achievement of the "epistemological" child is that she learns both to think and to learn, teaching a computer how it is done. In the first place children have to learn how to communicate with the computer. To this end an entire family of programming languages, called LOGO, has been developed, of which numerous variants exist, their specific characteristics depending on the particular aims of the end user and the area of knowledge dealt with. There are versions for 4-year-old children, for children of school age, for children with mental and physical handicaps, and even for autistic children. LOGO is not a game, but a high-level programming language, offering recursive and interactive opportunities, conceived to allow easy and immediate access to programming.

The best known method of access to LOGO is that subset containing the "tortoise" commands. The tortoise is a robot on wheels, dome shaped, and equipped with a pen tracing a line on the ground when it moves. A video tortoise also exists, but this is a more banal small triangular shaped tracer appearing on the monitor. LOGO's effectiveness is by no means separable from the physical nature of the robot—the mechanical tortoise offers immediate appeal to the child's sense of corporeity. At the same time it constitutes a stimulus to play rather than to boredom, and a suggestion as to how to tackle problems of a high degree of abstraction. The child controls the movements of the tortoise by means of the computer, first by drawing squares and other simple figures, and then increasingly complex drawings. In this way, the child learns both how to program and how to build a sort of concrete geometry, and may even reach an advanced state of mathematical knowledge, always taking as the starting point a personal interest in solving problems aimed at inducing the tortoise to achieve some extraordinary feat.

The approach adopted by Papert, which took careful heed of the proposals advanced by psychologists and which aimed at building an environment in which children could freely become the creators of their own learning, still retains its validity and vitality today.

CAI: Computer-aided instruction

In the 1960s the idea began to emerge that computers should be employed in education. The result was a series of programs that represented the computer equivalent of well-prepared pedagogic materials. The computer was utilised as if it was a book students had to follow, though they were free to jump from one topic to another or to skip those parts they were not interested in. The new technology was thus placed at the service of education—no pretence at methodological innovation was made.

CAI programs consist of structures that attempt to reproduce the knowledge of teachers who are experts in a given area. Everything is foreseen in advance, even though the student can select which particular pathway to follow within the network of knowledge available. Since everything is predetermined, even the most sophisticated CAI program cannot meet the challenge of even the slightest unforeseen variant. If the student asks a question the teachers had not thought of, the computer goes dead. If a mistake is made the teachers had not imagined possible, the computer is unable to correct it. Programs do not manage the knowledge they contain; they simply present it in a sophisticated fashion. Furthermore, they are incapable of drawing any autonomous inference, or of permitting any real interaction with the user. The basic criticism levelled at CAI programs is that they do no more than what a good book can do, apart from costing a hundred times the price. Nevertheless, they have been employed in special contexts with good results, and they have constituted the departure point for intelligent CAI, the difference between them demonstrating what path was to be followed.

ITS: Intelligent tutoring systems

The natural evolution of computer-assisted learning is constituted by intelligent CAI, now renamed ITS, namely intelligent tutoring systems. Since the 1970s, this field has become one of the privileged meeting grounds of artificial intelligence, cognitive science and pedagogy. This cross-over of industrial, theoretical and pedagogic interests has led to florid growth in the entire field, with books, conferences and journals devoted exclusively to this topic. A good reference text to go into the subject in greater depth is Etienne Wenger's *Artificial intelligence and tutoring systems* (1987).

The decisive turning point was reached by Jaime Carbonell (1970) with his program SCHOLAR, the first to achieve a partially interactive structure. Not only does the program answer the student's questions, but it is also capable of taking the initiative by asking the student questions that are pertinent from the didactic point of view, questions that demonstrate the level of knowledge attained. SCHOLAR employs semantic networks to represent its knowledge of the geography of South America. The means selected for teaching is so-called Socratic tutoring, which was later to be used by many other ITSs. First the computer tries to establish the knowledge of the user. Then it attempts to get students to correct their errors by presenting material contradicting students' mistaken assertions. SCHOLAR can make inferences that allow it to answer questions that the programmer did not foresee, if the student wishes to take the initiative by asking the computer questions. An example of a conversation between SCHOLAR and a student is reported in Fig. 12.4. Carbonell and the designers of programs of a similar nature to SCHOLAR identified what general features a good pedagogic system should possess, even though the way these specifications could be realised was left unsolved.

The other intelligent pedagogic system worth mentioning, in as much as it represents a significant turning point compared to the programs preceding it, is GUIDON, designed by William Clancey.

Clancey noted that an expert system might constitute a good departure point for an ITS, the main reason being that an expert system possesses a large quantity of knowledge that has already been formalised. One of the best known expert systems is MYCIN (see Chapter 3), an expert at diagnosing bacterial infections. Its designer, Shortliffe, moreover, had drawn his inspiration from SCHOLAR's conversations in implementing the interactive characteristics of his own program. Both Shortliffe and Clancey's groups work at Stanford, California, and their physical proximity has enabled in depth exchanges to take place between the two projects.

SCHOLAR:	Approximately what is the area of Brazil?
Student:	2,500,000 square miles.
SCHOLAR:	Wrong. Please indicate if the following statement is correct: The area of Paraguay is approximately 47,432 square miles.
Student:	What is the area of Brazil?
SCHOLAR:	The area of Brazil is approximately 3,600,000 square miles. Now answer the question you didn't answer before.
Student:	Incorrect.
SCHOLAR:	Very good. The area of Paraguay is approximately 157,000 square miles. Approximately what is the population of Brazil?
Student:	60,000,000
SCHOLAR:	Approximately correct, but a more exact value is 70,000,000. What is Brazil?
Student:	Tell me about Peru.

FIG. 12.4. Excerpt from a dialogue between SCHOLAR and a student.

In *Knowledge-based tutoring: The GUIDON program* (1987) Clancey stated the program's objectives as being the following:

(a) evaluating the pedagogic usefulness of the knowledge possessed by MYCIN;
(b) discovering the additional knowledge an ITS would require compared to that of the original expert system;
(c) expressing the didactic strategies in a general manner, independently of the specific domain in which it would be employed.

The undertaking was so ambitious as to be destined to partial failure from the outset. Suffice it to say that point c by itself is tantamount to a complete pedagogical theory. Notwithstanding this, it has led to significant advances in achieving full awareness of what constructing a pedagogic system with pretensions of intelligence means. Clancey implemented three different systems exhibiting increasing levels of efficacy: GUIDON, NEOMYCIN and GUIDON2, which in their turn influenced the later versions of the expert system MYCIN. Since the philosophy has remained constant, we shall ignore the variants here and refer to a single virtual program.

The didactic strategy adopted by GUIDON is one of mixed initiative—both the student and the computer may take the initiative in exploring a sector of knowledge. The Socratic approach employed by SCHOLAR has been softened to make way for possible clarifications the user might require. In fact teaching geography is different from teaching a branch of medicine—the structure of the subject changes, as does the type of student. In actual practice, a method has been adopted consisting in the presentation of clinical cases so that the student is faced with realistic problems. The cases are then discussed at the end, so the student will learn how a doctor acts when faced with a problem of that nature. It should be borne in mind that MYCIN's performance in the field of bacteriology is the equivalent of that of a specialist. When a student's performance differs from that of MYCIN, GUIDON comes on the scene with the correct diagnosis. At any point in the program the student may ask for help or for a more extended explanation. An example is shown in Fig. 12.5.

In Chapter 3 we saw that the program TEIRESIAS, designed to clarify for doctors the line of argument employed by MYCIN, was an interesting failure, since the production rules utilised by the program differed too greatly from the way humans reason to enable an effective exchange of information. GUIDON represents a great advance in this sense, but in the long run it comes up against the same difficulties. MYCIN's knowledge is represented in too artificial a manner for

After presentation of relevant case data . . .

Student: I think that the organisms present are Diplococcus-pneumoniae, Pseudomonas-aeruginosa and Neisseria-meningitidis.

GUIDON: Good. The age of the patient provides evidence for concluding that Diplococcus-pneumoniae might be present (Rule 507). The fact that the patient has been seriously burned is good reason to suspect Pseudomonas-aeruginosa (Rule 578) . . . The age of the patient also leads one to suspect Neisseria-meningitidis (Rule 507). Do the facts that the infection was acquired while the patient was hospitalised and information about diseases to which the patient has been exposed support this conclusion?

Student: No.

GUIDON: Right . . . What facts about this case tell you that the organisms which might be present are Escherichia-coli and Klebsiella-pneumoniae?

Student: Help.

GUIDON: *Help is provided by GUIDON with a list of all relevant factors for the current goal.*

FIG. 12.5. An excerpt from a dialogue between GUIDON and a student.

students to be able to acquire it with any ease. The individual rules of expert systems are a far cry from the coherent corpus a human being needs in order to memorise and efficaciously employ knowledge. Production rule number 507, which was quoted in the dialogue reported in Fig. 12.5, was shown in Fig. 3.2.

We shall now see, in Fig. 12.6, how GUIDON makes improper use of it to reject a perfectly plausible hypothesis advanced by a student. In this case the student has followed a diagnostic pathway that is different from top-down research in the hierarchical tree of illnesses, but perfectly consistent with the knowledge expressed in rule 507. Indeed the patient's age is the causal factor in the rule, and the other conditions are present purely as control conditions. Since, however, MYCIN first establishes the control conditions and then takes age into consideration, GUIDON judges the student's explanation unsatisfactory. While MYCIN's *modus operandi* is efficacious when employed on a computer, it turns out to be inappropriate when taken as a didactic model since its mode of reasoning is too distant from that of humans.

After presentation of relevant case data . . .

GUIDON: What are the organisms which might be causing the infection?

Student: Neisseria-meningitidis.

GUIDON: What facts about this case tell you that Neisseria-meningitidis is one of the organisms which might be causing the infection?

Student: Age.

GUIDON: The fact that the age of the patient is 34 years is not evidence for the conclusion that Neisseria-meningitidis is one of the organisms which might be causing the infection. Do you have a question?

FIG. 12.6. A plausible hypothesis rejected by GUIDON.

Having pointed out this insurmountable obstacle, the successive stage in GUIDON led to a separation between the knowledge possessed by the system and the procedures utilising this knowledge, with a corresponding organisation of knowledge into explicit abstract categories. In this way, the student is no longer given rigid and abstruse rules on how to proceed. Rather, specific information is inserted in the context of a strategy of comprehensible reasoning, concentrating on the handling of a set of hypotheses according to general problem-solving principles. In designing the didactic module, Clancey drew upon protocols of doctors and students engaged in diagnostic problems, attempting to make the machine repeat the inferential pathway followed by the human being, and not vice versa. Finally, it was the task of a set of subprograms to keep the student model up to date, continuously monitoring the student's learning in an unobtrusive fashion and exhibiting a greater willingness to concede alternatives.

What GUIDON proved was the necessity to separate domain-specific knowledge from the way it is taught. Furthermore, for an ITS to really work, all its components must be designed in such a way as to respect human cognitive processes. With GUIDON, born in artificial intelligence and growing up in cognitive science, the history of pedagogic systems returns towards psychology—to teach human beings, we must know what human beings are like, not simply what machines are like.

The essential components of an intelligent pedagogic system (Fig. 12.7) therefore comprise: domain knowledge; a model of the student; a didactic module. To these three must be added the communication interface, a problem common to all artificial systems interacting with humans in natural language.

Domain knowledge

This is the module that contains the knowledge to be transmitted to the student. In the case of SCHOLAR it is the geography of South America. Other well-known ITSs deal with electronics, logic, medical diagnosis, telephony. The knowledge component is often called the expert, by analogy with expert systems, since its content is equivalent to that of an expert in the field. Different programs have used different techniques to store knowledge, ranging from highly static methods—such as semantic networks utilised in the earliest systems—to the more dynamic—such as production rules and mental models utilised in the most modern systems.

The expert acts both as the source of knowledge to be conveyed and as the yardstick by which the level of competence attained by the student may be measured. Since all ITSs aspire to interactiveness, then the first problem that must be solved is to establish what the student knows, in order to avoid boredom through presenting known material, or shock through presenting questions or topics that are too advanced. As the student proceeds along the learning pathway, progress made must be constantly monitored to ensure gradualism. Yet again, the only yardstick is the (presumably optimal) knowledge possessed by the system. The student is not supposed to know everything at the end. Indeed the program might aim at a highly diversified set of pedagogic objectives. Nevertheless, the system must be capable of transmitting all it knows to a willing pupil.

The way in which knowledge is stored in this module is not neutral with respect to the overall functioning of the ITS. It must guarantee constant updating of the didactic objective, interacting with the pedagogic component. Its inferential capacities are the determining factor to overcome a rigid relationship with the user. The latter must be provided with the opportunity to ask questions, and furnish answers. To understand user's replies the system must rely not on its direct knowledge, but on inferences based on that knowledge. The choice of

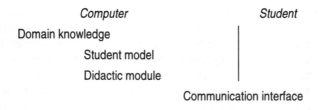

FIG. 12.7. Essential components of an intelligent pedagogic system.

how to represent knowledge also influences the manner in which that knowledge is transmitted to the student. Two major aspects must be borne in mind—transparency and psychological plausibility. With regard to the first aspect, expert modules may be classified into a spectrum varying from total opacity to total transparency. In opaque, or *black box* modules, knowledge representation cannot be accessed directly. Direct access can only be had to the results. Transparent, or *glass box*, modules, instead permit the observation and analysis both of knowledge and of the inferential steps that are taken. Psychological plausibility is connected to transparency, because the inferential chains followed by the ITS may differ quite radically from corresponding human reasoning, or may reproduce the latter within limits. The student will find it simpler to imitate a familiar mode of thought than one that is alien.

One problem ITSs have in common with every artificial system is that of the general knowledge the expert module is equipped with. There is an inverse relationship between general knowledge and domain-specific knowledge—the greater general knowledge is, the more limited domain-specific knowledge is, and vice versa. The ideal situation would be to have an extremely vast general knowledge base which contemporaneously serves as the foundation for different ITSs that are experts in specific sectors. Unfortunately, general knowledge is long and costly to formalise. Hence each ITS must reach a point of equilibrium grounded in the objectives the planners have set themselves. Let us take an example. In order for a medical ITS to explain nephritis, it must contain a set of notions concerning the kidneys and their functions in the human body. The same applies to a surgical ITS that has to explain the instructions for carrying out a nephrectomy. Is the information regarding the kidneys to be placed in the general information common to both ITSs or in the ITS-specific information? The concrete answer is that every case is decided on its merits, depending on the ambitions and the developmental potential of that particular system.

Student model

The module containing the student model furnishes a representation of how the student is modifying her knowledge and her behaviour as a result of the competence acquired. The need to provide this type of module emerged both because humans differ widely in their learning styles and because an ITS is used by humans possessing widely differing knowledge on the subject being treated. On what basis is a user model constructed? First, on explicitly requested information. The student may be asked questions such as:

[6]   With reference to topic x, how do you evaluate your competence?
       Zero   ☐
       Mediocre   ☐
       Intermediate   ☐
       Good   ☐
       Excellent   ☐

The student may be set crucial problems acting as a test enabling the ITS to evaluate the knowledge the subject has of the topic, or a human teacher may be asked to assign a student to a specific class, or, finally, the student may be asked what she personally prefers:

[7]   What level of explanation do you prefer?
       Elementary   ☐
       Intermediate   ☐
       Advanced   ☐

It is also possible to employ indirect information, which the ITS deduces from the type of question the student asks, from the type of answer the student gives and the type of mistakes committed. A good model of the student cannot be established once and for all, for the precise reason that it is expected that the student modify knowledge and behaviour as a result of the learning process. The model must therefore contain dynamic traits capable of representing internal developments. Ideally, at the end of a cycle of sessions with the ITS, the pupil should have reached a level of competence on the subject that is significantly different from her departure level. The only means the ITS has of capturing this increase is to keep a historical record of the student model.

The key to constructing a plausible model is not so much the correct answers provided as the wrong answers. The right answer represents the end of the thought process, and the only information it furnishes is whether the result the student has reached is correct or not. But the crucial point is: how did the student reach that result? Did the student follow a sensible path that will ensure a correct answer is always achieved? Or did luck play a vital part in solving this problem, but the student has adopted a potentially dangerous strategy to reach the solution, one that met with success in this particular case thanks to a series of fortuitous circumstances? The error furnishes information concerning the route followed by the subject, and hence is more informative than the correct answer when the aim is to plan learning in depth. Let us suppose an ITS on biology has the following exchange with a student:

[8]   ITS: Is a dolphin a fish?
       Student: No, it's a mammal.

The answer, which is one a CAI program would deem satisfactory, is insufficient. The student may have hazarded a guess, or she may have followed a mistaken inferential chain that can lay her open to potential errors. For example, the student may believe that a dolphin is a mammal because it is intelligent, whereas fish are stupid. But the size of the brain does not constitute a significant part of the definition of a mammal, which is grounded instead on reproductive features. When a right answer is given for a wrong reason, we can never modify any incorrect assumptions made by a student: In the case of 8, we should have lost an excellent opportunity for demonstrating to the pupil that the reply she furnished was based on an interesting correlation between type of reproduction and the evolution of intelligence. Brilliant errors lie at the root of deep learning, if they are diagnosed correctly. The challenge the student model thus launches us becomes: how does the student represent the problem situation, what knowledge does the student employ, how does the student reason on the basis of the data furnished? To provide a satisfactory answer to these questions, we must turn to a special sector of artificial intelligence, known as *qualitative reasoning*.

In his pioneering work, entitled "Naive physics I: Ontology for liquids" (1978) Patrick Hayes advances a revolutionary analysis of how people think liquids behave. For instance he explains how people predict what will happen if a cup of water is spilt on a table, with the water spreading over the table, dripping onto the floor and disappearing into some crack in the floor. Without carrying out the sophisticated mathematics quantitative physics has to utilise in order to comprehend the behaviour of the world around us and to capture its continual changes, humans employ an ingenuous, simplified form of physics, one that is nevertheless effective for their interaction with the world.

Let us reflect on how people imagine liquids progress from one state to another. A cube of solid ice, which may be taken with tongs, dissolves in a glass of whisky, mixing itself with the liquor; a kettle of water left on the burning gas will cause the water to evaporate and seemingly disappear. Even a child "knows" what is happening, but not because the child knows and applies the laws of physics. And if we adults had to teach the child those particular laws, it would be well to use as our starting point the child's own intuitive model of events.

Besides their pedagogic applications, the uses of a naïve knowledge of physics would be of benefit to human–machine interaction, a topic that will be taken up shortly. Anticipating one of the points to be made, however, how many mistakes are due to the fact that humans employ an incorrect model of the tools they use? Now, when the instrument is a television, then a blind eye can be turned to the fact. But if the tool is the sophisticated alarm system of a nuclear reactor or the complicated control panel of a helicopter, what then?

Hence the interest in the area of qualitative reasoning, an area that attempts to capture the way people reason with regard to simple physical domains, managing rather well, moreover, without deploying the complex tools of science. Examples that may be cited in support of the argument in addition to the naive theories on the physics of liquids, are those on mechanics and heat. In a fundamental work on the subject, *Mental models* (1983), Dedre Gentner and Albert Stevens make the point that if someone were tempted to think these domains are not that simple after all, then they might compare them with those really difficult domains (such as marriage!) which lack not only an accepted theory for making comparisons but also agreement as to what constitutes being an expert or a novice in the sector.

Qualitative models capture only a few of the aspects of the state of affairs they set out to explain. Sometimes their attention is focused simply on the description of a mechanism, but what they usually tend to do is to try to grasp the causal aspects of natural events or of the behaviour of a physical device. People use the electrical installations, the cooker, the coffee pot in a house successfully every day. Taking it for granted that people do not apply the theories of physics to predict that the light will go on by pressing a switch, what models do they employ to explain routine functioning? How do they reason when something does not go according to their expectations and requires them to act, as in the case when they turn on the stove and no gas comes out? Taking up Daniel Bobrow (1984), it may be said that the qualitative description of the behaviour of a system stems from a qualitative description of its structure, as schematised in Fig. 12.8.

A structural description corresponds to those variables characterising the system and to the interactions among them. It is based on the physical components of the system and on their connections. The behavioural description corresponds to the possible behaviours of the system. Finally, the functional description is couched in terms of the aims of the system itself or of some of its components. For example, the function of the safety valve in a pressure cooker is to prevent the cooker

Structural description ⟶ Behavioural description ⟶ Functional description

FIG. 12.8. Qualitative description of a physical system.

from exploding. Its behavioural description is that it makes pressure exit when it exceeds a given value by means of the physical structure of its components.

The fact that descriptions of the same event may be provided at different levels, but that each description is significant in relation to people's objectives, is what has induced scholars to adopt the concept of mental models. This particular use of the term is more limited than the meaning attributed to it in Chapter 5. It may here be defined as the mental representation of a physical system employed by a person or program to reason qualitatively about the processes involving that system. The model may be implemented on a computer, executing a sort of simulation of the events, for explanatory or predictive purposes.

Qualitative models place great emphasis on the causal aspects of the phenomenon being examined, and hence on the aspects concerning the way an event or a system evolves from one state to another. The fundamental notions of cause and time remain however elusive, one reason being the excessive attention devoted to the applicational aspects of the model, at the expense of those aspects that are more abstract and less tied to the specific context. This fact may be accounted for both by programmers of ITSs wishing to know how non-experts reason about a specific domain, and by the potential use qualitative models may be put to by expert systems, when such models are effective. In the volume edited recently by Johan de Kleer and Brian Williams (*Qualitative reasoning about physical systems II*, 1991), the interested reader will find the pioneering work in this area.

Didactic module

For this module to be really effective it should contain an exact model of the processes that the student it is interacting with employs to comprehend and learn. Unfortunately, the current state of the art in pedagogy is not sufficiently advanced to furnish precise details on these subjects, with the result that the pedagogic knowledge of ITSs is somewhat approximate and not particularly reliable. The pedagogic strategy adopted by the system is the result of the integration between domain knowledge and the student model. With regard to this topic Wenger distinguishes between a global level and a local level.

At the global level, didactic decisions influence the sequence of the specific teaching episodes. Using the information it already possesses on the general knowledge of the user and on the user's specific needs, the didactic module adapts the choice of topics and their order of presentation to the type of person utilising the program. At the local level, the module decides when it is necessary to intervene, whether the student has to be interrupted or not during an activity, and how much material has to be presented in a single work session. This includes guiding the student in the execution of the activities, explaining the phenomenon being treated, and correcting the student's mistakes.

All these decisions should be supported by solid knowledge of how to teach, and by a sufficient degree of flexibility permitting adaptation to different learning styles. This is exactly what an expert human teacher knows how to do intuitively, but that no one has yet managed to codify. ITSs may be distinguished along a broad spectrum ranging from diagnostic modules at one extreme, to mixed initiative modules, to training modules at the other extreme. *Diagnostic* modules (such as SCHOLAR) are so called because they emphasise what point the student has reached in the learning process by analysing the answers the student furnishes. Such modules maintain a constant check over each part of a session, adapting to the answers provided by the student but guiding the student along a specific track by means of the questions it sets.

In between the two extremes come the *mixed initiative* modules (such as GUIDON) where control is shared between system and student. Both parties may provide questions and answers. The computer can furnish efficacious answers to questions set it by the user, and hence can accept deviations from its initial didactic plan.

At the other extreme we find *training* modules. These leave the initiative totally to the student. They are not programmed to follow an exact didactic plan to teach a pre-established number of things within a certain length of time. Their objective is to encourage the acquisition of motor or thought abilities by involving the user in an activity resembling a game played on the computer. The training module is not intrusive, it rarely gives a lesson, and it directs learning exclusively by making the relevant modifications to the environment presented to the student. The Socratic teaching strategy, the mixed initiative strategy and the free strategy follow different routes, none of which can be defined as optimal *a priori*. Their respective validity varies according to whether the pedagogic system aims at teaching meteorology or piloting a jet plane.

The communication interface

Most didactic activity takes place through the medium of language—the system asks a question, the student answers, or asks other questions, and so on. ITSs must succeed in filtering their knowledge through the modes of communication available to them, and those usually consist of a video, a keyboard and perhaps a set of instruments the student can manipulate but which are under the computer's control.

This brings us up against all the problems encountered in human–machine communication, when we are far from possessing sufficient knowledge of the principles of communication in general. The following pages are devoted to these topics. Suffice it to say for now the ITSs oscillate between imposing a rigid artificial language, which is difficult to learn and use but which avoids any possible ambiguity in communication, and free natural language, with which the student feels far more comfortable but which causes comprehension difficulties for the computer, creating the risk of misunderstandings and consequently disillusionment on the part of the user.

HUMAN–MACHINE INTERACTION

The rise in the use of the computer as a tool at the work place is parallelled by the increase in the attention devoted to make the computer more friendly—easier even for non-experts to use. The direction we are moving in is one in which it is no longer the user who is obliged to put up with the difficulties inherent in the utilisation of the computer, such as having to memorise an abstruse set of commands, and above all having to learn complex programming languages. It is an opposite direction, one in which it should be the machine that is completely at our service, by becoming simple to use and able to talk in natural language.

This change in perspective corresponds to the fact that the identity of the typical user has changed over time. Up to the time of the explosion of the personal computer, the type of person employing this machine was an expert who had a good knowledge both of hardware and of software, otherwise that person would have found it impossible to work with the machines in general use up to the 1970s. Software designers could count on the fact that those employing the program were people with similar training and mentality to the designers themselves, with independent knowledge that could integrate the meagre and cryptic instructions contained in the instruction manual. The moment the user becomes a person—a doctor, a lawyer, an architect—who has no knowledge of computers, then the designer can no longer count on independent knowledge. The non-expert user employs the computer without knowing how it works, just as someone uses a washing machine perfectly well without having the slightest idea of what happens once the dials are set, and he or she would certainly not be able to repair the spin drier were it not to spin dry.

The usability of computers has a paradoxical feature: progress in computer technology makes increasingly complex systems available to users who have increasingly less knowledge of the subject. The human–machine interface thus acquires greater and greater importance if the computer is to be put to good use outside research centres. There is no point in producing an object with excellent properties if no one but the constructor is capable of exploiting it to the full. In this particular branch of computer science, which acts as a bottle-neck slowing down the spread and intelligent use of computer science, psychologists have greater expertise than engineers. Cognitive science has therefore an important social function in reconquering the privilege for humans of being served by intelligent as well as other types of machines. Two problems will now be identified—that of the use of the machine, and that of communicating with the machine itself. Finally, we will briefly comment on the social impact of the type of interaction we have hypothesised.

A machine for people

To know how to build good computers one should know how those people who will use the computers are built. While waiting for our psychological knowledge to increase, it is nevertheless possible to avoid at least the worst sources of inconvenience to the user. General consensus already exists on the elementary commonsense principles software designers should employ.

The principle of least surprise.   The actions taken by the computer should be predictable on the part of the user of the system. The more the mental model the user creates of the system corresponds to the actual behaviour of the machine, the more the machine can be employed with familiarity and efficaciousness, without inhibitions or superstitions.

The principle of retractability.   Each and every action the user calls on the computer to carry out may be retracted if the effects it produces do not correspond to the desired reactions. Hence if a command exists capitalising all letters, then a command must also exist that will convert all those letters into small letters. If there is a command taking one to the beginning of a text, then there must be one that takes one to the end of the text. Furthermore, it must be possible to countermand any order if the user has changed her mind.

The principle of equity.   With regard to actions bringing about definitive changes to the state of the system, the amount of time required to write the commands causing the change must be proportional to the magnitude of the change to be effected. Hence the command to delete a single line has to be simple and rapid, the command to delete an entire file has to be longer and contain a request for confirmation, and where a command for deleting all files on the disk exists, it should require a complex operation and should be executed only after a number of confirmations have been received, in order to avoid the disastrous consequences ensuing from an accidental typing error.

Two types of error may occur in human–machine interaction, one ascribable to each of the two parties in the interaction. Errors attributable to the machine do not concern us here. This sector is of great interest to computer science which makes enormous efforts to impede them taking place, with technical solutions which are continuously evolving. Bad design, bad production techniques, malfunctioning, unforeseen behaviour—engineers and computer scientists try to avoid every type of failure, whether the device be the automatic switch on the boiler or the automatic pilot on a plane.

Human errors are, instead, of greater interest to us, since preventing their occurrence is closely connected to how the brain works. Donald Norman (1986) classifies errors into two categories.

Involuntary slips

Slips are those errors not caused by conscious intention—the person intends doing one thing but he does another through the erroneous execution of the intended action. For instance, the driver wants to reverse but he puts the car into first gear; or the user wants to copy a file but deletes it instead.

Equipment design can do a great deal to increase or decrease the probability of involuntary errors. The most significant case is that of the computer keyboard which computers are guilty of inheriting exactly from the mechanical technology of the typewriter. As is common knowledge, the sequence of letters on the keyboard is such that it positively encourages making slips. Paul David (1986) explains that the reason for this treacherous design is that the earliest typewriters had to slow down typing speed to avoid the individual levers becoming entangled. It thus seemed a good idea to put the letters in the most inconvenient and unreasonable position, in order to oblige the typist to type as slowly as possible. Thanks to mechanical improvements and to the introduction of electronics, what had already been an insane solution at its conception has acquired an almost surrealist quality, unrivalled for the unhappiness it causes to the individual user and for the world-wide extension of the damage caused.

Even though it is hard to imagine someone today cynically setting out deliberately to encourage human error, designers should nevertheless devote some effort to foreseeing slips and trying to prevent them occurring. It is exactly their unintentional nature that renders them so difficult to inhibit voluntarily.

Mistakes

Mistakes are actions intentionally carried out by the user not knowing or not realising that these conflict with his long-term goals. For example, an inexperienced user might fail to save the file being worked on at regular intervals if the machine does not prompt him, and on some computers this entails running the risk of losing all the work done, if another user with higher priority needs more space. Or the security person at the control centre at a nuclear power station could switch off an alarm signal in the belief that a local circuit is faulty and that no breakdown has occurred, when instead the circuit is working perfectly and the breakdown really has taken place. This has been the official explanation furnished for at least a couple of extremely serious accidents at American and Russian nuclear power stations.

Avoiding mistakes is possible precisely because of their intentionality. The method to employ is that of ensuring the user has a working model of the machine that is reasonable in relation to the use that must be made of the machine itself, thereby enabling the user to predict and comprehend what that machine is capable of doing in its sphere of competence.

It is not essential that the user's mental model be realistic and the best possible in an abstract sense. Not knowing exactly how a computer works is not important if the user is able to use the machine efficaciously. If a fault arises, the user can always call in the technician. To utilise the machine in an optimal fashion, however, the user's mental model must bear a significant correspondence to the way the machine actually works, though limited to the level that is pertinent to the user's objectives. Thus a first rate flight controller may be blissfully ignorant of the nature of the silicon chip, but he has to possess a valid working model of the radar and the computer that processes the data received from airplanes.

It must further be remembered that no single mental model exists, either of a simple object such as a book, or even less so of a complex object such as a computer—it all depends on what purpose it is built for. What is important is that the model should really correspond to the competence of the machine; in other words, the competence should be neither overestimated or underestimated. The person at the control panel who ignores a warning signal thinking that the system is working badly may cause disaster. However, the same disaster could be brought about by assuming the machine is capable of understanding and sharing the long-term intentions of the user, and that it will therefore take those measures that are obvious to the human being but not to the computer.

Talking about bad human–machine cooperation, we may recall an important episode in the Falklands War between Great Britain and Argentina. A British destroyer was sunk by a French-built missile launched by the Argentinians, since the ship's defensive system failed to recognise the missile as being hostile. Indeed, it had been classified as being part of the equipment of an allied nation. The computer failed to raise the alarm, giving credit to the data stored in its memory, and the operator trusted the computer.

In order to avoid such uninspiring results in cooperation, current attention to human factors in computer science coincides largely with the attempt to ensure users have a model of the machine adequate to their objectives. Readers interested in the subject may refer to *Designing interface* (1991) edited by Carroll, and to the more technical *Handbook of human–computer interaction* by Helander (1988).

Human–machine communication

Communication between humans and machines comes about today almost exclusively through the medium of language because artificial systems of perception have not yet been connected up to the computers at present in use, nor will they be in the foreseeable future. Gestures and other types of non-verbal messages remain foreign to machines even if new communication environments like virtual reality are now developing. This limit lends extra weight to the language interface, and it is worthwhile trying to extend what we saw with regard to human communication in Chapter 4, to the case in which one of the two participants in a conversation is a machine. To do so, I will again refer to the cognitive pragmatics of Airenti, Bara, and Colombetti (1993a,b).

To be able to state a proper conversation is taking place, there has to be on the part of the participants reciprocal comprehension of their respective mental states, as well as mutual agreement on the relationship existing between them. In particular, it has to be possible to achieve the transmission and comprehension of the respective deep and conversational intentions of the interactants. In human–machine communication the decisive intentions are exclusively those of the user, since machines do not possess intentions. In addition, the sphere in which the conversation takes place is limited to a precise topic, namely that in which the computer is competent.

Human–machine communication may range between two extremes, from a language that is totally artificial to one that is totally natural. The former is the norm today—the user interacts with the machine through a formal language that is syntactically and semantically unambiguous. The brunt is borne completely by the human being, who is obliged to learn an abstract language from scratch, a task which requires an enormous investment in terms of time and energy. However, once the learning stage is overcome, a high level of interactional efficiency is achieved. The second type of communication, by means of natural language, would presuppose the machine possessing linguistic competence on a par with humans. This would do away with the learning period the user must go through before being able to interact with the computer. The consequence would be to provide access to the machine even to those people having no time or lacking the capacity to learn formal languages.

It is obvious what the optimal solution would be, had we the choice. However, no machine actually has the ability to interact employing an exclusively natural language. At best, we have programs capable of comprehending a limited number of simple expressions, but lacking adequate pragmatic comprehension that would enable the computer to work out the deep intentions of the user. It is precisely this limited degree of comprehension that besets interaction with a machine employing an intermediate language with pitfalls—when humans are not forced to express themselves in an artificial language, but can make use of their own language instead, they tend to attribute communicative capacities to the machine equal to those of a person. Stated differently, humans possess a communicative competence which, when employing natural language, leads them to attribute an identical level of competence to their partner. For this reason, if the computer appears to be capable of carrying on a normal conversation, it is immediately attributed with a capacity for linguistic comprehension that is equivalent to that of a human partner, and in particular with possessing intentions and the ability to reason about the intentions of others.

This type of attitude is compatible with the Chomskian approach to language, which assumes an innate linguistic competence which for this reason is activated to the full in human beings, no intermediate degrees being possible. To illustrate the concept, we shall analyse conversation [9], where a user U tries to obtain a simple piece of information on the time of a flight from computer C:

[9]   U: Do you know when the first flight for New York leaves?
       C: YES.
       U: When?
       C: WHEN WHAT?
       U: When does the plane leave?
       C: WHICH PLANE?

The computer answers the first question correctly, but literally. That is, it does not realise that the question is an indirect speech act corresponding to the request for information regarding the departure time of the plane for New York. The user's subsequent questions contain implicit reference to the previously mentioned plane, without however making explicit reference to it, and are therefore incomplete and incomprehensible to the computer program, despite the fact that it is capable of understanding each utterance separately. Let us now analyse the following conversation:

[10] U: Do you know when the train from Chicago arrives?
      C: AT 12.25, AT PLATFORM 9. IT IS 30 MINUTES LATE ON ITS SCHEDULED TIME OF ARRIVAL.

In this case the computer has understood the indirect speech act concerning the first train due to arrive, and cooperates fully with the user, completing the information with the platform the train is due to arrive at and news of the delay. It is unlikely the user will know the platform number, since he did not even know the time of the train, and knowledge of only the train time will be insufficient in the most obvious cases the information is required, such as catching the train or meeting someone when they arrive. The added specification of the delay is cooperative, aiming at helping the user to avoid wasting time. Here it may be said that human and machine really are communicating. Naturally, the user could have put the request to the machine in a complete and direct fashion, as in:

[11] U: Tell me at what time and from which platform the first train for Pittsburgh leaves.
      C: AT 9.44, FROM PLATFORM 7

This is exactly what humans cannot be expected to do spontaneously, unless they are forced to employ a formal language obliging them to specify all the necessary information in an unambiguous manner.

We have already seen that the user builds a functional model of the machine representing the type of performance the machine is expected to furnish within the particular applicational domain. Clearly, the initial functional model may be inadequate and a version having greater validity may be learned precisely through interaction. In the linguistic sphere, the most important aspect of the functional model is that this determines the number and content of games—both conversational and behavioural—the user believes may be carried out with the computer.

With regard to conversational games, the user may assume that if the machine is able to converse, then it is also cooperating with the user in accordance with Gricean maxims (see Chapter 8). We shall now analyse conversation 12, in which the machine does not exhibit linguistic problems, but does not have the norms of cooperation incorporated:

[12] U: Put me through to the head of maintenance.
　　　C: HE'S NOT IN.
　　　U: Put me through to his secretary.
　　　C: SHE'S NOT IN EITHER.
　　　U: Someone from maintenance then.
　　　C: NOBODY FROM THE MAINTENANCE DEPARTMENT IS IN TODAY.
　　　U: Is there a meeting or a strike or something?
　　　C: NOT THAT I KNOW OF.
　　　U: Well why is no one in at this time?
　　　C: THE FIREMEN EVACUATED THE BUILDING AFTER THE BLAZE LAST NIGHT.

Behavioural games are knowledge structures prescribing the actions each actor has to accomplish within a given context. For the game to be played, it must be known to both partners, and the conditions in which it applies agreed on. The game is activated by a specific intention, which in turn derives from the motivation of the actors. In the type of human–machine interaction that can be achieved today, the sole motivation the user may attribute the machine is what we may term the good professional, and it is therefore on this assumption that conversation is based. That is, to obtain the cooperation desired, the user will start out from the presupposition that the specific session is dominated by the intention on the part of the machine to act as a good professional: cooperative, competent, informed, reserved etc.

To obtain the desired answers, the user does not expect to have to call upon other forms of motivation, appealing for example to generosity or pity, in as much as deep goals of generosity or pity clearly cannot be attributed to the machine. In addition, the specific games that are acceptable in the context of the use of the machine should be limited to the field in which the machine is recognised as being experienced. This is analogous to what happens in interaction between human clients and professionals, where requests are confined to the expert's domain of competence. Indeed, in the legal professional sphere, one would refrain from asking a medical opinion from a lawyer.

In support of the thesis that the use of natural language leads to the attribution of exclusively human intentions, we may recall the extreme case of ELIZA, the program that simulates a psychotherapist. We have seen that its users did attribute stable intentions to ELIZA that were appropriate to a human therapist, attempting to arouse in the machine's reactions of likeableness or affect, solely on the grounds of the seeming richness of the linguistic interaction. The inference was of the kind "If it understands and produces language like a human being, then it must have the cognitive and emotional structures of a human being". But such a highly misleading functional model is unacceptable for any computer system whose goal is not simply amusement.

In any communicative process the performances achieved are limited *a priori*. This does not stem from any limitation to competence which, as we have seen, may only be exercised in an integral manner, but by restrictions to the area of knowledge to which those processes constituting competence are applied. In other terms, competence in possible games is always maintained, but the number of games actually playable is limited by the situation. This is true both in the case of communication among humans (a nurse cannot give orders to a surgeon) and in human–machine communication.

One precise prescription for human–machine interface designers is therefore not to arouse the desire in users to attribute to the machine behavioural games the machine is unable to handle. On the contrary, the user has to be guided to maintain interaction within the bounds of the aims of the conversation and within the limits of the computer being employed. If the partial use of natural language can lead the user into making errors, then it is far better, bearing in mind the objective is optimal interaction, to reject the employment of natural language in favour of artificial language, perhaps making it as simple as possible to learn, but rendering the nature and limits of type of relationship the machine can act in transparent. It will be possible to introduce natural language only when it becomes manageable by the computer without requiring the aid of any type of trick whatsoever.

The social computer

Another aspect of human interaction which could benefit from a simulative approach concerns social interaction among several people, or several machines, instead of only two partners which is the case we have examined so far. So-called *distributed artificial intelligence* (DAI) deals with how a problem may be solved in some cases directly through the exchange of messages among several participants, when all that is required is the pooling of knowledge already available to the participants individually but not yet structured at the level of the social system. In other cases the collective problem is first decomposed into various sub-problems which are then assigned to the various participants so that the general solution consists of a cooperative effort in which local solutions are pooled together. Although the reader may be referred to Huhn's work entitled *Distributed artificial intelligence* (1987) for the details, it is important to make a number of general points here.

The simulation of social behaviour means raising the difficulties encountered in individual simulation n-fold, if the participants are not to be reduced to the simplest of systems, devoid of independent intentionality. The basic idea is impeccable for its linearity and it may be summarised as follows: once a single system simulating the mind has been constructed, allowing it to communicate with one or more other versions of the same system will yield two or more artificial participants interacting between themselves, thereby enabling computer simulation not only of individual systems, but also of the relationships between systems.

As our knowledge of the mind broadens, there can be no doubt that it is important that this knowledge be utilised by all disciplines, from economics to sociology, which take the human beings as their departure point, then to analyse both the microrelationships existing between them, as is at bottom the communication between two persons, and the macrorelationships they enter into, which involves groups of individuals. A nonconformist model of social contexts as frameworks for cooperation and communication has been proposed by the social psychologist, Giuseppe Mantovani (1994).

The proposals we have examined represent the first steps towards a sort of computational sociology on a constructive and simulative basis and not just on an analytical basis, and the results might in principle turn out to be as rich as those achieved by cognitive science compared to traditional psychology.

Artificial sociology might become a specialised branch of computational sociology, with the assignment of the task of drawing up the rules of social interaction between machines, a possibility that is not remote. We have seen that an intelligent pedagogic system may call on an expert system for help, and when the studies on human–machine interfaces have made further progress then artificial systems able to communicate with humans will be available, at least in certain domains. The possibility of connecting up two computers arises naturally, together with the need to establish limits to communication, and also to interaction, between one machine and another.

CHAPTER THIRTEEN

Conclusion

At the outset of this work we asked ourselves whether cognitive science represented the best approach to the mind possible today. Providing an answer to the question has required an examination of the various themes treated in the course of this book. We are now in a position to take up the original question, having gathered the information that makes a more accurate evaluation possible.

What is cognitive science worth? As all the obscure points and uncertainties of the preceding chapters demonstrate, cognitive science is certainly no panacea for the scholars of human life, nor is it a sort of ultimate, final, unbeatable method for the analysis of the mind.

There are too many weak points, and I am not referring here to those aspects that have not yet been investigated or comprehended, since science is by definition an activity that is constantly in progress, one in which no one can hope to have the last word. It is therefore acceptable and natural that a field such as that of mental life should be made up of a greater number of aspects that are unknown compared to the number known. The important weak points are, instead, connected to the peculiarities of the method adopted by cognitive science, since it may be presumed that they will not be solved by linear progress in the acquisition of scientific knowledge. Artificial intelligence is a powerful tool, and like all efficacious instruments it also conditions the intentionality of its users—if you give me a sharp sword, then the desire to cut comes to me, but if you give me an elegant pen, then it is the wish to write that is aroused. Scientists are the living proof of Bruner's assertion that the infantile phase of enactive representation is always present in the human being.

In particular, the computational method has extreme difficulty in maintaining the global aspects of mental activity. One specific phenomenon may be simulated without undue hardship, but the interaction between different functions is another matter. The intrinsic limits of the computational method applied to the study of humans lie in its natural predisposition to reproduce phenomena analytically, and in the insurmountable hardship such synthetic reproductions create.

Scholars studying the human being have adopted artificial intelligence, but the key to understanding the mind lies in comprehending "stupid" performances, those full of errors, typical of living beings. Apparently banal, everyday phenomena are the really interesting parts. What is perfect and impeccable may safely be left to computer scientists. Nature builds adults from children, and what is really surprising is the capacity the latter possess to make creative mistakes, to misunderstand, to fail to comprehend what is obvious. And in so doing they illuminate the non-obviousness of obviousness, illustrating how what seems natural is instead something constructed. This is how cognitive science should produce a developmental reconstruction of adult functions—by means of programs capable of making the mistakes children make, and of learning something useful from any possible experience.

There is no doubt that computer simulation cannot achieve such complex reconstructions today. Nevertheless the decision as to what the goals of research are to be should not be left to the machine. If I want to write and I am given a sword, then I had better make a new request rather than allow myself to be convinced that what I really wanted to do was cut a sheet of paper. Employing artificial intelligence as a tool in the study of humankind does not mean applying it in an uncritical fashion—cognitive science maintains the privilege of not being satisfied, of complaining when a pair of tennis shoes are provided for a gala evening. Technology is at the service of science, and not vice versa.

Cognitive science is thus the best of all possible errors today. New, unpredictable scientific revolutions will provide even more efficacious methods in future, but the results already obtained are of a sufficiently high quality to be worthy of being included in any future mental science. Optimism on this point stems from the reflection that the real computer revolution is still to come—we may expect it when the people who were born at the same time as the personal computer start out on their scientific careers, hence at the very beginning of the third millenium.

Returning to the present, it may be stated with regard to any given mental state that it is not possible to decide *a priori* if it can be reproduced on computer or whether the attempt is destined to failure. If failure is the outcome, then only the intuition of scholars will be able to attribute the reasons to a surmountable difficulty or to a constraint that cannot be eliminated. The essential condition to achieve success is that the program reach a level of complexity comparable to the phenomenon under scrutiny, without simplifying it in any way. The fact that the necessary complexity is often well beyond the present development of cognitive science may be comforting to some and represent a challenge to others, but this does not alter the essential significance of the computational project.

Cognitive science does not cover every area of knowledge—some subjects we know a lot about (perception, thought), many others we know something about (language, emotion), others little or nothing (learning, development). But not investigated empirically does not correspond to unknowable. A precise distinction must be made between:

What has been achieved
Entire series of results have been obtained, and these may be judged of relatively high quality, especially when compared with those obtained by non-computational disciplines. This supports the thesis that to comprehend the human mind by far the most important tool today is still artificial intelligence—in both its forms, social and parallel—despite all its bothersome limits.

What can be achieved
What cognitive science has not yet done but can do with the means available to it. To this category belongs research currently being undertaken by cognitive scientists. Some problems seem as if they can be solved, and a group of scholars are investing human and technological resources into finding the solutions. As I pointed out earlier, that a problem can be solved is demonstrated only by the actual finding of the solution. A more brilliant and obstinate researcher may succeed where another colleague has failed, or he may simply provide a more expensive and spectacular failure. When one is engaged in science, future is a word that means hope, not certainty. The path trodden by cognitive science is nevertheless the most promising one available today, and advances in knowledge seem to be proceeding at a good rate.

What can be known
This refers to that which cognitive science has not done but what another science employing other methods might achieve. Some problems exist that have been recognised as important but that have escaped all attempts at dealing with them—perhaps our techniques are

unsuitable. Not even the cooks of Lucullus could pull out a joint of roast from two eggs and a drop of oil. In the near or distant future a revolutionary method might make simulations we find difficult today, such as those on synthesising and development, quite straightforward. Or perhaps a scientific revolution which goes well beyond the bounds of our imagination could change the type of approach, bringing with it excellent new results. Indeed the object of this book, a method allowing mental processes to be artificially reproduced, would have seemed madness or a dream only 50 years ago. Where cognitive science is destined to fail, there is absolutely no doubt that a better science—one equipped with more effective techniques—will take its place, proceeding along the road to knowledge.

What cannot be known

Some things are impossible to do. Whoever said humans can do everything, understand everything? Ethologists have pointed out the biological limits of the various animal species, but they are unable to tell us with equal accuracy what the limits of *Homo sapiens* are. A scholar from Venus with three times our intellectual capacities might perhaps explain to us we have no hope of ever comprehending a given phenomenon, just as we can explain to a gorilla that it will never be able to understand the theory of universal gravity, even though the primate is unquestionably aware of the fact that bananas falling from a tree finish up on the ground.

No scientist has ever received assurances that she is intelligent enough to comprehend everything about herself and about the surrounding world—nature has no obligations to us. That something exists beyond our limited capacities for comprehension is so probable that it may be considered a certainty. This something—unless another intelligent system, built by nature or by humans themselves, understands it and renders it intelligible to us—is destined to remain unknown, always beyond the bounds of our explanatory frameworks.

However, we can but try—this is what science is for.

References

Abelson, R.P. (1968). Computer simulation of social behavior. In G. Lindzey & E. Aronson (Eds.), *Handbook of social psychology*, Vol. II. Reading, MA: Addison-Wesley.

Agre, P. & Rosenschein, S.J. (Eds.) (1995). Special volume on computational research on interaction and agency, Parts 1 and 2. *Artificial Intelligence, 72–73*.

Airenti, G., Bara, B.G., & Colombetti, M. (1993a). Conversation and behavior games in the pragmatics of dialogue. *Cognitive Science, 17*, 197–256.

Airenti, G., Bara, B.G., & Colombetti, M. (1993b). Failures, exploitations and deceits in communication. *Journal of Pragmatics, 20*, 303–326.

Allen, J.F. (1984). Towards a general theory of action and time. *Artificial Intelligence, 23*, 123–154.

Allen, J.F. (1987). *Natural language understanding*. Menlo Park, CA: Benjamin/Cummings.

Anderson, J.R. (1983). *The architecture of cognition*. Cambridge, MA: Harvard University Press.

Appelt, D.E. (1985). *Planning English sentences*. Cambridge: Cambridge University Press.

Austin, J.L. (1962). *How to do things with words*. Oxford: Oxford University Press.

Baddeley, A.D. (1986). *Working memory*. Oxford: Oxford University Press.

Baddeley, A.D. (1990). *Human memory: Theory and practice*. Hillsdale, NJ: Lawrence Erlbaum Associates Inc.

Baillargeon, R. (1987). Object permanence in 3.5 and 4.5-month-old infants. *Developmental Psychology, 23*, 655–664.

Baillargeon, R. (1991). Reasoning about the height and location of a hidden object in 4.5 and 6.5-month-old infants. *Cognition, 38*, 13–42.

Bara, B.G. (1984). Modifications of knowledge by memory processes. In M.A. Reda & M.J. Mahoney (Eds.), *Cognitive psychotherapies* (pp. 47–64). Cambridge, MA: Ballinger, Harper & Row.

Bara, B.G., Bucciarelli, M., & Johnson-Laird, P.N. (1995). The development of syllogistic reasoning. *American Journal of Experimental Psychology, 108,* 157–193.

Bara, B.G., Bucciarelli, M., Johnson-Laird, P.N., & Lombardo, V. (1994). Mental models in propositional reasoning. *Proceedings XVI Conference of the Cognitive Science Society.* Hillsdale, NJ: Lawrence Erlbaum Associates Inc.

Bara, B.G., Carassa, A., & Geminiani, G. (1989). Il ragionamento: dal formale al quotidiano. In R. Viale (Ed.), *Mente umana, mente artificiale.* Milano: Feltrinelli, 299–326.

Bara B.G., Carassa A., & Geminiani G. (in press). Causality by contact. In A. Garnham & J. Oakhill (Eds.), *Mental models and interpretation of anaphora.*

Bara, B.G., & Guida, G. (Eds.) (1984). *Computational models of natural language processing.* Amsterdam: Elsevier Science.

Barr, A., Cohen, P.R., & Feigenbaum, E. (1982). *Handbook of artificial intelligence.* Los Altos, CA: Kaufmann.

Bartlett, F.C. (1932). *Remembering.* Cambridge: Cambridge University Press.

Bartlett, F.C. (1958). *Thinking: An experimental and social study.* London: Allen & Unwin.

Bateson, G. (1972). *Steps to an ecology of mind.* New York: Chandler.

Bateson, G. (1979). *Mind and nature: A necessary unity.* New York: Chandler.

Bechtel, W., & Abrahamsen, A. (1991). *Connectionism and the mind: An introduction to parallel processing in networks.* Oxford: Blackwell.

Bell, D., Raiffa, H., & Tversky, A. (Eds.), (1988). *Decision making: Descriptive, normative, and prescriptive interactions.* Cambridge: Cambridge University Press.

Bergson, H. (1889). *Essai sur les données immédiates de la conscience.* Paris: Felix Alcan.

Block, R.A. (1990). *Cognitive models of psychological time.* Hillsdale, NJ: Lawrence Erlbaum Associates Inc.

Bobrow, D.G. (Ed.) (1984). Special volume on qualitative reasoning about physical systems. *Artificial Intelligence, 24,* 1–3.

Boden, M.A. (1988). *Computer models of mind.* Cambridge: Cambridge University Press.

Boole, G. (1854). *An investigation of the laws of thought, on which are founded the mathematical theories of logic and probability.* Reprint (1951): New York.

Bornstein, B. (1963). Prosopagnosia. In L. Halpern (Ed.), *Problems of dynamic neurology.* Jerusalem: Hadasseh Medical Organization.

Bower, G.H. (1981). Mood and memory. *American Psychologist, 36,* 129–148.

Bowlby, J. (1969). *Attachment and loss: I—Attachment.* London: Hogarth Press.

Bowlby, J. (1973). *Attachment and loss: II—Separation: Anxiety and anger.* London: Hogarth Press.

Bowlby, J. (1980). *Attachment and loss: III—Loss: Sadness and depression.* London: Hogarth Press.

Bresnan, J. (1978). A realistic transformational grammar. In M. Halle, J. Bresnan, & G.A. Miller (Eds.), *Linguistic theory and psychological reality.* Cambridge, MA: MIT Press.

Bridgman, P.W. (1927). *The logic of modern physics.* New York: Macmillan.

Broadbent, D.E. (1958). *Perception and communication.* Oxford: Pergamon.

Broca, P. (1861). Remarques sur le siège de la faculté du langage articulé, suivies d'une observation d'aphémie (perte de la parole). *Bulletins de la Société Anatomique, XXXVI*, 330–357.

Bruce, V., & Green, P.R. (1992). *Visual perception: Physiology, psychology and ecology*, 2nd ed. Hillsdale, NJ: Lawrence Erlbaum Associates Inc.

Bruner, J.S., et al. (1966). *Studies in cognitive growth*. New York: Wiley.

Bruner, J.S. (1983a). *In search of mind: Essays in autobiography*. New York: Wiley.

Bruner, J.S. (1983b). *Child's talk: Learning to use language*. New York: Norton.

Bruner, J.S., & Goodman, C.L. (1947). Value and need as an organizing factor in perception. *Journal of Abnormal and Social Psychology, 42*, 33–44.

Bruner, J.S., Goodnow, J.J., & Austin, G.A. (1956). *A study of thinking*. New York: Wiley.

Buchanan, B.G., & Feigenbaum, E.A. (1978). DENDRAL and Meta-DENDRAL: Their applications dimension. *Artificial Intelligence, 11*, 5–24.

Cannon, W.B. (1927). The James–Lange theory of emotion: A critical examination and alternative theory. *American Journal of Psychology, 39*, 106–124.

Caramazza, A. (1986). On drawing inferences about the structure of normal cognitive systems from the analysis of patterns of impaired performance: The case for single-patient studies. *Brain and Cognition, 5*, 41–66.

Carbonell, J.R. (1970). AI in CAI: An artificial intelligence approach to computer-assisted instruction. *IEEE Transactions on Man-Machine Systems, 11*, 190–202.

Carroll, L.M. (Ed.). (1991). *Designing interface: Psychology at the human–computer interface*. Cambridge, MA: MIT Press.

Cesa-Bianchi, M., Beretta, A., & Luccio R. (1970). *La percezione*. Milano: Angeli.

Chess, S., & Thomas, A. (1987). *Know your child: An authoritative guide for today's parents*. New York: Basic Books.

Chomsky, N. (1957). *Syntactic structures*. The Hague: Mouton.

Chomsky, N. (1965). *Aspects of a theory of syntax*. Cambridge, MA: MIT Press.

Chomsky, N. (1980). *Rules and representations*. New York: Columbia University Press.

Christianson, S.A. (Ed.) (1992). *The handbook of emotion and memory: Research and theory*. Hillsdale, NJ: Lawrence Erlbaum Associates Inc.

Clancey, W.J. (1987). *Knowledge-based tutoring: The GUIDON program*. Cambridge, MA: MIT Press.

Clark, A. (1989). *Microcognition: Philosophy, cognitive science, and parallel distributed processing*. Boston, MA: MIT Press.

Clark, H.H. (1992). *Arenas of language use*. Chicago: Chicago University Press.

Cohen G. (1989). *Memory in the real world*. Hillsdale, NJ: Lawrence Erlbaum Associates Inc.

Cohen, P.R., Morgan, J., & Pollack, M.E. (Eds.) (1990). *Intentions in communication*. Cambridge, MA: MIT Press.

Cohen, P.R., & Perrault, C.R. (1979). Elements of a plan based theory of speech acts. *Cognitive Science, 3*, 177–212.

Colby, K.M. (1975). *Artificial paranoia*. New York: Pergamon.

Curnow, R., & Curran, S. (1983). *The Penguin computing book*. Harmondsworth, UK: Penguin.

David, P.A. (1986). Understanding the economics of QWERTY. In W.N. Parker (Ed.), *The necessity of history* (pp. 30–49). Oxford: Basil Blackwell.

Davies, D.J.M., & Isard, S.D. (1972). Utterances as programs. In D. Michie (Ed.), *Machine intelligence 7.* Edinburgh: Edinburgh University Press.

Davis, R. (1980). Meta-rules: Reasoning about control. *Artificial Intelligence. 15,* 179–222.

Dawkins, R. (1985). Creation and natural selection. *New Scientist,* 34–38, 25 September.

de Kleer, J., & Williams, B.C. (1991). Special volume on qualitative reasoning about physical systems II. *Artificial Intelligence, 51,* 1–3.

Dennett, D.C. (1991). *Consciousness explained.* London: Little, Brown & Co.

Descartes, R. (1663). *De homine.* Leiden. Reprinted in: F. Alquié (Ed.), *Oeuvres philosophiques.* Paris: Garnier.

Diderot, D., & D'Alembert, J.R. (1751–1780). *Encyclopédie, ou Dictionnaire raisonné des sciences, des arts et des métiers, par une société de gens de lettres.* Paris: Le Breton.

Dixon, N.F. (1981). *Preconscious processing.* New York: Wiley.

Duda, R., Gaschnig, J., & Hart, P.E. (1979). Model design in the PROSPECTOR consultant system for mineral exploration. In D. Michie (Ed.), *Expert systems in the micro-electronic age* (pp. 153–167). Edinburgh: Edinburgh University Press.

Dyer, M.G (1983). The role of affect in narratives. *Cognitive Science, 7,* 211–242.

Edelman, G.M. (1992). *Bright air, brilliant fire: On the matter of mind.* New York: Basic Books.

Ekman, P. (Ed.) (1982). *Emotion in the human face.* Cambridge: Cambridge University Press.

Einstein, A. (1917). *Uber die spezielle und die allgemeine Relativitatstheorie.* Braunschweig: Vieweg.

Evans, J.St.B.T. (1982). *The psychology of deductive reasoning.* London: Routledge & Kegan Paul.

Evans, J.St.B.T., Newstead, S.E., & Byrne, R.M.J. (1993). *Human reasoning: The psychology of deduction.* Hove, UK: Lawrence Erlbaum Associates Ltd.

Feigenbaum, E.A., & Feldman, J. (Eds.) (1963). *Computers and thought.* New York: McGraw-Hill.

Finke, R.A. (1989). *Principles of mental imagery.* Cambridge, MA: MIT Press.

Fodor, J.A. (1983). *The modularity of mind: An essay on faculty psychology.* Cambridge, MA: MIT Press.

Fraser, J.T. et al. (Eds.) (1972–1995). *The study of time I-VIII.* Bloomington, IN: International Society for the Study of Time.

Freud, S. (1901). *Zur Psychopathologie des Alltagslebens.* (Psychopathology of everyday life). In Standard Edition. London: Hogarth Press.

Frijda, N.H. (1986). *The emotions.* Cambridge: Cambridge University Press.

Gallup, G.G. (1970). Chimpanzees: Self-recognition. *Science, 167,* 86–87.

Garman, G. (1990). *Psycholinguistics.* Cambridge: Cambridge University Press.

Garnham, A., & Oakhill, J. (1994). *Thinking and reasoning.* Oxford: Basil Blackwell.

Gazzaniga, M.S. (1985). *The social brain: Discovering the networks of the mind.* New York: Basic Books.

Gelman, R. (1990). First principles organize attention to and learning about relevant data: Number and the animate–inanimate distinction as examples. *Cognitive Science, 14,* 79–106.

Gentner, D., & Stevens, A.L. (Eds.). (1983). *Mental models.* Hillsdale, NJ: Lawrence Erlbaum Associates Inc.

Gernsbacher, M.A. (Ed.) (1994). *Handbook of psycholinguistics.* San Diego, CA: Academic Press.

Geschwind, N. (1965). Disconnection syndromes in animals and man. *Brain, 88,* 237–294; 585–644.

Gibson, J.J. (1979). *The ecological approach to visual perception.* Boston: Houghton-Mifflin.

Gluck, M.A., & Rumelhart, D.E. (Eds.) (1990). *Neuroscience and connectionist theory.* Hillsdale, NJ: Lawrence Erlbaum Associates Inc.

Gordon, I.E. (1989). *Theories of visual perception.* New York: Wiley.

Grice, H.P. (1975). Logic and conversation: The William James Lectures, II. In P. Cole & J.L. Morgan (Eds.), *Syntax and semantics 3: Speech acts.* New York: Academic Press.

Grice, H.P. (1989). *Studies in the way of words.* Cambridge, MA: Harvard University Press.

Guidano, V.F. (1987). *Complexity of the self: A developmental approach to psychopathology and therapy.* New York: Guilford.

Guidano, V.F. (1991). *The self in process: Toward a post-rationalist cognitive therapy.* New York: Guilford.

Guidano, V.F., & Liotti G. (1983). *Cognitive processes and emotional disorders: A structural approach to psychotherapy.* New York: Guilford.

Gunderson, K. (1964). The imitation game. In A. Anderson (Ed.), *Minds and machines.* Englewood Cliffs, NJ: Prentice-Hall.

Hamilton, V., Bower, G.H., & Frijda, N.H. (Eds.) (1988). *Cognitive perspectives on emotion and motivation.* Dordrecht: Kluwer Academic Publishers.

Hayes, P.J. (1978). Naive physics 1 - Ontology for liquids. Memo 35, Institute pour les études semantiques et cognitives, Geneva. Reprinted in J.R. Hobbs & R.C. Moore (Eds.), *Formal theories of the commonsense world* (pp. 71–107). Norwood, NJ: Ablex

Hebb, D.O. (1949). *The organization of behavior.* New York: Wiley.

Helander, M. (Ed.) (1988). *Handbook of human–computer interaction.* Amsterdam: North Holland.

Hillis, D.W. (1985). *The connection machine.* Cambridge, MA: MIT Press.

Hobbs, J.R., & Moore, R.C. (Eds.) (1985). *Formal theories of the commonsense world.* Norwood, NJ: Ablex.

Holender, D. (1986). Semantic activation without conscious identification in dichotic listening, parafoveal vision, and visual masking: A survey and appraisal. *Behavioral and Brain Sciences, 9,* 1–66.

Holland, J.H., Holyoak, K.J., Nisbett, R.E., & Thagard, P.R. (1986). *Induction: Processes of inference, learning, and discovery.* Cambridge, MA: MIT Press.

Huhns, M.N. (Ed.) (1987). *Distributed artificial intelligence.* London: Pitman.

Hunt, E.B. (1962). *Concept learning: An information processing problem.* New York: Wiley.

Hutchins, E. (1994). *Cognition in the wild.* Cambridge, MA: MIT Press.

Izard, C.E. (1991). *The psychology of emotions.* New York: Plenum Press.

Jackendoff, R. (1986). *Semantics and cognition*. Cambridge, MA: MIT Press.
Jackson, P. (1990). *Introduction to expert systems*, 2nd ed. Reading, MA: Addison-Wesley.
Jacobs, W.J., & Nadel, L. (1985). Stress-induced recovery of fears and phobias. *Psychological Review, 92*, 512–531.
James, W. (1884). What is an emotion? *Mind, 9*, 188–205.
Johnson-Laird, P.N. (1983). *Mental models*. Cambridge: Cambridge University Press.
Johnson-Laird, P.N. (1993). *Human and machine thinking*. Hillsdale, NJ: Lawrence Erlbaum Associates Inc.
Johnson-Laird, P.N., & Bara, B.G. (1984). Syllogistic inference. *Cognition, 16*, 1–61.
Johnson-Laird, P.N., & Byrne, R.M.J. (1991). *Deduction*. Hillsdale, NJ: Lawrence Erlbaum Associates Inc.
Kandel, E.R,. & Schwartz, J.H. (Eds.) (1991). *Principles of neural science*, 3rd ed. Amsterdam: Elsevier Science.
Karmiloff-Smith, A. (1992). *Beyond modularity: A developmental perspective on cognitive science*. Cambridge, MA: MIT, Press.
Kenji Tokitsu. (1979). *La voie du karaté*. Paris: Editions du Seuil.
Koffka, K. (1935). *Principles of Gestalt psychology*. New York: Harcourt.
Konolige, K. (1988). Default and autoepistemic logic. *Artificial Intelligence, 35*, 343–382.
Kosslyn, S.M. (1980). *Image and mind*. Cambridge, MA: Harvard University Press.
Kosslyn, S.M. (1983). *Ghosts in the mind's machine: Creating and using images in the brain*. New York: Norton.
Kosslyn, S.M. (1994). *Image and brain: The resolution of the imagery debate*. Cambridge, MA: MIT Press.
Kuhn, T.S. (1962). *The structure of scientific revolutions*. Chicago: University of Chicago Press.
Kuhn, T.S. (1977). *The essential tension*. Chicago: University of Chicago Press.
Lakatos, I. (1978). The methodology of scientific research programmes. *Philosophical papers, vol. I*. Cambridge: Cambridge University Press.
La Mettrie, J.O. (1748). *L'homme machine*. Leyden.
Landauer, T.K. (1986). How much do people remember? Some estimates of the quantity of learned information in long-term memory. *Cognitive Science, 10*, 477–493.
Langley, P., Simon, H.A., Bradshaw, G.L., & Zytkow, J.M. (1987). *Scientific discovery: Computational explorations of the cognitive processes*. Cambridge, MA: MIT Press.
Lave, L. (1988). *Cognition in practice*. Cambridge: Cambridge University Press.
LeDoux, J.E. (1989). Cognitive–emotional interactions in the brain. *Cognition and Emotion, 3*, 267–299.
Lenat, D.B. (1982). Heuretics: The nature of heuristics. *Artificial Intelligence, 19*, 189–249.
Lenat, D.B. (1983a). Theory formation by heuristics search. The nature of heuristics II: Background and examples. *Artificial Intelligence, 20*, 31–59.
Lenat, D.B. (1983b). EURISKO: A program that learns new heuristics and domain concepts. The nature of heuristics III: Program design and results. *Artificial Intelligence. 20*, 61–98.

Leslie, A.M. (1994). Pretending and believing: Issues in the theory of ToMM. *Cognition, 50,* 211–238.

Lettvin, J.Y., Maturana, H.R., McCulloch, W.S., & Pitts, W.H. (1959). What the frog's eye tells the frog's mind. *Proceedings of the Institute for Radio Engineers, 47,* 1940–1951.

Levinson, S.C. (1983). *Pragmatics.* Cambridge: Cambridge University Press.

Lindsay, P.H., & Norman, D.A. (1972). *Human information processing: An introduction to psychology.* New York: Academic Press.

Longuet-Higgins, H.C. (1987). *Mental processes: Studies in cognitive science.* Cambridge, MA: MIT Press.

Mandler, G. (1984). *Mind and body: Psychology of emotion and stress.* New York: Norton.

Mantovani, G. (1994). Is computer-mediated communication intrinsically apt to enhance democracy in organizations? *Human Relations, 47,* 45–62.

Marcel, A.J. (1983a). Conscious and unconscious perception: Experiments on visual masking and word recognition. *Cognitive Psychology, 15,* 197–237.

Marcel, A.J. (1983b). Conscious and unconscious perception: An approach to the relations between phenomenal experience and perceptual processes. *Cognitive Psychology, 15,* 238–300.

Marcel, A.J., & Bisiach, E. (Eds.) (1988). *Consciousness in contemporary science.* Oxford: Oxford University Press.

Marr, D. (1977). Artificial intelligence: A personal view. *Artificial Intelligence, 9,* 37–48.

Marr, D. (1982). *Vision: A computational investigation into the human representation and processing of visual information.* San Francisco, CA: Freeman.

Marr, D., & Nishihara, H.K. (1978). Representation and recognition of the spatial organization of three-dimensional shape. *Proceedings of the Royal Society,* London, B, *200,* 269–294.

Marx, K. (1867). *Das Kapital.*

Maturana, H., & Varela, F.J. (1980). *Autopoiesis and cognition: The realization of the living.* Dordrecht: Reidel.

McCarthy, J., & Hayes, P.J. (1969). Some philosophical problems from the standpoint of artificial intelligence. In D. Michie & B. Meltzer (Eds.), *Machine intelligence 4* (pp. 463–502). Edinburgh: Edinburgh University Press.

McCarthy, R.A., & Warrington, E.K. (1990). *Cognitive neuropsychology: A clinical introduction.* New York: Academic Press.

McCloskey, M. (1991). Networks and theories: The place of connectionism in cognitive science. *Psychological Science, 6.*

McCorduck, P. (1979). *Machines who think.* New York: Freeman.

McCulloch, W.S., & Pitts, W.H. (1943). A logical calculus of the ideas imminent in nervous activity. *Bulletin of Mathematical Biophysics, 5,* 115–137.

McDermott, D. (1978). Planning and acting. *Cognitive Science, 2,* 71–109.

McDermott, D. (1982). A temporal logic for reasoning about processes and plans. *Cognitive Science, 6,* 101–155.

McLelland, J.L., & Rumelhart, D.E. (1988). *Explorations in parallel distributed processing: A handbook of models, programs, and exercises.* Cambridge, MA: MIT Press.

McLelland, J.L., Rumelhart, D.E., & the PDP Research Group. (1986). *Parallel distributed processing: Explorations in the microstructure of cognition. Vol. 2: Psychological and biological models.* Cambridge, MA: MIT Press.

Mehler, J., & Bertoncini, J. (1988). Development: A question of properties, not change? *Cognition, 115*, 121–133.

Miller, G.A. (1956). The magical number seven, plus or minus two: Some limits on our capacity for processing information. *Psychological Review, 63*, 81–97.

Miller, G.A., Galanter, E., & Pribram, K.H. (1960). *Plans and structure of behavior.* New York: Holt, Rinehart & Winston.

Miller, G.A., & Johnson-Laird, P.N. (1976). *Language and perception.* Cambridge: Cambridge University Press.

Miller, P.L. (1986). *Expert critiquing systems: Practice-based medical consultation by computer.* New York: Springer-Verlag.

Minsky, M.L. (1975). A framework for representing knowledge. In P.H. Winston (Ed.), *The psychology of computer vision.* New York: McGraw-Hill.

Minsky, M.L. (1986). *The society theory of mind.* New York: Simon & Schuster.

Moore, R.C. (1985). Semantical considerations on nonmonotonic knowledge. *Artificial Intelligence, 25*, 75–94.

Morrison, P., & Morrison, E. (Eds.) (1961). *Charles Babbage and his calculating engines.* New York: Dover.

Mussen, P.H., Conger, J.J., Kagan, J., & Huston, A. (1984). *Child development and personality.* Cambridge, MA: Harper & Row.

Neisser, U. (1967). *Cognitive psychology.* New York: Appleton-Century-Crofts.

Neisser, U. (1982). *Memory observed: Remembering in natural contexts.* San Francisco: Freeman.

Neisser, U., & Winograd, E. (Eds.) (1988). *Remembering reconsidered: Ecological and traditional approaches to the study of memory.* Cambridge: Cambridge University Press.

Newell, A. (1990). *Unified theories of cognition.* Cambridge, MA: Harvard University Press.

Newell, A., & Simon, H.A. (1956). The Logic Theory machine: A complex information processing system. *IRE Transactions on Information Theory,* IT-2, 61–79.

Newell, A., & Simon, H.A. (1972). *Human problem solving.* Englewood Cliffs, NJ: Prentice-Hall.

Newell, A., & Simon, H.A. (1976). Computer science as empirical enquiry: Symbols and search. *Communication of the ACM, 19*, 113–126.

Newton, I. (1687). *Philosophiae naturalis principia mathematica,* London.

Nilsson, N. (1971). *Problem solving methods in artificial intelligence.* New York: McGraw-Hill.

Norman, D.A. (1982). *Learning and memory.* San Francisco: Freeman.

Norman, D.A. (1986). Cognitive engineering. In D.A. Norman & S.W. Draper (Eds.), *User centered systems design.* Hillsdale, NJ: Lawrence Erlbaum Associates Inc.

Oatley, K. (1992). *Best laid schemes: The psychology of emotions.* Cambridge: Cambridge University Press.

Oatley, K., & Johnson-Laird, P.N. (1987). Towards a cognitive theory of emotions. *Cognition and Emotion, 1*, 29–50.

Ogden, C.K. (1930). *English basics: A general introduction with rules and grammar.* London: Kegan Paul, Trench & Trubner.

Ortony, A., & Turner, T.J. (1990). What's basic about basic emotions? *Psychological Review, 97*, 315–331.

Paivio, A. (1986). *Mental representations: A dual coding approach.* Oxford: Clarendon Press.

Papert, S. (1993). *Mindstorms: Children, computers, and powerful ideas* (Rev. ed.). New York: Basic Books.

Parker, S.T., & Gibson, K.R. (Eds.) (1990). *"Language" and intelligence in monkeys and apes.* Cambridge: Cambridge University Press.

Partridge, D., & Wilks, Y. (Eds.) (1990). *The foundations of artificial intelligence: A sourcebook.* Cambridge: Cambridge University Press.

Perner, J. (1991). *Understanding the representational mind.* Cambridge, MA: MIT Press.

Perrault, C.R., & Allen, J.F. (1980). A plan based analysis of indirect speech acts. *American Journal of Computational Linguistics, 6,* 167–182.

Piaget, J. (1937). *La construction du réel chez l'enfant.* Geneve: Neucâtel, Delachaux et Niestlé. (*The construction of reality in the child.* Translated 1954. New York: Basic Books).

Piaget, J. (1964). *Six études de psychologie.* Paris: Gonthier. (*Six psychological studies.* New York: Random House.)

Piattelli-Palmarini, M. (Ed.) (1979). *Théories du langage, théories de l'apprentissage.* Paris: Editions du Seuil. (*Language and learning.* Cambridge, MA: Harvard University Press).

Pinker, S. (1989). *Learnability and cognition.* Cambridge, MA: MIT Press.

Poggio, T. (1984). Vision by man and machine. *Scientific American,* April, 68–77.

Polanyi, M. (1966). *The tacit dimension.* Garden City, NY: Doubleday.

Poli, M.D. (1988). Species-specific differences in learning. In H.J. Jerison & I. Jerison (Eds.), *Intelligence and evolutionary biology* (pp. 277–297). Heidelberg: Springer-Verlag.

Pollack, M.E. (1992). The uses of plans. *Artificial Intelligence, 57,* 43–68.

Polya, G. (1945). *How to solve it.* Princeton, NJ: Princeton University Press.

Popper, K.R. (1959). *The logic of scientific discovery.* London: Hutchinson.

Pratt, V. (1987). *Thinking machines: The evolution of artificial intelligence.* Oxford: Basil Blackwell.

Prigogine, I. (1978). *From being to becoming: Time and complexity in the physical sciences.* San Francisco, CA: Freeman.

Prior, A.N. (1957). *Time and modality.* Oxford: Clarendon Press.

Prior, A.N. (1967). *Past, present and future.* Oxford: Clarendon Press.

Putnam, H. (1975). *Mind, language and reality. Philosophical papers, vol. II.* Cambridge: Cambridge University Press.

Pylyshyn, Z.W. (1984). *Computation and cognition: Toward a foundation for cognitive science.* Cambridge, MA: MIT Press.

Pylyshyn, Z.W. (Ed.) (1987). *The robot's dilemma: The frame problem in artificial intelligence.* Norwood, NJ: Ablex.

Quillian, M.R. (1968). Semantic memory. In M.L. Minsky (Ed.), *Semantic information processing* (pp. 216–260). Cambridge, MA: MIT Press.

Reiter, R. (1980). A logic for default reasoning. *Artificial Intelligence, 13,* 81–132.

Rich, E., & Knight, K. (1991). *Artificial intelligence.* New York: McGraw-Hill.

Rosch, E. (1973). Natural categories. *Cognitive Psychology, 4,* 328–350.

Rumelhart, D.E., McLelland, J.L., & the PDP Research Group. (1986). *Parallel distributed processing: Explorations in the microstructure of cognition. Vol. 1: Foundations.* Cambridge, MA: MIT Press.

Sacerdoti, E.D. (1977). *A structure for plans and behavior.* Amsterdam: Elsevier North-Holland.

Safran, J.D., & Segal, Z.V. (1990). *Interpersonal processes in cognitive therapy.* New York: Basic Books.
Sandoz, M. (1971). *Crazy Horse.* Lincoln, NE: University of Nebraska Press.
Schachter, S., & Singer, J. (1962). Cognitive, social and physiological determinants of emotional states. *Psychological Review, 69,* 379–399.
Schank, R.C. (1972). Conceptual dependency: A theory of natural language understanding. *Cognitive Psychology, 3,* 552–631.
Schank, R.C., & Abelson, R.P. (1977). *Scripts, plans, goals, and understanding: An inquiry into human knowledge structures.* Hillsdale, NJ: Lawrence Erlbaum Associates Inc.
Schubert, L.K. (1976). Extending the expressive power of networks. *Artificial Intelligence, 7,* 163–198.
Schusterman, R.J., Thomas, J., & Wood, F.G. (Eds.) (1986). *Dolphin cognition and behavior: A comparative approach.* Hillsdale, NJ: Lawrence Erlbaum Associates Inc.
Searle, J.R. (1979). *Expression and meaning: Studies in the theory of speech acts.* Cambridge: Cambridge University Press.
Searle, J.R. (1980). Minds, brains and programs. *The Behavioral and Brain Science, 3,* 417–457.
Searle, J.R. (1983). *Intentionality.* Cambridge: Cambridge University Press.
Searle, J.R. (1992). *The rediscovery of the mind.* Cambridge, MA: MIT Press.
Shallice, T. (1988). *From neuropsychology to mental structure.* Cambridge: Cambridge University Press.
Shannon, C.E., & Weaver, W. (1949). *The mathematical theory of communication.* Urbana, IL: University of Illinois Press.
Shapiro, S. (Ed.) (1990). *Encyclopedia of artificial intelligence.* New York: Wiley.
Shepard, R.N. (1980). *Internal representations: Studies in perception, imagery and cognition.* Montgomery, VE: Bradford.
Shortliffe, E.H. (1976). *Computer-based medical consultations: MYCIN.* New York: Elsevier.
Simon, H.A. (1969). *Sciences of the artificial.* Cambridge, MA: MIT Press.
Simon, H.A. (1983). *Reason in human affairs.* Stanford, CA: Stanford University Press.
Simon, H.A. (1991). *Models of my life.* New York: Basic Books.
Skinner, B.F. (1948). *Walden Two.* New York: Macmillan.
Smith, A. (1776). *An inquiry into the nature and causes of the wealth of nations.* London.
Smolensky, P. (1988). On the proper treatment of connectionism. *Behavioral and Brain Sciences, 11,* 1–23.
Spelke, E.S. (1990). Principles of object perception. *Cognitive Science, 14,* 29–56.
Sperber, D., & Wilson, D. (1986). *Relevance: Communication and cognition.* Oxford: Blackwell.
Sperry, R.W. (1961). Cerebral organization and behavior. *Science, 133,* 1749–1757.
Sroufe, L.A. (1984). The organization of emotional development. In K. Scherer & P. Ekman (Eds.), *Approaches to emotion.* Hillsdale, NJ: Lawrence Erlbaum Associates Inc.
Stalnaker, R. (1973). Presuppositions. *Journal of Philosophical Logic, 2,* 447–457.

Stefik, M. (1981a). Planning with constraints (MOLGEN: Part 1). *Artificial Intelligence, 16,* 111–139.

Stefik, M. (1981b). Planning and meta–planning (MOLGEN: Part 2). *Artificial Intelligence, 16,* 141–169.

Stein, N.L., & Oatley, K. (1992). Special issue on basic emotions. *Cognition and Emotion, 6,* 3–4.

Suchman, L.A. (1987). *Plans and situated actions: The problem of human–machine communication.* Cambridge: Cambridge University Press.

Toda, M. (Ed.) (1982). *Man, robot and society.* The Hague: Nijhoff.

Tranel, E., & Damasio, A.R. (1985). Knowledge without awareness: An autonomic index of facial recognition by prosopagnosics. *Science, 228,* 1453–1454.

Tulving, E. (1972). Episodic and semantic memory. In E. Tulving & W. Donaldson (Eds.), *Organization of memory.* New York: Academic Press.

Turing, A.M. (1936). On computable numbers, with an application to the Entscheidungsproblem. *Proceedings of the London Mathematical Society, Series 2, 42,* 230–265.

Turing, A.M. (1950). Computing machinery and intelligence. *Mind, 59,* 433–460.

Varela, F.J., Thompson, E., & Rosch, E. (1991). *The embodied mind: Cognitive science and human experience.* Cambridge, MA: MIT Press.

Verschueren, J., Ostman, J.-O., & Blommaert, J. (Eds.) (1995). *Handbook of pragmatics.* Amsterdam: John Benjamins.

Wanner, E., & Gleitman, L. (1982). *Language acquisition: The state of the art.* Cambridge: Cambridge University Press.

Watanabe, S. (1972). Creative time. In J.T. Fraser, F.C. Haber, & G.H. Muller (Eds.), *The study of time I,* 159–189. Berlin: Springer-Verlag.

Webster, N. (1979). *Webster's new universal unabridged dictionary.* Dorset & Baber.

Weiss, S.M., & Kulikowski, C.A. (1979). EXPERT: A system for developing consultation models. *Proceedings IJCAI 6,* 942–947.

Weizenbaum, J. (1975). *Computer power and human reason: From judgment to calculation.* San Francisco: Freeman.

Wenger, E. (1987). *Artificial intelligence and tutoring systems: Computational and cognitive approaches to the communication of knowledge.* Los Altos, CA: Morgan Kaufmann.

Wexler, K., & Culicover, P.W. (1980). *Formal principles of language acquisition.* Cambridge, MA: MIT Press.

Whitehead, A.N., & Russell, B. (1910). *Principia mathematica.* Cambridge: Cambridge University Press.

Wiener, N. (1948). *Cybernetics, or control and communication in the animal and the machine.* Cambridge, MA: MIT Press.

Wilensky, R. (1983). *Planning and understanding: A computational approach to human reasoning.* Reading, MA: Addison-Wesley.

Winograd, T. (1972). *Understanding natural language.* New York: Academic Press.

Winograd, T. (1983). *Language as a cognitive process. Vol. 1: Syntax.* Reading, MA: Addison-Wesley.

Winston, P.H (1975). Learning structural descriptions from examples. In P.H. Winston (Ed.), *The psychology of computer vision* (pp. 157–209). New York: McGraw-Hill.

Wittgenstein, L. (1922). *Tractatus logico-philosophicus*. London: Kegan Paul, Trench, Trubner & Co.

Wittgenstein, L. (1953). *Philosophische Untersuchungen*. Oxford: Basil Blackwell. (*Philosophical investigations*, Oxford: Basil Blackwell, 1958.)

Woods, W.A. (1970). Transition network grammars for natural language analysis.*Communications of the ACM, 13*, 591–606.

Woods, W.A. (1975). What's in a link: Foundations for semantic networks. In D.G. Bobrow & A. Collins (Eds.), *Representation and understanding* (pp. 35–82). New York: Academic Press.

Woods, W.A. (1981). Procedural semantics as a theory of meaning. In A. Joshi, B.N. Webber, & I.A. Sag (Eds.), *Elements of discourse understanding*. Cambridge: Cambridge University Press.

Zazzo, R. (1993). *Reflects de miroir et autres doubles*. Paris: Presse Universitaire de France.

Author index

Subject index